Green's
Functions
with
Applications

Studies in Advanced Mathematics

Titles Included in the Series

Green's
Functions
with
Applications

DEAN G. DUFFY

CHAPMAN & HALL/CRC

Boca Raton   London   New York   Washington, D.C.

Library of Congress Cataloging-in-Publication Data

Duffy, Dean G.
  Green's functions with applications / Dean Duffy
      p.   cm. (Studies in advanced mathematics)
  Includes bibliographical references and index.
  ISBN 1-58488-110-0
  1. Green's functions. I. Title. II. Series

QA371 .D74 2001
515'.35—dc21   2001028216
                                                                   CIP
          Catalog record is available from the Library of Congress

Visit the CRC Press Web site at www.crcpress.com

Dedicated to my friends

Contents

Introduction

This book had its origin in some electronic mail that I received from William S. Price a number of years ago. He needed to construct a Green's function and asked me if I knew a good book that might assist him. In suggesting several standard texts, I could not help but think that, based on my own experiences utilizing Green's functions alone and in conjunction with numerical solvers, I had my own ideas on how to present this material. It was this thought that ultimately led to the development of this monograph.

The purpose of this book is to provide applied scientists and engineers with a systematic presentation of the various methods available for deriving a Green's function. To this end, I have tried to make this book the most exhaustive source book on Green's function yet available, focusing on every possible analytic technique rather than theory.

After some introductory remarks, the material is classified according to whether we are dealing with an ordinary differential, wave, heat, or Helmholtz equation. Turning first to ordinary differential equations, we have either initial-value or boundary-value problems. After examining initial-value problems, I explore in depth boundary-value problems, both regular and singular. There are essentially two methods: piecing together a solution from the homogeneous solutions, and eigenfunction expansions. Both methods are presented.

Green's functions are particularly well suited for wave problems, as is shown in Chapter 3 with the detailed analysis of electromagnetic waves in surface waveguides and water waves. Before presenting this material, some of the classic solutions in one, two, and three dimensional free space are discussed.

The heat equation and Green's functions have a long association with each other. After discussing heat conduction in free space, the classic solutions of the heat equation in rectangular, cylindrical, and spherical coordinates are offered. The chapter concludes with an interesting application: The application of Green's functions in understanding the stability of fluids and plasmas.

It is not surprising that the final chapter on Poisson's and Helmholtz's equations is the longest. Finding solutions to Poisson's equation gave birth to this technique and Sommerfeld's work at the turn of the twentieth century spurred further development. For each equation, the techniques available for solving them as a function of coordinate system are presented. The final section deals with the computational efficiency of evaluating this class of Green's functions.

This book may be used in a class on boundary-value problems or as a source book for researchers, in which case I recommend that the reader not overlook the problems.

Most books are written with certain assumptions concerning the background of the reader. This book is no exception. The methods for finding Green's functions lean heavily on transform methods because they are particularly well suited for handling the Dirac delta function. For those unfamiliar with using transform methods to solve differential equations, I have summarized these techniques in Appendices A and B. In this sense, the present book is a continuation of my *Transform Methods for Solving Partial Differential Equations*. Because many of the examples and problems involve cylindrical coordinates, an appendix on Bessel functions has been included. In particular, I cover how to find Fourier-Bessel expansions.

A unique aspect of this book is its emphasis on the numerical evaluation of Green's functions. This has taken two particular forms. First, many of the Green's functions that are found in the text and problem sets are illustrated. The motivation here was to assist the reader in developing an intuition about the behavior of Green's functions in certain classes of problem. Second, Green's functions are of little value if they cannot be rapidly computed. Therefore, at several points in the book the question of the computational efficiency and possible methods to accelerate the process have been considered.

Special thanks go to Prof. Michael D. Marcozzi for his many useful suggestions for improving this book. Dr. Tim DelSole provided outstanding guidance in the section on convective/absolute instability. Dr. Chris Linton made several useful suggestions regarding Section 5.8. Finally, I would like to express my appreciation to all those authors and publishers who allowed me the use of their material from the scientific and engineering literature.

Definitions of the Most Commonly Used Functions

Function	Definition
$\delta(t-a)$	$= \begin{cases} \infty, & t=0 \\ 0, & t \neq 0 \end{cases}$, $\quad \int_{-\infty}^{\infty} \delta(t)\,dt = 1$
$H(t-a)$	$= \begin{cases} 1, & t>a, \\ 0, & t<a, \end{cases}$, $\quad a \geq 0$
$I_n(x)$	modified Bessel function of the first kind and order n
$J_n(x)$	Bessel function of the first kind and order n
$K_n(x)$	modified Bessel function of the second kind and order n
$r_<$	$= \min(r, \rho)$
$r_>$	$= \max(r, \rho)$
$x_<$	$= \min(x, \xi)$
$x_>$	$= \max(x, \xi)$
$Y_n(x)$	Bessel function of the second kind and order n
$y_<$	$= \min(y, \eta)$
$y_>$	$= \max(y, \eta)$
$z_<$	$= \min(z, \zeta)$
$z_>$	$= \max(z, \zeta)$

Chapter 1
Some Background Material

One of the fundamental problems of field theory[1] is the construction of solutions to linear differential equations when there is a specified source and the differential equation must satisfy certain boundary conditions. The purpose of this book is to show how Green's functions provide a method for obtaining these solutions. In this chapter, some of the mathematical aspects necessary for developing this technique are presented.

1.1 HISTORICAL DEVELOPMENT

In 1828 George Green (1793–1841) published an *Essay on the Application of Mathematical Analysis to the Theory of Electricity and Magnetism*. In this seminal work of mathematical physics, Green sought to determine the electric potential within a vacuum bounded by conductors with specified potentials. In today's notation we would say that he examined the solutions of $\nabla^2 u = -f$ within a volume V that satisfy certain boundary conditions along the boundary S.

---

[1] Any theory in which the basic quantities are fields, such as electromagnetic theory.

In modern notation, Green sought to solve the partial differential equation

$$\nabla^2 g(\mathbf{r}|\mathbf{r}_0) = -4\pi\delta(\mathbf{r} - \mathbf{r}_0), \qquad (1.1.1)$$

where $\delta(\mathbf{r} - \mathbf{r}_0)$ is the Dirac delta function. We now know that the solution to (1.1.1) is $g = 1/R$, where $R^2 = (x - \xi)^2 + (y - \eta)^2 + (z - \zeta)^2$. Although Green recognized the singular nature of g, he proceeded along a different track. First, he proved the theorem that bears his name:

$$\iiint_V \left(\varphi\nabla^2\chi - \chi\nabla^2\varphi\right) dV = \oiint_S (\varphi\nabla\chi - \chi\nabla\varphi) \cdot \mathbf{n}\, dS, \qquad (1.1.2)$$

where the outwardly pointing normal is denoted by $\mathbf{n}$ and χ and φ are scalar functions that possess bounded derivatives. Then, by introducing a small ball about the singularity at $\mathbf{r}_0$ because (1.1.2) cannot apply there and then excluding it from the volume V, Green obtained

$$\iiint_V g\nabla^2 u\, dV + \oiint_S g\nabla u \cdot \mathbf{n}\, dS$$
$$= \iiint_V u\nabla^2 g\, dV + \oiint_S u\nabla g \cdot \mathbf{n}\, dS - 4\pi u(\mathbf{r}_0), \quad (1.1.3)$$

because the surface integral over the small ball is $4\pi u(\mathbf{r}_0)$ as the radius of the ball tends to zero. Next, Green required that *both* g and u satisfy the homogeneous boundary condition $u = 0$ along the surface S. Since $\nabla^2 u = -f$ and $\nabla^2 g = 0$ within V (recall that the point $\mathbf{r}_0$ is excluded from V), he found

$$u(\mathbf{r}) = \frac{1}{4\pi} \oiint_S \overline{u}\,\nabla g \cdot \mathbf{n}\, dS, \qquad (1.1.4)$$

when $f = 0$ (Laplace's equation) for any point $\mathbf{r}$ within S, where $\overline{u}$ denotes the value of u on the boundary S. This solved the boundary-value problem once g was found. Green knew that g had to exist; it physically described the electrical potential from a point charge located at $\mathbf{r}_0$.

Green's essay remained relatively unknown until it was published[2] between 1850 and 1854. With its publication the spotlight shifted to the German school of mathematical physics. Although Green himself had not given a name for g, Riemann[3] (1826–1866) would subsequently call

---

[2] Green, G., 1850, 1852, 1854: An essay on the application of mathematical analysis to the theories of electricity and magnetism. *J. reine angewand. Math.*, **39**, 73–89; **44**, 356–374; **47**, 161–221.

[3] Riemann, B., 1869: *Vorlesungen über die partielle Differentialgleichungen der Physik*, §23; Burkhardt, H., and W. F. Meyer, 1900: *Potentialtheorie* in *Encyklop. d. math. Wissensch.*, **2**, **Part A**, 462–503. See §18.

Figure 1.1.1: Originally drawn to mathematics, Arnold Johannes Wilhelm Sommerfeld (1868–1951) migrated into physics due to Klein's interest in applying the theory of complex variables and other pure mathematics to a range of physical topics from astronomy to dynamics. Later on, Sommerfeld contributed to quantum mechanics and statistical mechanics. (Portrait, AIP Emilio Segrè Visual Archives, Margrethe Bohr Collection.)

it the "Green's function." Then, in 1877, Carl Neumann[4] (1832–1925) embraced the concept of Green's functions in his study of Laplace's equation, particularly in the plane. He found that the two-dimensional equivalent of the Green's function was not described by a singularity of the form $1/|\mathbf{r} - \mathbf{r}_0|$ as in the three-dimensional case but by a singularity of the form $\log(1/|\mathbf{r} - \mathbf{r}_0|)$.

With the function's success in solving Laplace's equation, other equations began to be solved using Green's functions. In the case of

---

[4] Neumann, C., 1877: *Untersuchungen über das Logarithmische und Newton'sche Potential*, Teubner, Leipzig.

the heat equation, Hobson[5] (1856–1933) derived the free-space Green's function for one, two and three dimensions and the French mathematician Appell[6] (1855–1930) recognized that there was a formula similar to Green's for the one-dimensional heat equation. However, it fell to Sommerfeld[7] (1868–1951) to present the modern theory of Green's function as it applies to the heat equation. Indeed, Sommerfeld would be the great champion of Green's functions at the turn of the twentieth century.[8]

The leading figure in the development of Green's functions for the wave equation was Kirchhoff[9] (1824–1887), who used it during his study of the three-dimensional wave equation. Starting with Green's second formula, he was able to show that the three-dimensional Green's function is

$$g(x, y, z, t | \xi, \eta, \zeta, \tau) = \frac{\delta(t - \tau - R/c)}{4\pi R}, \tag{1.1.8}$$

where $R = \sqrt{(x - \xi)^2 + (y - \eta)^2 + (z - \zeta)^2}$ (modern terminology). Although he did not call his solution a Green's function,[10] he clearly grasped the concept that this solution involved a function that we now call the Dirac delta function (see pg. 667 of his *Annalen d. Physik*'s paper). He used this solution to derive his famous *Kirchhoff's theorem*, which is the mathematical expression for Huygen's principle.

The application of Green's function to ordinary differential equations involving boundary-value problems began with the work of Burkhardt[11] (1861–1914). Using results from Picard's theory of ordinary differential equations, he derived the Green's function given by (1.5.35) as well as the properties listed in §2.3. Later on, Bôcher[12] (1867–1918) extended these results to nth order boundary-value problems.

---

[5] Hobson, E. W., 1888: Synthetical solutions in the conduction of heat. *Proc. London Math. Soc.*, **19**, 279–294.

[6] Appell, P., 1892: Sur l'équation $\frac{\partial^2 z}{\partial x^2} - \frac{\partial z}{\partial y} = 0$ et la théorie de la chaleur. *J. Math. pures appl.*, *4e série*, **8**, 187–216.

[7] Sommerfeld, A., 1894: Zur analytischen Theorie der Wärmeleitung. *Math. Ann.*, **45**, 263–277.

[8] Sommerfeld, A., 1912: Die Greensche Funktion der Schwingungsgleichung. *Jahresber. Deutschen Math.-Vereinung*, **21**, 309–353.

[9] Kirchhoff, G., 1882: Zur Theorie der Lichtstrahlen. *Sitzber. K. Preuss. Akad. Wiss. Berlin*, 641–669; reprinted a year later in *Ann. Phys. Chem., Neue Folge*, **18**, 663–695.

[10] This appears to have been done by Gutzmer, A., 1895: Über den analytischen Ausdruck des Huygens'schen Princips. *J. reine angewand. Math.*, **114**, 333–337.

[11] Burkhardt, H., 1894: Sur les fonctions de Green relatives a un domaine d'une dimension. *Bull. Soc. Math.*, **22**, 71–75.

[12] Bôcher, M., 1901: Green's function in space of one dimension. *Bull. Amer.*

Figure 1.1.2: Gustav Robert Kirchhoff's (1824–1887) most celebrated contributions to physics are the joint founding with Robert Bunsen of the science of spectroscopy, and the discovery of the fundamental law of electromagnetic radiation. Kirchhoff's work on light coincides with his final years as a professor of theoretical physics at Berlin. (Portrait taken from frontispiece of Kirchhoff, G., 1882: *Gesammelte Abhandlungen*. J. A. Barth, 641 pp.)

1.2 THE DIRAC DELTA FUNCTION

Since the 1950s, when Schwartz[13] (1915–) published his theory of distributions, the concept of generalized functions has had an enormous impact on many areas of mathematics, particularly on partial differential equations. In this section, we introduce probably the most important generalized function, the *Dirac delta function*. As we shall shortly see, the entire concept of Green's functions is intimately tied to this most "unusual" function.

Math. Soc., Ser. 2, **7**, 297–299; Bôcher, M., 1911/12: Boundary problems and Green's functions for linear differential and difference equations. *Annals Math., Ser. 2*, **13**, 71–88.

[13] Schwartz, L., 1973: *Théorie des distributions*. Hermann, 418 pp.

Figure 1.2.1: Laurent Schwartz' (1915–) work on distributions dates from the late 1940s. For this work he was awarded the 1950 Fields medal. (Portrait courtesy of the Ecole Polytechnique, France.)

For many, the Dirac delta function had its birth with the quantum mechanics of Dirac[14] (1902–1984). Modern scholarship[15] has shown, however, that this is simply not true. During the nineteenth century, both physicists and mathematicians used the delta function although physicists viewed it as a purely mathematical idealization that did not exist in nature, while mathematicians used it as an intuitive physical notion without any mathematical reality.

It was the work of Oliver Heaviside (1850–1925) and the birth of electrical engineering that brought the delta function to the attention of the broader scientific and engineering community. In his treatment of a cable that is grounded at both ends, Heaviside[16] introduced the delta function via its sifting property (1.2.9). Consequently, as Laplace transforms became a fundamental tool of electrical engineers, so too did

---

[14] Dirac, P., 1926-7: The physical interpretation of the quantum dynamics. *Proc. R. Soc. London*, **A113**, 621–641.

[15] Lützen, J., 1982: *The Prehistory of the Theory of Distributions*. Springer-Verlag, 232 pp. See chap. 4, part 2.

[16] Heaviside, O., 1950: *Electromagnetic Theory*. Dover Publications, Inc., §267. See Eq. 24.

Figure 1.2.2: Paul Adrien Maurice Dirac (1902–1984) ranks among the giants of twentieth-century physics. Awarded the 1933 Nobel Prize in physics for his relativistic quantum mechanics, Dirac employed the delta function during his work on quantum mechanics. In later years, Dirac also helped to formulate Fermi-Dirac statistics and contributed to the quantum theory of electromagnetic radiation. (Portrait reproduced by permission of the President and Council of the Royal Society.)

the use of the delta function.

Despite the delta function's fundamental role in electrical engineering and quantum mechanics, by 1945 there existed several schools of thought concerning its exact nature because Dirac's definition:

$$\delta(t) = \begin{cases} \infty, & t = 0, \\ 0, & t \neq 0, \end{cases} \tag{1.2.1}$$

such that

$$\int_{-\infty}^{\infty} \delta(t)\,dt = 1, \tag{1.2.2}$$

was unsatisfactory; no conventional function could be found that satisfied (1.2.2).

One approach, especially popular with physicists because it agreed with their physical intuition of a point mass or charge, sought to view the

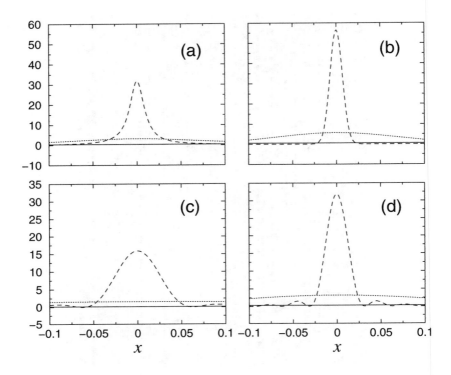

Figure 1.2.3: Frames (a)–(d) illustrate the delta sequences (1.2.4), (1.2.5), $[1 - \cos(nx)]/(n\pi x^2)$, and (1.2.6) as a function of x, respectively. The solid, dotted and dashed lines correspond to $n = 1$, $n = 10$ and $n = 100$, respectively.

delta function as the limit of the sequence of strongly peaked functions $\delta_n(t)$:

$$\delta(t) = \lim_{n\to\infty} \delta_n(t), \tag{1.2.3}$$

Candidates[17] included

$$\delta_n(t) = \frac{n}{\pi} \frac{1}{1 + n^2 t^2}, \tag{1.2.4}$$

$$\delta_n(t) = \frac{n}{\sqrt{\pi}} e^{-n^2 t^2}, \tag{1.2.5}$$

and

$$\delta_n(t) = \frac{1}{n\pi} \frac{\sin^2(nt)}{t^2}. \tag{1.2.6}$$

The difficulty with this approach was that the limits of these sequences may not exist.

---

[17] Kirchhoff [9] gave (1.2.5) in the limit of $n \to \infty$ as an example of a delta function.

Another approach, favored by electrical engineers, involved the Heaviside step function:

$$H(t) = \begin{cases} 1, & t > 0, \\ 0, & t < 0. \end{cases} \tag{1.2.7}$$

The delta function was now merely the derivative of $H(t)$,

$$\delta(t) = \frac{dH(t)}{dt}. \tag{1.2.8}$$

The difficulty here was that the derivative does not exist at $t = 0$.

Finally, some defined the delta function on the basis of its *sifting property*

$$\int_{-\infty}^{\infty} \delta(t-a)f(t)\, dt = f(a). \tag{1.2.9}$$

This property is given its name because $\delta(t-a)$ acts as a sieve, selecting from all possible values of $f(t)$ its value at the point $t = a$. Unfortunately, there is no conventional function $\delta(t-a)$ with this property.

• **Example 1.2.1**

Although viewing the delta function as the limit of a sequence of conventional functions lacks mathematical rigor, useful results can still be obtained by this method. For example, using the periodic set of pulses with period $2L$ and defined over the interval $[-L, L]$ by

$$\delta_n(x) = \begin{cases} n/2, & |x| < 1/n, \\ 0, & |x| > 1/n, \end{cases} \tag{1.2.10}$$

the Fourier series representation of $\delta_n(x)$ is

$$\delta_n(x) = \frac{1}{2L} + \sum_{m=1}^{\infty} \frac{n}{m\pi} \sin\left(\frac{m\pi}{nL}\right) \cos\left(\frac{m\pi x}{L}\right). \tag{1.2.11}$$

Consequently, the Fourier series representation for the delta function follows by letting $n \to \infty$ and we find that

$$\delta(x) = \frac{1}{2L} + \frac{1}{L} \sum_{n=1}^{\infty} \cos\left(\frac{n\pi x}{L}\right). \tag{1.2.12}$$

A quick check shows that this series is divergent, as it should be. If it were convergent, then $\delta(x)$ would be a conventional function.

• Example 1.2.2

Let us show that our Fourier series representation (1.2.12) of $\delta(x)$ possesses the sifting property (1.2.9).

Consider the integral $\int_{-L}^{L} f(x)\delta(x)\,dx$ where $f(x)$ is a conventional function. Thus,

$$\int_{-L}^{L} f(x)\delta(x)\,dx = \frac{1}{2L}\int_{-L}^{L} f(x)\,dx + \frac{1}{L}\sum_{n=1}^{\infty}\int_{-L}^{L} f(x)\cos\left(\frac{n\pi x}{L}\right)\,dx$$

$$(1.2.13)$$

$$= \frac{A_0}{2} + \sum_{n=1}^{\infty} A_n = f(0), \tag{1.2.14}$$

where A_0 and A_n are the Fourier coefficients of $f(x)$, and our demonstration is complete.

• Example 1.2.3

Let us use the delta sequence technique to compute the Fourier transform of the delta function.

As in the case of Fourier series, we first find the Fourier transform of $\delta_n(x)$ with $L \to \infty$ and then take the limit as $n \to \infty$. Therefore,

$$\Delta_n(\omega) = \int_{-\infty}^{\infty} \delta_n(t)e^{-i\omega t}\,dt = \int_{-1/n}^{1/n} \frac{n}{2}e^{-i\omega t}\,dt = \frac{\sin(\omega/n)}{\omega/n}. \tag{1.2.15}$$

Finally, the Fourier transform of the delta function follows by taking $n \to \infty$, or

$$\mathcal{F}[\delta(t)] = \lim_{n\to\infty}\Delta_n(\omega) = 1. \tag{1.2.16}$$

Although our use of delta sequences has netted several useful results, they are suspect because the limits of these sequences do not exist according to common definitions of convergence. It is the purpose of what is known as the *theory of distributions* or *generalized functions* to put $\delta(x)$ on a firm mathematical basis and to unify the many *ad hoc* mathematical approaches used by engineers and scientists. In the next few paragraphs we give an overview[18] of the arguments that lie behind this theory.

The theory of distributions is concerned with the problem of extending the definition of a function so that we may include expressions

---

[18] For more detail, see Hoskins, R. F., 1999: *Delta Functions: An Introduction to Generalized Functions*. Horwood Publishing, 262 pp.

Table 1.2.1: Some Useful Relationships Involving the Delta Function

---

$$\delta(t) = \begin{cases} \infty, & t = 0 \\ 0, & t \neq 0 \end{cases}, \qquad \int_{-\infty}^{\infty} \delta(t)\, dt = 1$$

$$\int_{-\infty}^{\infty} \delta(t-a)f(t)\, dt = f(a)$$

$$\delta(ct) = \delta(t)/|c|, \qquad \delta(-t) = \delta(t)$$

$$\delta(t) = -t\delta'(t)$$

$$f(t)\delta(t-a) = f(a)\delta(t-a)$$

$$\delta(t^2 - a^2) = \frac{\delta(t+a) + \delta(t-a)}{2|a|}$$

$$x^n \delta^{(m)}(x) = 0, \qquad \text{if} \qquad 0 \leq m < n$$

$$x^n \delta^{(m)}(x) = (-1)^n \frac{m!}{(m-n)!} \delta^{(m-n)}(x), \qquad \text{if} \qquad 0 \leq n \leq m$$

---

such as $\delta(x)$. This process might be thought of as akin to the manner in which natural numbers were extended to include integers, integers to rationals, and rationals to real number. Although there are several methods[19] open to us, we will use a sequential approach: We consider integrals of sequences of functions of the type

$$\int_{-\infty}^{\infty} g_n(x)\varphi(x)\, dx,$$

where $n = 1, 2, 3, \ldots$ A sequence of function $g_n(x)$, such as delta sequences (1.2.4)–(1.2.6), leads to a new mathematical concept, such as the delta function, provided such a sequence of integrals converges for *any* suitable function $\varphi(x)$.

---

[19] See, for example, Temple, G., 1953: Theories and applications of generalized functions. *J. Lond. Math. Soc.*, **28**, 134–148.

What do we mean by "suitable" for $\varphi(x)$? Because we want to define concepts such as $\delta'(x)$, $\delta''(x)$, and so forth, then $\varphi(x)$ should be infinitely differentiable (possess derivatives of all orders). Furthermore, it should vanish identically outside of a bounded region so that it behaves properly at infinity. A function $\varphi(x)$ that satisfies these requirements is called a *test function*. An example of a test function is

$$\varphi(x) = \begin{cases} e^{-a^2/(a^2-x^2)}, & |x| < a, \\ 0, & |x| \geq a, \end{cases} \qquad (1.2.17)$$

where $a > 0$.

Having introduced test functions, we now proceed to define the class of *admissible functions* from which the functions $g_n(x)$ will be selected. Although there is some choice in this matter, we require that these admissible functions be infinitely differentiable over the entire range $(-\infty, \infty)$, with their behavior at infinity left arbitrary. It is these functions that we are extending to encompass other (not infinitely differentiable) functions as well as distributions such as the delta function.

Having introduced test and admissible functions, we now define a *weakly convergent sequence* as one where the limit

$$\lim_{n \to \infty} \int_{-\infty}^{\infty} g_n(x)\varphi(x)\,dx$$

exists for all test functions $\varphi(x)$. A weakly convergent sequence may or may not be convergent in any of the conventional definitions such as pointwise convergent, uniformly convergent, convergent in the mean, and so forth. Although we could have extended the admissible functions by means of other types of convergence, the extension by weak convergence turns out to be very powerful.

We are now ready to give a rigorous definition for *distribution*: A distribution $g(x)$ is a "function" associated with a weakly convergent sequence of admissible functions for which the symbolic integral

$$\int_{-\infty}^{\infty} g(x)\varphi(x)\,dx$$

means

$$\int_{-\infty}^{\infty} g(x)\varphi(x)\,dx = \lim_{n \to \infty} \int_{-\infty}^{\infty} g_n(x)\varphi(x)\,dx. \qquad (1.2.18)$$

For example, the sequences given by (1.2.4)–(1.2.6) are equivalent because

$$\lim_{n \to \infty} \int_{-\infty}^{\infty} \delta_n(x)\varphi(x)\,dx = \varphi(0), \qquad (1.2.19)$$

Table 1.2.2: The Representation of the Dirac Delta Function $\delta(\mathbf{r} - \mathbf{r}_0)$ in Various Coordinate Systems

Coordinate	Three	Dimensions Two	One
Cartesian	$\delta(x - \xi)\delta(y - \eta)\delta(z - \zeta)$	$\delta(x - \xi)\delta(y - \eta)$	$\delta(x - \xi)$
Cylindrical	$\dfrac{\delta(r - \rho)\delta(\varphi - \varphi')\delta(z - \zeta)}{r}$	$\dfrac{\delta(r - \rho)\delta(z - \zeta)}{2\pi r}$	$\dfrac{\delta(r - \rho)}{2\pi r}$
Spherical	$\dfrac{\delta(r - \rho)\delta(\theta - \theta')\delta(\varphi - \varphi')}{r^2 \sin(\theta)}$	$\dfrac{\delta(r - \rho)\delta(\theta - \theta')}{2\pi r^2 \sin(\theta)}$	$\dfrac{\delta(r - \rho)}{4\pi r^2}$

for all $\varphi(x)$. The distribution $\delta(x)$ defined by any of these sequences is called the delta function.

• **Example 1.2.4**

Throughout this book we will solve differential equations that, in their simplest form, are similar to

$$\frac{dg}{dt} + ag = \delta(t - \tau). \tag{1.2.20}$$

What does (1.2.20) mean? On the left side we have a conventional differential operator; on the right side we have this peculiar delta function. Clearly $g(t)$ must be very strange in its own right.

How would the theory of distributions handle (1.2.20)? Because (1.2.20) involves a delta function, we should multiply it by a test function $\varphi(t)$ and integrate from $t = -\infty$ to $t = \infty$. From the definition of the delta function,

$$\int_{-\infty}^{\infty} \left(\frac{dg}{dt} + ag \right) \varphi(t)\, dt = \int_{-\infty}^{\infty} \delta(t - \tau)\varphi(t)\, dt = \varphi(\tau) \tag{1.2.21}$$

for every test function $\varphi(t)$. Integrating by parts, we have that

$$\int_{-\infty}^{\infty} g(t) \left[a\varphi(t) - \varphi'(t) \right] dt = \varphi(\tau) \tag{1.2.22}$$

for all functions $\varphi(t)$ because $g(t)\varphi(t)$ vanishes at infinity. Therefore, $g(t)$ is a distribution or generalized function such that (1.2.22) is satisfied for all sufficiently good test functions $\varphi(t)$. Although (1.2.22) is formally correct, it still does not give us a method for finding $g(t)$.

So far, we have dealt with the delta function only as it applies in one dimension. When it comes to partial differential equations, we need a corresponding definition of the multidimensional delta function:

$$\iiint_V f(\mathbf{r})\delta(\mathbf{r} - \mathbf{r}_0)\, dV = \begin{cases} f(\mathbf{r}_0), & \text{if } \mathbf{r}_0 \text{ inside } V, \\ 0, & \text{otherwise,} \end{cases} \qquad (1.2.23)$$

where $\mathbf{r}$ is the vector from the origin to some point $P(x, y, z)$ and $\mathbf{r}_0$ is the vector from the origin to another point $P(\xi, \eta, \zeta)$. Although there is no restriction on the number of dimensions involved and $f(\mathbf{r})$ can be a scalar or vector function, $f(\mathbf{r})$ must be defined at the point $P(\xi, \eta, \zeta)$. If $f(\mathbf{r})$ equals unity, then the delta function is normalized and is of unit magnitude in the sense that the integral of the delta function over the coordinates involved equals one. Following this convention, we list in Table 1.2.2 the representation for the Dirac delta function in the three most commonly used coordinate systems. For a three-dimensional orthogonal curvilinear coordinate system with elements of length $h_i u_i$, where h_i are the scale factors and u_i are the curvilinear coordinates, then the definition of the delta function can be written

$$\delta(\mathbf{r} - \mathbf{r}_0) = \frac{\delta(u_1 - \xi_1)}{h_1}\frac{\delta(u_2 - \xi_2)}{h_2}\frac{\delta(u_3 - \xi_3)}{h_3}. \qquad (1.2.24)$$

When the delta function contains only two dimensions, one of the terms on the right side of (1.2.24) vanishes. However, this does not allow us to arbitrarily omit one of the terms because the integral of the delta function must still equal unity. The proper procedure replaces the denominator of the right side of (1.2.24) by the integral of the three scale factors over the coordinate that is ignored. For example, in the case of spherical coordinates with no φ-dependence, the denominator becomes

$$\int_0^{2\pi} r^2 \sin(\theta)\, d\varphi = 2\pi r^2 \sin(\theta). \qquad (1.2.25)$$

If the problem involves spherical coordinates, but with no dependence on either φ or θ, the denominator becomes

$$\int_0^{\pi} \int_0^{2\pi} r^2 \sin(\theta)\, d\varphi\, d\theta = 4\pi r^2. \qquad (1.2.26)$$

1.3 GREEN'S FORMULAS

Many of the important results pertaining to Green's functions involve boundary conditions. To facilitate future analysis, we introduce two important vector identities that we will use in subsequent sections. These theorems are Green's first and second formulas:

$$\iiint_V \nabla\varphi \cdot \nabla\chi \, dV + \iiint_V \varphi\nabla^2\chi \, dV = \oiint_S \varphi(\nabla\chi \cdot \mathbf{n}) \, dS, \qquad (\mathbf{1.3.1})$$

and

$$\iiint_V \left(\varphi\nabla^2\chi - \chi\nabla^2\varphi\right) dV = \oiint_S \left(\varphi\nabla\chi - \chi\nabla\varphi\right) \cdot \mathbf{n} \, dS, \qquad (\mathbf{1.3.2})$$

where the volume V integral takes place over a closed surface S that has the outwardly pointing normal $\mathbf{n}$. Here ∇ denotes the three-dimensional gradient operator. Thus, Green's formulas establish the relationship between the sources (the Laplacian terms) and the fluxes of two scalar fields.

Green's first formula (1.3.1) follows from the vector identity

$$\nabla \cdot (\varphi\nabla\chi) = \nabla\varphi \cdot \nabla\chi + \varphi\nabla^2\chi. \qquad (\mathbf{1.3.3})$$

Integration of (1.3.3) over the closed volume yields

$$\iiint_V \nabla \cdot (\varphi\nabla\chi) \, dV = \iiint_V \left(\nabla\varphi \cdot \nabla\chi + \varphi\nabla^2\chi\right) dV. \qquad (\mathbf{1.3.4})$$

Applying Gauss's divergence theorem to the left side of (1.3.4) leads directly to (1.3.1).

To prove the second formula (1.3.2), we have from the first formula

$$\iiint_V \nabla\varphi \cdot \nabla\chi \, dV + \iiint_V \varphi\nabla^2\chi \, dV = \oiint_S \varphi(\nabla\chi \cdot \mathbf{n}) \, dS, \qquad (\mathbf{1.3.5})$$

and

$$\iiint_V \nabla\chi \cdot \nabla\varphi \, dV + \iiint_V \chi\nabla^2\varphi \, dV = \oiint_S \chi(\nabla\varphi \cdot \mathbf{n}) \, dS. \qquad (\mathbf{1.3.6})$$

Subtracting (1.3.6) from (1.3.5), we obtain the second formula (1.3.2).

• Example 1.3.1

Let us apply Green's second formula to a close surface S surrounding the point (ξ, η, ζ) where $\varphi = r^{-1}$ with $r^2 = (x-\xi)^2 + (y-\eta)^2 + (z-\zeta)^2$, and $\chi = u(x, y, z)$. Because we cannot use Green's formulas if the point (ξ, η, ζ) is included within the closed surface, we introduce a small sphere with surface area Σ, radius ϵ, and its center at (ξ, η, ζ). Then the volume integration V consists of the space enclosed by S *minus* the space inside Σ; the closed surface extends over the surface Σ as well as S.

We begin by substituting φ and χ into the second formula. Then

$$\iiint_V \frac{1}{r} \nabla^2 u \, dV = \oiint_{S+\Sigma} \left[u \nabla\left(\frac{1}{r}\right) - \frac{1}{r} \nabla u \right] \cdot \mathbf{n} \, dS \qquad (\mathbf{1.3.7})$$

$$= \oiint_S \left[u \nabla\left(\frac{1}{r}\right) - \frac{1}{r} \nabla u \right] \cdot \mathbf{n} \, dS$$

$$+ \oiint_\Sigma \left[u \nabla\left(\frac{1}{r}\right) - \frac{1}{r} \nabla u \right] \cdot \mathbf{n} \, dS, \qquad (\mathbf{1.3.8})$$

because $\nabla^2 \left(r^{-1} \right) = 0$ over the enclosed volume. Recalling that the surface integration about the space Σ involves a normal *into* the small sphere,

$$\oiint_\Sigma u \, \nabla\left(\frac{1}{r}\right) \cdot \mathbf{n} \, dS = -\int_0^{2\pi} \int_0^\pi u(\epsilon, \varphi, \theta) \, \frac{1}{\epsilon^2} \, \epsilon^2 \sin(\theta) \, d\theta \, d\varphi, \quad (\mathbf{1.3.9})$$

and

$$\oiint_\Sigma \frac{1}{r} \nabla u \cdot \mathbf{n} \, dS = \epsilon \int_0^{2\pi} \int_0^\pi \nabla u(\epsilon, \varphi, \theta) \, \sin(\theta) \, d\theta \, d\varphi. \qquad (\mathbf{1.3.10})$$

In the limit of $\epsilon \to 0$, (1.3.10) vanishes because $\nabla u(\epsilon, \varphi, \theta)$ is finite while (1.3.9) becomes

$$\oiint_\Sigma u \, \nabla\left(\frac{1}{r}\right) \cdot \mathbf{n} \, dS = -4\pi u(\xi, \eta, \zeta). \qquad (\mathbf{1.3.11})$$

Substituting (1.3.11) into (1.3.8), we have that

$$u(\xi, \eta, \zeta) = \frac{1}{4\pi} \oiint_S \left[u \nabla\left(\frac{1}{r}\right) - \frac{1}{r} \nabla u \right] \cdot \mathbf{n} \, dS - \frac{1}{4\pi} \iiint_V \frac{1}{r} \nabla^2 u \, dV.$$
$$(\mathbf{1.3.12})$$

This formula (1.3.12), first derived by Green, shows that a function that is continuous in a certain space inside a closed surface S is determined at any point (ξ, η, ζ) in the interior if we know:

- $\nabla^2 u$ at every point in V,

- the value of u at every point on the surface S, and

- the gradient of u along the normal to S at every point on S.

If u is harmonic so that $\nabla^2 u = 0$, then we need only know u and the gradient of u on the surface S.

• Example 1.3.2: The Dirichlet and Neumann problems

Let us examine (1.3.12) further under the condition that u is harmonic. Then (1.3.12) simplifies to

$$u(\xi, \eta, \zeta) = \frac{1}{4\pi} \oiint_S \left[u \nabla \left(\frac{1}{r} \right) - \frac{1}{r} \nabla u \right] \cdot \mathbf{n} \, dS. \qquad (1.3.13)$$

Consequently, (1.3.13) provides a means for *analytical continuation* of the surface values of u into the volume provided we know *both* u and $\partial u / \partial n$ on the surface S. What we wish to prove here is that we can specify *either* u *or* $\partial u / \partial n$ on the surface and have a unique answer, but *not* both.

We begin by assuming that there are two analytical continuations u_1 and u_2. Then the difference $v = u_1 - u_2$ is also harmonic because u_1 and u_2 are, and $v = 0$ on the surface S. From Green's first formula with $\varphi = \chi = v$,

$$\iiint_V \left[\left(\frac{\partial v}{\partial x} \right)^2 + \left(\frac{\partial v}{\partial y} \right)^2 + \left(\frac{\partial v}{\partial z} \right)^2 \right] dV = \oiint_S v \frac{\partial v}{\partial n} \, dS - \iiint_V v \nabla^2 v \, dV. \qquad (1.3.14)$$

Because $\nabla^2 v = 0$ in the volume and $v = 0$ on the surface, both integrals on the right side of (1.3.14) vanish. Consequently, the integrand of the left side of (1.3.14) must vanish for any arbitrary volume. Thus,

$$\frac{\partial v}{\partial x} = \frac{\partial v}{\partial y} = \frac{\partial v}{\partial z} = 0, \qquad (1.3.15)$$

and v is constant. Since $v = 0$ on S, it must equal zero everywhere within the volume. Thus, $u_1 = u_2$ and the continuation is unique for a harmonic function. This result is due to P. G. L. Dirichlet[20] (1805–1859). The problem of finding the harmonic continuation for given surface values is known as the *Dirichlet problem* or the *first boundary problem*.

The *second boundary problem* is associated with the name of C. G. Neumann (1832–1925). It differs from the Dirichlet problem by specifying the normal derivative of u, rather than u, along the boundary. The proof follows the one for the Dirichlet problem because (1.3.14) still vanishes when the normal derivative of u along the boundary equals zero.

---

[20] Dirichlet, P. G. L., 1850: Über einen neuen Ausdruck zur Bestimmung der Dichtigkeit einer unendlich dünnen Kugelschale, wenn der Werth des Potentials derselben in jedem Punkte ihrer Oberfläche gegeben ist. *Abh. Königlich. Preuss. Akad. Wiss.*, 99–116.

1.4 WHAT IS A GREEN'S FUNCTION?

An important aspect of mathematical physics is the solution of linear, nonhomogeneous differential equations. In the case of ordinary differential equations we can express this problem as $Lu = f$, where L is an ordinary linear differential operator involving the independent variable x such that $f(x)$ is a known function and $u(x)$ is the desired solution. Although we can solve this problem through the use of expansions of eigenfunctions given by the eigenvalue problem $Lu = \lambda u$, where λ is an arbitrary constant, this book develops an alternative. This method consists of finding solutions to the differential equation $Lg = \delta(x - \xi)$, where ξ is an arbitrary point of excitation. We will then show that the solution $u(x)$ is given by an integral involving the *Green's function* $g(x, \xi)$ [more commonly written $g(x|\xi)$] and $f(\xi)$.

In the following examples taken from engineering, we show how Green's functions naturally appear during the solution of initial-value and boundary-value problems. We also show that the solution $u(x)$ can be expressed as an integral involving the Green's function and $f(x)$.

Circuit theory

In electrical engineering one of the simplest electrical devices consists of a voltage source $v(t)$ connected to a resistor with resistance R and inductor with inductance L. See Figure 1.4.1. Denoting the current by $i(t)$, the equation that governs this circuit is

$$L\frac{di}{dt} + Ri = v(t). \tag{1.4.1}$$

Consider now the following experiment: With the circuit initially dead, we allow the voltage suddenly to become $V_0/\Delta\tau$ during a very short duration $\Delta\tau$ starting at $t = \tau$. Then, at $t = \tau + \Delta\tau$, we again turn off the voltage supply. Mathematically, for $t > \tau + \Delta\tau$, the circuit's performance obeys the homogeneous differential equation

$$L\frac{di}{dt} + Ri = 0, \qquad t > \tau + \Delta\tau, \tag{1.4.2}$$

whose solution is

$$i(t) = I_0 e^{-Rt/L}, \qquad t > \tau + \Delta\tau, \tag{1.4.3}$$

where I_0 is a constant and L/R is the *time constant* of the circuit. Because the voltage $v(t)$ during $\tau < t < \tau + \Delta\tau$ is $V_0/\Delta\tau$, then

$$\int_\tau^{\tau+\Delta\tau} v(t)\, dt = V_0. \tag{1.4.4}$$

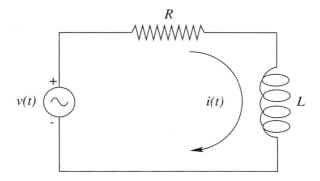

Figure 1.4.1: The *RL* electrical circuit driven by the voltage $v(t)$.

Therefore, over the interval $\tau < t < \tau + \Delta\tau$, (1.4.1) can be integrated to yield

$$L \int_\tau^{\tau + \Delta\tau} di + R \int_\tau^{\tau + \Delta\tau} i(t)\, dt = \int_\tau^{\tau + \Delta\tau} v(t)\, dt, \qquad (\mathbf{1.4.5})$$

or

$$L\left[i(\tau + \Delta\tau) - i(\tau)\right] + R \int_\tau^{\tau + \Delta\tau} i(t)\, dt = V_0. \qquad (\mathbf{1.4.6})$$

If $i(t)$ remain continuous as $\Delta\tau$ becomes small, then

$$R \int_\tau^{\tau + \Delta\tau} i(t)\, dt \approx 0. \qquad (\mathbf{1.4.7})$$

Finally, because

$$i(\tau) = 0, \qquad (\mathbf{1.4.8})$$

and

$$i(\tau + \Delta\tau) = I_0 e^{-R(\tau + \Delta\tau)/L} \approx I_0 e^{-R\tau/L}, \qquad (\mathbf{1.4.9})$$

for small $\Delta\tau$, (1.4.6) reduces to

$$L I_0 e^{-R\tau/L} = V_0, \qquad (\mathbf{1.4.10})$$

or

$$I_0 = \frac{V_0}{L} e^{R\tau/L}. \qquad (\mathbf{1.4.11})$$

Therefore, (1.4.3) can be written as

$$i(t) = \begin{cases} 0, & t < \tau, \\ V_0 e^{-R(t-\tau)/L}/L, & \tau \le t, \end{cases} \qquad (\mathbf{1.4.12})$$

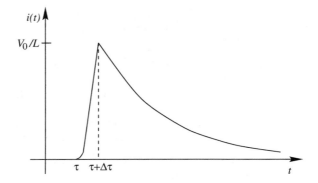

Figure 1.4.2: The current $i(t)$ within a RL circuit when the voltage $V_0/\Delta\tau$ is introduced between $\tau < t < \tau + \Delta\tau$.

after using (1.4.11). Equation (1.4.12) is plotted in Figure 1.4.2.

Consider now a new experiment with the same circuit where we subject the circuit to N voltage impulses, each of duration $\Delta\tau$ and amplitude $V_i/\Delta\tau$ with $i = 0, 1, \ldots, N$, occurring at $t = \tau_i$. See Figure 1.4.3. The current response is then

$$
i(t) = \begin{cases}
0, & t < \tau_0, \\[2ex]
V_0 e^{-R(t-\tau_0)/L}/L, & \tau_0 < t < \tau_1, \\[2ex]
V_0 e^{-R(t-\tau_0)/L}/L + V_1 e^{-R(t-\tau_1)/L}/L, & \tau_1 < t < \tau_2, \\[2ex]
\quad\vdots & \quad\vdots \\[2ex]
\displaystyle\sum_{i=0}^{N} V_i e^{-R(t-\tau_i)/L}/L, & \tau_N < t < \tau_{N+1}.
\end{cases}
$$

$$(1.4.13)$$

Finally, consider our circuit subjected to a continuous voltage source $v(t)$. Over each successive interval $d\tau$, the step change in voltage is $v(\tau)\,d\tau$. Consequently, from (1.4.13) the response $i(t)$ is now given by the *superposition integral*

$$
i(t) = \int_{\tau}^{t} \frac{v(\tau)}{L} e^{-R(t-\tau)/L}\, d\tau,
\tag{1.4.14}
$$

or

$$
i(t) = \int_{\tau}^{t} v(\tau) g(t|\tau)\, d\tau,
\tag{1.4.15}
$$

where

$$
g(t|\tau) = \frac{e^{-R(t-\tau)/L}}{L}, \qquad \tau < t.
\tag{1.4.16}
$$

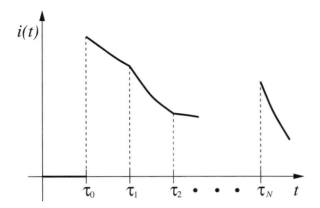

Figure 1.4.3: The current $i(t)$ within a RL circuit when the voltage is changed at $t = \tau_0$, $t = \tau_1$, and so forth.

Here we have assumed that $i(t) = v(t) = 0$ for $t < \tau$. In (1.4.15), $g(t|\tau)$ is called the *Green's function*. As (1.4.15) shows, given the Green's function to (1.4.1), the response $i(t)$ to any voltage source $v(t)$ can be obtained by convolving the voltage source with the Green's function.

We now show that we could have found the Green's function (1.4.16) by solving (1.4.1) subject to an impulse or delta forcing function. Mathematically, this corresponds to solving the following initial-value problem:

$$L\frac{dg}{dt} + Rg = \delta(t - \tau), \qquad g(0|\tau) = 0. \qquad (1.4.17)$$

Taking the Laplace transform[21] of (1.4.17), we find that

$$G(s|\tau) = \frac{e^{-s\tau}}{Ls + R}, \qquad (1.4.18)$$

or

$$g(t|\tau) = \frac{e^{-R(t-\tau)/L}}{L}H(t - \tau), \qquad (1.4.19)$$

where $H(\ )$ is the Heaviside step function (1.2.7). As our short derivation showed, the most direct route to finding a Green's function is solving the differential equation when its forcing equals the impulse or delta function. This is the technique that we will use throughout this book.

Statics

Consider a string of length L that is connected at both ends to supports and is subjected to a load (external force per unit length) of

---

[21] If the reader is unfamiliar with Laplace transform and their use in solving ordinary differential equations, Appendix B gives an overview of this technique.

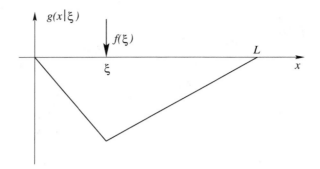

Figure 1.4.4: The response, commonly called a Green's function, of a string fixed at both ends to a point load at $x = \xi$.

$f(x)$. We wish to find the displacement $u(x)$ of the string. If the load $f(x)$ acts downward (negative direction), the displacement $u(x)$ of the string is given by the differential equation

$$T\frac{d^2 u}{dx^2} = f(x), \tag{1.4.20}$$

where T denotes the uniform tensile force of the string. Because the string is stationary at both ends, the displacement $u(x)$ satisfies the boundary conditions

$$u(0) = u(L) = 0. \tag{1.4.21}$$

Instead of directly solving for the displacement $u(x)$ of the string subject to the load $f(x)$, let us find the displacement that results from a load $\delta(x - \xi)$ concentrated at the point $x = \xi$. See Figure 1.4.4. For this load, the differential equation (1.4.20) becomes

$$T\frac{d^2 g}{dx^2} = \delta(x - \xi), \tag{1.4.22}$$

subject to the boundary conditions

$$g(0|\xi) = g(L|\xi) = 0. \tag{1.4.23}$$

In (1.4.22) $g(x|\xi)$ denotes the displacement of the string when it is subjected to an impulse load at $x = \xi$. In line with our circuit theory example, it gives the *Green's function* for our statics problem. Once found, the displacement $u(x)$ of the string subject to any arbitrary load $f(x)$ can be found by convolving the load $f(x)$ with the Green's function $g(x|\xi)$ as we did earlier.

Let us now find this Green's function. At any point $x \neq \xi$, (1.4.22) reduces to the homogeneous differential equation

$$\frac{d^2 g}{dx^2} = 0, \tag{1.4.24}$$

which has the solution

$$g(x|\xi) = \begin{cases} ax + b, & 0 \le x < \xi, \\ cx + d, & \xi < x \le L. \end{cases} \qquad (1.4.25)$$

Applying the boundary conditions (1.4.23), we find that

$$g(0|\xi) = a \cdot 0 + b = b = 0, \qquad (1.4.26)$$

and

$$g(L|\xi) = cL + d = 0, \quad \text{or} \quad d = -cL. \qquad (1.4.27)$$

Therefore, we can rewrite (1.4.25) as

$$g(x|\xi) = \begin{cases} ax, & 0 \le x < \xi, \\ c(x - L), & \xi < x \le L, \end{cases} \qquad (1.4.28)$$

where a and c are undetermined constants.

At $x = \xi$, the displacement $u(x)$ of the string must be continuous; otherwise, the string would be broken. Therefore, the Green's function given by (1.4.28) must also be continuous there. Thus,

$$a\xi = c(\xi - L), \quad \text{or} \quad c = \frac{a\xi}{\xi - L}. \qquad (1.4.29)$$

From (1.4.22) the second derivative of $g(x|\xi)$ must equal the impulse function. Therefore, the first derivative of $g(x|\xi)$, obtained by integrating (1.4.22), must be discontinuous by the amount $1/T$ or

$$\lim_{\epsilon \to 0} \left[\frac{dg(\xi + \epsilon|\xi)}{dx} - \frac{dg(\xi - \epsilon|\xi)}{dx} \right] = \frac{1}{T}, \qquad (1.4.30)$$

in which case

$$\frac{dg(\xi^+|\xi)}{dx} - \frac{dg(\xi^-|\xi)}{dx} = \frac{1}{T}, \qquad (1.4.31)$$

where ξ^+ and ξ^- denote points lying just above or below ξ, respectively. Using (1.4.28), we find that

$$\frac{dg(\xi^-|\xi)}{dx} = a, \qquad (1.4.32)$$

and

$$\frac{dg(\xi^+|\xi)}{dx} = c = \frac{a\xi}{\xi - L}. \qquad (1.4.33)$$

Thus, (1.4.31) leads to

$$\frac{a\xi}{\xi - L} - a = \frac{1}{T}, \quad \text{or} \quad \frac{aL}{\xi - L} = \frac{1}{T}, \quad \text{or} \quad a = \frac{\xi - L}{LT}, \qquad (1.4.34)$$

and the Green's function is

$$g(x|\xi) = \frac{1}{TL}(x_> - L)x_<. \tag{1.4.35}$$

To find the displacement $u(x)$ subject to the load $f(x)$, we proceed as we did in the previous example. The result of this analysis is

$$u(x) = \int_0^L f(\xi)g(x|\xi)\,d\xi \tag{1.4.36}$$

$$= \frac{x-L}{TL}\int_0^x f(\xi)\,\xi\,d\xi + \frac{x}{TL}\int_x^L f(\xi)\,(\xi - L)\,d\xi, \tag{1.4.37}$$

since $\xi < x$ in the first integral and $x < \xi$ in the second integral in (1.4.37).

Problems

In problems 1–9, you will show that the delta function can be re-expressed in terms of various eigenfunction expansions. We will use these results repeatedly in subsequent chapters.

1. Show that

$$\delta(x-\xi) = \frac{2}{L}\sum_{n=1}^{\infty}\sin\left(\frac{n\pi\xi}{L}\right)\sin\left(\frac{n\pi x}{L}\right), \quad 0 < x, \xi < L.$$

2. Show that

$$\delta(x-\xi) = \frac{1}{L} + \frac{2}{L}\sum_{n=1}^{\infty}\cos\left(\frac{n\pi\xi}{L}\right)\cos\left(\frac{n\pi x}{L}\right), \quad 0 < x, \xi < L.$$

3. Show that

$$\delta(x-\xi) = 2\sum_{n=1}^{\infty}\frac{k_n^2 + h^2}{L(k_n^2 + h^2) + h}\cos(k_n\xi)\cos(k_n x),$$

where $0 < x, \xi < L$, $\cos(k_n x)$ is the eigenfunction for the Sturm-Liouville problem $y'' + \lambda y = 0$ with $y'(0) = 0$ and $y'(L) + hy(L) = 0$, and k_n is the nth root of $k\tan(kL) = h$.

4. Show that

$$\delta(x-\xi) = \frac{2e^{x+\xi}}{e^{2L}-1} + 2L\sum_{n=1}^{\infty}\frac{\varphi_n(\xi)\varphi_n(x)}{n^2\pi^2 + L^2},$$

where $0 < x, \xi < L$, and $\varphi_n(x)$ is the eigenfunction for the Sturm-Liouville problem $\varphi'' + \lambda\varphi = 0$, $\varphi(0) - \varphi'(0) = 0$, and $\varphi'(L) - \varphi(L) = 0$.

5. Show that

$$\delta(x - \xi) = 4k^2 x e^{-k(x+\xi)} \sum_{n=0}^{\infty} \frac{L_n^1(2kx) L_n^1(2k\xi)}{n+1}, \quad 0 < x, \xi < \infty,$$

where $k > 0$ and $L_n^m(\ )$ is the (generalized) Laguerre polynomial:[22]

$$L_n^m(x) = e^x \frac{x^{-m}}{n!} \frac{d^n}{dx^n} \left(e^{-x} x^{n+m} \right), \quad n = 0, 1, 2, \ldots.$$

6. Show that the Fourier-Bessel expansion[23] for $\delta(r - \rho)/(2\pi r)$ is

$$\frac{\delta(r - \rho)}{2\pi r} = \frac{1}{\pi b^2} \sum_{n=1}^{\infty} \frac{J_0(k_n\rho/b) J_0(k_n r/b)}{J_1^2(k_n)}, \quad 0 < r, \rho < b,$$

where k_n is the nth root of $J_0(k) = 0$.

7. Show that the Fourier-Bessel expansion for $\delta(r - \rho)/(2\pi r)$ is

$$\frac{\delta(r - \rho)}{2\pi r} = \frac{1}{\pi} \sum_{n=0}^{\infty} \frac{J_0(k_n\rho) J_0(k_n r)}{J_0^2(k_n)}, \quad 0 < r, \rho < 1,$$

where k_n is the nth root of $J_0'(k) = -J_1(k) = 0$.

8. Show that the Fourier-Legendre expansion[24] for $\delta(\theta - \theta')/\sin(\theta')$ is

$$\frac{\delta(\theta - \theta')}{\sin(\theta')} = \sum_{n=0}^{\infty} \left(n + \tfrac{1}{2} \right) P_n[\cos(\theta)] P_n[\cos(\theta')],$$

where $0 < \theta, \theta' < \pi$.

9. Using double Fourier series, show that

$$\delta(x-\xi)\delta(y-\eta) = \frac{4}{ab} \sum_{n=1}^{\infty} \sum_{m=1}^{\infty} \sin\left(\frac{n\pi\xi}{a}\right) \sin\left(\frac{m\pi\eta}{b}\right) \sin\left(\frac{n\pi x}{a}\right) \sin\left(\frac{m\pi y}{b}\right),$$

---

[22] See Lebedev, N. N., 1972: *Special Functions and Their Applications*. Dover Publications, Inc., §§4.17–4.25.

[23] For those unfamiliar with Fourier-Bessel expansions, see §C.3 in Appendix C.

[24] A Fourier-Legendre expansion is a Fourier series where Legendre polynomials are used in place of sine and cosine. See the section on this in my *Advanced Engineering Mathematics* book.

where $0 < x, \xi < a$, and $0 < y, \eta < b$.

10. Prove that

$$\delta(x - x_0) = \int_{-\infty}^{\infty} \delta(x - \tau)\delta(\tau - x_0)\, d\tau.$$

11. Prove that

$$\int_a^b f(\tau)\, \delta'(\tau - t)\, d\tau = -f'(t)$$

if $a < t < b$.

12. Show that

$$\int_a^b \tau\, \delta(t - \tau)\, d\tau = t\left[H(t - a) - H(t - b)\right].$$

Hint: Use integration by parts.

13. Prove that

$$f(x)\delta(x - x_0) = \left[\frac{f(x) + f(x_0)}{2}\right]\delta(x - x_0).$$

Hint: Use $f(x)\delta(x - x_0) = f(x_0)\delta(x - x_0)$.

14. Prove that

$$\lim_{\epsilon \to 0} \frac{\epsilon}{(r^2 + \epsilon^2)^{3/2}} = \frac{\delta(r)}{r}.$$

15. Verify Green's formulas using the scalar functions $\varphi = xyz$ and $\chi = x + y + z$ and the cube defined by the planes $x = 0$, $x = 1$, $y = 0$, $y = 1$, $z = 0$, and $z = 1$.

Chapter 2
Green's Functions
for Ordinary Differential Equations

Having given a general overview of Green's functions, we provide over the next four chapters the Green's functions for a wide class of ordinary and partial differential equations. We begin with ordinary differential equations. In §§2.1 and 2.2 we show how Laplace transforms are used to find Green's functions for initial-value problems. In the case of boundary-value problems, there are two techniques. In the operator method (§§2.3 and 2.5), solutions are constructed for regions to the right and left of the point of excitation and then pieced together to give the complete Green's function. This is quite different from modal expansions presented in §2.4, which express the Green's function in terms of a superposition of eigenfunctions that are valid over the entire domain. We conclude the chapter by introducing the property of reciprocity: the response at x due to a delta function at ξ equals the response at ξ due to a delta function at x.

2.1 INITIAL-VALUE PROBLEMS

The simplest application of Green's functions to solving ordinary differential equations arises with the initial-value problem

$$\frac{d^n y}{dt^n} + a_1 \frac{d^{n-1}y}{dt^{n-1}} + \cdots + a_{n-1}\frac{dy}{dt} + a_n y = f(t), \quad 0 < t, \qquad (\mathbf{2.1.1})$$

where $a_1, a_2, \ldots$ are constants and all of the values of $y, y', \ldots, y^{(n-1)}$ at $t = 0$ are zero. Although there are several methods for solving (2.1.1), let us employ Laplace transforms. In that case, (2.1.1) becomes

$$\left(s^n + a_1 s^{n-1} + a_2 s^{n-2} + \cdots + a_n\right) Y(s) = F(s), \qquad (2.1.2)$$

where $Y(s)$ and $F(s)$ denote the Laplace transforms of $y(t)$ and $f(t)$, respectively. Consequently, once $F(s)$ is known, (2.1.2) is inverted to give $y(t)$.

What are the difficulties with this technique? The most obvious is the dependence of the solution on the forcing function $f(t)$. For each new $f(t)$ the inversion process must be repeated. Is there any way to avoid this problem?

We begin by rewriting (2.1.2) as

$$Y(s) = \frac{F(s)}{s^n + a_1 s^{n-1} + a_2 s^{n-2} + \cdots + a_n} = G(s)F(s), \qquad (2.1.3)$$

where $G(s) = (s^n + a_1 s^{n-1} + a_2 s^{n-2} + \cdots + a_n)^{-1}$. Thus, by the convolution theorem,

$$y(t) = g(t) * f(t) = \int_0^t g(x) f(t - x) \, dx, \qquad (2.1.4)$$

where $g(t)$ is the inverse Laplace transform of $G(s)$. Once we have $g(t)$, we can compute the response $y(t)$ to the forcing function $f(t)$ by an integration. Although we appear to have gained nothing—we have replaced an inversion by an integration—we actually have made progress. We have succeeded in decoupling the forcing from that portion of the solution that depends solely upon the differential equation. In this manner, we can explore the properties of the differential equation, and hence the mechanical or electrical properties of the corresponding physical system by itself. Indeed, this is what engineers do as they explore the *transfer function* $G(s)$ and its inverse, the *impulse response* or *Green's function* $g(t)$.

So far, we have assumed homogeneous initial conditions. What do we do if we have nonhomogeneous initial conditions? In that case, we would simply find a homogeneous solution that satisfies the initial conditions and add that solution to our nonhomogeneous solution (2.1.4).

Having shown the usefulness of Green's function, let us define it for the initial-value problem (2.1.1). Given a *linear* ordinary differential equation (2.1.1), the Green's function $g(t|\tau)$ of (2.1.1) is the solution to the differential equation

$$\frac{d^n g}{dt^n} + a_1 \frac{d^{n-1} g}{dt^{n-1}} + \cdots + a_{n-1} \frac{dg}{dt} + a_n g = \delta(t - \tau), \quad 0 < t, \tau. \quad (2.1.5)$$

The forcing occurs at time $t = \tau$, which is later than the initial time $t = 0$. This avoids the problem of the Green's function *not* satisfying all of the initial conditions.

Remark. Although we will use (2.1.5) as our fundamental definition of the Green's function as it applies to the ordinary differential equation, we can also find it by solving the initial-value problem:

$$\frac{d^n u}{dt^n} + a_1 \frac{d^{n-1} u}{dt^{n-1}} + \cdots + a_{n-1} \frac{du}{dt} + a_n u = 0, \qquad \tau < t, \qquad \textbf{(2.1.6)}$$

with the initial conditions

$$u(\tau) = u'(\tau) = \cdots = u^{(n-2)}(\tau) = 0, \quad \text{and} \quad u^{(n-1)}(\tau) = 1. \quad \textbf{(2.1.7)}$$

Equation (2.1.7) is an example of a *stationary* system, one where the differential equation and initial conditions are invariant under a translation in time. The Green's function is related to $u(t)$ via $g(t|\tau) = u(t - \tau)H(t - \tau)$. This can be seen by introducing a new time variable $t' = t - \tau$ and showing that the Laplace transform of (2.1.5) and (2.1.7) are identical.

Having defined the Green's function, our final task is to prove that

$$y(t) = \int_0^t g(t|\tau) f(\tau) \, d\tau. \qquad \textbf{(2.1.8)}$$

We begin by noting that

$$y'(t) = g(t|t)f(t) + \int_0^t g'(t|\tau)f(\tau) \, d\tau,$$

$$y''(t) = g'(t|t)f(t) + \int_0^t g''(t|\tau)f(\tau) \, d\tau, \qquad \textbf{(2.1.9)}$$

$$\vdots$$

$$y^{(n)}(t) = g^{(n-1)}(t|t)f(t) + \int_0^t g^{(n)}(t|\tau)f(\tau) \, d\tau.$$

We now require that the Green's function possess the properties that $g(t|t) = g'(t|t) = g''(t|t) = \cdots = g^{(n-2)}(t|t) = 0$ and $g^{(n-1)}(t|t) = 1$. Then, upon substituting (2.1.9) into (2.1.1), we find that

$$\frac{d^n y}{dt^n} + a_1 \frac{d^{n-1} y}{dt^{n-1}} + \cdots + a_{n-1} \frac{dy}{dt} + a_n y$$

$$= f(t) + \int_0^t \left[\frac{d^n g}{dt^n} + a_1 \frac{d^{n-1} g}{dt^{n-1}} + \cdots + a_{n-1} \frac{dg}{dt} + a_n g \right] f(\tau) \, d\tau$$

$$\textbf{(2.1.10)}$$

$$= f(t), \qquad \qquad \textbf{(2.1.11)}$$

because the bracketed term in (2.1.10) vanishes except at the point $t = \tau$. However, the contribution from this point is infinitesimally small.

• **Example 2.1.1**

Let us find the transfer and Green's functions for the system

$$y'' - 3y' + 2y = f(t), \qquad (2.1.12)$$

with $y(0) = y'(0) = 0$. To find the Green's function, we replace (2.1.12) with

$$g'' - 3g' + 2g = \delta(t - \tau), \qquad (2.1.13)$$

with $g(0|\tau) = g'(0|\tau) = 0$. Taking the Laplace transform of (2.1.13), we find that

$$G(s|\tau) = \frac{e^{-s\tau}}{s^2 - 3s + 2}, \qquad (2.1.14)$$

which is the transfer function for this system when $\tau = 0$. The Green's function equals the inverse of $G(s|\tau)$ or

$$g(t|\tau) = \left[e^{2(t-\tau)} - e^{t-\tau} \right] H(t - \tau), \qquad (2.1.15)$$

where $H(t)$ is the Heaviside step function

$$H(t - a) = \begin{cases} 1, & t > a, \\ 0, & t < a, \end{cases} \qquad (2.1.16)$$

where $a \geq 0$.

• **Example 2.1.2: The damped harmonic oscillator**

Second-order differential equations are ubiquitous in the physical sciences. A prototypical example involves a mass m attached to a spring with a spring constant k and damped with a velocity-dependent resistance. The governing equation is

$$my'' + cy' + ky = f(t), \qquad (2.1.17)$$

where $y(t)$ denotes the displacement of the oscillator from its equilibrium position, c denotes the damping coefficient, and $f(t)$ denotes the forcing.

Let us find the Green's function $g(t|\tau)$ for this *damped harmonic oscillator*. The Green's function is given by the equation

$$mg'' + cg' + kg = \delta(t - \tau), \qquad g(0|\tau) = g'(0|\tau) = 0, \qquad (2.1.18)$$

with $\tau > 0$.

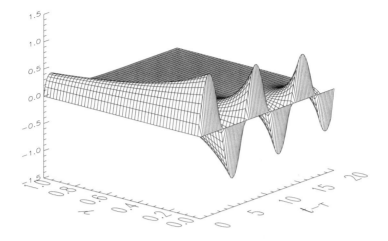

Figure 2.1.1: The Green's function for the damped harmonic oscillator as functions of time $t - \tau$ and γ when $\omega_0 = 0.9$.

Although we could solve (2.1.18) by Laplace transforms, an alternative approach employs Fourier transforms. To illustrate this method, let us use it here.

Taking the Fourier transform of both sides of (2.1.18),

$$G(\omega|\tau) = \frac{e^{-i\omega\tau}}{k + ic\omega - m\omega^2} = \frac{e^{-i\omega\tau}/m}{\omega_0^2 + ic\omega/m - \omega^2}, \qquad (2.1.19)$$

where $\omega_0^2 = k/m$ is the natural frequency of the system. From the definition of the inverse Fourier transform,

$$mg(t|\tau) = -\frac{1}{2\pi} \int_{-\infty}^{\infty} \frac{e^{i\omega(t-\tau)}}{\omega^2 - ic\omega/m - \omega_0^2} \, d\omega \qquad (2.1.20)$$

$$= -\frac{1}{2\pi} \int_{-\infty}^{\infty} \frac{e^{i\omega(t-\tau)}}{(\omega - \omega_1)(\omega - \omega_2)} \, d\omega, \qquad (2.1.21)$$

where

$$\omega_{1,2} = \pm\sqrt{\omega_0^2 - \gamma^2} + \gamma i, \qquad (2.1.22)$$

and $\gamma = c/(2m) > 0$.

We evaluate (2.1.21) by residues. If $t < 0$, we close the line integration in (2.1.21) along the real axis with a semicircle of infinite radius in the lower half of the ω-plane by Jordan's lemma. Because the poles in (2.1.21) always lie in the upper half of the ω-plane, the integrand is analytic within the closed contour and $g(t|\tau) = 0$ for $t < 0$. This is simply the causality condition,[1] the impulse forcing being the cause of the

---

[1] The principle stating that an event cannot precede its cause.

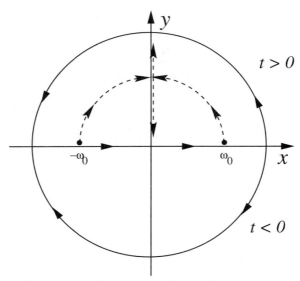

Figure 2.1.2: The migration of the poles of $G(\omega|\tau)$ of a damped harmonic oscillator as a function of γ.

excitation. Clearly, causality is closely connected with the analyticity of $G(\omega|\tau)$ in the lower half of the ω-plane.

If $t > 0$, we close (2.1.21) along the real axis with a semicircle of infinite radius in the upper half of the ω-plane. A useful method for understanding the inversion is to describe the migration of the poles $\omega_{1,2}$ in the complex ω-plane as γ increases from 0 to ∞. See Figure 2.1.2. For $\gamma \ll \omega_0$ (weak damping), the poles $\omega_{1,2}$ are very near to the real axis, above the points $\pm\omega_0$, respectively. As γ increases from 0 to ω_0, the poles approach the positive imaginary axis, moving along a semicircle of radius ω_0 centered at the origin. Thus, at any point along this semicircle except $\gamma = \omega_0$, we have simple poles and

$$mg(t|\tau) = 2\pi i \left(-\frac{1}{2\pi} \right) \left\{ \mathrm{Res}\left[\frac{e^{iz(t-\tau)}}{(z - \omega_1)(z - \omega_2)}; \omega_1 \right] \right.$$

$$\left. + \mathrm{Res}\left[\frac{e^{iz(t-\tau)}}{(z - \omega_1)(z - \omega_2)}; \omega_2 \right] \right\} \quad (2.1.23)$$

$$= \frac{-i}{\omega_1 - \omega_2} \left[e^{i\omega_1(t-\tau)} - e^{i\omega_2(t-\tau)} \right] \quad (2.1.24)$$

$$= \frac{e^{-\gamma(t-\tau)} \sin\left[(t - \tau)\sqrt{\omega_0^2 - \gamma^2} \right]}{\sqrt{\omega_0^2 - \gamma^2}} H(t - \tau). \quad (2.1.25)$$

Equation (2.1.25) gives the Green's function for an underdamped harmonic oscillator.

What happens when γ equals ω_0 and the poles coalesce? We have $\omega_1 = \omega_2$, a second-order pole in the integrand of (2.1.21) and a critically damped oscillator. In this case, the Green's function becomes

$$mg(t|\tau) = (t - \tau)e^{-\gamma(t-\tau)}H(t - \tau). \qquad (2.1.26)$$

Finally, for $\gamma > \omega_0$, the poles move in opposite directions along the positive imaginary axis; one of them approaches the origin, while the other tends to $i\infty$ as $\gamma \to \infty$. We again have two simple poles and two purely decaying, overdamped solutions. The Green's function in this case is then

$$mg(t|\tau) = \frac{e^{-\gamma(t-\tau)} \sinh\left[(t - \tau)\sqrt{\gamma^2 - \omega_0^2}\right]}{\sqrt{\gamma^2 - \omega_0^2}}H(t - \tau). \qquad (2.1.27)$$

2.2 THE SUPERPOSITION INTEGRAL

One of the earliest uses of Green's functions with ordinary differential equations involved electrical circuits. This theory developed independently of its evolution in physics and consequently employed the term impulse response in place of Green's function. Its roots lie in the superposition principle, which states that the response of a linear system to the simultaneous action of several causes is obtained by first determining the effects of the separate causes and then adding them in the proper manner.

To derive the superposition integral we need to introduce a new concept, namely the *indicial response* $a(t)$. The indicial response is defined as the response of a linear system initially at rest to a unit step function $H(t)$. Therefore, if the system is dormant until $t = \tau_1$ and we subject the system at that instant to a forcing $H(t - \tau_1)$, then the response will be zero if $t < \tau_1$ and will equal the indicial admittance $a(t - \tau_1)$ when $t > \tau_1$ because the time of excitation is $t = \tau_1$ and not $t = 0$.

Consider now a linear system that we force with the value $f(0)$ when $t = 0$ and hold that value until $t = \tau_1$. We then abruptly change the forcing by an amount $f(\tau_1) - f(0)$ to the value $f(\tau_1)$ at the time τ_1 and hold it at that value until $t = \tau_2$. Then we again abruptly change the forcing by an amount $f(\tau_2) - f(\tau_1)$ at the time τ_2, and so forth (see Figure 2.2.1). From the *linearity* of the problem, the response after the instant $t = \tau_n$ equals the sum

$$y(t) = f(0)a(t) + [f(\tau_1) - f(0)]a(t - \tau_1) + [f(\tau_2) - f(\tau_1)]a(t - \tau_2)$$
$$+ \cdots + [f(\tau_n) - f(\tau_{n-1})]a(t - \tau_n). \qquad (2.2.1)$$

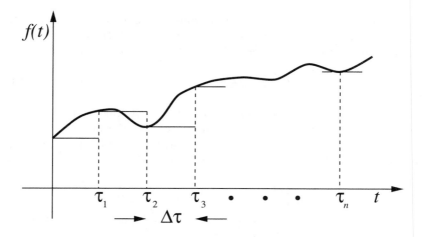

Figure 2.2.1: Diagram used in the derivation of Duhamel's integral.

If we write $f(\tau_k) - f(\tau_{k-1}) = \Delta f_k$, and $\tau_k - \tau_{k-1} = \Delta\tau_k$, (2.2.1) becomes

$$y(t) = f(0)a(t) + \sum_{k=1}^{n} a(t - \tau_k)\frac{\Delta f_k}{\Delta\tau_k}\,\Delta\tau_k. \qquad (2.2.2)$$

Finally, proceeding to the limit as the number n of jumps becomes infinite, in such a manner that all jumps and intervals between successive jumps tend to zero, this sum has the limit

$$y(t) = f(0)a(t) + \int_0^t f'(\tau)a(t - \tau)\,d\tau. \qquad (2.2.3)$$

Because the total response of the system equals the weighted sum [the weights being $a(t)$] of the forcing from the initial moment up to the time t, we refer to (2.2.3) as the *superposition integral*, or *Duhamel's integral*,[2] named after the French mathematical physicist Jean-Marie-Constant Duhamel (1797–1872) who first derived it in conjunction with heat conduction.

What does the superposition integral have to do with Green's functions? Integrating by parts transforms (2.2.3) into

$$y(t) = f(t)a(0) + \int_0^t f(\tau)a'(t - \tau)\,d\tau. \qquad (2.2.4)$$

---

[2] Duhamel, J.-M.-C., 1833: Mémoire sur la méthode générale relative au mouvement de la chaleur dans les corps solides plongés dans des milieux dont la température varié avec le temps. *J. École Polytech.*, **22**, 20–77.

However, from the definition of the Green's function and indicial response, we have that

$$g(t) = a'(t) + a(0)\delta(t).$$ (2.2.5)

Substituting (2.2.5) into (2.2.4), we obtain the final result that

$$y(t) = \int_0^t f(\tau)g(t-\tau)\,d\tau = \int_0^t f(t-\tau)g(\tau)\,d\tau.$$ (2.2.6)

2.3 REGULAR BOUNDARY-VALUE PROBLEMS

One of the purposes of this chapter is the solution of a wide class of nonhomogeneous ordinary differential equations of the form

$$\frac{d}{dx}\left[p(x)\frac{dy}{dx}\right] + s(x)y = -f(x), \qquad a \le x \le b,$$ (2.3.1)

with

$$\alpha_1 y(a) + \alpha_2 y'(a) = 0, \qquad \beta_1 y(b) + \beta_2 y'(b) = 0.$$ (2.3.2)

Although this may appear to be a fairly restrictive class of equations, it can be readily shown that any second-order, linear, nonhomogeneous ordinary differential equation can be written as (2.3.1). We placed a minus sign in front of $f(x)$ because we will encounter boundary-value problems of this particular form when we study Green's function for partial differential equations.

From the countless equations of the form (2.3.1), the most commonly encountered are those of the Sturm-Liouville type

$$\frac{d}{dx}\left[p(x)\frac{dy}{dx}\right] + [q(x) + \lambda r(x)]y = -f(x), \qquad a \le x \le b,$$ (2.3.3)

where λ is a parameter. This section is devoted to *regular Sturm-Liouville problems*. From the theory of differential equations, a regular Sturm-Liouville problem has three distinct properties: a finite interval $[a, b]$, continuous $p(x)$, $p'(x)$, $q(x)$ and $r(x)$, and strictly positive $p(x)$ and $r(x)$ on the *closed* interval $a \le x \le b$. If *any* of these conditions do not apply, then the problem is *singular*. We treat singular boundary-value problems in §2.5.

• Example 2.3.1

Consider the nonhomogeneous boundary-value problem

$$y'' + k^2 y = -f(x), \qquad y(0) = y(L) = 0. \qquad (2.3.4)$$

Because we can reexpress (2.3.4) as

$$\frac{d}{dx}\left(1 \cdot \frac{dy}{dx}\right) + (0 + k^2 \cdot 1)y = -f(x), \qquad (2.3.5)$$

(2.3.5) can be brought into the Sturm-Liouville form by choosing $p(x) = 1$, $q(x) = 0$, $r(x) = 1$, and $\lambda = k^2$.

Let us now determine the Green's function for the equation

$$\frac{d}{dx}\left[p(x)\frac{dg}{dx}\right] + s(x)g = -\delta(x - \xi), \qquad (2.3.6)$$

subject to yet undetermined boundary conditions. We know that such a function exists for the special case $p(x) = 1$ and $s(x) = 0$, and we now show that this is *almost always* true in the general case. Presently we construct Green's functions by requiring that they satisfy the following conditions:

> • $g(x|\xi)$ satisfies the *homogeneous* equation $f(x) = 0$ *except* at $x = \xi$,
>
> • $g(x|\xi)$ satisfies certain *homogeneous* conditions, and
>
> • $g(x|\xi)$ is continuous at $x = \xi$.

These homogeneous boundary conditions for a finite interval (a, b) will be

$$\alpha_1 g(a|\xi) + \alpha_2 g'(a|\xi) = 0, \qquad \beta_1 g(b|\xi) + \beta_2 g'(b|\xi) = 0, \qquad (2.3.7)$$

where g' denotes the x derivative of $g(x|\xi)$ and neither a nor b equals ξ. The coefficients α_1 and α_2 cannot both be zero; this also holds for β_1 and β_2. These conditions include the commonly encountered Dirichlet, Neumann, and Robin boundary conditions.

What about the value of $g'(x|\xi)$ at $x = \xi$? Because $g(x|\xi)$ is a continuous function of x, (2.3.6) dictates that there must be a discontinuity in $g'(x|\xi)$ at $x = \xi$. We now show that this discontinuity consists of a jump in the value $g'(x|\xi)$ at $x = \xi$. To prove this, we begin by integrating (2.3.6) from $\xi - \epsilon$ to $\xi + \epsilon$, which yields

$$p(x) \frac{dg(x|\xi)}{dx}\bigg|_{\xi-\epsilon}^{\xi+\epsilon} + \int_{\xi-\epsilon}^{\xi+\epsilon} s(x)g(x|\xi)\,dx = -1. \qquad (2.3.8)$$

Because $g(x|\xi)$ and $s(x)$ are both continuous at $x = \xi$,

$$\lim_{\epsilon \to 0} \int_{\xi-\epsilon}^{\xi+\epsilon} s(x)g(x|\xi)\,dx = 0. \qquad (2.3.9)$$

Applying the limit $\epsilon \to 0$ to (2.3.8), we have that

$$p(\xi)\left[\frac{dg(\xi^+|\xi)}{dx} - \frac{dg(\xi^-|\xi)}{dx}\right] = -1, \qquad (2.3.10)$$

where ξ^+ and ξ^- denote points just above and below $x = \xi$, respectively. Consequently, our last requirement on $g(x|\xi)$ will be that

- dg/dx must have a jump discontinuity of magnitude $-1/p(\xi)$ at $x = \xi$.

Similar conditions hold for higher-order ordinary differential equations.[3]

Consider now the region $a \le x < \xi$. Let $y_1(x)$ be a nontrivial solution of the *homogeneous* differential equation satisfying the boundary condition at $x = a$; then $\alpha_1 y_1(a) + \alpha_2 y_1'(a) = 0$. Because $g(x|\xi)$ must satisfy the same boundary condition, $\alpha_1 g(a|\xi) + \alpha_2 g'(a|\xi) = 0$. Since the set α_1, α_2 is nontrivial, then the Wronskian[4] of y_1 and g must vanish at $x = a$ or $y_1(a)g'(a|\xi) - y_1'(a)g(a|\xi) = 0$. However, for $a \le x < \xi$, both $y_1(x)$ and $g(x|\xi)$ satisfy the same differential equation, the homogeneous one. Therefore, their Wronskian is zero at all

---

[3] Ince, E. L., 1956: *Ordinary Differential Equations.* Dover Publications, Inc., §11.1.

[4] The Wronskian is defined by

$$W(y_1, y_2) = \begin{vmatrix} y_1(x) & y_2(x) \\ y_1'(x) & y_2'(x) \end{vmatrix} = y_1(x)y_2'(x) - y_1'(x)y_2(x).$$

points and $g(x|\xi) = c_1 y_1(x)$ for $a \leq x < \xi$, where c_1 is an arbitrary constant. In a similar manner, if the nontrivial function $y_2(x)$ satisfies the homogeneous equation and the boundary conditions at $x = b$, then $g(x|\xi) = c_2 y_2(x)$ for $\xi < x \leq b$. The continuity condition of g and the jump discontinuity of g' at $x = \xi$ imply

$$c_1 y_1(\xi) - c_2 y_2(\xi) = 0, \qquad c_1 y_1'(\xi) - c_2 y_2'(\xi) = 1/p(\xi). \qquad \textbf{(2.3.11)}$$

We can solve (2.3.11) for c_1 and c_2 provided the Wronskian of y_1 and y_2 does not vanish at $x = \xi$ or

$$y_1(\xi) y_2'(\xi) - y_2(\xi) y_1'(\xi) \neq 0. \qquad \textbf{(2.3.12)}$$

In other words, $y_1(x)$ must *not* be a multiple of $y_2(x)$. Is this always true? The answer is "generally yes." If the homogeneous equation admits no nontrivial solutions satisfying both boundary conditions at the same time,[5] then $y_1(x)$ and $y_2(x)$ must be linearly independent.[6] On the other hand, if the homogeneous equation possesses a single solution, say $y_0(x)$, which also satisfies $\alpha_1 y_0(a) + \alpha_2 y_0'(a) = 0$ and $\beta_1 y_0(b) + \beta_2 y_0'(b) = 0$, then $y_1(x)$ will be a multiple of $y_0(x)$ and so is $y_2(x)$. Then they are multiples of each other and their Wronskian vanishes. This would occur, for example, if the differential equation is a Sturm-Liouville equation, λ equals the eigenvalue, and $y_0(x)$ is the corresponding eigenfunction. No Green's function exists in this case.

- **Example 2.3.2**

Consider the problem of finding the Green's function for $g'' + k^2 g = -\delta(x - \xi)$, $0 < x < L$, subject to the boundary conditions $g(0|\xi) = g(L|\xi) = 0$ with $k \neq 0$. The corresponding homogeneous equation is $y'' + k^2 y = 0$. Consequently, $g(x|\xi) = c_1 y_1(x) = c_1 \sin(kx)$ for $0 \leq x \leq \xi$, while $g(x|\xi) = c_2 y_2(x) = c_2 \sin[k(L - x)]$ for $\xi \leq x \leq L$.

Let us compute the Wronskian. For our particular problem,

$$\begin{aligned} W(x) &= y_1(x) y_2'(x) - y_1'(x) y_2(x) & \textbf{(2.3.13)} \\ &= -k \sin(kx) \cos[k(L - x)] - k \cos(kx) \sin[k(L - x)] & \textbf{(2.3.14)} \\ &= -k \sin[k(x + L - x)] = -k \sin(kL), & \textbf{(2.3.15)} \end{aligned}$$

---

[5] In the theory of differential equations, this system would be called *incompatible*: one that admits no solution, save $y = 0$, which is also continuous for all x in the interval (a, b) and satisfies the homogeneous boundary conditions.

[6] Two functions f and g are said to be *linearly dependent* if there exist two constants c_1 and c_2, not both zero, such that $c_1 f(x) + c_2 g(x) = 0$ for all x in the interval. The functions f and g are said to be *linearly independent* on an interval if they are not linearly dependent.

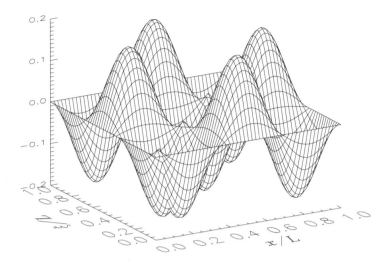

Figure 2.3.1: The Green's function (2.3.18), divided by L, as functions of x and ξ when $kL = 10$.

and $W(\xi) = -k\sin(kL)$. Therefore, the Green's function will exist as long as $kL \neq n\pi$. If $kL = n\pi$, $y_1(x)$ and $y_2(x)$ are linearly *dependent* with $y_0(x) = c_3\sin(n\pi x/L)$, the solution to the regular Sturm-Liouville problem $y'' + \lambda y = 0$, and $y(0) = y(L) = 0$.

Let us now proceed to find $g(x|\xi)$ when it does exist. The system (2.3.11) has the unique solution

$$c_1 = -\frac{y_2(\xi)}{p(\xi)W(\xi)}, \quad \text{and} \quad c_2 = -\frac{y_1(\xi)}{p(\xi)W(\xi)}, \qquad (\mathbf{2.3.16})$$

where $W(\xi)$ is the Wronskian of $y_1(x)$ and $y_2(x)$ at $x = \xi$. Therefore,

$$g(x|\xi) = -\frac{y_1(x_<)y_2(x_>)}{p(\xi)W(\xi)}. \qquad (\mathbf{2.3.17})$$

Clearly $g(x|\xi)$ is symmetric in x and ξ. It is also unique. The proof of the uniqueness is as follows: We can always choose a different $y_1(x)$, but it will be a multiple of the "old" $y_1(x)$, and the Wronskian will be multiplied by the same factor, leaving $g(x|\xi)$ the same. This is also true if we modify $y_2(x)$ in a similar manner.

- **Example 2.3.3**

Let us find the Green's function for $g'' + k^2g = -\delta(x-\xi)$, $0 < x < L$, subject to the boundary conditions $g(0|\xi) = g(L|\xi) = 0$. As we showed in the previous example, $y_1(x) = c_1 \sin(kx)$, $y_2(x) = c_2 \sin[k(L - x)]$, and $W(\xi) = -k \sin(kL)$. Substituting into (2.3.17), we have that

$$g(x|\xi) = \frac{\sin(kx_<) \sin[k(L - x_>)]}{k \sin(kL)}. \qquad (2.3.18)$$

- **Example 2.3.4**

Let us find the Green's function[7] for

$$g^{iv} - Pg'' - k^4 g = \delta(x - \xi), \qquad 0 < x, \xi < 1, \qquad (2.3.19)$$

subject to the boundary conditions

$$g(0|\xi) = A, \quad g'(0|\xi) = B, \quad g''(0|\xi) = C, \qquad (2.3.20)$$

and

$$g^{iii}(0|\xi) - Pg'(0|\xi) = D. \qquad (2.3.21)$$

In the region $0 \le x \le \xi$, the general solution to (2.3.19) is

$$g(x|\xi) = c_1 \cosh(\lambda x) + c_2 \sinh(\lambda x) + c_3 \cos(\omega x) + c_4 \sin(\omega x), \qquad (2.3.22)$$

where

$$\lambda = \sqrt{(Q + P)/2}, \quad \omega = \sqrt{(Q - P)/2}, \quad \text{and} \quad Q = \sqrt{P^2 + 4k^4}. \qquad (2.3.23)$$

Upon substituting (2.3.22) into the boundary conditions at $x = 0$, we have that

$$g(0|\xi) = c_1 + c_3 = A, \quad g'(0|\xi) = \lambda c_2 + \omega c_4 = B, \qquad (2.3.24)$$

$$g''(0|\xi) = \lambda^2 c_1 - \omega^2 c_3 = C, \qquad (2.3.25)$$

and

$$g^{iii}(0|\xi) - Pg'(0|\xi) = \lambda \omega^2 c_2 - \lambda^2 \omega c_4 = D. \qquad (2.3.26)$$

Solving for c_1 to c_4 and substituting these values into (2.3.22), we obtain

$$g(x|\xi) = A \frac{\omega^2 \cosh(\lambda x) + \lambda^2 \cos(\omega x)}{\lambda^2 + \omega^2} + B \frac{\lambda \sinh(\lambda x) + \omega \sin(\omega x)}{\lambda^2 + \omega^2}$$
$$+ C \frac{\cosh(\lambda x) - \cos(\omega x)}{\lambda^2 + \omega^2} + \frac{D}{\lambda^2 + \omega^2} \left[\frac{\sinh(\lambda x)}{\lambda} - \frac{\sin(\omega x)}{\omega} \right]. \qquad (2.3.27)$$

---

[7] Taken from Bergman, L. A., and J. E. Hyatt, 1989: Green functions for transversely vibrating uniform Euler-Bernoulli beams subject to constant axial preload. *J. Sound Vibr.*, **134**, 175–180. Published by Academic Press Ltd., London, U.K.

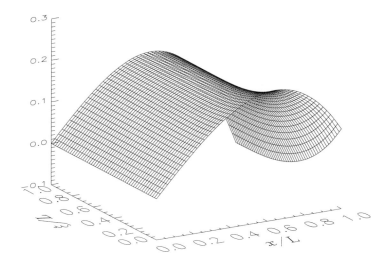

Figure 2.3.2: The Green's function (2.3.28), divided by L^3, as functions of x and ξ when $kL = 1$, $PL^2 = 0.1$, $A = C = 0$, $B = 0.5$, and $D = -3.5$.

Turning our attention to the interval $\xi \leq x \leq 1$, we know that g^{iii} increases by 1 in that interval while g, g' and g'' remain unchanged as we pass through $x = \xi$. Therefore, in this region,

$$
\begin{aligned}
g(x|\xi) &= A\,\frac{\omega^2\cosh(\lambda x) + \lambda^2\cos(\omega x)}{\lambda^2 + \omega^2} + B\,\frac{\lambda\sinh(\lambda x) + \omega\sin(\omega x)}{\lambda^2 + \omega^2} \\
&\quad + C\,\frac{\cosh(\lambda x) - \cos(\omega x)}{\lambda^2 + \omega^2} + \frac{D}{\lambda^2 + \omega^2}\left[\frac{\sinh(\lambda x)}{\lambda} - \frac{\sin(\omega x)}{\omega}\right] \\
&\quad + \frac{\sinh[\lambda(x - \xi)]}{\lambda(\lambda^2 + \omega^2)} - \frac{\sin[\omega(x - \xi)]}{\omega(\lambda^2 + \omega^2)}.
\end{aligned}
\tag{2.3.28}
$$

Using the symmetry property, we combine (2.3.27) and (2.3.28) into a single expression

$$
g(x|\xi) = \frac{F(x_>, x_<)}{\lambda^2 + \omega^2},
\tag{2.3.29}
$$

where

$$
\begin{aligned}
F(x, \xi) &= A[\omega^2\cosh(\lambda x) + \lambda^2\cos(\omega x)] + B[\lambda\sinh(\lambda x) + \omega\sin(\omega x)] \\
&\quad + C[\cosh(\lambda x) - \cos(\omega x)] + D[\sinh(\lambda x)/\lambda - \sin(\omega x)/\omega] \\
&\quad + \sinh[\lambda(x - \xi)]/\lambda - \sin[\omega(x - \xi)]/\omega.
\end{aligned}
\tag{2.3.30}
$$

So far, we showed that the Green's function for (2.3.6) exists, is symmetric and enjoys certain properties [see the material in the boxes after (2.3.6) and (2.3.10)]. But how does this help us solve (2.3.1)? We now prove that

$$y(x) = \int_a^b g(x|\xi)f(\xi)\,d\xi \qquad (2.3.31)$$

is the solution to the nonhomogeneous differential equation (2.3.1) and the homogeneous boundary conditions (2.3.7).

We begin by noting that in (2.3.31) x is a parameter while ξ is the dummy variable. As we perform the integration, we must switch from the form for $g(x|\xi)$ for $\xi \leq x$ to the second form for $\xi \geq x$ when ξ equals x; thus,

$$y(x) = \int_a^x g(x|\xi)f(\xi)\,d\xi + \int_x^b g(x|\xi)f(\xi)\,d\xi. \qquad (2.3.32)$$

Differentiation yields

$$\frac{d}{dx}\int_a^x g(x|\xi)f(\xi)\,d\xi = \int_a^x \frac{dg(x|\xi)}{dx}f(\xi)\,d\xi + g(x|x^-)f(x), \qquad (2.3.33)$$

and

$$\frac{d}{dx}\int_x^b g(x|\xi)f(\xi)\,d\xi = \int_x^b \frac{dg(x|\xi)}{dx}f(\xi)\,d\xi - g(x|x^+)f(x). \qquad (2.3.34)$$

Because $g(x|\xi)$ is continuous everywhere, we have that $g(x|x^+) = g(x|x^-)$ so that

$$\frac{dy}{dx} = \int_a^x \frac{dg(x|\xi)}{dx}f(\xi)\,d\xi + \int_x^b \frac{dg(x|\xi)}{dx}f(\xi)\,d\xi. \qquad (2.3.35)$$

Differentiating once more gives

$$\frac{d^2 y}{dx^2} = \int_a^x \frac{d^2 g(x|\xi)}{dx^2}f(\xi)\,d\xi + \frac{dg(x|x^-)}{dx}f(x)$$
$$+ \int_x^b \frac{d^2 g(x|\xi)}{dx^2}f(\xi)\,d\xi - \frac{dg(x|x^+)}{dx}f(x). \qquad (2.3.36)$$

The second and fourth terms on the right side will not cancel in this case; to the contrary,

$$\frac{dg(x|x^-)}{dx} - \frac{dg(x|x^+)}{dx} = -\frac{1}{p(x)}. \qquad (2.3.37)$$

To show this, we note that the term $dg(x|x^-)/dx$ denotes a differentiation of $g(x|\xi)$ with respect to x using the $x > \xi$ form and then letting $\xi \to x$. Thus,

$$\frac{dg(x|x^-)}{dx} = -\lim_{\substack{\xi \to x \\ \xi < x}} \frac{y_2'(x)y_1(\xi)}{p(\xi)W(\xi)} = -\frac{y_2'(x)y_1(x)}{p(x)W(x)}, \qquad (2.3.38)$$

while for $dg(x|x^+)/dx$ we use the $x < \xi$ form or

$$\frac{dg(x|x^+)}{dx} = -\lim_{\substack{\xi \to x \\ \xi > x}} \frac{y_1'(x)y_2(\xi)}{p(\xi)W(\xi)} = -\frac{y_1'(x)y_2(x)}{p(x)W(x)}. \qquad (2.3.39)$$

Upon introducing these results into the differential equation

$$p(x)\frac{d^2y}{dx^2} + p'(x)\frac{dy}{dx} + s(x)y = -f(x), \qquad (2.3.40)$$

we have

$$\int_a^x [p(x)g''(x|\xi) + p'(x)g'(x|\xi) + s(x)g(x|\xi)]f(\xi)\,d\xi$$

$$+ \int_x^b [p(x)g''(x|\xi) + p'(x)g'(x|\xi) + s(x)g(x|\xi)]f(\xi)\,d\xi$$

$$- p(x)\frac{f(x)}{p(x)} = -f(x). \qquad (2.3.41)$$

Because

$$p(x)g''(x|\xi) + p'(x)g'(x|\xi) + s(x)g(x|\xi) = 0, \qquad (2.3.42)$$

except for $x = \xi$, (2.3.41), and thus (2.3.1), is satisfied. Although (2.3.42) does not hold at the point $x = \xi$, the results are still valid because that one point does not affect the values of the integrals. As for the boundary conditions,

$$y(a) = \int_a^b g(a|\xi)f(\xi)\,d\xi, \qquad y'(a) = \int_a^b \frac{dg(a|\xi)}{dx}f(\xi)\,d\xi, \qquad (2.3.43)$$

and $\alpha_1 y(a) + \alpha_2 y'(a) = 0$ from (2.3.2).

Finally, let us consider the solution for the nonhomogeneous boundary conditions $\alpha_1 y(a) + \alpha_2 y'(a) = \alpha$, and $\beta_1 y(b) + \beta_2 y'(b) = \beta$. The solution in this case is

$$y(x) = \frac{\alpha y_2(x)}{\alpha_1 y_2(a) + \alpha_2 y_2'(a)} + \frac{\beta y_1(x)}{\beta_1 y_1(b) + \beta_2 y_1'(b)} + \int_a^b g(x|\xi)f(\xi)\,d\xi.$$

$$(2.3.44)$$

A quick check shows that (2.3.44) satisfies the differential equation and both nonhomogeneous boundary conditions.

2.4 EIGENFUNCTION EXPANSION FOR REGULAR BOUNDARY-VALUE PROBLEMS

In the previous section we showed how Green's functions can be used to solve the nonhomogeneous linear differential equation (2.1.1). The next question is how do you find the Green's function? In this section we present the most common method: *series expansion*.

Consider the nonhomogeneous problem

$$y'' = -f(x), \qquad \text{with} \qquad y(0) = y(L) = 0. \qquad (2.4.1)$$

The Green's function $g(x|\xi)$ must therefore satisfy

$$g'' = -\delta(x - \xi), \quad \text{with} \quad g(0|\xi) = g(L|\xi) = 0. \qquad (2.4.2)$$

Because $g(x|\xi)$ vanishes at the ends of the interval $(0, L)$, this suggests that it can be expanded in a series of suitably chosen orthogonal functions such as, for instance, the Fourier sine series

$$g(x|\xi) = \sum_{n=1}^{\infty} G_n(\xi) \sin\left(\frac{n\pi x}{L}\right), \qquad (2.4.3)$$

where the expansion coefficients G_n are dependent on the parameter ξ. Although we have chosen to use the orthogonal set of functions $\sin(n\pi x/L)$, we could have used other orthogonal functions as long as they vanish at the end points.

Because

$$g''(x|\xi) = \sum_{n=1}^{\infty} \left(-\frac{n^2\pi^2}{L^2}\right) G_n(\xi) \sin\left(\frac{n\pi x}{L}\right), \qquad (2.4.4)$$

and

$$\delta(x - \xi) = \sum_{n=1}^{\infty} A_n(\xi) \sin\left(\frac{n\pi x}{L}\right), \qquad (2.4.5)$$

where

$$A_n(\xi) = \frac{2}{L} \int_0^L \delta(x - \xi) \sin\left(\frac{n\pi x}{L}\right) dx = \frac{2}{L} \sin\left(\frac{n\pi \xi}{L}\right), \qquad (2.4.6)$$

we have that

$$-\sum_{n=1}^{\infty} \left(\frac{n^2\pi^2}{L^2}\right) G_n(\xi) \sin\left(\frac{n\pi x}{L}\right) = -\frac{2}{L} \sum_{n=1}^{\infty} \sin\left(\frac{n\pi \xi}{L}\right) \sin\left(\frac{n\pi x}{L}\right),$$

$$(2.4.7)$$

after substituting (2.4.4)–(2.4.6) into the differential equation (2.4.2). Since (2.4.7) must hold for any arbitrary x,

$$\left(\frac{n^2\pi^2}{L^2}\right) G_n(\xi) = \frac{2}{L}\sin\left(\frac{n\pi\xi}{L}\right). \qquad (2.4.8)$$

Thus, the Green's function is

$$g(x|\xi) = \frac{2L}{\pi^2}\sum_{n=1}^{\infty}\frac{1}{n^2}\sin\left(\frac{n\pi\xi}{L}\right)\sin\left(\frac{n\pi x}{L}\right). \qquad (2.4.9)$$

How might we use (2.4.9)? We can use this series to construct the solution of the nonhomogeneous equation (2.4.1) via the formula

$$y(x) = \int_0^L g(x|\xi)\,f(\xi)\,d\xi. \qquad (2.4.10)$$

This leads to

$$y(x) = \frac{2L}{\pi^2}\sum_{n=1}^{\infty}\frac{1}{n^2}\sin\left(\frac{n\pi x}{L}\right)\int_0^L f(\xi)\sin\left(\frac{n\pi\xi}{L}\right)\,d\xi, \qquad (2.4.11)$$

or

$$y(x) = \frac{L^2}{\pi^2}\sum_{n=1}^{\infty}\frac{a_n}{n^2}\sin\left(\frac{n\pi x}{L}\right), \qquad (2.4.12)$$

where a_n are the Fourier sine coefficients of $f(x)$.

• **Example 2.4.1**

Consider now the more complicated boundary-value problem

$$y'' + k^2 y = -f(x), \qquad \text{with} \qquad y(0) = y(L) = 0. \qquad (2.4.13)$$

The Green's function $g(x|\xi)$ must now satisfy

$$g'' + k^2 g = -\delta(x-\xi), \quad \text{and} \quad g(0|\xi) = g(L|\xi) = 0. \qquad (2.4.14)$$

Once again, we use the Fourier sine expansion

$$g(x|\xi) = \sum_{n=1}^{\infty} G_n(\xi)\sin\left(\frac{n\pi x}{L}\right). \qquad (2.4.15)$$

Direct substitution of (2.4.15) and (2.4.5) into (2.4.14) and grouping by corresponding harmonics yields

$$-\frac{n^2\pi^2}{L^2}G_n(\xi) + k^2 G_n(\xi) = -\frac{2}{L}\sin\left(\frac{n\pi\xi}{L}\right), \qquad (2.4.16)$$

or

$$G_n(\xi) = \frac{2}{L} \frac{\sin(n\pi\xi/L)}{n^2\pi^2/L^2 - k^2}. \qquad (2.4.17)$$

Thus, the Green's function is

$$g(x|\xi) = \frac{2}{L} \sum_{n=1}^{\infty} \frac{\sin(n\pi\xi/L)\sin(n\pi x/L)}{n^2\pi^2/L^2 - k^2}. \qquad (2.4.18)$$

Examining (2.4.18) more closely, we note that it enjoys the symmetry property that $g(x|\xi) = g(\xi|x)$.

• Example 2.4.2

Let us find the series expansion for the Green's function for

$$xg'' + g' + \left(k^2 x - \frac{m^2}{x}\right)g = -\delta(x - \xi), \quad 0 < x < L, \qquad (2.4.19)$$

where $m \geq 0$ and is an integer. The boundary conditions are

$$\lim_{x \to 0} |g(x|\xi)| < \infty, \quad \text{and} \quad g(L|\xi) = 0. \qquad (2.4.20)$$

To find this series, consider the Fourier-Bessel series[8]

$$g(x|\xi) = \sum_{n=1}^{\infty} G_n(\xi) J_m(k_{nm}x), \qquad (2.4.21)$$

where k_{nm} is the nth root of $J_m(k_{nm}L) = 0$. This series enjoys the advantage that it satisfies the boundary conditions and we will not have to introduce any homogeneous solutions so that $g(x|\xi)$ satisfies the boundary conditions.

Substituting (2.4.21) into (2.4.19) after we divide by x and using the Fourier-Bessel expansion for the delta function, we have that

$$(k^2 - k_{nm}^2)G_n(\xi) = -\frac{2k_{nm}^2 J_m(k_{nm}\xi)}{L^2[J_{m+1}(k_{nm}L)]^2} = -\frac{2J_m(k_{nm}\xi)}{L^2[J_m'(k_{nm}L)]^2}, \qquad (2.4.22)$$

so that

$$g(x|\xi) = \frac{2}{L^2} \sum_{n=1}^{\infty} \frac{J_m(k_{nm}\xi)J_m(k_{nm}x)}{(k_{nm}^2 - k^2)[J_m'(k_{nm}L)]^2}. \qquad (2.4.23)$$

---

[8] If you are unfamiliar with Fourier-Bessel expansions, see §C.3 in Appendix C.

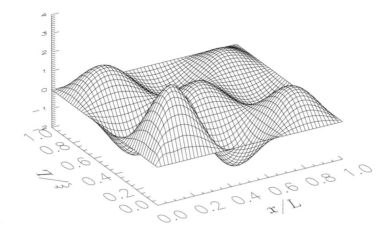

Figure 2.4.1: The Green's function (2.4.23) as functions of x/L and ξ/L when $kL = 10$ and $m = 1$.

Equation (2.4.23) is plotted in Figure 2.4.1.

We summarize the expansion technique as follows: Suppose that we want to solve the differential equation

$$Ly(x) = -f(x), \qquad (2.4.24)$$

with some condition $By(x) = 0$ along the boundary, where L now denotes the Sturm-Liouville differential operator

$$L = \frac{d}{dx}\left[p(x)\frac{d}{dx}\right] + [q(x) + \lambda r(x)], \qquad (2.4.25)$$

and B is the boundary condition operator

$$B = \begin{cases} \alpha_1 + \alpha_2\dfrac{d}{dx}, & \text{at } x = a, \\[2mm] \beta_1 + \beta_2\dfrac{d}{dx}, & \text{at } x = b. \end{cases} \qquad (2.4.26)$$

We begin by seeking a Green's function $g(x|\xi)$, which satisfies

$$Lg = -\delta(x - \xi), \qquad Bg = 0. \qquad (2.4.27)$$

To find the Green's function, we utilize the set of eigenfunctions $\varphi_n(x)$ associated with the regular Sturm-Liouville problem

$$\frac{d}{dx}\left[p(x)\frac{d\varphi_n}{dx}\right] + [q(x) + \lambda_n r(x)]\varphi_n = 0, \qquad (2.4.28)$$

where $\varphi_n(x)$ satisfies the same boundary conditions as $y(x)$. If g exists and if the set $\{\varphi_n\}$ is complete, then $g(x|\xi)$ can be represented by the series

$$g(x|\xi) = \sum_{n=1}^{\infty} G_n(\xi)\varphi_n(x). \qquad (2.4.29)$$

Applying L to (2.4.29),

$$Lg(x|\xi) = \sum_{n=1}^{\infty} G_n(\xi)L\varphi_n(x) = \sum_{n=1}^{\infty} G_n(\xi)(\lambda-\lambda_n)r(x)\varphi_n(x) = -\delta(x-\xi),$$
$$(2.4.30)$$

provided that λ does not equal any of the eigenvalues λ_n. Multiplying both sides of (2.4.30) by $\varphi_m(x)$ and integrating over x,

$$\sum_{n=1}^{\infty} G_n(\xi)(\lambda - \lambda_n) \int_a^b r(x)\varphi_n(x)\varphi_m(x)\, dx = -\varphi_m(\xi). \qquad (2.4.31)$$

If the eigenfunctions are *orthonormal*,

$$\int_a^b r(x)\varphi_n(x)\varphi_m(x)\, dx = \begin{cases} 1, & n = m, \\ 0, & n \neq m, \end{cases} \quad \text{and} \quad G_n(\xi) = \frac{\varphi_n(\xi)}{\lambda_n - \lambda}.$$
$$(2.4.32)$$

This leads directly to the *bilinear formula*:

$$g(x|\xi) = \sum_{n=1}^{\infty} \frac{\varphi_n(\xi)\varphi_n(x)}{\lambda_n - \lambda}, \qquad (2.4.33)$$

which permits us to write the Green's function at once if the eigenvalues and eigenfunctions of L are known.

One of the intriguing aspects of the bilinear formula is that the Green's function possesses poles, called the *point spectrum* of the Green's function, at λ_n. For regular second-order differential equations, there are an infinite number of simple poles that all lie along the real λ-axis according to Sturm's oscillation theorem.[9] This fact leads to the following important theorem:

---

[9] Ince, E. L., 1956: *Ordinary Differential Equations*. Dover Publications, Inc., §10.6.

Theorem:

$$\frac{1}{2\pi i} \oint_C g(x|\xi)\, d\lambda = -\frac{\delta(x-\xi)}{r(\xi)}, \qquad (2.4.34)$$

where C is a closed contour in the complex λ-plane that encloses all of the singularities of $g(x|\xi)$.

Proof: Starting with the bilinear formula (2.4.33), we apply Cauchy's integral formula and find that

$$\frac{1}{2\pi i} \oint_C g(x|\xi)\, d\lambda = -\frac{1}{2\pi i} \sum_{n=1}^{\infty} \varphi_n(\xi)\varphi_n(x) \oint_C \frac{d\lambda}{\lambda - \lambda_n} \quad (2.4.35)$$

$$= -\sum_{n=1}^{\infty} \varphi_n(\xi)\varphi_n(x). \qquad (2.4.36)$$

Next, let us expand $\delta(x-\xi)$ in terms of $\varphi_n(x)$. From Sturm-Liouville theory,

$$\delta(x-\xi) = \sum_{n=1}^{\infty} c_n \varphi_n(x), \qquad (2.4.37)$$

where

$$c_n = \frac{\int_a^b r(x)\delta(x-\xi)\varphi_n(x)\, dx}{\int_a^b r(x)\varphi_n^2(x)\, dx} = r(\xi)\varphi_n(\xi), \qquad (2.4.38)$$

because $\varphi_n(x)$ is an orthonormal eigenfunction. Thus,

$$\frac{\delta(x-\xi)}{r(\xi)} = \sum_{n=1}^{\infty} \varphi_n(\xi)\varphi_n(x), \qquad (2.4.39)$$

which completes the proof. $\qquad\qquad\qquad\qquad\qquad\qquad\qquad\qquad$ $\square$

• **Example 2.4.3**

In Example 2.3.3, we showed that the Green's function for the problem

$$g'' + \lambda g = -\delta(x-\xi), \qquad 0 < x < L, \qquad (2.4.40)$$

with $g(0|\xi) = g(L|\xi) = 0$, is

$$g(x|\xi) = \frac{\sin(\sqrt{\lambda}\, x_<) \sin[\sqrt{\lambda}\,(L - x_>)]}{\sqrt{\lambda}\, \sin(\sqrt{\lambda}\, L)}. \qquad (2.4.41)$$

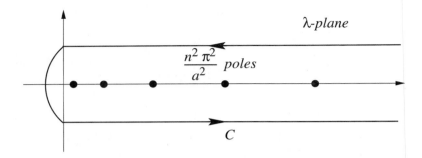

Figure 2.4.1: Closed contour C enclosing the poles of the Green's function (2.4.41).

Let us verify our theorem using the solution over the interval $[0, \xi]$. For the verification we must evaluate

$$\oint_C \frac{\sin(\sqrt{\lambda}\,x)\sin[\sqrt{\lambda}\,(L-\xi)]}{\sqrt{\lambda}\,\sin(\sqrt{\lambda}\,L)}\,d\lambda,$$

where the closed contour C is shown in Figure 2.4.1.

Our first concern is the presence of square roots in the integrand; we might have a multivalued function. However, when the Taylor expansion for sine is substituted into the integral, the square roots disappear and we have a single-valued integrand.

Next, we must find the location and nature of singularities. Clearly $\lambda = 0$ is one, while $\sqrt{\lambda_n}\,L = n\pi$, or $\lambda_n = n^2\pi^2/L^2$, $n = 1, 2, 3, \ldots$, is another. To discover the nature of the singularities, we use the infinite product formula for sine and find that

$$
\frac{\sin(\sqrt{\lambda}\,x)\sin[\sqrt{\lambda}\,(L-\xi)]}{\sqrt{\lambda}\,\sin(\sqrt{\lambda}\,L)}
$$
$$
= \frac{\sqrt{\lambda}\,x\left(1 - \frac{\lambda x^2}{\pi^2}\right)\left(1 - \frac{\lambda x^2}{9\pi^2}\right)\cdots\left[1 - \frac{\lambda(L-\xi)^2}{\pi^2}\right]\left[1 - \frac{\lambda(L-\xi)^2}{9\pi^2}\right]\cdots}{\sqrt{\lambda}\left(1 - \frac{\lambda L^2}{\pi^2}\right)\left(1 - \frac{\lambda L^2}{9\pi^2}\right)\cdots}.
$$

$$(2.4.42)$$

Consequently, $\lambda = 0$ is a removable singularity; $\lambda_n = n^2\pi^2/L^2$ are simple poles. Therefore,

$$\frac{1}{2\pi i}\oint_C g(x|\xi)\,d\lambda = \sum_{n=1}^{\infty}\operatorname{Res}\left\{\frac{\sin(\sqrt{\lambda}\,x)\sin[\sqrt{\lambda}\,(L-\xi)]}{\sqrt{\lambda}\,\sin(\sqrt{\lambda}\,L)}; \frac{n^2\pi^2}{L^2}\right\} \quad (2.4.43)$$

$$= \sum_{n=1}^{\infty}\lim_{\lambda\to\lambda_n}\frac{\lambda - \lambda_n}{\sin(\sqrt{\lambda}\,L)}\lim_{\lambda\to\lambda_n}\frac{\sin(\sqrt{\lambda}\,x)\sin[\sqrt{\lambda}\,(L-\xi)]}{\sqrt{\lambda}}$$

$$(2.4.44)$$

$$= \sum_{n=1}^{\infty} \frac{2\sqrt{\lambda_n}\sin(\sqrt{\lambda_n}x)\sin[\sqrt{\lambda_n}(L-\xi)]}{L\cos(\sqrt{\lambda_n}L)\sqrt{\lambda_n}} \qquad (2.4.45)$$

$$= \frac{2}{L}\sum_{n=1}^{\infty}(-1)^n\sin\left[\frac{n\pi}{L}(L-\xi)\right]\sin\left(\frac{n\pi x}{L}\right) \qquad (2.4.46)$$

$$= -\frac{2}{L}\sum_{n=1}^{\infty}\sin\left(\frac{n\pi\xi}{L}\right)\sin\left(\frac{n\pi x}{L}\right) = -\delta(x-\xi) \quad (2.4.47)$$

and the theorem holds because $r(\xi) = 1$ in the present example. A similar demonstration exists for $\xi \leq x \leq L$.

2.5 SINGULAR BOUNDARY-VALUE PROBLEMS

In the previous two sections we considered the Green's function associated with the regular Sturm-Liouville problem. When we turn to singular differential equations, most of the work centers on the singular Sturm-Liouville problem because of its importance in mathematical physics. In this section, we seek the Green's function for this particular class of boundary-value problems.

What is a singular Sturm-Liouville problem? Simply put, it is one that is not regular. For example, if $p(x) = 0$ or $p(x)$ becomes infinite at some point $x = c$ on the interval (a, b), then we have a singular Sturm-Liouville problem. From this wide class of problems we will focus on two: In the first class, (2.3.3) is defined on a finite interval $[a, b]$ while $p(x)$ may equal zero or become infinite at some point on this interval. In the second class, one or both endpoints lie at $x = \pm\infty$.

The difficulty with singular Sturm-Liouville problems becomes clear when we try to construct the Green's function. If we use homogeneous solutions to the second-order ordinary differential equation, say $y_1(x)$ and $y_2(x)$, as we did in §2.3, then one or both of these solutions may diverge to infinity at the singular boundary and we cannot hope to satisfy any boundary condition there. Similarly, a linear combination of these solutions may refuse to satisfy the boundary condition. On the other hand, *both* solutions may satisfy the boundary condition and it becomes ambiguous which one to choose.

Because most singular Sturm-Liouville problems involve a singular endpoint, the analysis of the problem begins by classifying these points along the lines introduced by Weyl.[10] As such, there are two types of singular endpoints:

---

[10] Weyl, H., 1910: Über gewöhnliche Differentialgleichungen mit Singularitäten und die zugehörigen Entwicklungen willkürlicher Funktionen. *Ann. Math.*, **68**, 220–269.

• *Limit-circle.* In this case, all solutions $y(x)$ from the singular Sturm-Liouville differential equation are *square-integrable*, i.e., $\int |y(x)|^2 r(x)\,dx$ converges at $x = b$. If this is so, we impose the boundary condition

$$\lim_{x \to b} [p(x)y'(x)f(x) - p(x)f'(x)y(x)] = 0$$

for a suitable function $f(x)$. This is an extension of the regular boundary condition to the singular boundary point.

• *Limit-point.* In this case, there is just one solution (up to a scalar factor) for which $\int |y(x)|^2 r(x)\,dx$ converges. If this is true, no boundary condition is required at $x = b$.

To illustrate these concepts, recall the boundary-value problem given by (2.4.13), namely

$$y'' + k^2 y = -f(x), \qquad \text{with} \qquad y(0) = y(L) = 0. \qquad \textbf{(2.5.1)}$$

A popular (although not strictly correct) method for converting (2.5.1) into a singular boundary-value problem is to consider the limiting process of $L \to \infty$. In that case, the eigenfunction is still $y(x) = \sin(kx)$. Because that is the only solution valid at $x = \infty$, we have a limit-point there.

Let us now find the Green's function for this singular Sturm-Liouville problem. We begin by using the results from Example 2.4.1, which were for the regular Sturm-Liouville problem

$$g'' + k^2 g = -\delta(x - \xi), \quad \text{with} \quad g(0|\xi) = g(L|\xi) = 0. \qquad \textbf{(2.5.2)}$$

In that example, we showed that

$$g(x|\xi) = \frac{2}{L} \sum_{n=1}^{\infty} \frac{\sin(n\pi\xi/L)\sin(n\pi x/L)}{n^2\pi^2/L^2 - k^2}. \qquad \textbf{(2.5.3)}$$

As before, we convert this problem into a singular Sturm-Liouville problem by taking the limit of $L \to \infty$.

One of the properties of bilinear expansions for regular Sturm-Liouville problems is that the Green's function possesses poles at $\lambda = \lambda_n$. What happens in the case of singular Sturm-Liouville problems? From our example we see that the poles move closer together. When $L = \infty$, the poles coalesce into a continuous branch cut. This branch cut in the λ-plane is called the *continuous spectrum*. Just as all of the discrete eigenvalues for the regular Sturm-Liouville problem lie along the real λ-axis, so does the continuous spectrum for the singular Sturm-Liouville problem. Not all singular differential equations will possess a continuous spectrum; some may contain only the point spectrum, and some may have both.

In §2.4 we also defined the completeness relationship (2.4.34). We formally extend this relationship in the case of singular Sturm-Liouville problems. For example, if we only have a continuous spectrum and no point spectrum (a common situation), the completeness relationship becomes

$$\frac{1}{2\pi i} \oint_{C_\infty} g(x|\xi) \, d\lambda = -\frac{\delta(x-\xi)}{r(\xi)}, \tag{2.5.4}$$

where the contour C_∞ denotes a circle taken in the positive direction with infinite radius so that it encloses every pole or branch cut in the complex λ-plane. In this way, (2.5.4) holds both for regular as well as singular problems.

For the singular problem, it is generally easier to find the Green's function by solving the ordinary differential equation as we did in §2.3 rather than calculate the equivalent bilinear expansion. Generalizing the completeness relation (2.4.34), we can derive the equivalent "eigenfunction" expansion for the Green's function that contains a branch cut. We do this by substituting (2.3.17) into (2.4.34) and then performing the contour integration over C. Upon simplifying the final integral, the completeness relation becomes

$$\delta(x-\xi) = \int \varphi_\omega(x) \widetilde{\varphi}_\omega(\xi) \, d\omega, \tag{2.5.5}$$

where $\varphi_\omega(x)$ and $\widetilde{\varphi}_\omega(x)$ are any two "eigenfunctions" of the differential equation (2.3.3). The limits of integration over ω will be from 0 to ∞ or from $-\infty$ to ∞, depending how the contour integral was simplified. An application of this completeness relationship is the definition of the integral transform pair:

$$G(\omega|\xi) = \int_0^\infty \widetilde{\varphi}_\omega(x) g(x|\xi) \, dx, \text{ and } g(x|\xi) = \int_C \varphi_\omega(x) G(\omega|\xi) \, d\omega. \tag{2.5.6}$$

For example, in our present example, $\varphi_\omega(x) = 2\sin(\omega x)/\pi$ and $\widetilde{\varphi}_\omega(x) = \sin(\omega x)$. If the integral over ω in (2.5.6) occurs between 0 and ∞, we have Fourier sine transforms.

• Example 2.5.1: Fokker-Planck equation

An important partial differential equation in stochastic differential equations and plasma physics is the Fokker-Planck equation. In finding its Green's function,[11] we must solve the boundary-value problem

$$\frac{d}{dx}\left(x^2\frac{dg}{dx}\right) - a\frac{d}{dx}(xg) - bg + \lambda g = -\delta(x - \xi), \quad 0 < x, \xi < \infty. \quad \textbf{(2.5.7)}$$

A quick check shows that this equation is singular for two reasons: First, the boundary point $x = \infty$ is singular by definition. Second, the point $x = 0$ is singular because the coefficients of the derivative terms of (2.5.7) vanish there.

Two independent solutions of the homogeneous form of (2.5.7) are

$$u_1(x) = x^{-1+\delta_+}, \qquad \text{and} \qquad u_2(x) = x^{-1+\delta_-}, \qquad \textbf{(2.5.8)}$$

where

$$\delta_\pm = \frac{a+1}{2} \pm \mu, \quad \mu = \sqrt{\lambda_0 - \lambda}, \quad \lambda_0 = \left(\frac{a+1}{2}\right)^2 + b. \quad \textbf{(2.5.9)}$$

From the classification of singular boundary conditions, $x = 0$ is a limit point and u_1 is the only possible solution for $x < \xi$. Similarly, $x = \infty$ is also a limit point and u_2 is the only possible solution for $x > \xi$.

Using the techniques outlined in §2.3, we find that

$$g(x|\xi) = \frac{1}{2\xi\mu}\begin{cases} (x/\xi)^{-1+\delta_+}, & x < \xi, \\ (x/\xi)^{-1+\delta_-}, & x > \xi. \end{cases} \qquad \textbf{(2.5.10)}$$

Equation (2.5.10) shows that this problem has a continuous spectrum but no discrete spectrum.

Upon substituting (2.5.10) into (2.5.4), we find that

$$\delta(x - \xi) = \frac{1}{2\pi\xi}\left(\frac{x}{\xi}\right)^{-(a+1)/2}\int_{-\infty}^{\infty}\left(\frac{x}{\xi}\right)^{i\omega}d\omega. \qquad \textbf{(2.5.11)}$$

Thus, the "eigenfunctions" in this case are

$$\varphi_\omega(x) = \frac{1}{2\pi}x^{(a-1)/2-i\omega}, \quad \text{and} \quad \widetilde{\varphi}_\omega(\xi) = \xi^{(a+1)/2+i\omega}, \qquad \textbf{(2.5.12)}$$

so that we obtain the well-known Mellon transform pair:

$$G(\omega|\xi) = \int_0^{\infty} x^{-(a+1)/2+i\omega}g(x|\xi)\,dx, \qquad \textbf{(2.5.13)}$$

---

[11] Taken from Park, B. T., and V. Petrosian, 1995: Fokker-Planck equations of stochastic acceleration: Green's functions and boundary conditions. *Astrophys. J.*, **446**, 699–716. Published by The University of Chicago Press.

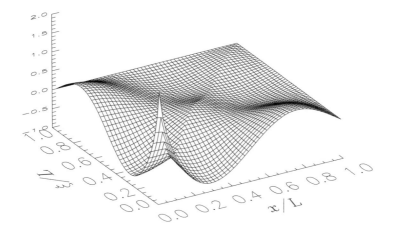

Figure 2.5.1: The Green's function (2.5.22) as functions of x/L and ξ/L when $kL = 7$ and $m = 0$.

and

$$g(x|\xi) = \frac{1}{2\pi} \int_{-\infty}^{\infty} x^{(a-1)/2-i\omega} G(\omega|\xi) \, d\omega, \qquad (2.5.14)$$

from (2.5.6).

• **Example 2.5.2: Bessel's equation**

Let us now find the Green's function for

$$xg'' + g' + \left(k^2 x - \frac{m^2}{x} \right) g = -\delta(x - \xi), \qquad 0 < x < L, \qquad (2.5.15)$$

where $m \geq 0$ and is an integer. The boundary conditions are

$$\lim_{x \to 0} |g(x|\xi)| < \infty, \quad \text{and} \quad g(L|\xi) = 0. \qquad (2.5.16)$$

The homogeneous solutions that satisfy the boundary conditions are

$$y_1(x) = J_m(kx), \quad y_2(x) = J_m(kx)Y_m(kL) - J_m(kL)Y_m(kx). \quad (2.5.17)$$

Computing the Wronskian,

$$\begin{aligned}
W(x) &= y_1(x)y_2'(x) - y_1'(x)y_2(x) & (2.5.18) \\
&= kJ_m(kx)[J_m'(kx)Y_m(kL) - J_m(kL)Y_m'(kx)] \\
&\quad - kJ_m'(kx)[J_m(kx)Y_m(kL) - J_m(kL)Y_m(kx)] & (2.5.19) \\
&= kJ_m(kL)[J_m'(kx)Y_m(kx) - J_m(kx)Y_m'(kx)] & (2.5.20) \\
&= -\frac{2}{\pi x} J_m(kL), & (2.5.21)
\end{aligned}$$

after using a well known result from Watson's book.[12]   Substituting (2.5.17) and (2.5.21) into (2.3.17) with $p(x) = x$,

$$g(x|\xi) = \frac{\pi J_m(kx_<)[J_m(kx_>)Y_m(kL) - J_m(kL)Y_m(kx_>)]}{2J_m(kL)}. \qquad (2.5.22)$$

Equation (2.5.22) is illustrated in Figure 2.5.1 when $kL = 7$, and $m = 0$.

What is the advantage of an expression like (2.5.22) over an eigenfunction expansion? First, there is no summation of a series outside of evaluating $J_m(\ )$ and $Y_m(\ )$. Second, you do not have to find the zeros of Bessel functions.

• Example 2.5.3: Confluent hypergeometric equation

The confluent hypergeometric equation arises in problems such as the three-dimensional harmonic oscillator.[13] Let us find the fundamental or free-space Green's function[14] governed by the confluent hypergeometric equation[15]

$$zg'' + (c - z)g' - ag = \delta(z - \zeta), \qquad 0 < z, \zeta < \infty. \qquad (2.5.23)$$

Motivated by (2.3.17), we guess that the Green's function is of the form

$$g(z|\zeta) = C\Phi(a, c; z_<)\Psi(a, c; z_>), \qquad (2.5.24)$$

where $\Phi(a, c; z)$ and $\Psi(a, c; z)$ are the homogeneous solutions to (2.5.23) and are known as confluent hypergeometric functions[16] of the first and second kind, respectively.

To compute C, we integrate (2.5.23) over the small interval $[\zeta^-, \zeta^+]$, where ζ^+ and ζ^- denote points just above and below $z = \zeta$, respectively.

---

[12] Watson, G. N., 1966: *A Treatise on the Theory of Bessel Functions.* Cambridge University Press, p. 76.

[13] See Negro, J., L. M. Nieto, and O. Rosas-Ortiz, 2000: Confluent hypergeometric equations and related solvable potentials in quantum mechanics. *J. Math. Phys.*, **41**, 7964–7996.

[14] In electromagnetic theory, a free-space Green's function is the particular solution of the differential equation valid over a domain of infinite extent, where the Green's function remains bounded as we approach infinity, or satisfies a radiation condition there.

[15] Taken with permission from Granovskii, I. Ya., and V. I. Nechet, 1974: Non-Hermitian nature of the degenerate hypergeometric equation. *Sov. Phys. J.*, **17**, 1085–1087. Publshed by Plenum Publishers.

[16] See Lebedev, N. N., 1972: *Special Functions and Their Applications.* Dover Publications, Inc., §§9.9 and 9.10.

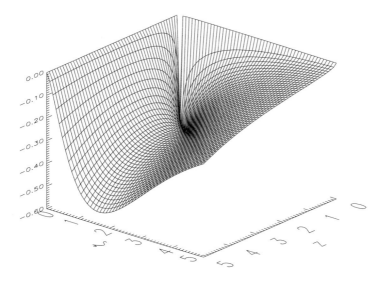

Figure 2.5.2: The Green's function (2.5.32) as functions of z and ζ when $a = \frac{3}{2}$, and $c = 3$.

Thus,

$$z \left.\frac{dg}{dz}\right|_{z=\zeta+} - z \left.\frac{dg}{dz}\right|_{z=\zeta-} = C\zeta \left[\Phi(a,c;\zeta)\Psi'(a,c;\zeta) - \Phi'(a,c;\zeta)\Psi(a,c;\zeta)\right]$$

$$(\mathbf{2.5.25})$$

$$= 1. \qquad (\mathbf{2.5.26})$$

Because

$$\Phi(a,c;z)\Psi'(a,c;z) - \Phi'(a,c;z)\Psi(a,c;z) = -\frac{\Gamma(c)}{\Gamma(a)}z^{-c}e^z, \qquad (\mathbf{2.5.27})$$

where $\Gamma(\ )$ is the gamma function,

$$C = -\frac{\Gamma(a)}{\Gamma(c)}\zeta^{c-1}e^{-\zeta}, \qquad (\mathbf{2.5.28})$$

and

$$g(z|\zeta) = -\frac{\Gamma(a)}{\Gamma(c)}\zeta^{c-1}e^{-\zeta}\Phi(a,c;z_<)\Psi(a,c;z_>). \qquad (\mathbf{2.5.29})$$

The Green's function may be re-expressed in terms of Whittaker functions since

$$\Phi(a,c;z) = z^{-c/2}e^{z/2}M_{\frac{c}{2}-a,\frac{c}{2}-\frac{1}{2}}(z), \qquad (\mathbf{2.5.30})$$

and

$$\Psi(a, c; z) = z^{-c/2} e^{z/2} W_{\frac{c}{2}-a, \frac{c}{2}-\frac{1}{2}}(z), \qquad (2.5.31)$$

so that

$$g(z|\zeta) = -\frac{\Gamma(a)}{\Gamma(c)} z^{-c/2} \zeta^{\frac{c}{2}-1} \exp\left(\frac{z-\zeta}{2}\right) M_{\frac{c}{2}-a, \frac{c}{2}-\frac{1}{2}}(z_<) W_{\frac{c}{2}-a, \frac{c}{2}-\frac{1}{2}}(z_>).$$
$$(2.5.32)$$

Finally, we use the integral formulas for the product of two Whittaker functions to obtain the following integral representation:

$$g(z|\zeta) = -\left(\frac{\zeta}{z}\right)^{\frac{c-1}{2}} \exp\left(\frac{z-\zeta}{2}\right) \int_1^\infty \left(\frac{\xi+1}{\xi-1}\right)^{\frac{c}{2}-a} \exp\left(-\frac{z+\zeta}{2}\xi\right)$$
$$\times I_{c-1}\left[\sqrt{z\zeta(\xi^2-1)}\right] \frac{d\xi}{\sqrt{\xi^2-1}},$$
$$(2.5.33)$$

when $\text{Re}(c) > \text{Re}(a) > 0$. Equation (2.5.33) is asymmetric in z and ζ; this asymmetry results from the non-self-adjoint nature of the confluent hypergeometric equation.

• **Example 2.5.4: Whittaker equation**

In this example, we find the free-space Green's function[15] governed by Whittaker's equation

$$\left(\frac{d^2}{dz^2} - \frac{1}{4} + \frac{\kappa}{z} + \frac{\frac{1}{4}-\mu^2}{z^2}\right) g = \delta(z-\zeta), \qquad 0 < z, \zeta < \infty. \quad (2.5.34)$$

Again, motivated by (2.3.17), we assume that the Green's function has the form

$$g(z|\zeta) = C M_{\kappa,\mu}(z_<) W_{\kappa,\mu}(z_>), \qquad (2.5.35)$$

where $M_{\kappa,\mu}(z)$ and $W_{\kappa,\mu}(z)$ are the homogeneous solutions and are known as Whittaker's functions.[17] To compute C, we integrate (2.5.34) over the small interval $[\zeta^-, \zeta^+]$, where ζ^+ and ζ^- denote points just above and below $z = \zeta$, respectively. Thus,

$$\left.\frac{dg}{dz}\right|_{z=\zeta^+} - \left.\frac{dg}{dz}\right|_{z=\zeta^-} = 1. \qquad (2.5.36)$$

Carrying out the calculation,

$$C = -\frac{\Gamma(\mu - \kappa + \frac{1}{2})}{\Gamma(2\mu + 1)}, \qquad (2.5.37)$$

---

[17] See Whittaker, E. T., and G. N. Watson, 1963: *A Course of Modern Analysis.* Cambridge University Press, Chapter 16.

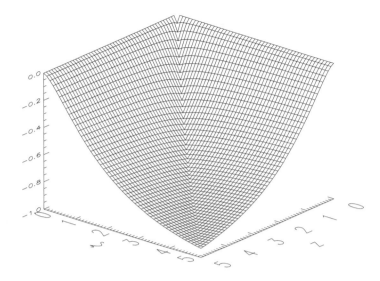

Figure 2.5.3: The Green's function (2.5.38) as functions of z and ζ when $\kappa = 0$, and $\mu = 1$.

and

$$g(z|\zeta) = -\frac{\Gamma(\mu - \kappa + \frac{1}{2})}{\Gamma(2\mu + 1)} M_{\kappa,\mu}(z_<)W_{\kappa,\mu}(z_>), \qquad (\mathbf{2.5.38})$$

where $\Gamma(\ )$ is the gamma function.

Finally, we use the integral formulas for the product of two Whittaker functions to obtain the integral representation

$$g(z|\zeta) = -\sqrt{z\zeta} \int_1^\infty \left(\frac{\xi + 1}{\xi - 1}\right)^\kappa \exp\left(-\frac{z + \zeta}{2}\,\xi\right)$$
$$\times I_{2\mu}\left[\sqrt{z\zeta(\xi^2 - 1)}\right] \frac{d\xi}{\sqrt{\xi^2 - 1}}, \qquad (\mathbf{2.5.39})$$

when $\mathrm{Re}(\kappa - \mu + \frac{1}{2}) > 0$. This Green's function is symmetric in z and ζ because Whittaker's equation is self-adjoint.

2.6 MAXWELL'S RECIPROCITY

From the theory of ordinary differential equations, we know that every linear differential equation of the form

$$L(u) = a_0(x)\frac{d^n u}{dx^n} + a_1(x)\frac{d^{n-1}u}{dx^{n-1}} + \cdots + a_{n-1}(x)\frac{du}{dx} + a_n(x)u = 0 \quad (\mathbf{2.6.1})$$

has associated with it an adjoint equation

$$\overline{L}(v) = (-1)^n \frac{d^n}{dx^n}[a_0(x)v] + (-1)^{n-1}\frac{d^{n-1}}{dx^{n-1}}[a_1(x)v] + \cdots$$

$$- \frac{d}{dx}[a_{n-1}(x)v] + a_n(x)v = 0. \quad (\mathbf{2.6.2})$$

The concept of adjoints is useful because

$$vL(u) - u\overline{L}(v) = \frac{d}{dx}P(u,v), \quad\quad\quad (\mathbf{2.6.3})$$

where $P(u,v)$ is known as the *bilinear concomitant*.[18] If we apply these theoretical considerations to Green's functions, we discover the important property of *reciprocity*. We prove it here for second-order differential equations; the results can be generalized to higher orders.

Consider the Green's function to the second-order differential equation

$$a_0(\xi)\frac{d^2 g(\xi|x)}{d\xi^2} + [2a_0'(\xi) - a_1(\xi)]\frac{dg(\xi|x)}{d\xi}$$

$$+ [a_0''(\xi) - a_1'(\xi) + a_2(\xi)]g(\xi|x) = \delta(\xi - x). \quad (\mathbf{2.6.4})$$

We assume that $g(\xi|x)$ satisfies homogeneous boundary conditions at $\xi = a$ and $\xi = b$. The associated adjoint problem of (2.6.4) is

$$a_0(\xi)\frac{d^2 g^*(\xi|x')}{d\xi^2} + a_1(\xi)\frac{dg^*(\xi|x')}{d\xi} + a_2(\xi)g^*(\xi|x') = \delta(\xi - x') \quad (\mathbf{2.6.5})$$

plus suitable homogeneous boundary conditions. Here we add a prime to x because x and x' are not necessarily the same.

We now compute (2.6.3) and find

$$g(\xi|x)\left[a_0(\xi)\frac{d^2 g^*(\xi|x')}{d\xi^2} + a_1(\xi)\frac{dg^*(\xi|x')}{d\xi} + a_2(\xi)g^*(\xi|x')\right]$$

$$- g^*(\xi|x')\left\{a_0(\xi)\frac{d^2 g(\xi|x)}{d\xi^2} + [2a_0'(\xi) - a_1(\xi)]\frac{dg(\xi|x)}{d\xi}\right.$$

$$\left. + [a_0''(\xi) - a_1'(\xi) + a_2(\xi)]g(\xi|x)\right\}$$

$$= \frac{d}{d\xi}\left\{a_0(\xi)g(\xi|x)\frac{dg^*(\xi|x')}{d\xi} + [a_1(\xi) - a_0'(\xi)]g(\xi|x)g^*(\xi|x')\right.$$

$$\left. - a_0(\xi)g^*(\xi,x')\frac{dg(\xi|x)}{d\xi}\right\}. \quad\quad (\mathbf{2.6.6})$$

---

[18] See Ince, E. L., 1956: *Ordinary Differential Equations*. Dover Publications, Inc., §5.3.

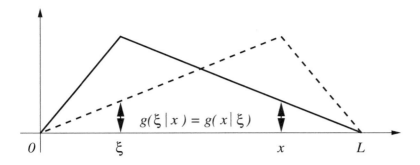

Figure 2.6.1: Maxwell's reciprocity.

Upon integrating from a to b, we obtain

$$\int_a^b g(\xi|x) \left[a_0(\xi) \frac{d^2 g^*(\xi|x')}{d\xi^2} + a_1(\xi) \frac{dg^*(\xi|x')}{d\xi} + a_2(\xi) g^*(\xi|x') \right] d\xi$$

$$= \int_a^b g^*(\xi|x') \left\{ a_0(\xi) \frac{d^2 g(\xi|x)}{d\xi^2} + [2a_0'(\xi) - a_1(\xi)] \frac{dg(\xi|x)}{d\xi} \right.$$

$$\left. + [a_0''(\xi) - a_1'(\xi) + a_2(\xi)] g(\xi|x) \right\} d\xi, \qquad (2.6.7)$$

because both $g(\xi|x)$ and $g^*(\xi|x')$ satisfy homogeneous boundary conditions. Upon substituting (2.6.4) and (2.6.5), we have that

$$\int_a^b \delta(\xi - x') g(\xi|x) \, d\xi = \int_a^b g^*(\xi|x') \delta(\xi - x) \, d\xi, \qquad (2.6.8)$$

or

$$g(x'|x) = g^*(x|x'). \qquad (2.6.9)$$

If the differential operators in (2.6.4) and (2.6.5) are *self-adjoint*, i.e., $L( ) = \overline{L}( )$, then $g(x|\xi) = g(\xi|x)$.

● **Example 2.6.1**

In §1.5 we showed that the Green's function for the differential equation

$$g''(x|\xi) = \delta(x - \xi), \qquad g(0|\xi) = g(L|\xi) = 0, \qquad (2.6.10)$$

is

$$g(x|\xi) = (L - x_>)x_<. \qquad (2.6.11)$$

A quick check shows that $g(x|\xi) = g(\xi|x)$ is symmetric because the differential operator $L = d^2/dx^2$ is self-adjoint. In Figures 2.6.1 and 2.6.2 we illustrate this result graphically.

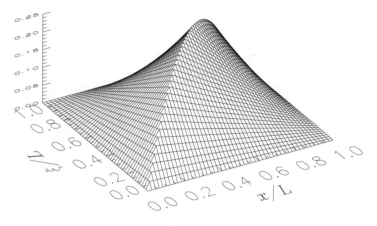

Figure 2.6.2: The Green's function (2.6.11) as functions of x and ξ. The symmetry of the solution in x and ξ is quite apparent.

Problems

For the following initial-value problems, find the Green's function. Assume that the initial conditions are zero and $\tau > 0$.

1. $g' + kg = \delta(t - \tau)$

2. $g'' - 2g' - 3g = \delta(t - \tau)$

3. $g'' + 4g' + 3g = \delta(t - \tau)$

4. $g'' - 2g' + 5g = \delta(t - \tau)$

5. $g'' - 3g' + 2g = \delta(t - \tau)$

6. $g'' + 4g' + 4g = \delta(t - \tau)$

7. $g'' - 9g = \delta(t - \tau)$

8. $g'' + g = \delta(t - \tau)$

9. $g'' - g' = \delta(t - \tau)$

Find the Green's function and the corresponding bilinear expansion [using eigenfunctions from the regular Sturm-Liouville problem $y_n'' + k^2 y_n = 0$] for

$$g'' = -\delta(x - \xi), \qquad 0 < x, \xi < L,$$

which satisfy the following boundary conditions:

10. $g(0|\xi) - \alpha g'(0|\xi) = 0, \alpha \neq 0, -L, \qquad g(L|\xi) = 0,$

11. $g(0|\xi) - g'(0|\xi) = 0, \qquad\qquad g(L|\xi) - g'(L|\xi) = 0,$

12. $g(0|\xi) - g'(0|\xi) = 0, \qquad\qquad g(L|\xi) + g'(L|\xi) = 0.$

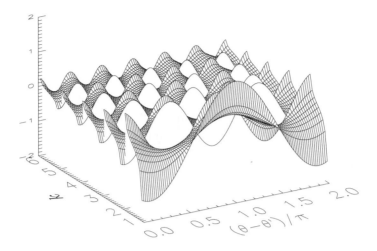

Problem 20

13. Find the Green's function for

$$g'' - \frac{1}{a} = -\delta(x - \xi), \qquad 0 < x, \xi < L,$$

with the boundary conditions $g'(0|\xi) = g'(a|\xi) = 0$. Is there a bilinear expansion? Why or why not?

Find the Green's function[19] and the corresponding bilinear expansion [using eigenfunctions from the regular Sturm-Liouville problem $y_n'' + k^2 y_n = 0$] for

$$g'' - k^2 g = -\delta(x - \xi), \qquad 0 < x, \xi < L,$$

which satisfy the following boundary conditions:

14. $g(0|\xi) = 0,$ $\qquad\qquad\qquad g(L|\xi) = 0,$

15. $g'(0|\xi) = 0,$ $\qquad\qquad\qquad g'(L|\xi) = 0,$

16. $g(0|\xi) = 0,$ $\qquad\qquad\qquad g(L|\xi) + g'(L|\xi) = 0,$

17. $g(0|\xi) = 0,$ $\qquad\qquad\qquad g(L|\xi) - g'(L|\xi) = 0,$

18. $ag(0|\xi) + g'(0|\xi) = 0,$ $\qquad g'(L|\xi) = 0,$

19. $g(0|\xi) + g'(0|\xi) = 0,$ $\qquad g(L|\xi) - g'(L|\xi) = 0.$

---

[19] Problem 18 taken from Chakrabarti, A., and T. Sahoo, 1996: Reflection of water waves by a nearly vertical porous wall. *J. Austral. Math. Soc., Ser. B*, **37**, 417–429.

20. Show that the Green's function[20] for

$$g'' + k^2 g = -\delta(\theta - \theta'), \qquad -\pi < \theta, \theta' < \pi,$$

subject to 2π periodicity is

$$g(\theta|\theta') = -\frac{\cos[k(\pi - |\theta - \theta'|)]}{2k\sin(k\pi)}.$$

Then use (2.4.34) to prove that

$$\delta(\theta - \theta') = \frac{1}{2\pi} \sum_{n=-\infty}^{\infty} e^{in(\theta - \theta')}.$$

This Green's function is illustrated in the figure captioned Problem 20. The gaps are located where k is an integer and the Green's function does not exist.

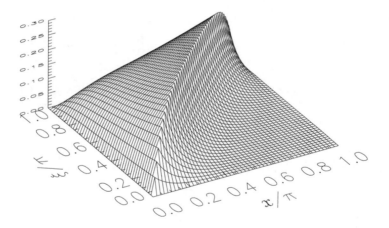

Problem 21

21. Find the Green's function and corresponding bilinear expansion for

$$g'' + ag' - k^2 g = -\delta(x - \xi), \qquad 0 < x, \xi < \pi,$$

subject to the boundary conditions $g(0|\xi) = g(\pi|\xi) = 0$. This Green's function is illustrated in the figure captioned Problem 21 as functions of x and ξ when $a = 3$, and $k = 1$.

---

[20] See Felsen, F. B., and N. Marcuvitz, 1973: *Radiation and Scattering of Waves.* Prentice-Hall, Inc., p. 310.

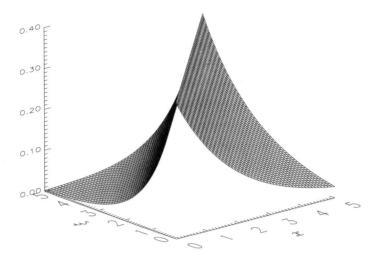

Problem 22

22. Find the free-space Green's function for

$$\nu g'' - g' - sg = -\delta(x - \xi), \qquad -\infty < x, \xi < \infty,$$

where $s, \nu > 0$. This Green's function is illustrated under the figure captioned Problem 22 as functions of x and ξ when $\nu = 1$, and $\nu s = 2$.

23. Show that the free-space Green's function for

$$\frac{1}{r}\frac{d}{dr}\left(r\frac{dg}{dr}\right) - k^2 g = -\frac{\delta(r - \rho)}{r}, \qquad 0 < r, \rho < \infty,$$

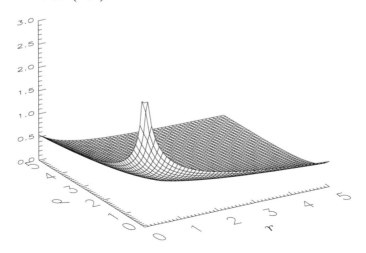

Problem 23

is

$$g(r|\rho) \doteq I_0(kr_<)K_0(kr_>).$$

This Green's function is illustrated in the figure captioned Problem 23 as functions of r and ρ when $k = 1$.

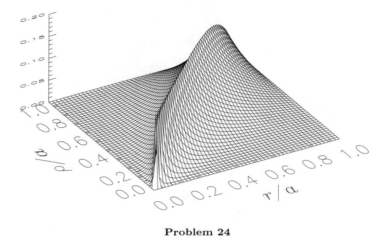

Problem 24

24. Show[21] that the Green's function for

$$\frac{1}{r}\frac{d}{dr}\left(r\frac{dg}{dr}\right) - \frac{m^2}{r^2}g = -\frac{\delta(r-\rho)}{r}, \qquad 0 < r, \rho < a,$$

with the boundary conditions

$$\lim_{r\to 0}|g(r|\rho)| < \infty, \qquad \text{and} \qquad g(a|\rho) = 0,$$

is

$$g(r|\rho) = \frac{1}{2m}\left(\frac{r_<}{r_>}\right)^m\left[1 - \left(\frac{r_>}{a}\right)^{2m}\right].$$

This Green's function is illustrated in the figure captioned Problem 24 as functions of r/a and ρ/a when $m = 3$.

25. Show that the Green's function for

$$\frac{d^2g}{dr^2} + \frac{1}{r}\frac{dg}{dr} + \alpha^2 g = \frac{\delta(r-\rho)}{r}, \qquad 0 < r, \rho < 1,$$

---

[21] Reprinted with permission from Schecter, D. A., D. H. E. Dubin, A. C. Cass, C. F. Driscoll, I. M. Lansky, and T. M. O'Neil, 2000: Inviscid damping of asymmetries on a two-dimensional vortex. *Phys. Fluids*, **12**, 2397–2412. ©American Institute of Physics, 2000.

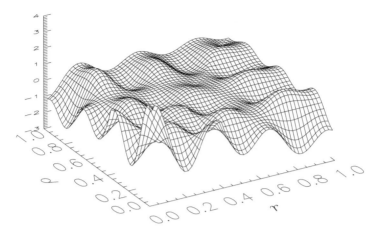

Problem 25

with the boundary conditions $g'(0|\rho) = g'(1|\rho) = 0$, is

$$g(r|\rho) = [aJ_0\left(\alpha r_>\right) + bY_0\left(\alpha r_>\right)] J_0\left(\alpha r_<\right),$$

where $aJ_1(\alpha) + bY_1(\alpha) = 0$, and $b = \pi/2$. This Green's function is illustrated in the figure captioned Problem 25 as functions of r and ρ when $\alpha = 20$.

26. Find the Green's function[22] and its corresponding bilinear expansion for the singular differential equation

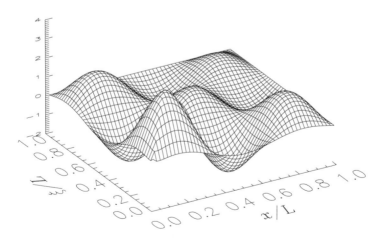

Problem 26

[22] Taken from Cockran, J. A., 1988: Unusual identities for special functions from waveguide propagation analyses. *IEEE Trans. Microwave Theory Tech.*, **36**, 611–614. ©1988 IEEE

$$xg'' + g' + \left(k^2 x - \frac{m^2}{x}\right) g = -\delta(x - \xi), \quad 0 < x < L,$$

where $m > -1$. The boundary conditions are

$$\lim_{x \to 0} |g'(x|\xi)| < \infty, \quad \text{and} \quad g'(L|\xi) = 0.$$

This Green's function (divided by $k^2 L^2$) is illustrated in the figure captioned Problem 26 as functions of x/L and ξ/L when $kL = 10$.

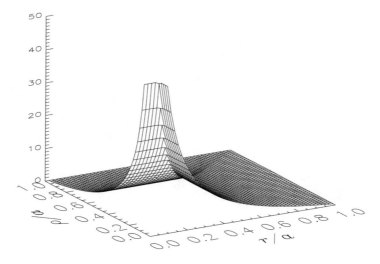

Problem 27

27. Find the eigenfunction expansion for the Green's function within a sphere of radius a:

$$\frac{1}{r^2} \frac{d}{dr} \left(r^2 \frac{dg}{dr}\right) = \frac{1}{r} \frac{d^2}{dr^2} (rg) = -\frac{\delta(r - \rho)}{4\pi \rho r}, \quad 0 < r, \rho < a,$$

with the boundary conditions

$$\lim_{r \to 0} |g(r|\rho)| < \infty, \quad \text{and} \quad g(a|\rho) = 0.$$

This Green's function (multiplied by $2\pi^3 a$) is illustrated in the figure captioned Problem 27 for $0.02 \le r/a, \rho/a \le 1$. No value greater than 50 is plotted.

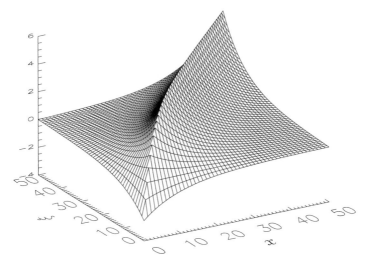

Problem 28

28. Find the Green's function[23] for the singular differential equation

$$g'' - k^2 g = -\delta(x - \xi), \qquad 0 < x, \xi < \infty,$$

with the boundary conditions

$$ag(0|\xi) + bg'(0|\xi) = 0, \qquad \lim_{x \to \infty} |g'(x|\xi)| < \infty.$$

Assume that $k \neq 0$ and $kb \neq a$. This Green's function is illustrated in the figure captioned Problem 28 as functions of x and ξ when $a = 2$, $b = 4$, and $k = 0.1$.

29. Show that the Green's function[24] for the singular differential equation

$$a^2 x^2 g'' + 2a^2 xg' - g = -\delta(x - \xi), \qquad 0 < x, \xi < \infty,$$

with the boundary conditions

$$g(0|\xi) = 0, \qquad \lim_{x \to \infty} |g(x|\xi)| < \infty,$$

---

[23] Taken from Chakrabarti, A., and T. Sahoo, 1998: Reflection of water waves in the presence of surface tension by a nearly vertical porous wall. *J. Austral. Math. Soc., Ser. B*, **39**, 308–317.

[24] Taken from Kamb, B., and K. A. Echelmeyer, 1986: Stress-gradient coupling in glacier flow: I. Longitudinal averaging of the influence of ice thickness and surface slope. *J. Glaciol.*, **32**, 267–284. Reprinted from the *Journal of Glaciology* with permission of the International Glaciological Society.

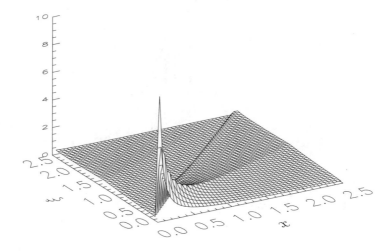

Problem 29

where $a > 0$, is

$$g(x|\xi) = \frac{(\xi/x)^{p_\pm}}{ax\sqrt{a^2 + 4}},$$

with

$$p_\pm = \tfrac{1}{2}\left(-1 \pm \sqrt{1 + 4/a^2}\right).$$

The negative sign applies for $0 < x \le \xi$ while the positive sign holds for $\xi \le x < \infty$. This Green's function is illustrated in the figure captioned Problem 29 when $0.1 \le x, \xi \le 2.5$, and $a = 1$.

30. Find the free-space Green's function[25] governed by

$$g'' - [s + \rho H(-x)]g = -\delta(x - \xi).$$

Hint: Consider the four separate cases: (1) $x \ge 0$, $\xi \ge 0$; (2) $x \le 0$, $\xi \ge 0$; (3) $x \ge 0$, $\xi \le 0$; and (4) $x \le 0$, $\xi \le 0$.

31. Find the Green's function for

$$g^{iv} = \delta(x - \xi), \qquad 0 < x, \xi < L,$$

subject to the boundary conditions

$$g(0|\xi) = g''(0|\xi) = g(L|\xi) = g''(L|\xi) = 0.$$

---

[25] Adapted from Linetsky, V., 1999: Step options. *Math. Finance*, **9**, 55–96. Published by Blackwell Science Ltd, Oxford, UK.

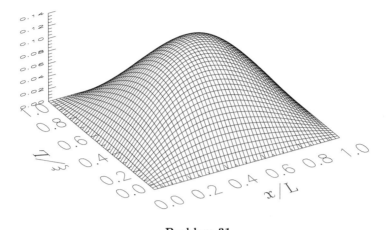

Problem 31

This Green's function (multiplied by $6/L^3$) is illustrated in the figure captioned Problem 31 as functions of x/L and ξ/L.

32. Use Fourier transforms to find the Green's function for

$$g^{iv} - k^4 g = -\delta(x - \xi), \qquad -\infty < x, \xi < \infty.$$

33. Find the Green's function[26] and its corresponding bilinear expansion for

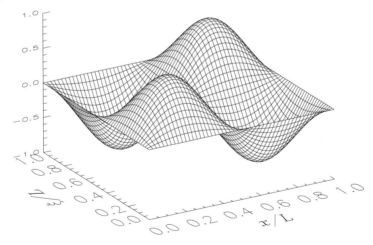

Problem 33

[26] Taken from Nicholson, J. W., and L. A. Bergman, 1985: On the efficacy of the modal series representation for Green functions of vibrating continuous structures. *J. Sound Vibr.*, **98**, 299–304. Published by Academic Press Ltd., London, U.K.

$$g^{iv} - k^4 g = \delta(x - \xi), \qquad 0 < x, \xi < L,$$

with the boundary conditions

$$g(0|\xi) = g(L|\xi) = g''(0|\xi) = g''(L|\xi) = 0.$$

This Green's function (multiplied by $100/L^3$) is illustrated in the figure captioned Problem 33 as functions of x/L and ξ/L when $kL = 4$.

34. Find the Green's function[27] for

$$g^{iv} - a^4 g + R\delta(x - x_0)g = \delta(x - \xi), \qquad -\infty < x, x_0, \xi < \infty,$$

where $a > 0$, and $x_0 \neq \xi$.

Step 1: Taking the Fourier transform of the ordinary differential equation, show that

$$(k^4 - a^4)G(k|\xi) + \frac{R}{2\pi} \int_{-\infty}^{\infty} e^{-ix_0(k-\tau)} G(\tau|\xi)\,d\tau = e^{-ik\xi},$$

where k is the transform variable, or

$$G(k|\xi) = \frac{e^{-ik\xi}}{k^4 - a^4} - \frac{Rf(\xi)e^{-ikx_0}}{k^4 - a^4},$$

where

$$f(\xi) = \frac{1}{2\pi} \int_{-\infty}^{\infty} G(\tau|\xi)e^{ix_0\tau}\,d\tau.$$

Step 2: Show that

$$f(\xi) = \frac{g_0(x_0|\xi)}{1 + Rg_0(x_0|x_0)},$$

where

$$g_0(x|y) = \frac{1}{2\pi} \int_{-\infty}^{\infty} \frac{e^{i(x-y)\tau}}{\tau^4 - a^4}\,d\tau = \frac{1}{4a^3}\left(ie^{ia|x-y|} - e^{-a|x-y|}\right).$$

Hint: Set $k = \tau$ in the second equation in Step 1. Then multiply the equation by $e^{ix_0\tau}\,d\tau$ and integrate from $-\infty$ to ∞.

---

[27] Reprinted with permission from DiPerna, D. T., and D. Feit, 2000: A perturbation technique for the prediction of the displacement of a line-driven plate with discontinuities. *J. Acoust. Soc. Am.*, **107**, 2004–2010. ©Acoustical Society of America, 2000.

Step 3: Show that

$$g(x|\xi) = g_0(x|\xi) - \frac{Rg_0(x_0|\xi)}{1 + Rg_0(x_0|x_0)} g_0(x|x_0).$$

35. Find the Green's function for the singular differential equation

$$\frac{1}{r}\frac{d}{dr}\left(r\frac{dg}{dr}\right) - \frac{m^2}{r^2}g - \omega\delta(r-1)g = -\frac{\delta(r-\rho)}{r}, \quad 0 < r, \rho < \infty,$$

with the boundary conditions

$$\lim_{r\to 0}|g(r|\rho)| < \infty, \quad \text{and} \quad \lim_{r\to\infty}|g(r|\rho)| < \infty.$$

Step 1: If $g(r|\rho) = g_1(r|\rho) + g_2(r|\rho)$, show that g_1 and g_2 are governed by

$$\frac{1}{r}\frac{d}{dr}\left(r\frac{dg_1}{dr}\right) - \frac{m^2}{r^2}g_1 = -\frac{\delta(r-\rho)}{r},$$

and

$$\frac{1}{r}\frac{d}{dr}\left(r\frac{dg_2}{dr}\right) - \frac{m^2}{r^2}g_2 - \omega\delta(r-1)g_2 = \omega\delta(r-1)g_1.$$

Step 2: Show that

$$g_1(r|\rho) = \frac{1}{2m}\left(\frac{r_<}{r_>}\right)^m.$$

Step 3: Show that

$$g_2(r|\rho) = -\frac{\omega f(\rho)}{2m(2m+\omega)}\begin{cases} r^m, & r < 1, \\ r^{-m}, & r > 1, \end{cases}$$

where

$$f(\rho) = \begin{cases} \rho^{-m}, & \rho < 1, \\ \rho^m, & \rho > 1. \end{cases}$$

Hint: Show that

$$\frac{dg_2}{dr}\Big|_{r=1^-}^{r=1^+} - \omega g_2(1|\rho) = \frac{\omega}{2m}f(\rho).$$

36. Show that any nonhomogeneous, linear ordinary differential equation

$$A(x)\frac{d^2y}{dx^2} + B(x)\frac{dy}{dx} + C(x)y = f(x)$$

can be written as (2.3.1) by first multiplying it by $\exp\left[\int^x B(\xi)\,d\xi/A(\xi)\right]$ $/A(x)$ and then factoring.

Chapter 3
Green's Functions
for the Wave Equation

In Chapter 2, we showed how Green's functions could be used to solve initial and boundary-value problems involving ordinary differential equations. When we approach partial differential equations, similar considerations hold, although the complexity increases. In the next three chapters, we work through the classic groupings of the wave, heat and Helmholtz's equations.

Of these three groups, we start with the wave equation

$$\nabla^2 u - \frac{1}{c^2} \frac{\partial^2 u}{\partial t^2} = -q(\mathbf{r}, t), \qquad (3.0.1)$$

where ∇ is the three-dimensional gradient operator, t denotes time, $\mathbf{r}$ is the position vector, c is the phase velocity of the wave, and $q(\mathbf{r}, t)$ is the source density. In addition to (3.0.1) it is necessary to state boundary and initial conditions to obtain a unique solution. The condition on the boundary can be either Dirichlet or Neumann or a linear combination of both (Robin condition). The conditions in time must be Cauchy—that is, we must specify the value of u and its time derivative at $t = t_0$ for each point of the region under consideration.

The purpose of this introductory section is to prove that we can express the solution to (3.0.1) in terms of boundary conditions, initial conditions and the Green's function, which is found by solving

$$\nabla^2 g - \frac{1}{c^2}\frac{\partial^2 g}{\partial t^2} = -\delta(\mathbf{r} - \mathbf{r}_0)\delta(t - \tau), \qquad (3.0.2)$$

where $\mathbf{r}_0$ denotes the position of the source. Equation (3.0.2) expresses the effect of an impulse as it propagates from $\mathbf{r} = \mathbf{r}_0$ as time increases from $t = \tau$. For $t < \tau$, causality requires that $g(\mathbf{r}, t|\mathbf{r}_0, \tau) = g_t(\mathbf{r}, t|\mathbf{r}_0, \tau) = 0$ if the impulse is the sole source of the disturbance. We also require that g satisfies the homogeneous form of the boundary condition satisfied by u.

Remark. Although we will use (3.0.2) as our fundamental definition of the Green's function as it applies to the wave equation, we can also find it by solving the initial-value problem:

$$\nabla^2 u - \frac{1}{c^2}\frac{\partial^2 u}{\partial t^2} = 0, \quad t > \tau, \quad u(\mathbf{r}, \tau) = 0, \quad u_t(\mathbf{r}, \tau) = \delta(\mathbf{r} - \mathbf{r}_0). \quad (3.0.3)$$

Then $g(\mathbf{r}, t|\mathbf{r}_0, \tau) = u(\mathbf{r}, t - \tau)H(t - \tau)$. This is most easily seen by introducing a new time variable $t' = t - \tau$ into (3.0.2)–(3.0.3) and noting that the Laplace transform of (3.0.2) and (3.0.3) are identical.

Before we turn to the solution of (3.0.2), we first deal with *reciprocity*, namely that $g(\mathbf{r}, t|\mathbf{r}_0, \tau) = g(\mathbf{r}_0, -\tau|\mathbf{r}, -t)$. Our analysis starts with the two equations:

$$\nabla^2 g(\mathbf{r}, t|\mathbf{r}_0, \tau_0) - \frac{1}{c^2}\frac{\partial^2 g(\mathbf{r}, t|\mathbf{r}_0, \tau_0)}{\partial t^2} = -\delta(\mathbf{r} - \mathbf{r}_0)\delta(t - \tau_0), \qquad (3.0.4)$$

and

$$\nabla^2 g(\mathbf{r}, -t|\mathbf{r}_1, -\tau_1) - \frac{1}{c^2}\frac{\partial^2 g(\mathbf{r}, -t|\mathbf{r}_1, -\tau_1)}{\partial t^2} = -\delta(\mathbf{r} - \mathbf{r}_1)\delta(t - \tau_1). \quad (3.0.5)$$

Equation (3.0.5) holds because the delta function is an even function. Multiplying (3.0.4) by $g(\mathbf{r}, -t|\mathbf{r}_1, -\tau_1)$ and (3.0.5) by $g(\mathbf{r}, t|\mathbf{r}_0, \tau_0)$, subtracting, and integrating over the volume V and over the time t from $-\infty$ to t', where t' is greater than either τ_0 or τ_1, then

$$\int_{-\infty}^{t'} \iiint_V \left[g(\mathbf{r}, t|\mathbf{r}_0, \tau_0)\nabla^2 g(\mathbf{r}, -t|\mathbf{r}_1, -\tau_1) \right.$$
$$- g(\mathbf{r}, -t|\mathbf{r}_1, -\tau_1)\nabla^2 g(\mathbf{r}, t|\mathbf{r}_0, \tau_0)$$
$$+ \frac{1}{c^2}g(\mathbf{r}, t|\mathbf{r}_0, \tau_0)\frac{\partial^2 g(\mathbf{r}, -t|\mathbf{r}_1, -\tau_1)}{\partial t^2}$$
$$\left. - \frac{1}{c^2}g(\mathbf{r}, -t|\mathbf{r}_1, -\tau_1)\frac{\partial^2 g(\mathbf{r}, t|\mathbf{r}_0, \tau_0)}{\partial t^2} \right] dV\, dt$$
$$= g(\mathbf{r}_0, -\tau_0|\mathbf{r}_1, -\tau_1) - g(\mathbf{r}_1, \tau_1|\mathbf{r}_0, \tau_0). \quad (3.0.6)$$

The left side of (3.0.6) can be transformed by Green's second formula and the identity

$$\frac{\partial}{\partial t}\left[g(\mathbf{r},t|\mathbf{r}_0,\tau_0)\frac{\partial g(\mathbf{r},-t|\mathbf{r}_1,-\tau_1)}{\partial t} - g(\mathbf{r},-t|\mathbf{r}_1,-\tau_1)\frac{\partial g(\mathbf{r},t|\mathbf{r}_0,\tau_0)}{\partial t}\right]$$

$$= g(\mathbf{r},t|\mathbf{r}_0,\tau_0)\frac{\partial^2 g(\mathbf{r},-t|\mathbf{r}_1,-\tau_1)}{\partial t^2} - g(\mathbf{r},-t|\mathbf{r}_1,-\tau_1)\frac{\partial^2 g(\mathbf{r},t|\mathbf{r}_0,\tau_0)}{\partial t^2} \tag{3.0.7}$$

to become

$$\int_{-\infty}^{t'} \oiint_S \left[g(\mathbf{r},t|\mathbf{r}_0,\tau_0) \, \nabla g(\mathbf{r},-t|\mathbf{r}_1,-\tau_1) \right.$$

$$\left. - g(\mathbf{r},-t|\mathbf{r}_1,-\tau_1) \, \nabla g(\mathbf{r},t|\mathbf{r}_0,\tau_0) \right] \cdot \mathbf{n} \, dS \, dt$$

$$+ \frac{1}{c^2} \iiint_V \left[g(\mathbf{r},t|\mathbf{r}_0,\tau_0) \frac{\partial g(\mathbf{r},-t|\mathbf{r}_1,-\tau_1)}{\partial t} \right.$$

$$\left. - g(\mathbf{r},-t|\mathbf{r}_1,-\tau_1)\frac{\partial g(\mathbf{r},t|\mathbf{r}_0,\tau_0)}{\partial t} \right]_{t=-\infty}^{t=t'} dV. \tag{3.0.8}$$

The surface integral in (3.0.8) vanishes because both Green's functions satisfy the same homogeneous boundary conditions on S. The volume integral in (3.0.8) also vanishes for the following reasons: (1) At the lower limit both $g(\mathbf{r},-\infty|\mathbf{r}_0,\tau_0)$ and its time derivative vanish from the causality condition. (2) At the time $t = t'$, $g(\mathbf{r},-t'|\mathbf{r}_1,\tau_1)$ and its time derivative vanish because $-t'$ is earlier than $-\tau_1$. Thus, the left side of (3.0.6) equals zero and $g(\mathbf{r},t|\mathbf{r}_0,\tau) = g(\mathbf{r}_0,-\tau|\mathbf{r},-t)$.

We now establish that the solution to the nonhomogeneous wave equation can be expressed in terms of the Green's function, boundary conditions and initial conditions. We begin with the equations

$$\nabla_0^2 u(\mathbf{r}_0,t_0) - \frac{1}{c^2}\frac{\partial^2 u(\mathbf{r}_0,t_0)}{\partial t_0^2} = -q(\mathbf{r}_0,t_0), \tag{3.0.9}$$

and

$$\nabla_0^2 g(\mathbf{r},t|\mathbf{r}_0,t_0) - \frac{1}{c^2}\frac{\partial^2 g(\mathbf{r},t|\mathbf{r}_0,t_0)}{\partial t_0^2} = -\delta(\mathbf{r}-\mathbf{r}_0)\delta(t-t_0), \tag{3.0.10}$$

where we obtain (3.0.9) from a combination of (3.0.2) plus reciprocity. As we did above, we multiply (3.0.9) by $g(\mathbf{r},t|\mathbf{r}_0,t_0)$ and (3.0.8) by $u(\mathbf{r}_0,t_0)$ and subtract. Integrating over the volume V_0 and over t_0 from 0 to t^+, where t^+ denotes a time slightly later than t so that we avoid ending the integration exactly at the peak of the delta function, we obtain

$$
\int_0^{t^+} \iiint_{V_0} \left[g \, \nabla_0^2 u - u \, \nabla_0^2 g + \frac{1}{c^2} \left(u \, \frac{\partial^2 g}{\partial t_0^2} - g \, \frac{\partial^2 u}{\partial t_0^2} \right) \right] dV_0 \, dt_0
$$

$$
= u(\mathbf{r}, t) - \int_0^{t^+} \iiint_{V_0} q(\mathbf{r}_0, t_0) \, g(\mathbf{r}, t | \mathbf{r}_0, t_0) \, dV_0 \, dt_0. \quad \textbf{(3.0.11)}
$$

Again employing Green's second formula, Gauss's divergence theorem and rewriting some terms, we find that

$$
\int_0^{t^+} \oiint_{S_0} (g \, \nabla_0 u - u \, \nabla_0 g) \cdot \mathbf{n} \, dS_0 \, dt_0 + \frac{1}{c^2} \iiint_{V_0} \left[u \, \frac{\partial g}{\partial t_0} - g \frac{\partial u}{\partial t_0} \right]_0^{t^+} dV_0
$$

$$
+ \int_0^{t^+} \iiint_{V_0} q(\mathbf{r}_0, t_0) \, g(\mathbf{r}, t | \mathbf{r}_0, t_0) \, dV_0 \, dt_0 = u(\mathbf{r}, t). \quad \textbf{(3.0.12)}
$$

The integrand in the first integral is specified by the boundary conditions. In the second integral, the integrand vanishes at $t = t^+$ from the initial conditions on g. The limit at $t = 0$ is determined by the initial conditions. Hence,

$$
u(\mathbf{r}, t) = \int_0^{t^+} \iiint_{V_0} q(\mathbf{r}_0, t_0) \, g(\mathbf{r}, t | \mathbf{r}_0, t_0) \, dV_0 \, dt_0
$$

$$
+ \int_0^{t^+} \oiint_{S_0} \bigg[g(\mathbf{r}, t | \mathbf{r}_0, t_0) \, \nabla_0 u(\mathbf{r}_0, t_0)
$$

$$
- u(\mathbf{r}_0, t_0) \, \nabla_0 g(\mathbf{r}, t | \mathbf{r}_0, t_0) \bigg] \cdot \mathbf{n} \, dS_0 \, dt_0
$$

$$
- \frac{1}{c^2} \iiint_{V_0} \left\{ u(\mathbf{r}_0, 0) \left[\frac{\partial g(\mathbf{r}, t | \mathbf{r}_0, 0)}{\partial t_0} \right] \right.
$$

$$
\left. - g(\mathbf{r}, t | \mathbf{r}_0, 0) \left[\frac{\partial u(\mathbf{r}_0, 0)}{\partial t_0} \right] \right\} dV_0. \quad \textbf{(3.0.13)}
$$

Equation (3.0.13) gives the complete solution of the inhomogeneous problem including initial conditions. In the case of the surface integrals, the surface value equals the limit of the value of the function as the surface is approached from the interior. The first two integrals on the right side of (3.0.13) represent the effect of the source and the boundary conditions, respectively. The last term involves the initial conditions; it can be interpreted as asking what sort of source is needed so that the function u starts in the desired manner.

3.1 ONE-DIMENSIONAL WAVE EQUATION IN AN UNLIMITED DOMAIN

The simplest possible example of Green's functions for the wave equation is the one-dimensional vibrating string problem.[1] In this problem the Green's function is given by the equation

$$\frac{\partial^2 g}{\partial t^2} - c^2 \frac{\partial^2 g}{\partial x^2} = c^2 \delta(x - \xi)\delta(t - \tau), \qquad (3.1.1)$$

where $-\infty < x, \xi < \infty$, and $0 < t, \tau$. If the initial conditions equal zero, the Laplace transform of (3.1.1) is

$$\frac{d^2 G}{dx^2} - \frac{s^2}{c^2} G = -\delta(x - \xi)e^{-s\tau}, \qquad (3.1.2)$$

where $G(x, s|\xi, \tau)$ is the Laplace transform of $g(x, t|\xi, \tau)$. To solve (3.1.2) we take its Fourier transform and obtain the algebraic equation

$$\overline{G}(k, s|\xi, \tau) = \frac{\exp(-ik\xi - s\tau)}{k^2 + s^2/c^2}. \qquad (3.1.3)$$

Having found the joint Laplace-Fourier transform of $g(x, t|\xi, \tau)$, we must work our way back to the Green's function. From the definition of the Fourier transform, we have that

$$G(x, s|\xi, \tau) = \frac{e^{-s\tau}}{2\pi} \int_{-\infty}^{\infty} \frac{e^{ik(x-\xi)}}{k^2 + s^2/c^2} \, dk. \qquad (3.1.4)$$

To evaluate the Fourier-type integral (3.1.4), we apply the residue theorem. Performing the calculation,

$$G(x, s|\xi, \tau) = \frac{c \, \exp(-s\tau - s|x - \xi|/c)}{2s}. \qquad (3.1.5)$$

Finally, applying (B.1.3),

$$g(x, t|\xi, \tau) = \frac{c}{2} H\left(t - \tau - |x - \xi|/c\right), \qquad (3.1.6)$$

or

$$g(x, t|\xi, \tau) = \frac{c}{2} H\left[c(t - \tau) + (x - \xi)\right] H\left[c(t - \tau) - (x - \xi)\right], \qquad (3.1.7)$$

---

[1] See also Graff, K. F., 1991: *Wave Motion in Elastic Solids*. Dover Publications, Inc., §1.1.8.

where $H(\ )$ is the Heaviside step function

$$H(t-a) = \begin{cases} 1, & t > a, \\ 0, & t < a, \end{cases} \qquad (3.1.8)$$

where $a \geq 0$.

Let us show that (3.1.6) satisfies the initial conditions and the differential equation. We begin by noting that

$$g(x, 0|\xi, \tau) = \frac{c}{2} H(-\tau - |x - \xi|/c) = 0, \qquad (3.1.9)$$

since the argument of the Heaviside step function is always negative. Similarly,

$$g_t(x, 0|\xi, \tau) = \frac{c}{2} \delta(-\tau - |x - \xi|/c) = 0, \qquad (3.1.10)$$

as the argument of the delta function is nonzero.

To show that the solution satisfies the differential equation, we use (3.1.7) and find that

$$\frac{\partial^2 g}{\partial x^2} - \frac{1}{c^2} \frac{\partial^2 g}{\partial t^2} = -2c\, \delta[c(t-\tau) + (x-\xi)]\, \delta[c(t-\tau) - (x-\xi)] \quad (3.1.11)$$

$$= -2c\, \delta[c(t-\tau) + (x-\xi)]\, \delta[2(x-\xi)] \qquad (3.1.12)$$

$$= -c\, \delta[c(t-\tau) + (x-\xi)]\, \delta(x-\xi) \qquad (3.1.13)$$

$$= -\delta(t-\tau)\, \delta(x-\xi) . \qquad (3.1.14)$$

We can use (3.1.6) and the *method of images* to obtain the Green's function for

$$\frac{\partial^2 g}{\partial x^2} - \frac{1}{c^2} \frac{\partial^2 g}{\partial t^2} = \delta(x-\xi)\delta(t-\tau), \qquad 0 < x, t, \xi, \tau, \qquad (3.1.15)$$

subject to the boundary condition $g(0, t|\xi, \tau) = 0$.

We begin by noting that the free-space Green's function[2] (3.1.6) is the particular solution to (3.1.15). Therefore, we need only find a homogeneous solution $f(x, t|\xi, \tau)$ so that

$$g(x, t|\xi, \tau) = \frac{c}{2} H\left(t - \tau - |x - \xi|/c\right) + f(x, t|\xi, \tau) \qquad (3.1.16)$$

satisfies the boundary condition at $x = 0$.

---

[2] In electromagnetic theory, a free-space Green's function is the particular solution of the differential equation valid over a domain of infinite extent, where the Green's function remains bounded as we approach infinity, or satisfies a radiation condition there.

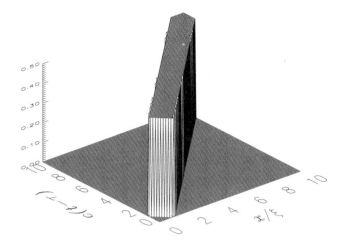

Figure 3.1.1: The Green's function $g(x,t|\xi,\tau)/c$ given by (3.1.17) for the one-dimensional wave equation for $x > 0$ at different distances x/ξ and times $c(t - \tau)$ subject to the boundary condition $g(0,t|\xi,\tau) = 0$.

To find $f(x,t|\xi,\tau)$, let us introduce a source at $x = -\xi$ at $t = \tau$. The corresponding free-space Green's function is $H(t - \tau - |x + \xi|/c)$. If, along the boundary $x = 0$ for any time t, this Green's function destructively interferes with the free-space Green's function associated with the source at $x = \xi$, then we have our solution. This will occur if our new source has a negative sign, resulting in the combined Green's function

$$g(x,t|\xi,\tau) = \frac{c}{2}\left[H(t - \tau - |x - \xi|/c) - H(t - \tau - |x + \xi|/c)\right].$$
$$(3.1.17)$$

See Figure 3.1.1. Because (3.1.17) satisfies the boundary condition, we need no further sources.

Let us check and see that (3.1.17) is the solution to our problem. First, direct substitution into (3.1.15) yields $\delta(x - \xi)\delta(t - \tau) - \delta(x + \xi)\delta(t - \tau)$, which equals $\delta(x - \xi)\delta(t - \tau)$ if $x > 0$. Second, an evaluation of (3.1.17) at $x = 0$ gives $g(0,t|\xi,\tau) = 0$. Thus, (3.1.17) is the Green's function for the problem (3.1.15) in the half-plane $x > 0$ because it (1) satisfies the wave equation in the domain $x > 0$, (2) remains finite as $x \to \infty$, and (3) satisfies the boundary condition $g(0,t|\xi,\tau) = 0$.

In a similar manner, we can use (3.1.6) and the method of images to find the Green's function for

$$\frac{\partial^2 g}{\partial x^2} - \frac{1}{c^2}\frac{\partial^2 g}{\partial t^2} = \delta(x - \xi)\delta(t - \tau), \quad 0 < x, t, \xi, \tau, \quad (3.1.18)$$

subject to the boundary condition $g_x(0,t|\xi,\tau) = 0$.

We begin by examining the related problem

$$\frac{\partial^2 g}{\partial x^2} - \frac{1}{c^2}\frac{\partial^2 g}{\partial t^2} = \delta(x-\xi)\delta(t-\tau) + \delta(x+\xi)\delta(t-\tau), \qquad \textbf{(3.1.19)}$$

where $-\infty < x, \xi < \infty$, and $0 < t, \tau$. In this particular case, we have chosen an image that is the mirror reflection of $\delta(x-\xi)$. This was dictated by the fact that the Green's function must be an even function of x along $x = 0$ for any time t. In line with this argument,

$$g(x,t|\xi,\tau) = \frac{c}{2}\left[H\left(t-\tau-|x-\xi|/c\right) + H\left(t-\tau-|x+\xi|/c\right)\right].$$
$$\textbf{(3.1.20)}$$

To show that this is the correct solution, we compute $g_x(x,t|\xi,\tau)$ and find that

$$g_x(x,t|\xi,\tau) = -\tfrac{1}{2}\mathrm{sgn}(x-\xi)\delta\left(t-\tau-|x-\xi|/c\right)$$
$$-\tfrac{1}{2}\mathrm{sgn}(x+\xi)\delta\left(t-\tau-|x+\xi|/c\right), \quad \textbf{(3.1.21)}$$

where the signum function $\mathrm{sgn}(t)$ is

$$\mathrm{sgn}(t) = \begin{cases} 1, & t > 0, \\ -1, & t < 0. \end{cases} \qquad \textbf{(3.1.22)}$$

Consequently, $g_x(0,t|\xi,\tau) = 0$. Thus, (3.1.20) is the correct Green's function because it satisfies (3.1.18) if $x,\xi > 0$ and the boundary condition $g_x(0,t|\xi,\tau) = 0$.

• Example 3.1.1: One-dimensional Klein-Gordon equation

The Klein-Gordon equation is a form of the wave equation that arose in particle physics as the relativistic scalar wave equation describing particles with nonzero rest mass. In this example, we find its Green's function when there is only one spatial dimension:

$$\frac{\partial^2 g}{\partial x^2} - \frac{1}{c^2}\left(\frac{\partial^2 g}{\partial t^2} + a^2 g\right) = -\delta(x-\xi)\delta(t-\tau), \qquad \textbf{(3.1.23)}$$

where $-\infty < x, \xi < \infty$, $0 < t, \tau$, c is a real, positive constant (the wave speed) and a is a real, nonnegative constant. The corresponding boundary conditions are

$$\lim_{|x|\to\infty}|g(x,t|\xi,\tau)| < \infty, \qquad \textbf{(3.1.24)}$$

and the initial conditions are

$$g(x,0|\xi,\tau) = g_t(x,0|\xi,\tau) = 0. \qquad \textbf{(3.1.25)}$$

We begin by taking the Laplace transform of (3.1.23) and find that

$$\frac{d^2G}{dx^2} - \left(\frac{s^2 + a^2}{c^2}\right)G = -\delta(x - \xi)e^{-s\tau}. \qquad (3.1.26)$$

Applying Fourier transforms to (3.1.26), we obtain

$$G(x, s|\xi, \tau) = \frac{c^2}{2\pi}e^{-s\tau}\int_{-\infty}^{\infty}\frac{e^{ik(x-\xi)}}{s^2 + a^2 + k^2c^2}\,dk \qquad (3.1.27)$$

$$= \frac{c^2}{\pi}e^{-s\tau}\int_0^{\infty}\frac{\cos[k(x - \xi)]}{s^2 + a^2 + k^2c^2}\,dk. \qquad (3.1.28)$$

Inverting the Laplace transform and employing the second shifting theorem,

$$g(x, t|\xi, \tau) = \frac{c^2}{\pi}H(t - \tau)\int_0^{\infty}\frac{\sin\left[(t - \tau)\sqrt{a^2 + k^2c^2}\,\right]\cos[k(x - \xi)]}{\sqrt{a^2 + k^2c^2}}\,dk$$

$$(3.1.29)$$

$$= \frac{c^2}{2\pi}H(t - \tau)\int_0^{\infty}\frac{\sin\left[(t - \tau)\sqrt{a^2 + k^2c^2} - k|x - \xi|\right]}{\sqrt{a^2 + k^2c^2}}\,dk$$

$$+ \frac{c^2}{2\pi}H(t - \tau)\int_0^{\infty}\frac{\sin\left[(t - \tau)\sqrt{a^2 + k^2c^2} + k|x - \xi|\right]}{\sqrt{a^2 + k^2c^2}}\,dk$$

$$(3.1.30)$$

$$= \frac{c}{2\pi}H(t - \tau)\int_a^{\infty}\frac{\sin\left[\omega(t - \tau) - \sqrt{\omega^2 - a^2}\,|x - \xi|/c\right]}{\sqrt{\omega^2 - a^2}}\,d\omega$$

$$+ \frac{c}{2\pi}H(t - \tau)\int_a^{\infty}\frac{\sin\left[\omega(t - \tau) + \sqrt{\omega^2 - a^2}\,|x - \xi|/c\right]}{\sqrt{\omega^2 - a^2}}\,d\omega.$$

$$(3.1.31)$$

Equation (3.1.29) represents a superposition of homogeneous solutions (normal modes) to (3.1.23). An intriguing aspect of (3.1.30)–(3.1.31) is that this solution occurs everywhere after $t > \tau$. If $|x - \xi| > c(t - \tau)$, these wave solutions destructively interfere so that we have zero there while they constructively interfere at those times and places where the physical waves are present.

Applying integral tables to (3.1.29), the final result is

$$g(x, t|\xi, \tau) = \frac{c}{2}J_0\left[a\sqrt{(t - \tau)^2 - (x - \xi)^2/c^2}\,\right]$$

$$\times H[c(t - \tau) - |x - \xi|]. \qquad (3.1.32)$$

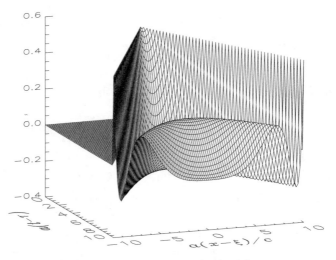

Figure 3.1.2: The free-space Green's function $g(x, t|\xi, \tau)/c$ for the one-dimensional Klein-Gordon equation at different distances $a(x - \xi)/c$ and times $a(t - \tau)$.

This Green's function is illustrated in Figure 3.1.2.

Let us return to (3.1.27) and invert the Fourier transform via the residue theorem. We obtain

$$G(x, s|\xi, \tau) = \frac{\exp[-|x - \xi|\sqrt{(s^2 + a^2)/c^2} - s\tau]}{2\sqrt{(s^2 + a^2)/c^2}}. \qquad (3.1.33)$$

Instead of using tables to invert the Laplace transform, we employ Bromwich's integral with the contour running just to the right of the imaginary axis. The square root is computed by taking the branch cuts to run from $s = ai$ out to $+\infty i$ and from $s = -ai$ out to $-\infty i$; the cuts are along the imaginary axis. Setting $s = \omega i$,

$$g(x, t|\xi, \tau) = \frac{c}{2\pi} H(t - \tau) \int_a^\infty \frac{\sin[\omega(t - \tau) - \sqrt{\omega^2 - a^2}\,|x - \xi|/c]}{\sqrt{\omega^2 - a^2}} \, d\omega$$

$$+ \frac{c}{2\pi} H(t - \tau) \int_0^a \frac{\exp[-|x - \xi|\sqrt{a^2 - \omega^2}\,/c] \cos[\omega(t - \tau)]}{\sqrt{a^2 - \omega^2}} \, d\omega.$$

$$(3.1.34)$$

Comparing (3.1.31) and (3.1.34), we see that the backward propagating waves in (3.1.31) have become evanescent waves in (3.1.34). This suggests that in certain transient problems there exists a correspondence between the evanescent modes and backward propagating solutions.

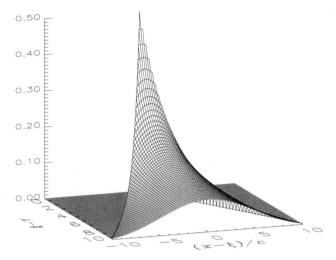

Figure 3.1.3: The free-space Green's function $g(x, t|\xi, \tau)/c$ for the one-dimensional equation of telegraphy with $\gamma = 1$ at different distances $(x - \xi)/c$ and times $t - \tau$.

• Example 3.1.2: Equation of telegraphy

When the vibrating string problem includes the effect of air resistance, (3.1.1) becomes

$$\frac{\partial^2 g}{\partial t^2} + 2\gamma \frac{\partial g}{\partial t} - c^2 \frac{\partial^2 g}{\partial x^2} = c^2 \delta(x - \xi)\delta(t - \tau), \qquad (3.1.35)$$

where $-\infty < x, \xi < \infty$, and $0 < t, \tau$, with the boundary conditions (3.1.24) and the initial conditions (3.1.25). Let us find the Green's function.

Our analysis begins by introducing an intermediate dependent variable $w(x, t|\xi, \tau)$, where $g(x, t|\xi, \tau) = e^{-\gamma t} w(x, t|\xi, \tau)$. Substituting for $g(x, t|\xi, \tau)$, we now have

$$\frac{\partial^2 w}{\partial t^2} - \gamma^2 w - c^2 \frac{\partial^2 w}{\partial x^2} = c^2 \delta(x - \xi)\delta(t - \tau)e^{\gamma \tau}. \qquad (3.1.36)$$

Taking the Laplace transform of (3.1.36), we obtain

$$\frac{d^2 W}{dx^2} - \left(\frac{s^2 - \gamma^2}{c^2}\right) W = -\delta(x - \xi)e^{\gamma \tau - s\tau}. \qquad (3.1.37)$$

Using Fourier transforms as in Example 3.1.1, the solution to (3.1.37) is

$$W(x, s|\xi, \tau) = \frac{\exp[-|x - \xi|\sqrt{(s^2 - \gamma^2)/c^2} + \gamma \tau - s\tau]}{2\sqrt{(s^2 - \gamma^2)/c^2}}. \qquad (3.1.38)$$

Employing tables to invert the Laplace transform and the second shifting theorem, we have that

$$w(x,t|\xi,\tau) = \frac{c}{2}e^{\gamma\tau}I_0\left[\gamma\sqrt{(t-\tau)^2 - (x-\xi)^2/c^2}\right]$$
$$\times H[c(t-\tau) - |x-\xi|], \qquad (3.1.39)$$

or

$$g(x,t|\xi,\tau) = \frac{c}{2}e^{-\gamma(t-\tau)}I_0\left[\gamma\sqrt{(t-\tau)^2 - (x-\xi)^2/c^2}\right]$$
$$\times H[c(t-\tau) - |x-\xi|]. \qquad (3.1.40)$$

• Example 3.1.3: D'Alembert's solution

Consider the case of an unbounded space that is free of sources. If the initial values are

$$u(x_0,0) = F(x_0), \qquad \text{and} \qquad \frac{\partial u(x_0,0)}{\partial t_0} = G(x_0), \qquad (3.1.41)$$

then by (3.0.13),

$$u(x,t) = \frac{1}{2c}\int_{-\infty}^{\infty} G(x_0)H(t - |x-x_0|/c)\,dx_0$$
$$+ \frac{1}{2c}\int_{-\infty}^{\infty} F(x_0)\,\delta(t - |x-x_0|/c)\,dx_0 \qquad (3.1.42)$$
$$= \frac{1}{2c}\int_{x-ct}^{x+ct} G(x_0)\,dx_0$$
$$+ [cF(x+ct) - (-c)F(x-ct)]/(2c) \qquad (3.1.43)$$
$$= \tfrac{1}{2}[F(x+ct) + F(x-ct)] + \frac{1}{2c}\int_{x-ct}^{x+ct} G(x_0)\,dx_0. \qquad (3.1.44)$$

Equation (3.1.44) is known as *d'Alembert's solution*. Note that $u(x,t)$ depends on *all* of the values of $u_t(x,0)$ within a distance ct of x but depends only on the value of $u(x,0)$ at the two points exactly at $x_0 = x \pm ct$.

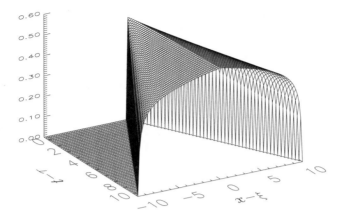

Figure 3.1.4: The free-space Green's function $g(x, t|\xi, \tau)$ for the one-dimensional wave equation in a visco-elastic media at different distances $x - \xi$ and times $t - \tau$ when $a = 0.05$

• **Example 3.1.4: Wave propagation in a visco-elastic media**

Visco-elastic media exhibit the important property of memory that is expressed mathematically in the dynamical equations by time convolution. In this example, we examine the effect of visco-elasticity by finding the free-space Green's function governed by the protypical equation[3]

$$[1 + K(t)*]\frac{\partial^2 g}{\partial t^2} - \frac{\partial^2 g}{\partial x^2} = [1 + K(t)*]\delta(t - \tau)\delta(x - \xi), \qquad (3.1.45)$$

with $-\infty < x, \xi < \infty$, and $0 < t, \tau$, where $K(t) = (a^2 + 2a/\sqrt{\pi t})H(t)$.
We begin by taking the Laplace transform of (3.1.45) or

$$(s + a\sqrt{s})^2 G - \frac{d^2 G}{dx^2} = \left(1 + \frac{a^2}{s} + \frac{2a}{\sqrt{s}}\right) e^{-s\tau}\delta(x - \xi). \qquad (3.1.46)$$

Solving (3.1.46), we have that

$$G(x, s|\xi, \tau) = \left(\frac{1}{2s} + \frac{a}{2s^{3/2}}\right) e^{-(s+a\sqrt{s})|x-\xi|-s\tau}. \qquad (3.1.47)$$

---

[3] See Hanyga, A., and M. Seredyńska, 1999: Some effects of the memory kernel singularity on wave propagation and inversion in poroelastic media–I. Forward problems. *Geophys. J. Int.*, **137**, 319–335. Published by Blackwell Science Ltd., Oxford, UK.

Taking the inverse Laplace transform and using the second shifting theorem, we obtain the Green's function

$g(x, t|\xi, \tau)$

$$= \tfrac{1}{2}\left(1 - a^2|x - \xi|\right) \operatorname{erfc}\left(\frac{a|x - \xi|}{2\sqrt{t - \tau - |x - \xi|}}\right) H(t - \tau - |x - \xi|)$$

$$+ \frac{a}{2}\sqrt{\frac{t - \tau - |x - \xi|}{\pi}} \exp\left[-\frac{a^2(x - \xi)^2}{4(t - \tau - |x - \xi|)}\right] H(t - \tau - |x - \xi|).$$

$$(3.1.48)$$

Let us compare our solution (3.1.48) with the solution when the visco-elastic effect is absent, namely (3.1.6). As Figure 3.1.4 shows, the sharp wave front at $t - \tau = |x - \xi|$ becomes progressively smoother as time increases.

● **Example 3.1.5: Wave propagation into a plasma**

So far, we treated the case where the same wave equation holds over the entire domain. Here, we consider the problem of the wave equation

$$\frac{\partial^2 g_1}{\partial t^2} - \frac{\partial^2 g_1}{\partial x^2} = \delta(t - \tau)\delta(x + \xi), \qquad \xi > 0, \qquad (3.1.49)$$

governing the semi-infinite domain $x < 0$ while the Klein-Gordon equation

$$\frac{\partial^2 g_2}{\partial t^2} - \frac{\partial^2 g_2}{\partial x^2} + g_2 = 0 \qquad (3.1.50)$$

holds for $x > 0$. At the interface, we have the conditions that

$$g_1(0, t|\xi, \tau) = g_2(0, t|\xi, \tau), \quad \text{and} \quad g_{1x}(0, t|\xi, \tau) = g_{2x}(0, t|\xi, \tau).$$

$$(3.1.51)$$

Of course, we also have the boundary conditions at infinity that

$$\lim_{x \to -\infty} |g_1(x, t|\xi, \tau)| < \infty, \quad \text{and} \quad \lim_{x \to \infty} |g_2(x, t|\xi, \tau)| < \infty, \quad (3.1.52)$$

and all of the initial conditions equal zero. This problem[4] models the propagation of progressive waves into a region where a plasma is present.

---

[4] See Nodland, B., and C. J. McKinstrie, 1997: Propagation of a short laser pulse in a plasma. *Phys. Review E*, **56**, 7174–7178; Khrushchinskii, A. A., and D. Yu. Churmakov, 2000: Passage of an electromagnetic pulse through a layer of homogeneous plasma (exact solution). *J. Engng. Phys. Thermophys.*, **73**, 554–562.

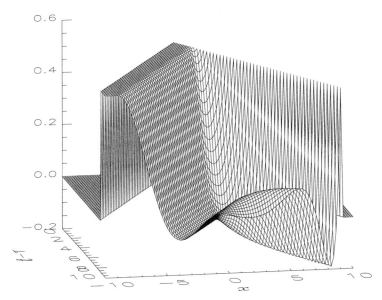

Figure 3.1.5: The free-space Green's function $g(x,t|\xi,\tau)$ governed by the one-dimensional wave equation for $x < 0$ and the one-dimensional Klein-Gordon equation for $x > 0$. The source of excitation at time $t = \tau$ occurs at $x = -\xi$.

As usual, we begin by taking the Laplace transform of the governing equations and find that

$$\frac{d^2 G_1}{dx^2} - s^2 G_1 = -\delta(x + \xi)e^{-s\tau}, \tag{3.1.53}$$

and

$$\frac{d^2 G_2}{dx^2} - (s^2 + 1)G_2 = 0, \tag{3.1.54}$$

with

$$G_1(0, s|\xi, \tau) = G_2(0, s|\xi, \tau), \quad \text{and} \quad G_1'(0, s|\xi, \tau) = G_2'(0, s|\xi, \tau). \tag{3.1.55}$$

The solution to (3.1.53)–(3.1.54) is

$$G_1(x, s|\xi, \tau) = \frac{e^{-s|x+\xi|-s\tau}}{2s} + Ae^{sx}, \tag{3.1.56}$$

and

$$G_2(x, s|\xi, \tau) = Be^{-x\sqrt{s^2+1}}. \tag{3.1.57}$$

The first term in (3.1.56) represents the particular solution to (3.1.53) while the remaining terms in (3.1.56)–(3.1.57) are necessary to satisfy the interfacial conditions. The solution e^{-sx} in (3.1.56) and $e^{x\sqrt{s^2+1}}$ in (3.1.57) are discarded because they violate the boundary conditions at infinity.

After satisfying the interfacial condition at $x = 0$, (3.1.56)–(3.1.57) become

$$G_1(x, s|\xi, \tau) = \frac{e^{-s|x+\xi|-s\tau}}{2s} + \frac{s - \sqrt{s^2 + 1}}{2s\left(s + \sqrt{s^2 + 1}\right)} e^{-s(\xi - x + \tau)}, \quad (3.1.58)$$

and

$$G_2(x, s|\xi, \tau) = \frac{e^{-x\sqrt{s^2+1} - s(\xi + \tau)}}{s + \sqrt{s^2 + 1}}. \quad (3.1.59)$$

The physical interpretation of (3.1.58)–(3.1.59) is as follows: The leading term in (3.1.58) represents the direct wave radiating out from the source point $x = -\xi$. At the interface, some of the energy passes into the region $x > 0$; this is given by (3.1.59). Some of the energy is reflected and this reflected wave is given by the second term in (3.1.58). Taking the inverse of (3.1.58)–(3.1.59), the corresponding Green's functions are

$$g_1(x, t|\xi, \tau) = \tfrac{1}{2}H(t - \tau - |x + \xi|) - H[t - \tau - (\xi - x)]\int_0^{t-\tau-(\xi-x)} \frac{J_2(z)}{z}\, dz,$$
$$(3.1.60)$$

and

$$g_2(x, t|\xi, \tau) = \frac{(t - \tau - \xi)\, J_1\left[\sqrt{(t - \tau - \xi)^2 - x^2}\right]}{(t - \tau + x - \xi)\, \sqrt{(t - \tau - \xi)^2 - x^2}} H(t - \tau - x - \xi)$$
$$+ \frac{x J_1'\left[\sqrt{(t - \tau - \xi)^2 - x^2}\right]}{t - \tau + x - \xi} H(t - \tau - x - \xi).$$
$$(3.1.61)$$

Equations (3.1.60)–(3.1.61) are illustrated in Figure 3.1.5 when $\xi = 1$. The flip side of this problem, the source located in the plasma, has been left as problem 7.

3.2 ONE-DIMENSIONAL WAVE EQUATION ON THE INTERVAL $0 < x < L$

One of the classic problems of mathematical physics involves finding the displacement of a taut string between two supports when an external force is applied. The governing equation is

$$\frac{\partial^2 u}{\partial t^2} - c^2 \frac{\partial^2 u}{\partial x^2} = f(x, t), \quad 0 < x < L, \quad 0 < t, \quad (3.2.1)$$

where c is the constant phase speed.

In this section, we find the Green's function for this problem by considering the following problem:

$$\frac{\partial^2 g}{\partial t^2} - c^2 \frac{\partial^2 g}{\partial x^2} = \delta(x - \xi)\delta(t - \tau), \quad 0 < x, \xi < L, \quad 0 < t, \tau, \quad \textbf{(3.2.2)}$$

with the boundary conditions

$$\alpha_1 g(0, t|\xi, \tau) + \beta_1 g_x(0, t|\xi, \tau) = 0, \quad 0 < t, \quad \textbf{(3.2.3)}$$

and

$$\alpha_2 g(L, t|\xi, \tau) + \beta_2 g_x(L, t|\xi, \tau) = 0, \quad 0 < t, \quad \textbf{(3.2.4)}$$

and the initial conditions

$$g(x, 0|\xi, \tau) = g_t(x, 0|\xi, \tau) = 0, \quad 0 < x < L. \quad \textbf{(3.2.5)}$$

We start by taking the Laplace transform of (3.2.2) and find that

$$\frac{d^2 G}{dx^2} - \frac{s^2}{c^2} G = -\frac{\delta(x - \xi)}{c^2} e^{-s\tau}, \quad 0 < x < L, \quad \textbf{(3.2.6)}$$

with

$$\alpha_1 G(0, s|\xi, \tau) + \beta_1 G'(0, s|\xi, \tau) = 0, \quad \textbf{(3.2.7)}$$

and

$$\alpha_2 G(L, s|\xi, \tau) + \beta_2 G'(L, s|\xi, \tau) = 0. \quad \textbf{(3.2.8)}$$

Problems similar to (3.2.6)–(3.2.8) were considered in the previous chapter. There, solutions were developed in terms of an eigenfunction expansion. Applying the same technique here,

$$G(x, s|\xi, \tau) = e^{-s\tau} \sum_{n=1}^{\infty} \frac{\varphi_n(\xi)\varphi_n(x)}{s^2 + c^2 k_n^2}, \quad \textbf{(3.2.9)}$$

where $\varphi_n(x)$ is the nth *orthonormal* eigenfunction to the regular Sturm-Liouville problem

$$\varphi''(x) + k^2 \varphi(x) = 0, \quad 0 < x < L, \quad \textbf{(3.2.10)}$$

subject to the boundary conditions

$$\alpha_1 \varphi(0) + \beta_1 \varphi'(0) = 0, \quad \textbf{(3.2.11)}$$

and

$$\alpha_2 \varphi(L) + \beta_2 \varphi'(L) = 0. \quad \textbf{(3.2.12)}$$

Taking the inverse of (3.2.9), we have that the Green's function is

$$
g(x, t | \xi, \tau) = \left\{ \sum_{n=1}^{\infty} \varphi_n(\xi) \varphi_n(x) \frac{\sin[k_n c(t - \tau)]}{k_n c} \right\} H(t - \tau). \quad \textbf{(3.2.13)}
$$

Let us verify that (3.2.13) is indeed the solution to (3.2.2). We begin by computing

$$
\frac{\partial g}{\partial t} = \left\{ \sum_{n=1}^{\infty} \varphi_n(\xi) \varphi_n(x) \cos[k_n c(t - \tau)] \right\} H(t - \tau)
$$

$$
+ \left\{ \sum_{n=1}^{\infty} \varphi_n(\xi) \varphi_n(x) \frac{\sin[k_n c(t - \tau)]}{k_n c} \right\} \delta(t - \tau) \quad \textbf{(3.2.14)}
$$

$$
= \left\{ \sum_{n=1}^{\infty} \varphi_n(\xi) \varphi_n(x) \cos[k_n c(t - \tau)] \right\} H(t - \tau), \quad \textbf{(3.2.15)}
$$

because the bracketed term multiplying the delta function equals zero since $f(t)\delta(t - a) = f(a)\delta(t - a)$. Next,

$$
\frac{\partial^2 g}{\partial t^2} = - \left\{ \sum_{n=1}^{\infty} \varphi_n(\xi) \varphi_n(x) k_n c \sin[k_n c(t - \tau)] \right\} H(t - \tau)
$$

$$
+ \left\{ \sum_{n=1}^{\infty} \varphi_n(\xi) \varphi_n(x) \cos[k_n c(t - \tau)] \right\} \delta(t - \tau) \quad \textbf{(3.2.16)}
$$

$$
= - \left\{ \sum_{n=1}^{\infty} \varphi_n(\xi) \varphi_n(x) k_n c \sin[k_n c(t - \tau)] \right\} H(t - \tau)
$$

$$
+ \left\{ \sum_{n=1}^{\infty} \varphi_n(\xi) \varphi_n(x) \right\} \delta(t - \tau) \quad \textbf{(3.2.17)}
$$

$$
= - \left\{ \sum_{n=1}^{\infty} \varphi_n(\xi) \varphi_n(x) k_n c \sin[k_n c(t - \tau)] \right\} H(t - \tau)
$$

$$
+ \delta(x - \xi)\delta(t - \tau). \quad \textbf{(3.2.18)}
$$

Therefore,

$$
\frac{\partial^2 g}{\partial t^2} - c^2 \frac{\partial^2 g}{\partial x^2}
$$

$$
= \left\{ \sum_{n=1}^{\infty} \varphi_n(\xi) \left[-k_n^2 c^2 \varphi_n(x) - c^2 \varphi_n''(x) \right] \frac{\sin[k_n c(t - \tau)]}{k_n c} \right\} H(t - \tau)
$$

$$
+ \delta(x - \xi)\delta(t - \tau) = \delta(x - \xi)\delta(t - \tau) \quad \textbf{(3.2.19)}
$$

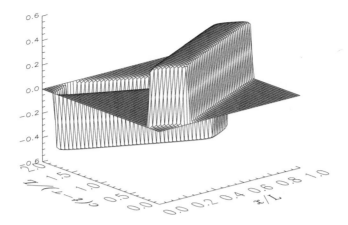

Figure 3.2.1: The Green's function $cg(x,t|\xi,\tau)$ given by (3.2.25) for the one-dimensional wave equation over the interval $0 < x < L$ as a function of location x/L and time $c(t-\tau)/L$ with $\xi/L = 0.2$. The boundary conditions are $g(0,t|\xi,\tau) = g(L,t|\xi,\tau) = 0$.

and (3.2.13) satisfies the differential equation (3.2.2).

- **Example 3.2.1**

Let us use our results to find the Green's function for

$$\frac{\partial^2 g}{\partial t^2} - c^2 \frac{\partial^2 g}{\partial x^2} = \delta(x-\xi)\delta(t-\tau), \tag{3.2.20}$$

with the boundary conditions

$$g(0,t|\xi,\tau) = g(L,t|\xi,\tau) = 0, \qquad 0 < t, \tag{3.2.21}$$

and the initial conditions

$$g(x,0|\xi,\tau) = g_t(x,0|\xi,\tau) = 0, \qquad 0 < x < L. \tag{3.2.22}$$

For this example, the Sturm-Liouville problem is

$$\varphi''(x) + k^2\varphi(x) = 0, \qquad 0 < x < L, \tag{3.2.23}$$

with the boundary conditions $\varphi(0) = \varphi(L) = 0$. The nth *orthonormal* eigenfunction for this problem is

$$\varphi_n(x) = \sqrt{\frac{2}{L}} \, \sin\left(\frac{n\pi x}{L}\right). \tag{3.2.24}$$

Consequently, from (3.2.13), the Green's function is

$$g(x,t|\xi,\tau) = \frac{2}{\pi c} \left\{ \sum_{n=1}^{\infty} \frac{1}{n} \sin\left(\frac{n\pi\xi}{L}\right) \sin\left(\frac{n\pi x}{L}\right) \sin\left[\frac{n\pi c(t-\tau)}{L}\right] \right\}$$
$$\times H(t-\tau). \tag{3.2.25}$$

● **Example 3.2.2**

Let us find the Green's function for

$$\frac{\partial^2 g}{\partial t^2} - c^2 \frac{\partial^2 g}{\partial x^2} = \delta(x-\xi)\delta(t-\tau), \tag{3.2.26}$$

with the boundary conditions

$$g_x(0,t|\xi,\tau) = g_x(L,t|\xi,\tau) = 0, \qquad 0 < t, \tag{3.2.27}$$

and the initial conditions

$$g(x,0|\xi,\tau) = g_t(x,0|\xi,\tau) = 0, \qquad 0 < x < L. \tag{3.2.28}$$

We begin by taking the Laplace transform of (3.2.26) and (3.2.27). This yields

$$\frac{d^2 G}{dx^2} - \frac{s^2}{c^2} G = -\frac{\delta(x-\xi)}{c^2} e^{-s\tau}, \tag{3.2.29}$$

with the boundary conditions

$$G'(0,s|\xi,\tau) = G'(L,s|\xi,\tau) = 0. \tag{3.2.30}$$

There are two ways that we can express the solution to (3.2.29)–(3.2.30). Applying the technique given in §2.3, we have

$$G(x,s|\xi,\tau) = \frac{\cosh[s(L-|x-\xi|)/c] + \cosh[s(L-x-\xi)/c]}{2sc \sinh(sL/c)} e^{-s\tau}.$$
$$\tag{3.2.31}$$

At this point, we note that

$$\frac{\cosh[s(L-\chi)/c]}{\sinh(sL/c)} = \frac{e^{s(L-\chi)/c} + e^{-s(L-\chi)/c}}{e^{sL/c} - e^{-sL/c}}$$

$$= e^{-s\chi/c}\left[1 + e^{-2sL/c} + e^{-4sL/c} + \cdots\right]$$

$$+ e^{-s(2L-\chi)/c}\left[1 + e^{-2sL/c} + e^{-4sL/c} + \cdots\right]. \tag{3.2.32}$$

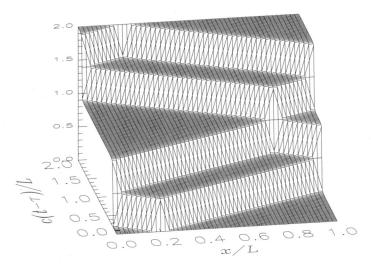

Figure 3.2.2: The Green's function $cg(x, t|\xi, \tau)$ given by (3.2.33) or (3.2.35) for the one-dimensional wave equation over the interval $0 < x < L$ as functions of location x/L and time $c(t - \tau)/L$ with $\xi/L = 0.2$. The boundary conditions are $g_x(0, t|\xi, \tau) = g_x(L, t|\xi, \tau) = 0$.

Applying this result to each of the hyperbolic terms in (3.2.31) and then inverting each term separately, we obtain

$$
\begin{aligned}
g(x, t|\xi, \tau) = \frac{1}{2c} \sum_{n=0}^{\infty} \Big\{ & H(t - \tau - |x - \xi|/c - 2nL/c) \\
& + H(t - \tau - (x + \xi)/c - 2nL/c) \\
& + H[t - \tau + |x - \xi|/c - 2(n + 1)L/c] \\
& + H[t - \tau + (x + \xi)/c - 2(n + 1)L/c] \Big\}. \quad \textbf{(3.2.33)}
\end{aligned}
$$

Equation (3.2.33) can be interpreted in terms of multiple reflections of the initial wave at the boundaries.

We can also express the solution to (3.2.29)–(3.2.30) as the eigenfunction expansion

$$
G(x, s|\xi, \tau) = \frac{1}{L} e^{-s\tau} \sum_{n=0}^{\infty} \epsilon_n \frac{\cos(n\pi\xi/L)\cos(n\pi x/L)}{s^2 + n^2\pi^2c^2/L^2}, \quad \textbf{(3.2.34)}
$$

where $\epsilon_0 = 1$ and $\epsilon_n = 2$ if $n > 0$. Equation (3.2.34) follows from (3.2.31) by employing the Mittag-Leffler expansion theorem from complex variable theory. A term-by-term inversion of the Laplace transform yields the Green's function

$$g(x, t|\xi, \tau) = \frac{(t - \tau)H(t - \tau)}{L}$$
$$+ \frac{2H(t - \tau)}{c\pi} \sum_{n=1}^{\infty} \frac{1}{n} \cos\left(\frac{n\pi\xi}{L}\right) \cos\left(\frac{n\pi x}{L}\right) \sin\left[\frac{n\pi c}{L}(t - \tau)\right].$$

$$(3.2.35)$$

3.3 AXISYMMETRIC VIBRATIONS OF A CIRCULAR MEMBRANE

An important one-dimensional problem involves the oscillations of a circular membrane that is clamped along the boundary $r = a$. To find its Green's function, we must solve the partial differential equation

$$\frac{1}{c^2}\left(\frac{\partial^2 g}{\partial t^2} + 2b\frac{\partial g}{\partial t} + \alpha^2 g\right) - \frac{\partial^2 g}{\partial r^2} - \frac{1}{r}\frac{\partial g}{\partial r} = \frac{\delta(r - \rho)\delta(t - \tau)}{2\pi r}, \quad (3.3.1)$$

with $0 < r, \rho < a$, and $0 < t, \tau$, subject to the boundary conditions

$$\lim_{r \to 0} |g(r, t|\rho, \tau)| < \infty, \quad g(a, t|\rho, \tau) = 0, \qquad 0 < t, \quad (3.3.2)$$

and the initial conditions

$$g(r, 0|\rho, \tau) = g_t(r, 0|\rho, \tau) = 0, \qquad 0 < r < a. \quad (3.3.3)$$

We begin by taking the Laplace transform of (3.3.1) and find that

$$\frac{d^2 G}{dr^2} + \frac{1}{r}\frac{dG}{dr} - \frac{s^2 + 2bs + \alpha^2}{c^2} G = -\frac{\delta(r - \rho)}{2\pi r} e^{-s\tau}. \quad (3.3.4)$$

To solve (3.3.4), we first expand the right side of (3.3.4) as a Fourier-Bessel series or

$$\frac{\delta(r - \rho)}{2\pi r} = \frac{1}{\pi a^2} \sum_{n=1}^{\infty} \frac{J_0(k_n\rho/a)J_0(k_n r/a)}{J_1^2(k_n)}, \quad (3.3.5)$$

where k_n is the nth root of $J_0(k) = 0$. This series has the advantage that it vanishes along $r = a$.

Because the delta function (3.3.5) is a Fourier-Bessel series, we anticipate that $G(r, s|\rho, \tau)$ will be of the form

$$G(r, s|\rho, \tau) = \sum_{n=1}^{\infty} A_n J_0(k_n r/a). \quad (3.3.6)$$

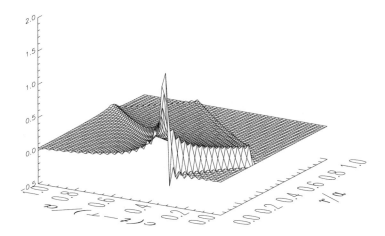

Figure 3.3.1: The Green's function $ag(x, t|\xi, \tau)/c$ for a circular membrane as functions of location r/a and time $c(t-\tau)/a$ when $\rho/a = 0.3$, $b = 1$, and $a^2(\alpha^2 - b^2)/c^2 = 1$.

Substitution of (3.3.5) and (3.3.6) into (3.3.1) yields

$$G(r, s|\rho, \tau) = \frac{c^2 e^{-s\tau}}{\pi a^2} \sum_{n=1}^{\infty} \frac{J_0(k_n\rho/a)J_0(k_nr/a)}{[(s + b)^2 + \alpha^2 - b^2 + c^2 k_n^2/a^2]J_1^2(k_n)}. \tag{3.3.7}$$

A straightforward application of tables and the second shifting theorem yields

$$g(r, t|\rho, \tau) = \frac{c^2}{\pi a^2} e^{-b(t-\tau)} H(t - \tau)$$
$$\times \sum_{n=1}^{\infty} \frac{J_0(k_n\rho/a)J_0(k_nr/a)\sin[(t - \tau)\sqrt{c^2 k_n^2/a^2 + \alpha^2 - b^2}]}{J_1^2(k_n)\sqrt{c^2 k_n^2/a^2 + \alpha^2 - b^2}}, \tag{3.3.8}$$

where we have assumed that the damping is sufficiently weak so that $c^2 k_n^2/a^2 + \alpha^2 > b^2$ for all n.

3.4 TWO-DIMENSIONAL WAVE EQUATION IN AN UNLIMITED DOMAIN

Consider the two-dimensional wave equation[5]

$$\frac{\partial^2 g}{\partial x^2} + \frac{\partial^2 g}{\partial y^2} - \frac{1}{c^2}\frac{\partial^2 g}{\partial t^2} = \delta(x - \xi)\delta(y - \eta)\delta(t - \tau), \tag{3.4.1}$$

---

[5] Patterned after Gopalsamy, K., and B. D. Aggarwala, 1972: Propagation of disturbances from randomly moving sources. *Zeit. angew. Math. Mech.*, **52**, 31–35. See also Graff, K. F., 1991: *Wave Motion in Elastic Solids*. Dover Publications, Inc., pp. 285–288.

where $-\infty < x, y, \xi, \eta < \infty$, and $0 < t, \tau$. If all of the initial conditions are zero, the transformed form of (3.4.1) is

$$\frac{d^2\overline{G}}{dy^2} - s^2\left(\alpha^2 + \frac{1}{c^2}\right)\overline{G} = -\delta(y - \eta)e^{-ik\xi}e^{-s\tau}, \qquad (3.4.2)$$

where we have taken the Laplace transform with respect to time and the Fourier transform with respect to x and replaced the Fourier transform parameter k with αs. The solution of (3.4.2) that tends to zero as $|y - \eta| \to \infty$ is

$$\overline{G}(\alpha, y, s|\xi, \eta, \tau) = \frac{1}{2s\beta}e^{-i\alpha s\xi - s\beta|y-\eta| - s\tau}, \qquad (3.4.3)$$

where

$$\beta(\alpha) = \left(\alpha^2 + \frac{1}{c^2}\right)^{1/2}, \qquad (3.4.4)$$

with $\text{Re}(\beta) \geq 0$.

Inverting the Fourier transform yields

$$G(x, y, s|\xi, \eta, \tau) = \frac{e^{-s\tau}}{4\pi}\int_{-\infty}^{\infty}\frac{e^{-s[\beta|y-\eta| - i\alpha(x-\xi)]}}{\beta}\, d\alpha. \qquad (3.4.5)$$

With the substitution $\alpha = iw$, (3.4.5) becomes

$$G(x, y, s|\xi, \eta, \tau) = \frac{e^{-s\tau}}{4\pi i}\int_{-\infty i}^{\infty i}\frac{e^{-s[w(x-\xi) + \beta|y-\eta|]}}{\beta}\, dw, \qquad (3.4.6)$$

with

$$\beta = \left(\frac{1}{c^2} - w^2\right)^{1/2}. \qquad (3.4.7)$$

To evaluate (3.4.6), we deform the line integration from the imaginary axis in the w-plane to one where

$$w(x - \xi) + \beta|y - \eta| = t \qquad (3.4.8)$$

is real and positive. This change of variable and deformation of the original contour results in (3.4.6) becoming a forward Laplace transform so that we can determine the inversion by inspection—the so-called Cagniard-de Hoop technique.[6] In the present case, we must deform our integral to the new Cagniard contour with great care. There are branch points at $w = 1/c$ and $w = -1/c$. To ensure that $\text{Re}(\beta) \geq 0$, we take

---

[6] De Hoop, A. T., 1960: A modification of Cagniard's method for solving seismic pulse problems. *Appl. Sci. Res.*, **B8**, 349–356.

the branch cuts to lie along the real axis from $1/c$ to ∞ and from $-1/c$ to $-\infty$.

From (3.4.8), a little algebra gives

$$w = \frac{(x - \xi)t}{r^2} \pm \frac{i|y - \eta|}{r^2}\sqrt{t^2 - \frac{r^2}{c^2}}, \qquad \frac{r}{c} < t < \infty, \qquad (3.4.9)$$

where $r^2 = (x - \xi)^2 + (y - \eta)^2$. Furthermore, along this hyperbola

$$\beta = \frac{|y - \eta|t}{r^2} \mp \frac{(x - \xi)i}{r^2}\sqrt{t^2 - \frac{r^2}{c^2}}, \qquad (3.4.10)$$

and

$$\frac{\partial w}{\partial t} = \pm \frac{i\beta}{\sqrt{t^2 - r^2/c^2}}. \qquad (3.4.11)$$

Upon using the symmetry of the path of integration, (3.4.6) becomes the t-integral

$$G(x, y, s|\xi, \eta, \tau) = \frac{e^{-s\tau}}{2\pi} \int_{r/c}^{\infty} \frac{e^{-st}}{\sqrt{t^2 - r^2/c^2}}\, dt = \frac{e^{-s\tau}}{2\pi} K_0\left(\frac{rs}{c}\right). \qquad (3.4.12)$$

Consequently, by inspection and the second shifting theorem,

$$g(x, y, t|\xi, \eta, \tau) = \frac{1}{2\pi} \frac{H(t - \tau - r/c)}{\sqrt{(t - \tau)^2 - r^2/c^2}}, \qquad (3.4.13)$$

the Green's function of the two-dimensional wave equation.

An interesting aspect of (3.4.13) follows from a comparison of it with the Green's function that we found in one dimension, (3.1.6). Here our Green's function decays as $(t - \tau)^{-1}$. Consequently, in addition to the abrupt change at $t - \tau = r/c$, we also have a tail that theoretically continues forever. In the next section, when we continue to the three-dimensional case, we again have a sharp propagation boundary.

3.5 THREE-DIMENSIONAL WAVE EQUATION IN AN UNLIMITED DOMAIN

In this section, we find the Green's function for the three-dimensional wave equation resulting from a point source in a domain of infinite extent. Mathematically, the problem is

$$\frac{\partial^2 g}{\partial x^2} + \frac{\partial^2 g}{\partial y^2} + \frac{\partial^2 g}{\partial z^2} - \frac{1}{c^2}\frac{\partial^2 g}{\partial t^2} = \delta(x - \xi)\delta(y - \eta)\delta(z - \zeta)\delta(t - \tau), \quad (3.5.1)$$

where $-\infty < x, y, z, \xi, \eta, \zeta < \infty$, $0 < t, \tau$, and c is the phase speed, with

$$g(x, y, z, 0|\xi, \eta, \zeta, \tau) = g_t(x, y, z, 0|\xi, \eta, \zeta, \tau) = 0. \qquad (3.5.2)$$

It arises in such diverse fields as acoustics and seismology.

We begin by taking the Laplace transform of (3.5.1). This yields

$$\frac{\partial^2 G}{\partial x^2} + \frac{\partial^2 G}{\partial y^2} + \frac{\partial^2 G}{\partial z^2} - \frac{s^2}{c^2} G = -\delta(x - \xi)\delta(y - \eta)\delta(z - \zeta)e^{-s\tau}. \quad (3.5.3)$$

Next, we introduce the three-dimensional Fourier transform

$$
\begin{aligned}
&G(x, y, z, s|\xi, \eta, \zeta, \tau) \\
&= \frac{1}{(2\pi)^3} \int_{-\infty}^{\infty} \int_{-\infty}^{\infty} \int_{-\infty}^{\infty} \overline{G}(k, \ell, m, s|\xi, \eta, \zeta, \tau) e^{i(kx + \ell y + mz)} \, dk \, d\ell \, dm.
\end{aligned}
$$

$$(3.5.4)$$

Direct substitution of (3.5.4) into (3.5.3) gives

$$\overline{G}(k, \ell, m, s|\xi, \eta, \zeta, \tau) = \frac{c^2}{s^2 + c^2 \kappa^2} e^{-ik\xi - i\ell\eta - im\zeta - s\tau}, \qquad (3.5.5)$$

so that

$$
\begin{aligned}
&G(x, y, z, s|\xi, \eta, \zeta, \tau) \\
&= \frac{e^{-s\tau}}{(2\pi)^3} \int_{-\infty}^{\infty} \int_{-\infty}^{\infty} \int_{-\infty}^{\infty} \frac{e^{i[k(x-\xi) + \ell(y-\eta) + m(z-\zeta)]}}{\kappa^2 + s^2/c^2} \, dk \, d\ell \, dm,
\end{aligned}
$$

$$(3.5.6)$$

where $\kappa^2 = k^2 + \ell^2 + m^2$.

To evaluate (3.5.6) we introduce spherical coordinates in such a manner that the polar axis ($\theta = 0$) lies along the half line from the origin to $(x - \xi, y - \eta, z - \zeta)$ so that $x - \xi = 0$, $y - \eta = 0$, $z - \zeta = R$, $k = \kappa \cos(\varphi) \sin(\theta)$, $\ell = \kappa \sin(\varphi) \sin(\theta)$, and $m = \kappa \cos(\theta)$. Then, (3.5.6) becomes

$$
\begin{aligned}
&G(x, y, z, s|\xi, \eta, \zeta, \tau) \\
&= \frac{e^{-s\tau}}{(2\pi)^3} \int_0^{\infty} \int_0^{\pi} \int_0^{2\pi} \frac{\kappa^2 \, e^{i\kappa R \cos(\theta)}}{\kappa^2 + s^2/c^2} \sin(\theta) \, d\varphi \, d\theta \, d\kappa
\end{aligned}
$$

$$(3.5.7)$$

$$= \frac{e^{-s\tau}}{(2\pi)^2} \int_0^{\infty} \int_0^{\pi} \frac{\kappa^2 \, e^{i\kappa R \cos(\theta)}}{\kappa^2 + s^2/c^2} \sin(\theta) \, d\theta \, d\kappa \qquad (3.5.8)$$

$$= \frac{e^{-s\tau}}{(2\pi)^2 iR} \int_{-\infty}^{\infty} \frac{\kappa \, e^{i\kappa R}}{\kappa^2 + s^2/c^2} \, d\kappa, \qquad (3.5.9)$$

where $R = \sqrt{(x-\xi)^2 + (y-\eta)^2 + (z-\zeta)^2}$. Finally, we use contour integration to evaluate (3.5.9) and find that

$$G(x,y,z,s|\xi,\eta,\zeta,\tau) = \frac{e^{-sR/c-s\tau}}{4\pi R} \qquad (3.5.10)$$

for $R \neq 0$. Taking the inverse of (3.5.10) and applying the second shifting theorem, the Green's function arising from a point source is

$$g(x,y,z,t|\xi,\eta,\zeta,\tau) = \frac{\delta(t-\tau-R/c)}{4\pi R}. \qquad (3.5.11)$$

- **Example 3.5.1**

Our first application of (3.5.11) is to verify the Green's function that we found for the two-dimensional wave equation (3.4.1). If we integrate (3.5.11) over ζ, the two-dimensional Green's function is

$$g(x,y,t|\xi,\eta,\tau) = \frac{1}{4\pi} \int_{-\infty}^{\infty} \frac{\delta[c(t-\tau)-R]}{R} d\zeta. \qquad (3.5.12)$$

Introducing the change of variable $s = z - \zeta$, we have that

$$g(x,y,t|\xi,\eta,\tau) = \frac{1}{4\pi} \int_{-\infty}^{\infty} \frac{\delta[c(t-\tau)-R]}{R} ds \qquad (3.5.13)$$

$$= \frac{1}{2\pi} \int_{0}^{\infty} \frac{\delta[c(t-\tau)-R]}{R} ds, \qquad (3.5.14)$$

where $R = \sqrt{(x-\xi)^2 + (y-\eta)^2 + s^2}$ and (3.5.14) follows from the fact that the integrand is an even function of s. If we now change to $R^2 = \rho^2 + s^2$, where $\rho^2 = (x-\xi)^2 + (y-\eta)^2$, so that

$$\frac{ds}{R} = \frac{dR}{s} = \frac{dR}{\sqrt{R^2 - \rho^2}}, \qquad (3.5.15)$$

then

$$g(x,y,t|\xi,\eta,\tau) = \frac{1}{2\pi} \int_{\rho}^{\infty} \frac{\delta[c(t-\tau)-R]}{\sqrt{R^2 - \rho^2}} dR. \qquad (3.5.16)$$

If $c(t - \tau) < \rho$, the integral vanishes because the argument of the delta function is always negative. If $c(t - \tau) > \rho$, the sifting property of the delta function yields

$$g(x, y, t | \xi, \eta, \tau) = \frac{1}{2\pi} \frac{H[c(t - \tau) - \rho]}{\sqrt{c^2(t - \tau)^2 - \rho^2}}. \tag{3.5.17}$$

The Heaviside function allows us to combine everything into a single expression. This agrees with the result (3.4.13) that we found earlier.

• **Example 3.5.2: Kirchhoff's formula**

Consider the wave equation (3.0.1) where there are no source terms and the initial conditions are zero. Then, by (3.0.13),

$$u(\mathbf{r}, t) = \int_0^{t^+} \oiint_{S_0} \Big[g(\mathbf{r}, t | \mathbf{r}_0, t_0) \, \nabla_0 u(\mathbf{r}_0, t_0)$$
$$- u(\mathbf{r}_0, t_0) \, \nabla_0 g(\mathbf{r}, t | \mathbf{r}_0, t_0) \Big] \cdot \mathbf{n} \, dS_0 \, dt_0. \tag{3.5.18}$$

If we now use (3.5.11) in (3.5.18),

$$u(\mathbf{r}, t) = \frac{1}{4\pi} \int_0^{t^+} \oiint_{S_0} \left\{ \left[\frac{\delta(t_0 - t + R/c)}{R} \right] \nabla_0 u(\mathbf{r}_0, t_0) \right.$$
$$\left. - u(\mathbf{r}_0, t_0) \, \nabla_0 \left[\frac{\delta(t_0 - t + R/c)}{R} \right] \right\} \cdot \mathbf{n} \, dS_0 \, dt_0, \tag{3.5.19}$$

where $R = |\mathbf{r} - \mathbf{r}_0|$. Carrying out the t_0 integration, we find that

$$\int_0^{t^+} \frac{\delta(t_0 - t + R/c)}{R} \nabla_0 u(\mathbf{r}_0, t_0) \, dt_0 = \frac{\nabla_0 u(\mathbf{r}_0, t - R/c)}{R}. \tag{3.5.20}$$

On the other hand,

$$\int_0^{t^+} u(\mathbf{r}_0, t_0) \, \nabla_0 \left[\frac{\delta(t_0 - t + R/c)}{R} \right] dt_0$$
$$= \int_0^{t^+} u(\mathbf{r}_0, t_0) \frac{\partial}{\partial R} \left[\frac{\delta(t_0 - t + R/c)}{R} \right] \nabla_0 R \, dt_0 \tag{3.5.21}$$
$$= \int_0^{t^+} u(\mathbf{r}_0, t_0) \left(\frac{\mathbf{R}}{R^3} \right) \left[\frac{R}{c} \delta'(t_0 - t + R/c) - \delta(t_0 - t + R/c) \right] dt_0 \tag{3.5.22}$$
$$= -\frac{\mathbf{R}}{R^3} \left\{ u(\mathbf{r}_0, t - R/c) + \frac{R}{c} \left[\frac{\partial u(\mathbf{r}_0, t_0)}{\partial t_0} \right]_{t_0 = t - R/c} \right\}, \tag{3.5.23}$$

where $\mathbf{R} = \mathbf{r} - \mathbf{r}_0$. Substituting (3.5.20) and (3.5.23) into (3.5.19), the final result is

$$
u(\mathbf{r}, t) = \frac{1}{4\pi} \oiint_{S_0} \left[\frac{1}{R} \nabla u(\mathbf{r}_0, t_0) \bigg|_{t_0 = t - R/c} - \frac{\mathbf{R}}{cR^2} \frac{\partial u(\mathbf{r}_0, t_0)}{\partial t_0} \bigg|_{t_0 = t - R/c} \right.
$$

$$
\left. + \left(\frac{\mathbf{R}}{R^3} \right) u(\mathbf{r}_0, t_0) \bigg|_{t_0 = t - R/c} \right] \cdot \mathbf{n} \, dS_0, \qquad (3.5.24)
$$

where the normal vector points outward. Equation (3.5.24) is the general form of *Kirchhoff's theorem*;[7] it states that if u and u_t are known at all points on a closed surface at any time, then the value of u can be calculated for all points at all times.

Consider now the simple case when S_0 is a sphere of radius $R = ct$ centered at $\mathbf{r}$. Because $dS_0 = R^2 \sin(\theta) \, d\theta \, d\varphi$, (3.5.24) becomes

$$
u(\mathbf{r}, t) = \frac{1}{4\pi} \left\{ \frac{\partial}{\partial(ct)} \left[ct \int_0^\pi \int_0^{2\pi} u(\mathbf{r}_0, t_0) \big|_{t_0 = 0} \sin(\theta) \, d\varphi \, d\theta \right] \right.
$$

$$
\left. + t \int_0^\pi \int_0^{2\pi} \frac{\partial u(\mathbf{r}_0, t_0)}{\partial t_0} \bigg|_{t_0 = 0} \sin(\theta) \, d\varphi \, d\theta \right\}. \qquad (3.5.25)
$$

This result (3.5.25) is known as *Poisson's formula*.[8] An alternative derivation[9] obtains this result by solving the radially symmetric wave equation in an unlimited domain using Laplace transforms.

Equation (3.5.25) says that the value of $u(\mathbf{r}, t)$ depends only upon the values of $u(\mathbf{r}_0, 0)$ and $u_{t_0}(\mathbf{r}_0, 0)$ on the spherical surface $R = ct$ and not on these quantities *inside* or *outside* of this sphere. This statement can be inverted to say that the values of u and u_{t_0} at a spatial point $\mathbf{r}$ influence the solution only on the surface $R = ct$ of the light or sound cone that emanates from $(\mathbf{r}, 0)$. Consequently, light or sound are carried through the medium at exactly a fixed speed c without "echoes."

• Example 3.5.3: Weyl and Sommerfeld integrals

For our third application, let us now invert the Fourier transform (3.5.6) without using spherical coordinates. We perform the m integration by closing the line integral along the real axis as dictated by

---

[7] Kirchhoff, G., 1882: Zur Theorie der Lichtstrahlen. *Sitzber. K. Preuss. Akad. Wiss. Berlin*, 641–669; reprinted a year later in *Ann. Phys. Chem., Neue Folge*, **18**, 663–695.

[8] Poisson, S. D., 1818: Sur l'intégration de quelques équations linéares aux différences partielles, et particulièrement de l'équation générale du mouvement des fluides élastiques. *Mém. Acad. Sci. Paris*, **3**, 121–176.

[9] Bromwich, T. J. I'A., 1927: Some solutions of the electromagnetic equations, and of the elastic equations, with applications to the problem of secondary waves. *Proc. London Math. Soc., Ser. 2*, **28**, 438–475. See §5.

Jordan's lemma. For $z > 0$, this requires an infinite semicircle in the upper half-plane; for $z < 0$, an infinite semicircle in the lower half-plane. One source of difficulty is the presence of singularities along the real axis at $m = \pm\sqrt{k^2 + l^2 + s^2/c^2}$. If we set $s = -i\omega$ so that the temporal behavior is $e^{-i\omega t}$, we must pass underneath the singularity at $m = \sqrt{k^2 + l^2 + s^2/c^2}$ and over the singularity at $m = -\sqrt{k^2 + l^2 + s^2/c^2}$ so that we have radiating waves. This is clearly seen after applying the residue theorem; we find that the Laplace transform of the Green's function is

$$G(x, y, z, s|\xi, \eta, \zeta, \tau)$$
$$= \frac{e^{-s\tau}}{8\pi^2} \int_{-\infty}^{\infty} \int_{-\infty}^{\infty} \frac{e^{ik(x-\xi)+il(y-\eta)-|z-\zeta|\sqrt{k^2+l^2+s^2/c^2}}}{\sqrt{k^2 + l^2 + s^2/c^2}} \, dk \, dl \quad \textbf{(3.5.26)}$$

with the condition that the square root has a real part ≥ 0. Because the $G(x, y, z, s|\xi, \eta, \zeta, \tau)$'s given by (3.5.10) and (3.5.26) must be equivalent, we immediately obtain the *Weyl integral*:[10]

$$\frac{e^{i\omega R/c}}{R} = \frac{1}{2\pi} \int_{-\infty}^{\infty} \int_{-\infty}^{\infty} \frac{e^{ik(x-\xi)+il(y-\eta)-|z-\zeta|\sqrt{k^2+l^2-\omega^2/c^2}}}{\sqrt{k^2 + l^2 - \omega^2/c^2}} \, dk \, dl,$$
$$\textbf{(3.5.27)}$$

if we substitute $s = -i\omega$. We can further simplify (3.5.27) by introducing $k = \rho\cos(\varphi)$, $l = \rho\sin(\varphi)$, $x - \xi = r\cos(\theta)$, and $y - \eta = r\sin(\theta)$ in which case

$$\frac{e^{i\omega R/c}}{R} = \frac{1}{2\pi} \int_0^{\infty} \int_0^{2\pi} \frac{e^{i\rho r\cos(\theta-\varphi)+i|z-\zeta|\sqrt{\omega^2/c^2-\rho^2}}}{-i\sqrt{\omega^2/c^2 - \rho^2}} \, d\varphi \, \rho \, d\rho. \quad \textbf{(3.5.28)}$$

If we now carry out the φ integration, we obtain the *Sommerfeld integral*:[11]

$$\frac{e^{i\omega R/c}}{R} = i \int_0^{\infty} \frac{J_0(\rho r)e^{i|z-\zeta|\sqrt{\omega^2/c^2-\rho^2}}}{\sqrt{\omega^2/c^2 - \rho^2}} \, \rho \, d\rho, \quad \textbf{(3.5.29)}$$

where the imaginary part of the square root must be positive. Both the Weyl and Sommerfeld integrals are used extensively in electromagnetism and elasticity as an integral representation of spherical waves propagating from a point source.

---

[10] Weyl, H., 1919: Ausbreitung elektromagnetischer Wellen über einem ebenen Leiter. *Ann. Phys., 4te Folge*, **60**, 481–500.

[11] Sommerfeld, A., 1909: Über die Ausbreitung der Wellen in der draftlosen Telegraphie. *Ann. Phys., 4te Folge*, **28**, 665–736.

• Example 3.5.4: Dispersive wave equation

Let us find the Green's function[12] for the three-dimensional wave equation

$$\frac{\partial^2 g}{\partial x^2} + \frac{\partial^2 g}{\partial y^2} + \frac{\partial^2 g}{\partial z^2} - \frac{1}{c^2}\left(\frac{\partial^2 g}{\partial t^2} + a^2 g\right) = -\delta(x - \xi)\delta(y - \eta)\delta(z - \zeta)\delta(t - \tau),$$
(3.5.30)

where $-\infty < x, y, z, \xi, \eta, \zeta < \infty$, $0 < t, \tau$, c is a real, positive constant (the wave speed), and a is a real, nonnegative constant. This equation arises in plasma physics.

We begin by taking the Laplace transform of (3.5.30) and find that

$$\frac{\partial^2 G}{\partial x^2} + \frac{\partial^2 G}{\partial y^2} + \frac{\partial^2 G}{\partial z^2} - \left(\frac{s^2 + a^2}{c^2}\right)G = -\delta(x - \xi)\delta(y - \eta)\delta(z - \zeta)e^{-s\tau}.$$
(3.5.31)

Using the techniques that gave (3.5.10), the solution to (3.5.31) is

$$G(x, y, z, s|\xi, \eta, \zeta, \tau) = \frac{e^{-R\sqrt{s^2 + a^2}/c - s\tau}}{4\pi R}, \qquad R \neq 0,$$
(3.5.32)

where $R = \sqrt{(x - \xi)^2 + (y - \eta)^2 + (z - \zeta)^2}$.

Consider now the transform

$$U(x, y, z, s|\xi, \eta, \zeta, \tau) = \frac{\exp(-R\sqrt{s^2 + a^2}/c)}{\sqrt{s^2 + a^2}}.$$
(3.5.33)

which has the inverse

$$u(x, y, z, t|\xi, \eta, \zeta, \tau) = J_0\left(a\sqrt{t^2 - R^2/c^2}\right)H(t - R/c).$$
(3.5.34)

Consequently, the transform

$$\frac{\partial U}{\partial R} = -\frac{\exp(-R\sqrt{s^2 + a^2}/c)}{c}$$
(3.5.35)

has the inverse

$$\frac{\partial u}{\partial R} = \frac{aR\, J_1(a\sqrt{t^2 - R^2/c^2})}{c^2\sqrt{t^2 - R^2/c^2}}H(t - R/c) - \frac{\delta(t - R/c)}{c}.$$
(3.5.36)

---

[12] An alternative derivation is given by Chambers, Ll. G., 1966: Derivation of solutions of the Klein-Gordon equation from solutions of the wave equation. *Proc. Edin. Math. Soc., Ser. 2*, **15**, 125–129.

so that

$$g(x, y, z, t | \xi, \eta, \zeta, \tau) = \frac{\delta(t - \tau - R/c)}{4\pi R}$$
$$- \frac{a\, J_1[a\sqrt{(t - \tau)^2 - R^2/c^2}]}{4\pi c\sqrt{(t - \tau)^2 - R^2/c^2}} H(t - \tau - R/c).$$

(3.5.37)

Thus, after the initial passage of the wave front at $c(t - \tau) = R$, there remains a contribution from the tail due to the dispersive effects introduced by the a^2 term in (3.5.30).

• **Example 3.5.5: Dissipative wave equation**

As sound waves propagate through a medium they may suffer frictional and viscous losses. A prototypical equation describing this phenomenon is

$$\frac{\partial^2 u}{\partial x^2} + \frac{\partial^2 u}{\partial y^2} + \frac{\partial^2 u}{\partial z^2} - \frac{1}{c^2}\left(\frac{\partial}{\partial t} + \alpha\right)\left(\frac{\partial}{\partial t} + \beta\right) u = -f(x, y, z, t), \quad \textbf{(3.5.38)}$$

where $-\infty < x, y, z, \xi, \eta, \zeta < \infty$, and $0 < t, \tau$, and α and β are real, non-negative constants. Thus, the Green's function for this *dissipative wave equation* is

$$\frac{\partial^2 g}{\partial x^2} + \frac{\partial^2 g}{\partial y^2} + \frac{\partial^2 g}{\partial z^2} - \frac{1}{c^2}\left(\frac{\partial}{\partial t} + \alpha\right)\left(\frac{\partial}{\partial t} + \beta\right) g$$
$$= -\delta(x - \xi)\delta(y - \eta)\delta(z - \zeta)\delta(t - \tau), \quad \textbf{(3.5.39)}$$

with the initial condition $g(x, y, z, 0 | \xi, \eta, \zeta, \tau) = g_t(x, y, z, 0 | \xi, \eta, \zeta, \tau) = 0$.

We begin by taking the Laplace transform of (3.5.39) or

$$\frac{\partial^2 G}{\partial x^2} + \frac{\partial^2 G}{\partial y^2} + \frac{\partial^2 G}{\partial z^2} - \gamma^2 G = -\delta(x - \xi)\delta(y - \eta)\delta(z - \zeta)e^{-s\tau}, \quad \textbf{(3.5.40)}$$

where $\gamma = \sqrt{(s + \alpha)(s + \beta)}/c$. Applying the same technique used in the derivation of (3.5.10), the solution to (3.5.40) is

$$G(x, y, z, s | \xi, \eta, \zeta, \tau) = \frac{e^{-\gamma R - s\tau}}{4\pi R}, \quad R \neq 0, \quad \textbf{(3.5.41)}$$

where $R = \sqrt{(x - \xi)^2 + (y - \eta)^2 + (z - \zeta)^2}$. If we introduce $a \equiv (\alpha + \beta)/2$ and $b \equiv |\beta - \alpha|/2$, we may write (3.5.41) as

$$G(x, y, z, s | \xi, \eta, \zeta, \tau) = \frac{\exp[-s\tau - R\sqrt{(s + a)^2 - b^2}/c]}{4\pi R}. \qquad (3.5.42)$$

Consider now the transform

$$U(x, y, z, s | \xi, \eta, \zeta, \tau) = \frac{\exp(-R\sqrt{s^2 - b^2}/c)}{\sqrt{s^2 - b^2}}. \qquad (3.5.43)$$

Its inverse is

$$u(x, y, z, t | \xi, \eta, \zeta, \tau) = I_0\left(b\sqrt{t^2 - R^2/c^2}\right) H(t - R/c). \qquad (3.5.44)$$

Consequently, the transform

$$\frac{\partial U}{\partial R} = -\frac{\exp(-R\sqrt{s^2 - b^2}/c)}{c} \qquad (3.5.45)$$

has the inverse

$$\frac{\partial u}{\partial R} = -\frac{bR \, I_1(b\sqrt{t^2 - R^2/c^2})}{c^2 \sqrt{t^2 - R^2/c^2}} H(t - R/c) - \frac{\delta(t - R/c)}{c}. \qquad (3.5.46)$$

Therefore,

$$\mathcal{L}^{-1}\left(e^{-R\sqrt{s^2-b^2}/c}\right) = \delta(t - R/c) + \frac{bR}{c} \frac{I_1(b\sqrt{t^2 - R^2/c^2})}{\sqrt{t^2 - R^2/c^2}} H(t - R/c), \qquad (3.5.47)$$

and

$$g(x, y, z, t | \xi, \eta, \zeta, \tau) = \frac{e^{-a(t-\tau)}}{4\pi R}\delta(t - \tau - R/c)$$
$$+ be^{-a(t-\tau)} \frac{I_1[b\sqrt{(t - \tau)^2 - R^2/c^2}]}{4\pi c\sqrt{(t - \tau)^2 - R^2/c^2}} H(t - \tau - R/c),$$

$$(3.5.48)$$

after applying the second shifting theorem. In the lossless case[13] when $a = b = 0$, we recover (3.5.11).

---

[13] In telegraphy, this is the case when the telegraphic line has no resistance or leakage.

The Green's function for the general case of $a, b \neq 0$ and $a \neq b$ consists of two parts. The first term in (3.5.48) is an attenuated delta function that is propagating outward at the speed c. The second term represents a tail or residue that will remain after the pulse and wave front pass the point $R = c(t - \tau)$.

3.6 ASYMMETRIC VIBRATIONS OF A CIRCULAR MEMBRANE

A problem of considerable practical importance involves the oscillations of a circular membrane that is clamped along the boundary $r = a$. To find its Green's function, we must solve the partial differential equation

$$\frac{1}{c^2}\left(\frac{\partial^2 g}{\partial t^2} + 2b\frac{\partial g}{\partial t} + \alpha^2 g\right) - \frac{\partial^2 g}{\partial r^2} - \frac{1}{r}\frac{\partial g}{\partial r} - \frac{1}{r^2}\frac{\partial^2 g}{\partial \theta^2}$$
$$= \frac{\delta(r - \rho)\delta(\theta - \theta')\delta(t - \tau)}{r}, \quad (3.6.1)$$

with $0 < r, \rho < a$, $0 \leq \theta - \theta' \leq 2\pi$, and $0 < t, \tau$, subject to the boundary conditions

$$\lim_{r \to 0} |g(r, \theta, t|\rho, \theta', \tau)| < \infty, \quad g(a, \theta, t|\rho, \theta', \tau) = 0, \qquad 0 < t, \quad (3.6.2)$$

and the initial conditions

$$g(r, \theta, 0|\rho, \theta', \tau) = g_t(r, \theta, 0|\rho, \theta', \tau) = 0, \qquad 0 \leq r < a. \quad (3.6.3)$$

We begin by taking the Laplace transform of (3.6.1) and find that

$$\frac{\partial^2 G}{\partial r^2} + \frac{1}{r}\frac{\partial G}{\partial r} + \frac{1}{r^2}\frac{\partial^2 G}{\partial \theta^2} - \frac{s^2 + 2bs + \alpha^2}{c^2}G = -\frac{\delta(r - \rho)\delta(\theta - \theta')}{r}e^{-s\tau}.$$
$$(3.6.4)$$

To solve (3.6.4), we first expand its right side as double Fourier-Bessel series or

$$\frac{\delta(r - \rho)\delta(\theta - \theta')}{r} = \frac{1}{\pi a^2}\sum_{m=0}^{\infty}\sum_{n=1}^{\infty}\epsilon_m\frac{J_m(k_{nm}\rho/a)J_m(k_{nm}r/a)}{J_{m+1}^2(k_{nm})}$$
$$\times \cos[m(\theta - \theta')], \quad (3.6.5)$$

where $\epsilon_0 = 1$, $\epsilon_m = 2$ for $m > 0$, and k_{nm} is the nth root of $J_m(k) = 0$. This series has the advantage that it vanishes along $r = a$.

With this Fourier-Bessel series representation of the delta functions, we anticipate that

$$G(r, \theta, s|\rho, \theta', \tau) = \frac{1}{\pi a^2}\sum_{m=0}^{\infty}\sum_{n=1}^{\infty}\epsilon_m G_{nm}(s|\tau)\frac{J_m(k_{nm}\rho/a)J_m(k_{nm}r/a)}{J_{m+1}^2(k_{nm})}$$
$$\times \cos[m(\theta - \theta')]. \quad (3.6.6)$$

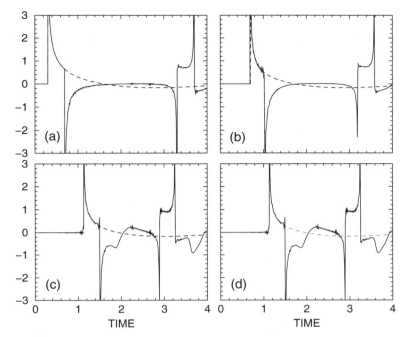

Figure 3.6.1: The Green's function $ag(x, t|\xi, \tau)/(\pi c)$ for a circular membrane with asymmetric vibrations as a function of time $c(t - \tau)/a$ when $\theta - \theta'$ has the values of (a) 0, (b) $\pi/3$, (c) $2\pi/3$, and (d) π when $b = 0$, $a^2\beta^2/c^2 = 1$, $r = 0.8$, and $\rho = 0.5$. The dashed line gives the free-space Green's function [the first term in (3.6.12)] for this particular problem.

Substitution of (3.6.5) and (3.6.6) into (3.6.4) yields

$$G_{nm}(s|\tau) = \frac{c^2\, e^{-s\tau}}{(s+b)^2 + \alpha^2 - b^2 + c^2 k_n^2/a^2}. \qquad (3.6.7)$$

A straightforward application of tables and the shifting theorems gives

$$g(r, \theta, t|\rho, \theta', \tau) = \frac{c^2}{\pi a^2} e^{-b(t-\tau)} H(t - \tau)$$

$$\times \sum_{m=0}^{\infty} \sum_{n=1}^{\infty} \epsilon_m \frac{J_m(k_{nm}\rho/a) J_m(k_{nm}r/a)}{J_{m+1}^2(k_{nm})} \cos[m(\theta - \theta')]$$

$$\times \frac{\sin[(t - \tau)\sqrt{c^2 k_{nm}^2/a^2 + \alpha^2 - b^2}]}{\sqrt{c^2 k_{nm}^2/a^2 + \alpha^2 - b^2}}, (3.6.8)$$

where we have assumed that the damping is sufficiently weak so that $c^2 k_{nm}^2/a^2 + \alpha^2 > b^2$ for all n and m.

Although (3.6.8) solves the problem, what is occurring physically? To address this question, let us retrace our steps and approach the problem in a different manner.

Once again, we take the Laplace transform of (3.6.1) and use the Fourier cosine expansion for the delta function. This suggests that our Green's function has the form

$$G(r, \theta, s|\rho, \theta', \tau) = e^{-s\tau} \sum_{n=0}^{\infty} G_n(r|\rho)\epsilon_n \cos[n(\theta - \theta')], \qquad (\textbf{3.6.9})$$

where $G_n(r|\rho)$ is given by

$$\frac{d^2 G_n}{dr^2} + \frac{1}{r}\frac{dG_n}{dr} - \frac{n^2}{r^2}G_n - \frac{s'^2 + \beta^2}{c^2}G_n = -\frac{\delta(r - \rho)}{2\pi r}, \qquad (\textbf{3.6.10})$$

where $s' = s + b$, and $\beta^2 = \alpha^2 - b^2$. We now solve (3.6.10) in a manner similar to that shown in Example 2.5.2 and find that

$$G_n(r|\rho) = \frac{1}{2\pi} I_n\left(\frac{r_<}{c}\sqrt{s'^2 + \beta^2}\right) K_n\left(\frac{r_>}{c}\sqrt{s'^2 + \beta^2}\right)$$

$$- \frac{K_n(a\sqrt{s'^2 + \beta^2}/c) I_n(\rho\sqrt{s'^2 + \beta^2}/c) I_n(r\sqrt{s'^2 + \beta^2}/c)}{2\pi I_n(a\sqrt{s'^2 + \beta^2}/c)}.$$

$$(\textbf{3.6.11})$$

The interpretation of (3.6.11) is straightforward. The first term is the free-space Green's function and the particular solution to (3.6.10) while the second term is a homogeneous solution of (3.6.10) required so that the Green's function satisfies the boundary condition (3.6.2).

Upon substituting (3.6.11) into (3.6.9), it remains for us to invert the Laplace transform and find $g(r, \theta, t|\rho, \theta', \tau)$. The combination of (3.6.9) and (3.6.11) consists of two terms. The first term is

$$\frac{e^{-s\tau}}{2\pi} \sum_{n=0}^{\infty} \epsilon_n I_n\left(\frac{r_<}{c}\sqrt{s'^2 + \beta^2}\right) K_n\left(\frac{r_>}{c}\sqrt{s'^2 + \beta^2}\right) \cos[n(\theta - \theta')].$$

Using the result from (4.3.41), this summation equals

$$\frac{e^{-s\tau}}{2\pi} K_0\left(R\sqrt{s'^2 + \beta^2}\right),$$

where $R^2 = r^2 + \rho^2 - 2r\rho \cos(\theta - \theta')$. Finally, using both shifting theorems and tables, we find that the free-space portion of $g(r, \theta, t|\rho, \theta', \tau)$ is

$$e^{-b(t-\tau)} \frac{\cos\left[\beta\sqrt{(t - \tau)^2 - R^2/c^2}\right]}{2\pi\sqrt{(t - \tau)^2 - R^2/c^2}} H(t - \tau - R/c).$$

We invert the second term using residues. From the asymptotic representation for modified Bessel functions, this term will "turn on" for $c(t - \tau) > r + \rho - 2a$ and corresponds to the minimum time needed for a wave emanating from the source to reflect from the boundary at $r = a$ and the reflected wave to travel back to the observational point. Carrying out the inversion, we find that an alternative to (3.6.8) is

$$
g(r, \theta, t | \rho, \theta', \tau) = e^{-b(t-\tau)} \frac{\cos\left[\beta\sqrt{(t-\tau)^2 - R^2/c^2}\right]}{2\pi\sqrt{(t-\tau)^2 - R^2/c^2}} H(t - \tau - R/c)
$$

$$
+ \frac{c}{2a} e^{-b(t-\tau)} H[c(t-\tau) + r + \rho - 2a]
$$

$$
\times \sum_{m=0}^{\infty} \epsilon_m \cos[m(\theta - \theta')]
$$

$$
\times \left\{ \sum_{n=1}^{\infty} \frac{k_{nm} Y_m(k_{nm}) J_m(k_{nm}\rho/a) J_m(k_{nm}r/a)}{J_{m+1}(k_{nm})} \right.
$$

$$
\left. \times \frac{\sin[c(t-\tau)\sqrt{k_{nm}^2 + a^2\beta^2/c^2}/a]}{\sqrt{k_{nm}^2 + a^2\beta^2/c^2}} \right\}, (3.6.12)
$$

where k_{nm} is the nth root of $J_m(k_{nm}) = 0$.

• **Example 3.6.1: Vibration of a circular plate**

Another problem of keen practical interest is the vibration of a circular plate with a clamped edge. To find the Green's function for this problem, we must solve

$$
\left(\frac{\partial^2}{\partial r^2} + \frac{1}{r}\frac{\partial}{\partial r} + \frac{1}{r^2}\frac{\partial^2}{\partial \theta^2} \right)^2 g + \frac{1}{c^2}\frac{\partial^2 g}{\partial t^2} = \frac{\delta(r - \rho)\delta(\theta - \theta')\delta(t - \tau)}{r},
$$
(3.6.13)

with $\alpha < r, \rho < 1$, $0 \le \theta - \theta' \le 2\pi$, and $0 < t, \tau$, subject to boundary conditions at $r = \alpha$ and $r = 1$, where $0 < \alpha < 1$.

The exact form of the boundary conditions vary considerably. If the plate is clamped along the boundary at $r = \alpha$, then

$$
g(\alpha, \theta, t | \rho, \theta', \tau) = g_r(\alpha, \theta, t | \rho, \theta', \tau) = 0.
$$
(3.6.14)

On the other hand, if the plate is allowed to vibrate freely there, then the boundary conditions become

$$
\frac{\partial}{\partial r}\left(\frac{\partial^2 g}{\partial r^2} + \frac{1}{r}\frac{\partial g}{\partial r} + \frac{1}{r^2}\frac{\partial^2 g}{\partial \theta^2} \right) + \frac{1 - \nu}{r^2}\frac{\partial^2}{\partial \theta^2}\left(\frac{\partial g}{\partial r} - \frac{g}{r} \right) = 0, \quad (3.6.15)
$$

and

$$
\frac{\partial^2 g}{\partial r^2} + \nu\left(\frac{1}{r}\frac{\partial g}{\partial r} + \frac{1}{r^2}\frac{\partial^2 g}{\partial \theta^2} \right) = 0
$$
(3.6.16)

at $r = \alpha$, where ν denotes Poisson's ratio. The initial conditions are the familiar

$$g(r, \theta, 0 | \rho, \theta', \tau) = g_t(r, \theta, 0 | \rho, \theta', \tau) = 0, \qquad \alpha < r < 1. \qquad \textbf{(3.6}.17)$$

We begin by writing the solution as

$$g(r, \theta, t | \rho, \theta', \tau) = \sum_{n=0}^{\infty} \sum_{m=1}^{\infty} g_{nm}(t) R_{nm}(r) \cos[n(\theta - \theta_n)]. \qquad \textbf{(3.6}.18)$$

Here

$$R_{nm}(r) = c_1 J_n(k_{nm}r) + c_2 Y_n(k_{nm}r) + c_3 I_n(k_{nm}r) + c_4 K_n(k_{nm}r). \qquad \textbf{(3.6}.19)$$

The coefficients c_i and k_{nm} are computed from substituting $R_{nm}(r)$ into the boundary conditions. This results in four homogeneous linear equations whose determinant must equal zero.[14] Because these computations must be done numerically for even the simplest boundary conditions, we presently assume that they have been computed by some numerical algorithm.

Upon substituting (3.6.18) into (3.6.13), we have that

$$\sum_{n=0}^{\infty} \sum_{m=1}^{\infty} \left(\frac{g''_{nm}}{c^2} + k_{nm}^4 g_{nm} \right) R_{nm}(r) \cos[n(\theta - \theta_n)]$$

$$= \frac{\delta(r - \rho)\delta(\theta - \theta')\delta(t - \tau)}{r}. \qquad \textbf{(3.6}.20)$$

We now multiply (3.6.20) by $r R_{NM}(r) \cos[N(\theta - \theta_N)]$ and integrate over r and θ. If we choose the normalization constant of $R_{nm}(r)$ so that

$$\int_{\alpha}^{1} R_{nm}(r) R_{n\ell}(r) \, r \, dr = \frac{\delta_{m\ell}}{\pi}, \qquad \textbf{(3.6}.21)$$

where $\delta_{m\ell}$ is the Kronecker delta function, (3.6.20) simplifies to

$$g''_{NM} + \omega_{NM}^2 g_{NM} = c^2 R_{NM}(\rho) \cos[N(\theta' - \theta_N)]\delta(t - \tau), \qquad \textbf{(3.6}.22)$$

where $c^2 k_{NM}^4 = \omega_{NM}^2$, and N, M are arbitrary integers such that $0 \leq N < \infty$ and $1 \leq M < \infty$. The solution to (3.6.22) is

$$g_{nm}(t) = c^2 R_{NM}(\rho) \cos[N(\theta' - \theta_N)] \frac{\sin[\omega_{nm}(t - \tau)]}{\omega_{nm}} H(t - \tau). \qquad \textbf{(3.6}.23)$$

---

[14] See, for example, Vogel, S. M., and D. W. Skinner, 1965: Natural frequencies of transversely vibrating uniform annular plates. *J. Appl. Mech.*, **32**, 926–931 and Anderson, G., 1969: On the determination of finite integral transforms for forced vibrations of circular plates. *J. Sound Vib.*, **9**, 126–144.

Therefore,

$$g(r, \theta, t | \rho, \theta', \tau) = c^2 H(t - \tau) \sum_{n=0}^{\infty} \sum_{m=1}^{\infty} R_{nm}(\rho) R_{nm}(r) \cos[n(\theta' - \theta_n)]$$
$$\times \cos[n(\theta - \theta_n)] \frac{\sin[\omega_{nm}(t - \tau)]}{\omega_{nm}}. \quad (3.6.24)$$

Quite often θ_n is set to zero for all n.

For the special case when $\alpha = 0$, so that there is no longer an annulus, the problem simplifies because $c_2 = c_4 = 0$, and

$$R_{nm}(r) = c_1 J_n(k_{nm} r) + c_3 I_n(k_{nm} r). \quad (3.6.25)$$

If the edge at $r = 1$ is clamped, k_{nm} is given by solution to

$$J_m(k_{nm}) I'_m(k_{nm}) - I_m(k_{nm}) J'_m(k_{nm}) = 0. \quad (3.6.26)$$

3.7 THERMAL WAVES

In the classical theory of heat conduction, Fourier's law implies that knowledge about changes in a temperature field propagates infinitely fast. Although this is clearly unrealistic, classical theory works well in most situations because the thermal diffusivity is usually 10 orders of magnitude slower than the speed at which this information travels.[15]

One case where diffusivity and the speed of these thermal waves becomes comparable is in superfluids such as liquid helium. Another example is short-pulse laser technology used in modern microfabrication technologies. In these cases the heat equation becomes

$$\frac{1}{c^2} \frac{\partial^2 u}{\partial t^2} + \frac{1}{\alpha} \frac{\partial u}{\partial t} - \nabla^2 u = f(\mathbf{r}, t), \quad (3.7.1)$$

where α is the thermal conductivity and c is the velocity of the thermal wave. The function $f(\mathbf{r}, t)$ represents an internal heat source. In this section we find[16] the Green's function for (3.7.1).

The mathematical problem is

$$\frac{1}{c^2} \frac{\partial^2 g}{\partial t^2} + \frac{1}{\alpha} \frac{\partial g}{\partial t} - \nabla^2 g = \delta(\mathbf{r} - \mathbf{r}_0) \delta(t - \tau). \quad (3.7.2)$$

---

[15] Nettleton, R. E., 1960: Relaxation theory of thermal conduction in liquids. *Phys. Fluids*, **3**, 216–225; Francis, P. H., 1972: Thermo-mechanical effects in elastic wave propagation: A survey. *J. Sound Vibr.*, **21**, 181–192.

[16] For a slightly different derivation, see Haji-Sheikh, A., and J. V. Beck, 1994: Green's function solution for thermal wave equation in finite bodies. *Int. J. Heat Mass Transfer*, **37**, 2615–2626.

We begin by assuming that we can express the Green's function in terms of the set of eigenfunction

$$g(\mathbf{r}, t|\mathbf{r}_0, \tau) = \sum_{n=1}^{\infty} C_n(t) G_n(\mathbf{r}), \qquad (\mathbf{3.7.3})$$

where $G_n(\mathbf{r})$ satisfies $\nabla^2 G_n = -\lambda_n^2 G_n$ as well as time-independent homogeneous boundary conditions, and $C_n(t)$ are the presently unknown, time-dependent Fourier coefficients. Although $G_n(\mathbf{r})$ depends upon the domain's geometry and boundary conditions, there is always an orthogonality condition that we can write as

$$\iiint_V G_n(\mathbf{r}) G_m(\mathbf{r}) \, dV = \begin{cases} N_n(\lambda_n), & n = m, \\ 0, & \text{otherwise.} \end{cases} \qquad (\mathbf{3.7.4})$$

Upon substituting (3.7.3) into (3.7.2), we obtain

$$\sum_{n=1}^{\infty} \left(\frac{1}{c^2} \frac{d^2 C_n}{dt^2} + \frac{1}{\alpha} \frac{dC_n}{dt} \right) G_n(\mathbf{r}) - C_n(t) \nabla^2 G_n(\mathbf{r}) = \delta(\mathbf{r} - \mathbf{r}_0) \delta(t - \tau),$$

$$(\mathbf{3.7.5})$$

or

$$\sum_{n=1}^{\infty} \left(\frac{1}{c^2} \frac{d^2 C_n}{dt^2} + \frac{1}{\alpha} \frac{dC_n}{dt} + \lambda_n^2 \right) G_n(\mathbf{r}) = \delta(\mathbf{r} - \mathbf{r}_0) \delta(t - \tau), \quad (\mathbf{3.7.6})$$

since $\nabla^2 G_n = -\lambda_n^2 G_n$. At this point, we multiply each side of (3.7.6) by $G_m(\mathbf{r})$ and integrate over the volume. Applying (3.7.4), (3.7.6) simplifies to

$$\frac{1}{c^2} \frac{d^2 C_n}{dt^2} + \frac{1}{\alpha} \frac{dC_n}{dt} + \lambda_n^2 C_n = \frac{G_n(\mathbf{r}_0) \delta(t - \tau)}{N_n(\lambda_n)}, \qquad (\mathbf{3.7.7})$$

where we have substituted n for m. Solving (3.7.7) by Laplace transforms, we find that

$$C_n(t) = c^2 e^{-c^2(t-\tau)/(2\alpha)} \frac{G_n(\mathbf{r}_0) \sin[(t - \tau)\sqrt{\lambda_n^2 - c^4/(4\alpha^2)}]}{N_n(\lambda_n)\sqrt{\lambda_n^2 - c^4/(4\alpha^2)}} H(t - \tau).$$

$$(\mathbf{3.7.8})$$

Consequently, the Green's function for (3.7.2) is

$$g(\mathbf{r}, t|\mathbf{r}_0, \tau) = c^2 e^{-c^2(t-\tau)/2\alpha} H(t - \tau)$$

$$\times \sum_{n=1}^{\infty} \frac{G_n(\mathbf{r}_0) G_n(\mathbf{r}) \sin[(t - \tau)\sqrt{\lambda_n^2 - c^4/(4\alpha^2)}]}{N_n(\lambda_n)\sqrt{\lambda_n^2 - c^4/(4\alpha^2)}}. \quad (\mathbf{3.7.9})$$

An application of (3.7.9) is left as problem 4.

3.8 DISCRETE WAVENUMBER REPRESENTATION

So far, we have found Green's functions through the application of transform methods. Often the inversion of a transform by analytic techniques fails and we must invert it numerically. A particularly straightforward method for regions near the source is the *discrete wavenumber technique*, which is used to invert the *Fourier spatial* transform. Away from the source, normal mode solutions are commonly used.

The numerical computation of the Fourier transform is often difficult because the integral is improper. Thus, if the transform remains finite at larger wavenumbers, the numerical approximation will require many grid points to achieve a desired accuracy. The problem is further exacerbated by the rapid oscillation of the integrand at larger wavenumbers. A method introduced by Bouchon and Aki[17] employs the discrete Fourier transform to approximate the Fourier transform.[18] In this approximation we replace the single impulse forcing at $x = 0$ with an infinite series of impulses located at $x = \pm nL$, where $n = 0, 1, 2, \ldots$ and L is the distance between the impulses. Because these extraneous source functions will produce disturbances that will eventually propagate near $x = 0$, the times for which this solution is valid are always limited.

The discrete wavenumber method can be applied in two different ways. In the first method, the joint transform of Green's function, say $\overline{G}(k, s | \xi, \tau)$, is found as before. Then the inverse is given by

$$G(x, s | \xi, \tau) = \frac{1}{L} \sum_{n=-\infty}^{\infty} \overline{G}(k_n, s | \xi, \tau) e^{i k_n x}, \qquad (3.8.1)$$

where $k_n = 2\pi n / L$. Of course, the number of terms retained in the Fourier expansion and L are dictated by the computational resources that are available and the desired accuracy of the solution.

The second method of using the discrete wavenumber method can be illustrated by finding the Green's function for the one-dimensional damped wave equation[19] (Klein-Gordon equation)

$$\frac{\partial^2 g}{\partial x^2} - \frac{1}{c^2} \frac{\partial^2 g}{\partial t^2} - \beta^2 g = -\delta(x - \xi)\delta(t - \tau). \qquad (3.8.2)$$

---

[17] Bouchen, M., and K. Aki, 1977: Discrete wave-number representation of seismic-source wave fields. *Bull. Seism. Soc. Am.*, **67**, 259–277.

[18] See, for example, Davis, P. J., and P. Rabinowitz, 1984: *Methods of Numerical Integration*. Academic Press, §3.9.5.

[19] Taken from Touhei, T., 1993: Impulse response of an elastic layered medium in the anti-plane wave field based on a thin-layered element and discrete wave number method. *Proc. JSCE—Struct. Engng. Earthq. Engng. (in Japanese)*, **No. 459/I-22**, 119–128.

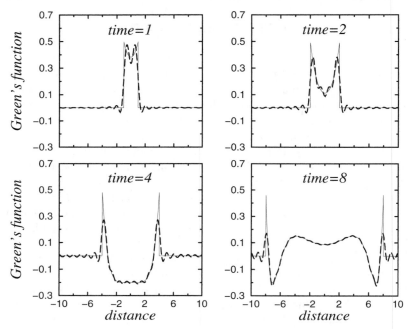

Figure 3.8.1: Comparison of the exact free-space Green's function $g(x, t|\xi, \tau)/c$ for a one-dimensional Klein-Gordon equation (solid line) and the approximate Green's function found by the discrete wavenumber method (dashed line) at different distances $x - \xi$ and times $c(t - \tau)$.

In Example 3.1.1 we showed that the exact solution is

$$g(x, t|\xi, \tau) = \frac{c}{2} J_0 \left[\beta \sqrt{c^2(t - \tau)^2 - (x - \xi)^2} \right] H[c(t - \tau) - |x - \xi|].$$
$$(3.8.3)$$

In the process of finding the analytic solution (3.8.3), we would have utilized the result that

$$\delta(x - \xi) = \frac{1}{2\pi} \int_{-\infty}^{\infty} e^{ik(x-\xi)} \, dk. \qquad (3.8.4)$$

In the discrete wavenumber method, the representation of the delta function is

$$\delta_c(x - \xi) = \frac{1}{L} \sum_{n=-\infty}^{\infty} e^{ik_n(x-\xi)}, \qquad (3.8.5)$$

where $k_n = 2\pi n/L$. This suggests that we should seek solutions of the form

$$g(x, t|\xi, \tau) = \frac{1}{L} \sum_{n=-\infty}^{\infty} g_n(t|\tau) e^{ik_n(x-\xi)}, \qquad (3.8.6)$$

where $g_n(t|\tau)$ is governed by

$$\frac{1}{c^2}\frac{d^2 g_n}{dt^2} + \left(\beta^2 + k_n^2\right) g_n = \delta(t - \tau). \tag{3.8.7}$$

A simple application of Laplace transform yields

$$g_n(t|\tau) = \frac{c\,\sin\left[c(t-\tau)\sqrt{\beta^2 + k_n^2}\right]}{\sqrt{\beta^2 + k_n^2}} H(t - \tau). \tag{3.8.8}$$

Upon substituting (3.8.8) into (3.8.6) and simplifying,

$$g(x,t|\xi,\tau) = \frac{cH(t-\tau)}{L}\sum_{n=0}^{\infty}\epsilon_n\frac{\sin\left[c(t-\tau)\sqrt{\beta^2 + k_n^2}\right]}{\sqrt{\beta^2 + k_n^2}}\cos[k_n(x - \xi)], \tag{3.8.9}$$

where $\epsilon_0 = 1$ and $\epsilon_n = 2$ if $n > 0$. Of course, the infinite number of terms in (3.8.9) must be truncated to just N terms during the actual computation of $g(x,t|\xi,\tau)$. Figure 3.8.1 compares the exact solution with the approximate solution containing 21 terms at different times $c(t - \tau)$ and positions $x - \xi$ with $L = 20$ and $\beta = 1$.

In this example we were able to find the temporal behavior exactly via Laplace transforms. If we had been unable to invert the Laplace transform, we would first sum the N terms of the discrete wavenumber inversion and then evaluate the Laplace inversion numerically.

3.9 LEAKY MODES

One of the most common structures in nature is a stratified medium which consists of homogeneous layers. Disturbances originating in one of these layers will reflect off interfaces as well as leak some of their energy into adjacent layers.

A particularly interesting case occurs when one of the adjacent layers has a phase speed larger than that of the layer of the incident wave. As a wave enters the layer with higher phase speed, it can race ahead of its counterpart in the source layer and reappear as a precursor to the direct wave. These "leaky modes"[20] are essential in such disciplines as acoustics and electromagnetics. Green's functions play a fundamental role[21] in understanding leaky modes. In this section we highlight

---

[20] Tamir, T., and A. A. Oliner, 1963: Guided complex waves. Part 1. Fields at an interface. *Proc. IEE*, **110**, 310–324.

[21] Marcuvitz, N., 1956: On field representations in terms of leaky modes or eigenmodes. *IRE Trans. Antennas Propag.*, **4**, 192–194.

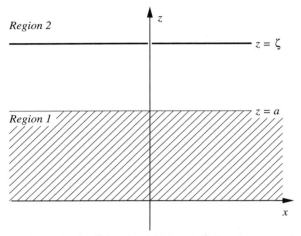

Figure 3.9.1: Schematic of a dielectric of thickness a lying above a perfect conductor at $z = 0$ and below some external medium $a < z < \infty$.

their use in explaining the reflection and transmission of electromagnetic waves in various stratified media.

A simple example involves electromagnetic wave propagation within an one-dimensional, open waveguide. Our model consists of a one-dimensional slab[22], as shown in Figure 3.9.1. Within the slab $0 < z < a$, waves travel at the speed of c/n, where c is the speed of light in region 2.

The governing equations for each layer are

$$\frac{\partial^2 g_1}{\partial z^2} - \frac{n^2}{c^2}\frac{\partial^2 g_1}{\partial t^2} = 0, \quad 0 < z < a, \quad 0 < t, \tag{3.9.1}$$

and

$$\frac{\partial^2 g_2}{\partial z^2} - \frac{1}{c^2}\frac{\partial^2 g_2}{\partial t^2} = -\delta(z - \zeta)\delta(t - \tau), \quad a < z < \infty, \quad 0 < t, \tag{3.9.2}$$

where $\zeta > a$, z is the vertical distance and t is time. At $z = 0$, we assume a perfect conductor and

$$g_1(0, t | \zeta, \tau) = 0, \qquad 0 < t. \tag{3.9.3}$$

At the interface $z = a$, we require continuity of tangential components or

$$g_1(a, t | \zeta, \tau) = g_2(a, t | \zeta, \tau), \quad 0 < t, \tag{3.9.4}$$

---

[22] Taken from §2 of Veselov, G. I., A. I. Kirpa and N. I. Platonov, 1986: Transient-field representation in terms of steady-improper solutions. *IEE Proc., Part H*, **133**, 21–25.

and

$$\frac{\partial g_1(a,t|\zeta,\tau)}{\partial z} = \frac{\partial g_2(a,t|\zeta,\tau)}{\partial z}, \quad 0 < t. \tag{3.9.5}$$

Initially, the system is undisturbed so that

$$g_1(z,0|\zeta,\tau) = \frac{\partial g_1(z,0|\zeta,\tau)}{\partial t} = 0, \quad 0 < z < a, \tag{3.9.6}$$

and

$$g_2(z,0|\zeta,\tau) = \frac{\partial g_2(z,0|\zeta,\tau)}{\partial t} = 0, \quad a < z < \infty. \tag{3.9.7}$$

At infinity, $\lim_{z\to\infty} |g_2(z,t|\zeta,\tau)| < \infty$.

Taking the Laplace transform of (3.9.1)–(3.9.5),

$$\frac{d^2 G_1}{dz^2} - \frac{n^2 s^2}{c^2} G_1 = 0, \tag{3.9.8}$$

$$\frac{d^2 G_2}{dz^2} - \frac{s^2}{c^2} G_2 = -\delta(z - \zeta)e^{-s\tau}, \tag{3.9.9}$$

with

$$G_1(0, s|\zeta,\tau) = 0, \tag{3.9.10}$$

$$G_1(a, s|\zeta,\tau) = G_2(a, s|\zeta,\tau), \tag{3.9.11}$$

and

$$G_1'(a, s|\zeta,\tau) = G_2'(a, s|\zeta,\tau). \tag{3.9.12}$$

The difficulty in solving this set of differential equations is the presence of the delta function in (3.9.9). We solve this problem by further breaking down the region $a < z < \infty$ into two separate regions $a < z < \zeta$ and $\zeta < z < \infty$. At $z = \zeta$, we integrate (3.9.9) across a very narrow strip from ζ^- to ζ^+. This gives the additional conditions that

$$G_2(\zeta^-, s|\zeta,\tau) = G_2(\zeta^+, s|\zeta,\tau), \tag{3.9.13}$$

and

$$G_2'(\zeta^+, s|\zeta,\tau) - G_2'(\zeta^-, s|\zeta,\tau) = -e^{-s\tau}. \tag{3.9.14}$$

Solutions that satisfy the differential equations (3.9.8)–(3.9.9) plus the boundary conditions (3.9.10)–(3.9.14) are

$$G_1(z, s|\zeta,\tau) = \frac{\sinh(nsz/c)}{\Delta(s)} e^{-s(\zeta-a)/c - s\tau}, \quad 0 < z < a, \tag{3.9.15}$$

and

$$G_2(z, s|\zeta,\tau) = \frac{c}{2s} \exp\left(-\frac{s}{c}|z - \zeta| - s\tau\right)$$
$$- \frac{c}{2s} \exp\left[-\frac{s(\zeta - a)}{c} - \frac{s(z - a)}{c} - s\tau\right] \tag{3.9.16}$$
$$+ \frac{\sinh(nsa/c)}{\Delta(s)} \exp\left[-\frac{s(\zeta - a)}{c} - \frac{s(z - a)}{c} - s\tau\right],$$

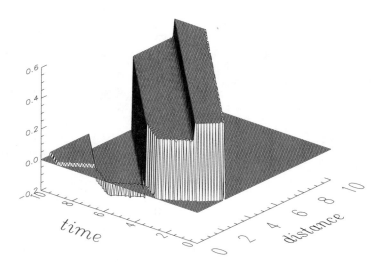

Figure 3.9.2: Plot of the Green's function (divided by c) at various distances z/a and times $c(t - \tau)/a$ for $\zeta = 2a$, and $n = 1.5811$.

where $a < z < \infty$, and

$$\Delta(s) = \frac{s}{c} \sinh\left(\frac{nsa}{c}\right) + \frac{ns}{c} \cosh\left(\frac{nsa}{c}\right). \tag{3.9.17}$$

We invert the first term of (3.9.16) by inspection and discover that

$$g_2^{(d)}(z, t|\zeta, \tau) = \frac{c}{2} H\left(t - \tau - \frac{|z - \zeta|}{c}\right). \tag{3.9.18}$$

This portion of the solution represents the direct wave emitted from $z = \zeta$. The second term in $G_2(z, s|\zeta, \tau)$ contains a simple pole at $s = 0$. Performing the inversion by inspection,

$$g_2^{(s)}(z, t|\zeta, \tau) = -\frac{c}{2} H\left(t - \tau - \frac{z - a}{c} - \frac{\zeta - a}{c}\right). \tag{3.9.19}$$

This is a steady-state field set up in region 2 due to the reflection of the direct wave from the interface at $z = a$.

The inversion of the last term in (3.9.16) depends upon the value of n. We will do the case for $n > 1$ while the case $n < 1$ is reserved as problem 18. In either case, this term contains simple poles that are the zeros of $s^{-1}\Delta(s)$. For $n > 1$, a little algebra gives

$$\frac{a s_m}{c} = \frac{1}{2n} \ln\left(\frac{n - 1}{n + 1}\right) + \frac{m\pi}{2n} i, \tag{3.9.20}$$

where $m = \pm 1, \pm 3, \ldots$. These poles at $s = s_m$ give the transient solutions to the problem. Applying the residue theorem in conjunction with Bromwich's integral,

$$
g_1^{(t)}(z, t|\zeta, \tau) = \frac{c}{n^2 - 1} H\left(t - \tau + \frac{n(z - a)}{c} - \frac{\zeta - a}{c}\right)
$$

$$
\times \left\{ \sum_m \frac{\sinh(n s_m z/c) \exp[s_m(t - \tau) - s_m(\zeta - a)/c]}{(a s_m/c) \sinh(n s_m a/c)} \right\},
$$

$$(3.9.21)$$

and

$$
g_2^{(t)}(z, t|\zeta, \tau) = \frac{c}{n^2 - 1} H\left(t - \tau - \frac{z - a}{c} - \frac{\zeta - a}{c}\right)
$$

$$
\times \left\{ \sum_m \frac{\exp[s_m(t - \tau) - s_m(\zeta - a)/c - s_m(z - a)/c]}{a s_m/c} \right\}.
$$

$$(3.9.22)$$

The solution $g_1^{(t)}(z, t|\zeta, \tau)$ gives the total field in region 1 while the total field within region 2 equals the sum of $g_2^{(d)}(z, t|\zeta, \tau)$, $g_2^{(s)}(z, t|\zeta, \tau)$ and $g_2^{(t)}(z, t|\zeta, \tau)$.

Let us examine the transient solution $g_2^{(t)}(z, t|\zeta, \tau)$ more closely. Its origins are in (3.9.16), which contain terms that behave as $e^{-s_m z/c}$. Because $\text{Re}(s_m) < 0$ for $n > 1$, this solution grows exponentially as $z \to \infty$. Have we made a mistake in deriving (3.9.22)? No, because there is an exponential *decay* e^{st} in time for each fixed space point at which the wave front has passed. This exponential time decay is sufficiently large to counteract the exponential growth in z so that the net effect yields solutions that are always finite. We see this in a plot of the solution given in Figure 3.9.2.

Individual episodes are illustrated in Figure 3.9.3. Initially, a square pulse emanates from $z = 2a$, propagating both inward and outward from the source. When the inwardly propagating wave strikes the interface, a portion enters the dielectric while the remaining energy is reflected, as shown in frame $c(t - \tau)/a = 2$. After reflecting from the conductor at $z = 0$, a portion of the energy escapes the dielectric while the remaining portion is reflected back into the dielectric, as frame $c(t - \tau)/a = 5$ shows. This repeated process of reflection and transmission continues until all of the energy escapes the dielectric.

Although our one-dimensional problem illustrates internal reflections, the addition of another spatial dimension introduces us to a new phenomenon: *head waves*.

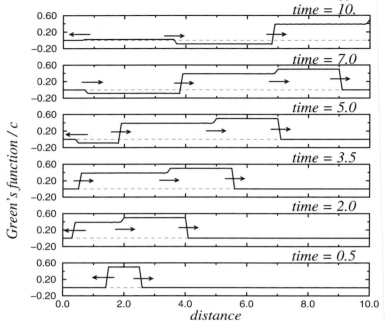

Figure 3.9.3: Snapshots of the Green's function (divided by c) at various distances z/a and times $c(t - \tau)/a$ for $\zeta = 2a$ and $n = 1.5811$.

Consider[23] two dielectric half-spaces where the speed of light is c/n in region 1 and c in region 2. (See Figure 3.9.4.) We locate the impulse forcing at $x = 0$ and $z = \zeta$. The Green's functions then satisfy the equations

$$\frac{\partial^2 g_1}{\partial x^2} + \frac{\partial^2 g_1}{\partial z^2} - \frac{n^2}{c^2} \frac{\partial^2 g_1}{\partial t^2} = 0, \qquad z < 0, \qquad (3.9.23)$$

and

$$\frac{\partial^2 g_2}{\partial x^2} + \frac{\partial^2 g_2}{\partial z^2} - \frac{1}{c^2} \frac{\partial^2 g_2}{\partial t^2} = -\delta(z - \zeta)\delta(x)\delta(t), \qquad z > 0, \qquad (3.9.24)$$

where x is the horizontal distance. Continuity in the electric and magnetic fields at the interface $z = 0$ implies

$$g_1(x, 0, t | 0, \zeta, 0) = g_2(x, 0, t | 0, \zeta, 0), \qquad (3.9.25)$$

---

[23] Taken from §2.1 of: Transient scattering responses from a plane interface between dielectric half-spaces, H. Shirai, *Electron. Comm. Japan, Part 2*, ©1995 Scripta Technica. Reprinted by permission of John Wiley & Sons, Inc. See also De Hoop, A. T., 1979: Pulsed electromagnetic radiation from a line source in a two-media configuration. *Radio Sci.*, **14**, 253–268.

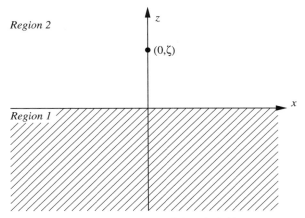

Figure 3.9.4: A line impulse lying above a dielectric half-space.

and

$$\frac{\partial g_1(x, 0, t|0, \zeta, 0)}{\partial z} = \frac{\partial g_2(x, 0, t|0, \zeta, 0)}{\partial z}. \tag{3.9.26}$$

At infinity, $\lim_{|z| \to \infty} |g_j(x, z, t|0, \zeta, 0)| < \infty$.

We solve (3.9.23)–(3.9.26) by the joint application of Laplace and Fourier transforms. We denote this joint transform of $g_j(x, z, t|0, \zeta, 0)$ by $\mathcal{G}_j(k, z, s|0, \zeta, 0)$ with $j = 1, 2$, where s and k are the transform variables of the Laplace and Fourier transforms, respectively. Assuming that the system is initially at rest,

$$\frac{d^2 \mathcal{G}_1}{dz^2} - \nu_1^2 \mathcal{G}_1 = 0, \qquad z < 0, \tag{3.9.27}$$

and

$$\frac{d^2 \mathcal{G}_2}{dz^2} - \nu_2^2 \mathcal{G}_2 = -\delta(z - \zeta), \qquad z > 0, \tag{3.9.28}$$

with the boundary conditions

$$\mathcal{G}_1(k, 0, s|0, \zeta, 0) = \mathcal{G}_2(k, 0, s|0, \zeta, 0), \tag{3.9.29}$$

and

$$\mathcal{G}_1'(k, 0, s|0, \zeta, 0) = \mathcal{G}_2'(k, 0, s|0, \zeta, 0), \tag{3.9.30}$$

where $\nu_1 = \sqrt{k^2 + n^2 s^2/c^2}$, and $\nu_2 = \sqrt{k^2 + s^2/c^2}$.

Solutions that satisfy (3.9.27)–(3.9.30) are

$$\mathcal{G}_1(k, z, s|0, \zeta, 0) = \frac{2\nu_2}{\nu_1 + \nu_2} \frac{e^{-\nu_2 \zeta + \nu_1 z}}{2\nu_2}, \tag{3.9.31}$$

and

$$\mathcal{G}_2(k, z, s|0, \zeta, 0) = \frac{e^{-\nu_2|z - \zeta|}}{2\nu_2} + \frac{\nu_2 - \nu_1}{\nu_1 + \nu_2} \frac{e^{-\nu_2(z + \zeta)}}{2\nu_2}. \tag{3.9.32}$$

The physical interpretation of (3.9.31)-(3.9.32) is straightforward. Equation (3.9.31) gives the wave that has been transmitted into region 1 from region 2. On the other hand, (3.9.32) gives the direct wave emanating from the source (the first term) and a reflected wave (the second term).

(a) Direct wave

Let us now invert (3.9.31)–(3.9.32). We begin with the first term in (3.9.32). If we denote this term by $G_2^{(d)}(k, z, s|0, \zeta, 0)$, then the inverse Fourier transform gives

$$G_2^{(d)}(x, z, s|0, \zeta, 0) = \frac{1}{4\pi} \int_{-\infty}^{\infty} \frac{e^{ikx - \nu_2|z - \zeta|}}{\nu_2}\, dk. \qquad (3.9.33)$$

Setting $x = r \sin(\varphi)$, $|z - \zeta| = r \cos(\varphi)$ where $r^2 = x^2 + (z - \zeta)^2$ and $k = s \sinh(w)/c$, then (3.9.33) becomes

$$G_2^{(d)}(x, z, s|0, \zeta, 0) = \frac{1}{4\pi} \int_{-\infty}^{\infty} e^{-sr \cosh(w - \varphi i)/c}\, dw, \qquad (3.9.34)$$

with the requirement that $\mathrm{Re}[\cosh(w - \varphi i)] > 0$. Introducing $w - \varphi i = \beta$,

$$G_2^{(d)}(x, z, s|0, \zeta, 0) = \frac{1}{4\pi} \int_{-\infty - \varphi i}^{\infty - \varphi i} e^{-sr \cosh(\beta)/c}\, d\beta \qquad (3.9.35)$$

$$= \frac{1}{4\pi} \int_{-\infty}^{\infty} e^{-sr \cosh(\beta)/c}\, d\beta \qquad (3.9.36)$$

$$= \frac{1}{2\pi} \int_{0}^{\infty} e^{-sr \cosh(\beta)/c}\, d\beta \qquad (3.9.37)$$

$$= \frac{1}{2\pi} \int_{r/c}^{\infty} \frac{e^{-st}}{\sqrt{t^2 - r^2/c^2}}\, dt, \qquad (3.9.38)$$

where $t = r \cosh(\beta)/c$. We can deform the contour in (3.9.35) to the one in (3.9.36) because the integrand has no singularities.

Examining the integrand in (3.9.38) closely, we note that it is the Laplace transform of the function $H(t - r/c)/\sqrt{t^2 - r^2/c^2}$. Consequently, we immediately have that

$$g_2^{(d)}(x, z, t|0, \zeta, 0) = \frac{H(t - r/c)}{2\pi \sqrt{t^2 - r^2/c^2}}. \qquad (3.9.39)$$

Equation (3.9.39) agrees with the two-dimensional, free-space Green's function (3.4.13) that we found for a line source.

This technique of deforming the inversion contour of the Fourier transform into the definition of the Laplace transform so that we can invert the Laplace transform by inspection was first developed by Cagniard[24] and then subsequently improved by De Hoop.[6]

(b) Reflected wave

We now turn our attention to the waves that are reflected as the result of the interface at $z = 0$. If we denote the Laplace transform of that portion of the Green's function by $G^{(r)}(x, z, s|0, \zeta, 0)$, it is governed by the equation

$$G_2^{(r)}(x, z, s|0, \zeta, 0) = \frac{1}{4\pi} \int_{-\infty}^{\infty} \left(\frac{\nu_2 - \nu_1}{\nu_2 + \nu_1} \right) \frac{e^{ikx - \nu_2(z+\zeta)}}{\nu_2} \, dk. \quad (\mathbf{3.9.40})$$

Using the same substitution that we used to evaluate the direct wave, we have that

$$G_2^{(r)}(x, z, s|0, \zeta, 0) = \frac{1}{4\pi} \int_{-\infty}^{\infty} \left[\frac{\cosh(w) - \sqrt{n^2 + \sinh^2(w)}}{\cosh(w) + \sqrt{n^2 + \sinh^2(w)}} \right]$$
$$\times e^{-s\rho \cosh(w - \varphi_r i)/c} \, dw, \quad (\mathbf{3.9.41})$$

where $\rho^2 = x^2 + (z + \zeta)^2$, $x = \rho \sin(\varphi_r)$ and $z + \zeta = \rho \cos(\varphi_r)$. Because the integrand of (3.9.41) has branch points at $\pm \sin^{-1}(n)i$, two branch cuts must be introduced. We take one to run from $\sin^{-1}(n)i$ out to ∞i while the other runs from $-\sin^{-1}(n)i$ out to $-\infty i$. We will do the analysis for $n < 1$; the case $n > 1$ is left as problem 19.

When $n < 1$, we introduce $w = \beta + \varphi_r i$ so that

$$G_2^{(r)}(x, z, s|0, \zeta, 0) = \frac{1}{4\pi} \int_{-\infty - \varphi_r i}^{\infty - \varphi_r i} e^{-s\rho \cosh(\beta)/c} \, d\beta$$
$$\times \left[\frac{\cosh(\beta + \varphi_r i) - \sqrt{n^2 + \sinh^2(\beta + \varphi_r i)}}{\cosh(\beta + \varphi_r i) + \sqrt{n^2 + \sinh^2(\beta + \varphi_r i)}} \right].$$
$$(\mathbf{3.9.42})$$

Turning to the case $\sin^{-1}(n) > |\varphi_r|$ first, we can still deform the contour $\beta - \varphi_r i$ to the real axis because we will not cross any branch

---

[24] The most accessible version is Cagniard, L., E. A. Flinn and C. H. Dix, 1962: *Reflection and Refraction of Progressive Seismic Waves.* McGraw-Hill, 282 pp.

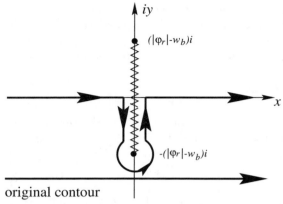

original contour

Figure 3.9.5: Schematic of the deformation of the contour used in (3.9.42) to the one used in (3.9.50).

points. Then

$$G_2^{(r)}(x, z, s|0, \zeta, 0) = \frac{1}{4\pi} \int_{-\infty}^{\infty} F(\beta) e^{-s\rho \cosh(\beta)/c} \, d\beta \qquad (3.9.43)$$

$$= \frac{1}{4\pi} \int_0^{\infty} [F(\beta) + F(-\beta)] e^{-s\rho \cosh(\beta)/c} \, d\beta \quad (3.9.44)$$

$$= \frac{1}{4\pi} \int_{\rho/c}^{\infty} \left\{ F[\cosh^{-1}(ct/\rho)] + F[-\cosh^{-1}(ct/\rho)] \right\}$$

$$\times \frac{e^{-st}}{\sqrt{t^2 - \rho^2/c^2}} \, dt, \qquad (3.9.45)$$

where

$$F(\beta) = \frac{\cosh(\beta + \varphi_r i) - \sqrt{n^2 + \sinh^2(\beta + \varphi_r i)}}{\cosh(\beta + \varphi_r i) + \sqrt{n^2 + \sinh^2(\beta + \varphi_r i)}}, \qquad (3.9.46)$$

$$\rho^2 = x^2 + (z + \zeta)^2, \qquad \varphi_r = \tan^{-1}\left(\frac{x}{z + \zeta}\right), \qquad (3.9.47)$$

$t = \rho \cosh(\beta)/c$, and the real part of the radical must be positive.

Because we recognize (3.9.45) as the definition of the Laplace transform, we can immediately write down

$$g_2^{(r)}(x, z, t|0, \zeta, 0) = \left\{ F[\cosh^{-1}(ct/\rho)] + F[-\cosh^{-1}(ct/\rho)] \right\}$$

$$\times \frac{H(t - \rho/c)}{4\pi \sqrt{t^2 - \rho^2/c^2}} \qquad (3.9.48)$$

$$= \text{Re}\left\{ F\left[\cosh^{-1}\left(\frac{ct}{\rho}\right)\right] \right\} \frac{H(t - \rho/c)}{2\pi \sqrt{t^2 - \rho^2/c^2}}, \qquad (3.9.49)$$

since $F(-\beta) = F^*(\beta)$.

Consider now the case when $|\varphi_r| > w_b = \sin^{-1}(n)$. Let us take the branch cut associated with the branch points $\pm(|\varphi_r| - w_b)i$ to run along the imaginary axis between the two branch points. Consequently, when we deform the original contour from $\beta - \varphi_r i$ to the real axis, we have additional contour integrals arising from integrations along the branch cut. For example, if $\varphi_r > 0$, then the branch cut integrals would be: (1) a line integral running from the real axis (and just to the left of the branch cut) down to the branch point, (2) an integration around the branch point, and (3) a line integral from the branch point (and just to the right of the branch cut) running up to the real axis. See Figure 3.9.5. Since the integration around the branch point equals zero, we have that

$$G_2^{(r)}(x, z, s|0, \zeta, 0) = \frac{1}{4\pi} \int_{-\infty}^{0} F(\beta) e^{-s\rho \cosh(\beta)/c} \, d\beta$$

$$- \frac{1}{4\pi} \int_{0}^{|\varphi_r|-w_b} \frac{\cos(|\varphi_r| - y) - \sqrt{n^2 - \sin^2(|\varphi_r| - y)}}{\cos(|\varphi_r| - y) + \sqrt{n^2 - \sin^2(|\varphi_r| - y)}}$$

$$\times e^{-s\rho \cos(y)/c} \, dy$$

$$- \frac{1}{4\pi} \int_{|\varphi_r|-w_b}^{0} \frac{\cos(|\varphi_r| - y) + \sqrt{n^2 - \sin^2(|\varphi_r| - y)}}{\cos(|\varphi_r| - y) - \sqrt{n^2 - \sin^2(|\varphi_r| - y)}}$$

$$\times e^{-s\rho \cos(y)/c} \, dy$$

$$+ \frac{1}{4\pi} \int_{0}^{\infty} F(\beta) e^{-s\rho \cosh(\beta)/c} \, d\beta \qquad (3.9.50)$$

$$= \frac{1}{4\pi} \int_{0}^{\infty} [F(\beta) + F(-\beta)] e^{-s\rho \cosh(\beta)/c} \, d\beta$$

$$+ \frac{1}{4\pi} \int_{0}^{|\varphi_n|-w_b} \frac{\cos(|\varphi_r| - y)\sqrt{n^2 - \sin^2(|\varphi_r| - y)}}{1 - n^2}$$

$$\times e^{-s\rho \cos(y)/c} \, dy \qquad (3.9.51)$$

$$= \frac{1}{4\pi} \int_{\rho/c}^{\infty} \left\{ F[\cosh^{-1}(ct/\rho)] + F[-\cosh^{-1}(ct/\rho)] \right\}$$

$$\times \frac{e^{-st}}{\sqrt{t^2 - \rho^2/c^2}} \, dt$$

$$+ \frac{1}{\pi} \int_{\rho/c}^{\rho \cos(|\varphi_r|-w_b)/c} G[|\varphi_r| - \cos^{-1}(ct/\rho)]$$

$$\times \frac{e^{-st}}{\sqrt{\rho^2/c^2 - t^2}} \, dt, \qquad (3.9.52)$$

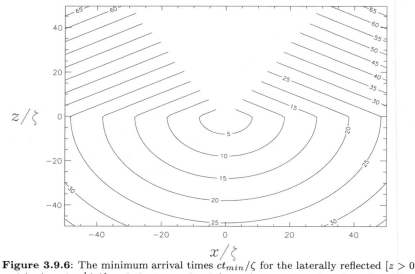

Figure 3.9.6: The minimum arrival times ct_{min}/ζ for the laterally reflected [$z > 0$ and $|\varphi_r| > \sin^{-1}(n)$] and the transmitted ($z < 0$) waves as a function of position when $n = 0.5$.

where

$$G(\beta) = \frac{\cos(\beta)|n^2 - \sin^2(\beta)|^{1/2}}{1 - n^2}. \tag{3.9.53}$$

In particular, we recognize (3.9.52) as the definition of the Laplace transform, in which case we can immediately write down

$$
g_2^{(r)}(x, z, t|0, \zeta, 0) = \text{Re}\left\{ F\left[\cosh^{-1}\left(\frac{ct}{\rho}\right)\right]\right\} \frac{H(t - \rho/c)}{2\pi\sqrt{t^2 - \rho^2/c^2}}
$$
$$
+ \frac{G[|\varphi_r| - \cos^{-1}(ct/\rho)]H(|\varphi_r| - w_b)}{\pi\sqrt{\rho^2/c^2 - t^2}}
$$
$$
\times \left\{ H[t - \rho\cos(w_b - |\varphi_r|)/c] - H(t - \rho/c)\right\}. \tag{3.9.54}
$$

Let us examine (3.9.54) more closely. The first term is merely the reflected wave from the interface that is felt at any point in region 2 after the time $t = \rho/c$. The second term is new and radically different. First, it exists *before* and *disappears when* the conventional reflected wave arrives at those points where $|\varphi_r| > \sin^{-1}(n)$; it is a *precursor* to the arrival of the reflected wave. Second, this wave occurs only for sufficiently large $|\varphi_r|$; this corresponds to large $|x|$. Consequently, these waves occur near the lateral sides of the expanding reflected wave field. For this reason, they are called *lateral waves*. In geophysics, similar waves occur due to different elastic properties within the solid earth.

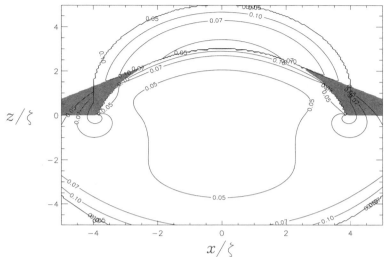

Figure 3.9.7: The Green's function $\zeta g_j(x,z,t|0,\zeta,0)/c$ for various values of x/ζ and z/ζ when $ct/\zeta = 4$. The shaded region denotes that region where head waves are present.

Because these lateral waves are recorded at the beginning or head of a seismographic record, they are often called *head waves*.

Why do these lateral or head waves exist? Recall that the case $n < 1$ corresponds to light traveling faster in region 1 than in region 2. Consequently, some of the wave energy that enters region 1 near $x = 0$ can outrun its counterpart in region 2 and then reemerge into region 2 where it is observed as head waves. Thus lateral waves can reach an observer sooner than the direct wave much as a car may reach a destination sooner by taking an indirect route involving an expressway rather than a direct route that consists of residential streets: the higher speeds of the highway more than compensate for the increased distance traveled.

(c) Transmitted wave

We complete our analysis by finding the Green's function for the waves that pass through the interface into region 1. If $G^{(t)}(x, y, s|0, \zeta, 0)$ denotes the Laplace transform of the Green's function, it is found by evaluating

$$G_1^{(t)}(x,z,s|0,\zeta,0) = \frac{1}{2\pi} \int_{-\infty}^{\infty} \left(\frac{\nu_2}{\nu_2 + \nu_1}\right) \frac{e^{ikx - \nu_2\zeta + \nu_1 z}}{\nu_2}\, dk. \quad \textbf{(3.9.55)}$$

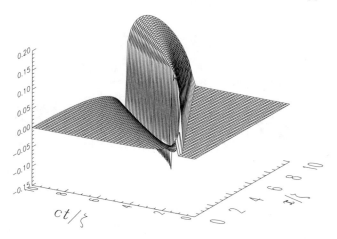

Figure 3.9.8: The reflected wave $\sqrt{|t^2 - \rho^2/c^2|}\, g_2^{(r)}(x, z, t|0, \zeta, 0)$ for various values of x/ζ and ct/ζ when $z/\zeta = 0.05$.

At this point, we introduce the new variable p such that $k = isp$ and

$$G_1^{(t)}(x, z, s|0, \zeta, 0) = \frac{1}{2\pi i} \int_{-\infty i}^{\infty i} \frac{e^{-s(px + \zeta\sqrt{1/c^2 - p^2} - z\sqrt{n^2/c^2 - p^2})}}{\sqrt{n^2/c^2 - p^2} + \sqrt{1/c^2 - p^2}}\, dp. \tag{3.9.56}$$

We next change to the variable

$$t = px + \zeta\sqrt{\frac{1}{c^2} - p^2} - z\sqrt{\frac{n^2}{c^2} - p^2}. \tag{3.9.57}$$

Just as we have done twice before, we would like to deform the original contour (along the imaginary axis) into a contour C along which t is real so that

$$G_1^{(t)}(x, z, s|0, \zeta, 0) = \frac{1}{2\pi i} \int_C \frac{e^{-st}}{\sqrt{n^2/c^2 - p^2} + \sqrt{1/c^2 - p^2}}\, dp \tag{3.9.58}$$

$$= \frac{1}{\pi} \int_{t_{min}}^{\infty} \mathrm{Im}\left(\frac{1}{\sqrt{n^2/c^2 - p^2} + \sqrt{1/c^2 - p^2}}\, \frac{dp}{dt} \right)$$

$$\times\, e^{-st}\, dt, \tag{3.9.59}$$

and

$$\frac{dp}{dt} = \frac{1}{x - \zeta pc/\sqrt{1 - p^2 c^2} + zpc/\sqrt{n^2 - p^2 c^2}}. \tag{3.9.60}$$

What is the nature of C? From a detailed analysis of (3.9.57), we find the following properties:

• C must remain on the same Riemann surface as the original contour. We take the branch cuts along the real axis from the branch points $(\pm 1/c, 0)$ and $(\pm n/c, 0)$ out to infinity.

• For $x > 0$, the contour C lies in the first and fourth quadrants of the complex p-plane.

• For $x > 0$, the contour C crosses the real axis with $0 < p < 1/c$ or n/c, depending upon which is smaller. At this crossing, t assumes its smallest value, t_{min}, which corresponds to the minimum travel time for a transmitted wave to arrive at a point in region 1. At that time, dp/dt is of infinite magnitude and

$$x - \frac{\zeta pc}{\sqrt{1 - p^2 c^2}} + \frac{zpc}{\sqrt{n^2 - p^2 c^2}} = 0. \qquad (3.9.61)$$

Combining this result with (3.9.57), we have

$$\frac{ct_{min}}{\zeta} = \frac{1}{\sqrt{1 - p^2 c^2}} - \frac{n^2(z/\zeta)}{\sqrt{n^2 - p^2 c^2}}. \qquad (3.9.62)$$

Figure 3.9.6 gives this minimum arrival time for the reflected lateral and transmitted waves when $n = 0.5$. Because the head waves have their origin in the leaking of energy from region 1 into region 2, the arrival times must match at the interface.

Having found t_{min}, we now find the transmitted wave solution for any time $t > t_{min}$, given x, z, ζ, c, and n. To do this, we must find the corresponding value of p that is the (complex) zero of the analytic function (3.9.57). Once the value of p is found, then

$$g_1^{(t)}(x, z, t|0, \zeta, 0) = \frac{1}{\pi} \mathrm{Im} \left(\frac{c}{\sqrt{n^2 - p^2 c^2} + \sqrt{1 - p^2 c^2}} \frac{dp}{dt} \right) H\left(t - t_{min}\right).$$

$$(3.9.63)$$

In Figure 3.9.7 the Green's function $\zeta g_j(x, z, t|0, \zeta, 0)/c$ has been plotted for various values of x and z when $ct/\zeta = 4$. The shaded region of the wave field denotes that region where head waves are present. We clearly see how these wave are at the lateral sides of the reflected wave field.

A similar, but much more difficult, problem arises if we wish to find the Green's function for the wave equation due a point source within one of two semi-infinite media. Towne[25] and Volodicheva and Lopukhov[26]

[25] Towne, D. H., 1968: Pulse shapes of spherical waves reflected and refracted at a plane interface separating two homogeneous fluids. *J. Acoust. Soc. Am.*, **44**, 65–76.

[26] Volodicheva, M. I., and K. V. Lopukhov, 1994: Effect of the sphericity of an acoustic wave on its reflection coefficient from a plane interface of two liquid media. *Acoust. Phys.*, **40**, 681–684.

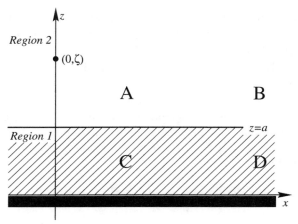

Figure 3.9.9: A dielectric slab on a perfectly conducting half-space with a line source of current above it. The four labels A, B, C, D give the locations at which the Green's function will be subsequently displayed.

have presented solutions to this problem using Cagniard's method for inverting the joint transform.

The reflection and transmission of electromagnetic waves within two semi-infinite dielectrics differs from our first problem because of an additional spatial dimension and the replacement of the slab in Region 1 with a semi-infinite region. We now combine the two problems together by replacing one of the semi-infinite regions (region 1) with one of finite depth a. See Figure 3.9.9. Because the dielectric in region 1 now lies on a conducting half-plane, we have the new boundary condition $g_1(x, 0, t|0, \zeta, 0) = 0$ at $z = 0$. We anticipate not only leaky modes but also internally reflecting waves.

This problem is nearly identical to the one that we just solved. The governing equations are the same, (3.9.23)–(3.9.26), except that the interface occurs at $z = a$ rather than $z = 0$. The mathematical analysis begins as before. After a joint application of Laplace and Fourier transforms, we have that

$$\frac{d^2 \mathcal{G}_1}{dz^2} - \nu_1^2 \mathcal{G}_1 = 0, \qquad 0 < z < a, \tag{3.9.64}$$

and

$$\frac{d^2 \mathcal{G}_2}{dz^2} - \nu_2^2 \mathcal{G}_2 = -\delta(z - \zeta), \qquad a < z, \tag{3.9.65}$$

with the boundary conditions $\mathcal{G}_1(k, 0, s|0, \zeta, 0) = 0$,

$$\mathcal{G}_1(k, a, s|0, \zeta, 0) = \mathcal{G}_2(k, a, s|0, \zeta, 0), \tag{3.9.66}$$

and

$$\mathcal{G}_1'(k, a, s|0, \zeta, 0) = \mathcal{G}_2'(k, a, s|0, \zeta, 0), \tag{3.9.67}$$

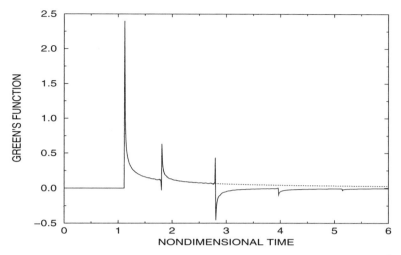

Figure 3.9.10: The temporal evolution of the Green's function at the point $(a, \frac{3a}{2})$. The dashed line corresponds to the Green's functions (3.9.39) and (3.9.54) for the problem of two half-spaces.

where $\nu_1 = \sqrt{k^2 + n^2 s^2/c^2}$, and $\nu_2 = \sqrt{k^2 + s^2/c^2}$. Solutions that satisfy (3.9.64)–(3.9.67) are

$$\mathcal{G}_1 = \frac{\sinh(\nu_1 z)e^{-\nu_2(\zeta-a)}}{\nu_1 \cosh(\nu_1 a) + \nu_2 \sinh(\nu_1 a)}, \quad 0 < z < a, \qquad (\textbf{3.9.68})$$

and

$$\mathcal{G}_2 = \frac{e^{-\nu_2|z-\zeta|}}{2\nu_2} - \frac{e^{-\nu_2(z+\zeta-2a)}}{2\nu_2}\left[\frac{\nu_1 \cosh(\nu_1 a) - \nu_2 \sinh(\nu_1 a)}{\nu_1 \cosh(\nu_1 a) + \nu_2 \sinh(\nu_1 a)}\right], \quad a < z.$$
$$(\textbf{3.9.69})$$

Because (3.9.68)–(3.9.69) are even functions of ν_1, no branch points are associated with ν_1, but there are branch points and cuts with ν_2. To satisfy the radiation condition as $z \to \infty$, $\text{Re}(\nu_2) \geq 0$.

Some physical insight concerning (3.9.68)–(3.9.69) can be achieved by rewriting the hyperbolic functions in terms of exponentials and then expanding the resulting expressions as a geometric series in $e^{-2\nu_1 a}$:

$$\mathcal{G}_1 = \frac{e^{\nu_1(z-a)-\nu_2(\zeta-a)}}{\nu_1 + \nu_2}\left(1 - Re^{-2\nu_1 a} + R^2 e^{-4\nu_1 a} - R^3 e^{-6\nu_1 a} + \ldots\right)$$
$$- \frac{e^{-\nu_1(z+a)-\nu_2(\zeta-a)}}{\nu_1 + \nu_2}\left(1 - Re^{-2\nu_1 a} + R^2 e^{-4\nu_1 a} - R^3 e^{-6\nu_1 a} + \ldots\right),$$
$$(\textbf{3.9.70})$$

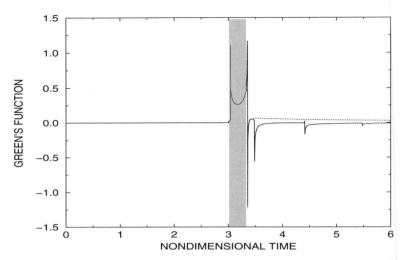

Figure 3.9.11: The temporal evolution of the Green's function at the point $(3a, \frac{3a}{2})$. The dashed line corresponds to the Green's functions (3.9.39) and (3.9.54) for the problem of two half-spaces. The shaded area corresponds to the period of time when leaky modes are present.

and

$$\mathcal{G}_2 = \frac{e^{-\nu_2|z-\zeta|}}{2\nu_2} - \frac{e^{-\nu_2(z+\zeta-2a)}}{2\nu_2}\left[R + (1-R^2)e^{-2\nu_1 a} - R(1-R^2)e^{-4\nu_1 a}\right.$$
$$\left. + R^2(1-R^2)e^{-6\nu_1 a} - \cdots\right], \quad (3.9.71)$$

where $R = (\nu_1 - \nu_2)/(\nu_1 + \nu_2)$. The first term in (3.9.70) and the first two terms in (3.9.71) equal (3.9.31)–(3.9.32), respectively, when the displacement of the interface is taken into account, and correspond to the solutions when there are simply two semi-infinite dielectrics present. Consequently, the remaining terms give the contribution from internal reflections within region 1. This illustrates how Green's functions within a finite domain can be constructed from an infinite series of free-space Green's functions.

Let us now find $g_j(x, z, t|0, \zeta, 0)$. Equations (3.9.68)–(3.9.69) cannot be inverted in closed form except for the leading term in (3.9.69). Therefore, we must invert these two transforms numerically via the inversion integrals

$$g_j(x, z, t) = \frac{1}{4\pi^2 i}\int_{\gamma-\infty i}^{\gamma+\infty i} e^{st}\left\{\int_{-\infty}^{\infty} \mathcal{G}_j(k, z, s)e^{ikx}\, dk\right\} ds, \quad (3.9.72)$$

where $\gamma > 0$ so that the Laplace transform converges. If we introduce $s = \omega i$, we may rewrite (3.9.72) as

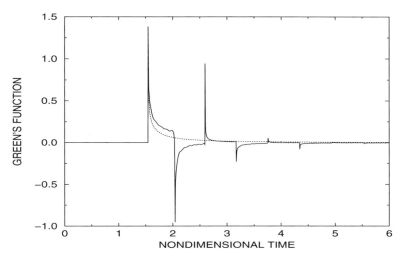

Figure 3.9.12: The temporal evolution of the Green's function at the point $(a, \frac{a}{2})$. The dashed line corresponds to the Green's function (3.9.63) for the problem of two half-spaces.

$$
g_j(x, z, t|0, \zeta, 0)
$$
$$
= \frac{1}{2\pi^2} \mathrm{Re} \left\{ \int_{0-\gamma i}^{\infty-\gamma i} \left[\int_{-\infty}^{\infty} \mathcal{G}_j(k, z, \omega|0, \zeta, 0) e^{i(kx+\omega t)} dk \right] d\omega \right\}.
$$
$$
(3.9.73)
$$

The numerical inversion was performed in a two-step procedure. First, for fixed ω, the inner integral was computed using Simpson's rule with $\Delta k = 0.01$ and k running from 0 to 20,000. Then the frequency integral was done with $\gamma = 0.05$, $\Delta\omega = 0.1$ and ω running from 0 to 3840 using Weddle's rule.

To illustrate the inversion of (3.9.68)–(3.9.69), the temporal behavior of the Green's function is given at four points A–D labeled on Figure 3.9.9. Point A is the closest to the source and its behavior with time is shown in Figure 3.9.10. The first peak corresponds to the direct wave and the second, negative peak is the reflected wave off the interface. Not surprisingly, the solution is in complete agreement with our earlier solution for the two dielectric half spaces (included as a dotted line) because the effect of the boundary at $z = 0$ has not yet been felt. After $ct/a = 2.8$, we see a series of gradually diminishing pulses. These are the portion of the energy that has passed through the interface from internal reflections off the boundaries at $z = 0$ and $z = a$.

Point B is still outside of the slab but located farther away from the source than point A. The temporal evolution of the Green's function

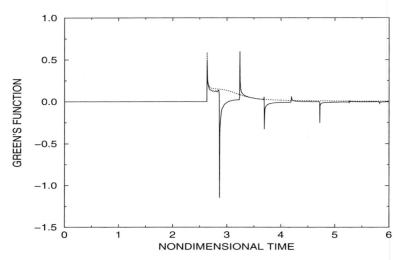

Figure 3.9.13: Same as Figure 3.9.12 except for the point $(3a, \frac{a}{2})$.

there is shown in Figure 3.9.11. This point was chosen because leaky waves are present. Indeed, the leaky waves arrive at $ct/a = 3$, while the direct wave does not arrive until $ct/a = 3.04$. After that time, they coexist (highlighted with shading) until $ct/a = 3.35$ when the reflected wave off the interface arrives. Again, we have our train of pulses from the internal reflections. However, the amplitudes are smaller than those shown in Figure 3.9.10 because some of the energy has already left the slab during earlier internal reflections.

Both points C and D are located within the slab and directly below points A and B, respectively. The temporal evolution of the Green's function is given in Figures 3.9.12 and 3.9.13, respectively. At both points, we first note the passage of the transmitted wave that we found in the case of two dielectric half spaces. After that, we have a train of pulses as internal reflections propagate through the slab.

3.10 WATER WAVES

If classical electrostatics gave birth to the concept of Green's function, it has enjoyed great success in the equally classic field of irrotational water waves. This section illustrates its use.

The theory of surface water waves stands on two assumptions: First, the fluid is irrotational. This allows us to introduce a velocity potential φ such that the Eulerian three-dimensional velocity $\mathbf{v}$ is given by $\mathbf{v} = \nabla\varphi$. Second, the fluid is incompressible so that $\nabla \cdot \mathbf{v} = 0$, or

$$\frac{\partial^2\varphi}{\partial x^2} + \frac{\partial^2\varphi}{\partial y^2} + \frac{\partial^2\varphi}{\partial z^2} = 0. \qquad (3.10.1)$$

In addition to the field equation (3.10.1), we must include boundary conditions. Assuming that the fluid lies on an infinitely flat surface, we have that

$$\frac{\partial \varphi}{\partial z} = 0, \qquad z = -h. \qquad (\mathbf{3.10.2})$$

Finally, from the linearization of Bernoilli's equation along the top surface, we find that

$$\frac{\partial^2 \varphi}{\partial t^2} + g \frac{\partial \varphi}{\partial z} = -\frac{1}{\rho} \frac{\partial p}{\partial t} = f(x, y, t), \qquad z = 0, \qquad (\mathbf{3.10.3})$$

where g temporarily denotes the acceleration due to gravity. Here $p(x, y, 0, t)$ is a prescribed surface pressure. From (3.10.3), we immediately see that φ and $\partial \varphi / \partial t$ must be known at $t = 0$.

The method that we will use to solve this initial-boundary-value problem involves the time-dependent Green's function. This Green's function is the velocity potential that gives the wave motions arising from an impulse forcing at (ξ, η, ζ) in a fluid that is initially at rest. After the Green's function is constructed, then we can write the solution $\varphi(x, y, z, t)$ in terms of any arbitrary initial conditions.[27] We now consider several classic problems involving (3.10.1)–(3.10.3).

- **Example 3.10.1: The Cauchy-Poisson problem**

One of the simplest problems of surface waves involves finding the waves that result from the impulse forcing of the surface. Solved at the beginning of the nineteenth century by Cauchy[28] and Poisson,[29] the Green's function method provides an insightful means of presenting their results. Although the original problem was for two spatial dimensions, we will solve it here in three dimensions, reserving the two-dimensional problem as problem 20.

We wish to determine the water waves that arise from an impulse forcing[30] at the point $x = \xi$, $y = \eta$ and at the depth $z = \zeta$. The water is at rest at $t = \tau$ in its equilibrium position and the surface pressure equals zero.

---

[27] See Wehausen, J. V., and E. V. Laitone, 1960: Surface waves in *Handbuch der Physik*. Springer-Verlag, §22α.

[28] Cauchy, A.-L., 1827: Théorie de la propagation des ondes à la surface d'un fluide pesant d'une profondeur indéfinie. *Mém. présentés par divers savans à l'Acad. Roy. Sci. Inst. France*, **1**, 3–123.

[29] Poisson, S. D., 1818: Mémoire sur la théorie des ondes. *Mém. Acad. Roy. Sci. Inst. France, Ser. 2*, **1**, 71–186.

[30] Taken from: The initial value problem for transient water waves, A. B. Finkelstein, *Commun. Pure Appl. Math.*, ©1957 John Wiley & Sons. Reprinted by permission of John Wiley & Sons, Inc.

To construct our Green's function we must find the solution to the following problem: The Green's function satisfies the partial differential equations

$$\frac{\partial^2 g}{\partial x^2} + \frac{\partial^2 g}{\partial y^2} + \frac{\partial^2 g}{\partial z^2} = -\delta(x - \xi)\delta(y - \eta)\delta(z - \zeta), \qquad (3.10.4)$$

where $-\infty < x, y, \xi, \eta < \infty$, $-h < z, \zeta < 0$, and $0 \leq \tau < t$, while at the free surface

$$\frac{\partial^2 g}{\partial t^2} + \frac{\partial g}{\partial z} = 0, \qquad z = 0, \qquad (3.10.5)$$

and at the bottom

$$\frac{\partial g}{\partial z} = 0, \qquad z = -h. \qquad (3.10.6)$$

Here, we have nondimensionalized the equations so that g again denotes the Green's function. At infinity, we require that g and $\partial g / \partial t$ tend to zero. As initial conditions we take

$$g(x, y, 0, \tau | \xi, \eta, \zeta) = g_t(x, y, 0, \tau | \xi, \eta, \zeta) = 0. \qquad (3.10.7)$$

Introducing the Fourier transform G of g,

$$G(k, \ell, z, t | \xi, \eta, \zeta) = \int_{-\infty}^{\infty} \int_{-\infty}^{\infty} g(x, y, z, t | \xi, \eta, \zeta) \, e^{-i(kx + \ell y)} \, dx \, dy,$$
$$(3.10.8)$$

or

$$g(x, y, z, t | \xi, \eta, \zeta) = \frac{1}{4\pi^2} \int_{-\infty}^{\infty} \int_{-\infty}^{\infty} G(k, \ell, z, t | \xi, \eta, \zeta) \, e^{i(kx + \ell y)} \, dk \, d\ell,$$
$$(3.10.9)$$

(3.10.4)–(3.10.7) becomes

$$\frac{\partial^2 G}{\partial z^2} - \kappa^2 G = -e^{-i(k\xi + \ell\eta)}\delta(z - \zeta), \qquad (3.10.10)$$

$$\frac{\partial^2 G}{\partial t^2} + \frac{\partial G}{\partial z} = 0, \qquad z = 0, \qquad (3.10.11)$$

$$\frac{\partial G}{\partial z} = 0, \qquad z = -h, \qquad (3.10.12)$$

and

$$G = \frac{\partial G}{\partial t} = 0, \quad t = \tau, \quad z = 0, \qquad (3.10.13)$$

with $\kappa^2 = k^2 + \ell^2$.

To solve (3.10.10)–(3.10.13), we apply techniques from §2.3. The general solution to (3.10.10) can be written

$$\widetilde{G} = e^{i(k\xi+\ell\eta)}G = \begin{cases} A(t)\cosh[\kappa(z+h)], & z < \zeta, \\ B(t)\cosh(\kappa z) - B''(t)\sinh(kz)/\kappa, & z > \zeta. \end{cases}$$
(3.10.14)

Requiring that the Green's function is continuous at $z = \zeta$,

$$\widetilde{G} = \cosh[\kappa(z_< + h)]\left[C(t)\cosh(\kappa z_>) - C''(t)\sinh(\kappa z_>)/\kappa\right]. \quad (3.10.15)$$

Because

$$\left.\frac{\partial \widetilde{G}}{\partial z}\right|_{z=\zeta^-}^{z=\zeta^+} = -1, \quad (3.10.16)$$

$C(t)$ is governed by

$$\cosh(\kappa h)C''(t) + \kappa\sinh(\kappa h)C(t) = 1. \quad (3.10.17)$$

Solving (3.10.17) with $C(\tau) = C'(\tau) = 0$,

$$C(t) = \frac{1 - \cos\left[(t-\tau)\sqrt{\kappa\tanh(\kappa h)}\right]}{\kappa\,\sinh(\kappa h)}. \quad (3.10.18)$$

Substituting (3.10.14) into (3.10.9),

$$g = \frac{1}{4\pi^2}\int_{-\infty}^{\infty}\int_{-\infty}^{\infty}\widetilde{G}(k,\ell,z,t|\xi,\eta,\zeta)\,e^{i[k(x-\xi)+\ell(y-\eta)]}\,dk\,d\ell. \quad (3.10.19)$$

Upon introducing $k = \kappa\cos(\alpha)$, $\ell = \kappa\sin(\alpha)$, $x - \xi = r\cos(\theta)$, $y - \eta = r\sin(\theta)$, with $r^2 = (x-\xi)^2 + (y-\eta)^2$, we find that

$$g = \frac{1}{4\pi^2}\int_0^{\infty}\widetilde{G}(k,\ell,z,t|\xi,\eta,\zeta)\left[\int_{-\pi+\theta}^{\pi+\theta}e^{i\kappa r\cos(\alpha-\theta)}\,d\alpha\right]\kappa\,d\kappa. \quad (3.10.20)$$

Using the integral definition for $J_0(\ )$,

$$J_0(x) = \frac{1}{2\pi}\int_{-\pi}^{\pi}e^{ix\cos(\alpha)}\,d\alpha, \quad (3.10.21)$$

(3.10.20) simplifies to

$$g = \frac{1}{2\pi}\int_0^{\infty}\widetilde{G}(k,\ell,z,t|\xi,\eta,\zeta)J_0(\kappa r)\,\kappa\,d\kappa. \quad (3.10.22)$$

or

$$g = -\frac{1}{2\pi} \int_0^\infty \frac{\sinh(\kappa z_>) \cosh[\kappa(z_< + h)]}{\cosh(\kappa h)} J_0(\kappa r) \, d\kappa$$

$$+ \frac{1}{2\pi} \int_0^\infty \frac{\cosh[\kappa(z + h)] \cosh[\kappa(\zeta + h)]}{\cosh^2(\kappa h)}$$

$$\times \left\{ \frac{1 - \cos\left[(t - \tau)\sqrt{\kappa \tanh(\kappa h)}\right]}{\kappa \tanh(\kappa h)} \right\} J_0(\kappa r) \, \kappa \, d\kappa. \quad (\mathbf{3.10.23})$$

Since $\cosh[\kappa(\zeta + h)] = e^{\kappa\zeta} \cosh(\kappa h) - e^{-\kappa h} \sinh(\kappa\zeta)$,

$$g = -\frac{1}{2\pi} \int_0^\infty \sinh(\kappa z_<) e^{\kappa z_>} J_0(\kappa r) \, d\kappa$$

$$+ \frac{1}{2\pi} \int_0^\infty e^{-\kappa h} \frac{\sinh(\kappa z) \sinh(\kappa\zeta)}{\cosh(\kappa h)} J_0(\kappa r) \, d\kappa$$

$$+ \frac{1}{2\pi} \int_0^\infty \frac{\cosh[\kappa(z + h)] \cosh[\kappa(\zeta + h)]}{\cosh^2(\kappa h)}$$

$$\times \left\{ \frac{1 - \cos\left[(t - \tau)\sqrt{\kappa \tanh(\kappa h)}\right]}{\kappa \tanh(\kappa h)} \right\} J_0(\kappa r) \, \kappa \, d\kappa. \quad (\mathbf{3.10.24})$$

Finally, we have that

$$\int_0^\infty e^{\kappa a} J_0(\kappa b) \, d\kappa = \left(a^2 + b^2\right)^{-1/2}, \quad\quad\quad (\mathbf{3.10.25})$$

$$g = \frac{1}{4\pi} \left[\frac{1}{R} - \frac{1}{R'} + 2 \int_0^\infty \frac{e^{-\kappa h} \sinh(\kappa z) \sinh(\kappa\zeta)}{\cosh(\kappa h)} J_0(\kappa r) \, d\kappa \right]$$

$$+ \frac{1}{2\pi} \int_0^\infty \frac{\cosh[\kappa(z + h)] \cosh[\kappa(\zeta + h)]}{\cosh^2(\kappa h)}$$

$$\times \left\{ \frac{1 - \cos\left[(t - \tau)\sqrt{\kappa \tanh(\kappa h)}\right]}{\kappa \tanh(\kappa h)} \right\} J_0(\kappa r) \, \kappa \, d\kappa, \quad (\mathbf{3.10.26})$$

where

$$R^2 = (x-\xi)^2 + (y-\eta)^2 + (z-\zeta)^2, \quad \text{and} \quad R'^2 = (x-\xi)^2 + (y-\eta)^2 + (z+\zeta)^2.$$
$$(\mathbf{3.10.27})$$

This is not the Green's function in the conventional sense because the source, although localized in position, is left "turned on." Consequently, another version of the problem is

$$\frac{\partial^2 g}{\partial x^2} + \frac{\partial^2 g}{\partial y^2} + \frac{\partial^2 g}{\partial z^2} = -\delta(x - \xi)\delta(y - \eta)\delta(z - \zeta)\delta(t - \tau), \quad (\mathbf{3.10.28})$$

with

$$g(x, y, z, 0|\xi, \eta, \zeta, \tau) = g_t(x, y, z, 0|\xi, \eta, \zeta, \tau) = 0, \qquad (3.10.29)$$

and (3.10.5)–(3.10.6). Upon taking the Fourier transform in both the x and y directions and the Laplace transform in time, we now have

$$\frac{d^2 G}{dz^2} - \kappa^2 G = -e^{-i(k\xi + \ell\eta) - s\tau} \delta(z - \zeta), \qquad (3.10.30)$$

$$s^2 G(k, \ell, 0, s|\xi, \eta, \zeta, \tau) + G'(k, \ell, 0, s|\xi, \eta, \zeta, \tau) = 0, \qquad (3.10.31)$$

and

$$G'(k, \ell, -h, s|\xi, \eta, \zeta, \tau) = 0, \qquad (3.10.32)$$

with $\kappa^2 = k^2 + \ell^2$.

We now solve (3.10.30)–(3.10.32) as we did earlier by piecing together solutions valid for $z < \zeta$ and $\zeta < z$ and choosing the constants so that $\widetilde{G}$ is continuous and possesses a jump in the first derivative equal to -1 at $z = \zeta$. This yields

$$\widetilde{G} = e^{i(k\xi + \ell\eta) + s\tau} G = \frac{\cosh[\kappa(z_< + h)] \left[\cosh(\kappa z_>) - s^2 \sinh(\kappa z_>)/\kappa\right]}{s^2 \cosh(\kappa h) + \kappa \sinh(\kappa h)}.$$
$$(3.10.33)$$

Taking the inverse Fourier transform in k and ℓ, the Laplace transform of $g(x, y, z, t|\xi, \eta, \zeta, \tau)$ is

$$\mathcal{L}(g) = \frac{e^{-s\tau}}{4\pi^2} \int_{-\infty}^{\infty} \int_{-\infty}^{\infty} \frac{\cosh[\kappa(z_< + h)] \left[\cosh(\kappa z_>) - s^2 \sinh(\kappa z_>)/\kappa\right]}{\cosh(\kappa h)[s^2 + \kappa \tanh(\kappa h)]}$$
$$\times e^{i[k(x-\xi) + \ell(y-\eta)]} \, dk \, d\ell. \qquad (3.10.34)$$

Upon introducing polar coordinates, we find that

$$\mathcal{L}(g) = -\frac{e^{-s\tau}}{2\pi} \int_0^{\infty} \frac{\sinh(\kappa z_>) \cosh[\kappa(z_< + h)]}{\cosh(\kappa h)} J_0(\kappa r) \, d\kappa$$

$$+ \frac{e^{-s\tau}}{2\pi} \int_0^{\infty} \frac{\cosh[\kappa(z + h)] \cosh[\kappa(\zeta + h)]}{\cosh^2(\kappa h)[s^2 + \kappa \tanh(\kappa h)]} J_0(\kappa r) \, \kappa \, d\kappa. \qquad (3.10.35)$$

Finally, taking the inverse Laplace transform and simplifying the first integral in (3.10.35), we obtain

$$g = \frac{\delta(t - \tau)}{4\pi} \left[\frac{1}{R} - \frac{1}{R'} + 2 \int_0^{\infty} e^{-\kappa h} \frac{\sinh(\kappa z) \sinh(\kappa \zeta)}{\cosh(\kappa h)} J_0(\kappa r) \, d\kappa\right]$$

$$+ \frac{H(t - \tau)}{2\pi} \int_0^{\infty} \frac{\cosh[\kappa(z + h)] \cosh[\kappa(\zeta + h)]}{\cosh^2(\kappa h)} J_0(\kappa r) \, \kappa \, d\kappa$$

$$\times \frac{\sin\left[(t - \tau)\sqrt{\kappa \tanh(\kappa h)}\right]}{\sqrt{\kappa \tanh(\kappa h)}}, \qquad (3.10.36)$$

where R and R' are defined by (3.10.27).

• Example 3.10.2: Time harmonic water waves

In the previous example, we found transient water waves arising from an impulse source that "turned on" at $t = \tau$ or acted for only an instant. In many cases, we are not interested in the transients and seek the "steady-state" solution when harmonic water waves have established themselves. In the case of three spatial dimensions, the Green's function is given by $\operatorname{Re}\{g(x, y, z|\xi, \eta, \zeta)e^{-i\omega t}\}$, where g is governed by

$$\frac{\partial^2 g}{\partial x^2} + \frac{\partial^2 g}{\partial y^2} + \frac{\partial^2 g}{\partial z^2} = -\delta(x - \xi)\delta(y - \eta)\delta(z - \zeta), \qquad (3.10.37)$$

with $-\infty < x, y, \xi, \eta < \infty$, and $-h < z, \zeta < 0$, while the free surface condition becomes

$$\frac{\partial g}{\partial z} = \omega^2 g, \qquad z = 0, \qquad (3.10.38)$$

and at the bottom

$$\frac{\partial g}{\partial z} = 0, \qquad z = -h. \qquad (3.10.39)$$

Finally, we require that

$$\lim_{(x,y,z)\to(\xi,\eta,\zeta)} \left| g(x, y, z|\xi, \eta, \zeta) - \frac{1}{R} \right| < \infty, \qquad (3.10.40)$$

where R is defined by (3.10.27).

In the previous example, we applied Fourier transforms in both the x and y directions. Due to axial symmetry these Fourier transforms became a Hankel transform in r. Because we have similar axial symmetry here, we anticipate that a similar process will occur and apply Hankel transforms from the outset. This leads to

$$\frac{d^2 G}{dz^2} - k^2 G = -\frac{\delta(z - \zeta)}{2\pi}, \qquad (3.10.41)$$

with

$$G'(k, 0|0, \zeta) = \omega^2 G(k, 0|0, \zeta), \qquad (3.10.42)$$

and

$$G'(k, -h|0, \zeta) = 0, \qquad (3.10.43)$$

where

$$G(k, z|0, \zeta) = \int_0^\infty g(r, z|0, \zeta)\, J_0(kr)\, r\, dr, \qquad (3.10.44)$$

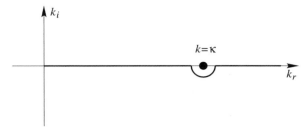

Figure 3.10.1: The line integral used in the evaluation of (3.10.46).

and $r^2 = (x - \xi)^2 + (y - \eta)^2$. The solution to (3.10.41)–(3.10.43) is

$$G(k, z|0, \zeta) = \frac{\omega^2 \sinh(kz_>) + k \cosh(kz_>)}{k \sinh(kh) - \omega^2 \cosh(kh)} \frac{\cosh[k(z_< + h)]}{2\pi k}. \quad (\mathbf{3.10.45})$$

Taking the inverse Hankel transform, the Green's function is

$$g(r, z|0, \zeta) = \frac{1}{2\pi} \oint_0^\infty \cosh[k(z_< + h)] \frac{\omega^2 \sinh(kz_>) + k \cosh(kz_>)}{k \sinh(kh) - \omega^2 \cosh(kh)} J_0(kr) \, dk,$$
$$(\mathbf{3.10.46})$$

where the special integral sign denotes an integration along the real axis except for a small semicircle *below* the singularity at $k = \kappa$, where $\kappa \tanh(\kappa h) = \omega^2$. See Figure 3.10.1.

Equation (3.10.46) does not lend itself to numerical evaluation. An alternative form is as follows: Because the integrand is an even function of k, we convert it to an integral along the entire real k axis except for passing *over* the singularity at $k = -\kappa$ and *below* the singularity at $k = \kappa$. We then close the line integration with a semicircular path of infinite radius joining $k = +\infty$ to $k = -\infty$. Because the contribution from the arc at infinity vanishes, the Green's function is given by a sum of the residues and equals

$$g(r, z|0, \zeta) = \frac{1}{2\pi} \sum_{n=0}^\infty \frac{K_0(\kappa_n r)}{N_n} \cos[\kappa_n(\zeta + h)] \cos[\kappa_n(z + h)], \quad (\mathbf{3.10.47})$$

where

$$N_n = \frac{h}{2} \left[1 + \frac{\sin(2\kappa_n h)}{2\kappa_n h} \right], \quad (\mathbf{3.10.48})$$

and κ_n is the nth positive root of $\kappa \tan(\kappa h) = -\omega^2$.

Another alternative to (3.10.46) rewrites the integrand so that we separate out explicitly the singular nature of the Green's function. When this is done, we obtain

$$g(r, z | 0, \zeta) = \frac{1}{4\pi} \left(\frac{1}{R} + \frac{1}{R''} \right)$$
$$- \frac{1}{2\pi} \fint_0^\infty \frac{(\omega^2 + k)e^{-kh}}{\omega^2 \cosh(kh) - k \sinh(kh)}$$
$$\times \cosh[k(z + h)] \cosh[k(\zeta + h)] J_0(kr) \, dk,$$

$$\text{(3.10.49)}$$

where $R''^2 = (x - \xi)^2 + (y - \eta)^2 + (z + \zeta + 2h)^2$.

In the limit of infinitely deep water, (3.10.46) and (3.10.49) become

$$g(r, z | 0, \zeta) = \frac{1}{4\pi R} - \frac{1}{4\pi} \fint_0^\infty \frac{\omega^2 + k}{\omega^2 - k} e^{k(z+\zeta)} J_0(kr) \, dk, \quad \text{(3.10.50)}$$

and

$$g(r, z | 0, \zeta) = \frac{1}{4\pi R} - \frac{1}{4\pi R'} + \frac{1}{2\pi} \fint_0^\infty \frac{k}{k - \omega^2} e^{k(z+\zeta)} J_0(kr) \, dk,$$

$$\text{(3.10.51)}$$

respectively.

• Example 3.10.3: Rapidly converging Green's function representations

In the previous examples, we found several Green's function representations without considering how difficult they would be to evaluate. For example, if we were to compute the Green's function given by (3.10.46) and (3.10.47), we would find that direct integration is difficult when $|z|$ and $|\zeta|$ are both small and that the eigenfunction expansion converges slowly when r is small.

Recently, Linton and McIver[31] developed a method for finding efficient representations of the Green's functions for water waves. To understand their technique, consider the initial-value problem

$$u_t = \nabla^2 u, \qquad 0 < t, \qquad \text{(3.10.52)}$$

where ∇ is the three-dimensional gradient operator. We take as our initial condition

$$u(\mathbf{x}, 0) = \delta(\mathbf{x} - \mathbf{x}_0), \qquad \text{(3.10.53)}$$

where $\mathbf{x}$ is the position vector and $\mathbf{x}_0$ is the point of excitation. The solution satisfies time-independent boundary conditions.

---

[31] Linton, C. M., and P. McIver, 2000: Green's functions for water waves in porous structures. *Appl. Ocean Res.*, **22**, 1–12.

Linton and McIver's first insight was to observe that the Green's function governed by

$$\nabla^2 g = -\delta(\mathbf{x} - \mathbf{x}_0) \tag{3.10.54}$$

for the same domain and boundary conditions that govern (3.10.52) equals

$$g(\mathbf{x}|\mathbf{x}_0) = \int_0^\infty u(\mathbf{x}, t)\, dt. \tag{3.10.55}$$

This technique depends, of course, upon the integral (3.10.55) existing. In point of fact, it does not in our particular problem and we must choose a $\tilde{u}$ so that $\int_0^\infty (u + \tilde{u})\, dt$ does exist. With this correction, the Green's function equals

$$g(\mathbf{x}|\mathbf{x}_0) = \int_0^\infty [u(\mathbf{x}, t) + \tilde{u}(\mathbf{x}, t)]\, dt - \tilde{g}(\mathbf{x}|\mathbf{x}_0), \tag{3.10.56}$$

where $\nabla^2 \tilde{g}(\mathbf{x}) = -\tilde{u}(\mathbf{x}, 0)$.

Why have we introduced (3.10.52)–(3.10.53)? This new initial-boundary-value problem is generally more difficult to solve than (3.10.54). This leads us to Linton and McIver's second insight. Let us split (3.10.55) into two parts:

$$g(\mathbf{x}|\mathbf{x}_0) = \int_0^a u(\mathbf{x}, t)\, dt + \int_a^\infty u(\mathbf{x}, t)\, dt, \tag{3.10.57}$$

where a is a free parameter. If we could find two representations of u such that first one, u_1, converges rapidly for small values of t and the second one, u_2, converges rapidly for large t, then the resulting expansion would converge more rapidly than using a single u for all t. As Linton and McIver showed, there is an optimal value for a that minimizes the total number of terms needed for a desired accuracy.

Let us now apply these theoretical considerations to find the Green's function for water waves. We start by solving the initial-boundary-value problem

$$\frac{\partial u}{\partial t} = \frac{\partial^2 u}{\partial x^2} + \frac{\partial^2 u}{\partial y^2} + \frac{\partial^2 u}{\partial z^2}, \tag{3.10.58}$$

where $-\infty < x, y < \infty$, and $-h < z < 0$, with the boundary conditions

$$u_z(x, y, 0, t) = \omega^2 u(x, y, 0, t), \tag{3.10.59}$$

and

$$u_z(x, y, -h, t) = 0, \tag{3.10.60}$$

and the initial condition

$$u(x, y, z, 0) = \delta(x - \xi)\delta(y - \eta)\delta(z - \zeta). \tag{3.10.61}$$

As we will show in Chapter 4, (3.10.58)–(3.10.61) correspond to finding the Green's function for a three-dimensional heat equation. One of the techniques for solving this problem is product solutions of the form

$$u(x, y, z, t) = u_1(x, y, t)w(z, t)H(t), \tag{3.10.62}$$

where $u_1(x, y, t)$ is the free-space Green's function for the two-dimensional problem

$$\frac{\partial u_1}{\partial t} = \frac{\partial^2 u_1}{\partial x^2} + \frac{\partial^2 u_1}{\partial y^2}, \tag{3.10.63}$$

with $u_1(x, y, 0) = \delta(x - \xi)\delta(y - \eta)$, and $w(z, t)$ is the Green's function for the one-dimensional problem

$$\frac{\partial w}{\partial t} = \frac{\partial^2 w}{\partial z^2}, \tag{3.10.64}$$

with the boundary conditions

$$w_z(0, t) = \omega^2 w(0, t), \quad \text{and} \quad w_z(-h, t) = 0, \tag{3.10.65}$$

and the initial condition $w(z, 0) = \delta(z - \zeta)$.

The free-space Green's function is found in §4.1 and equals

$$u_1(x, y, t) = \frac{H(t)}{4\pi t} \exp\left[-\frac{(x - \xi)^2 + (y - \eta)^2}{4t}\right]. \tag{3.10.66}$$

To find $w(z, t)$, we take the Laplace transform of (3.10.64)–(3.10.65) and find that

$$\frac{d^2 W}{dz^2} - sW = -\delta(z - \zeta), \tag{3.10.67}$$

with $W'(0, s) = \omega^2 W(0, s)$, and $W'(-h, s) = 0$. The solution to (3.10.67) is

$$W(z, s) = -\frac{[\omega^2 \sinh(qz_>) + q\cosh(qz_>)]\cosh[q(z_< + h)]}{q[\omega^2 \cosh(qh) - q\sinh(qh)]} \tag{3.10.68}$$

$$= \frac{e^{-q|z-\zeta|}}{2q} + \frac{e^{-q\chi_{04}}}{2q} + \frac{1}{2q}\sum_{n=1}^{\infty}\left(\frac{q + \omega^2}{q - \omega^2}\right)^n \sum_{m=1}^{4} e^{-q\chi_{nm}}, \tag{3.10.69}$$

where

$$\chi_{n1} = 2(n - 1)h - \zeta - z, \qquad \chi_{n2} = 2nh - \zeta + z, \tag{3.10.70}$$

$$\chi_{n3} = 2nh + \zeta - z, \qquad \chi_{n4} = 2(n + 1)h + \zeta + z, \tag{3.10.71}$$

and $q^2 = s$.

Our next step is to invert the Laplace transforms. From (3.10.68) the inversion yields

$$w(z,t) = \sum_{n=0}^{\infty} \frac{e^{-\mu_n^2 t}}{N_n} \cos[\mu_n(z+h)] \cos[\mu_n(\zeta+h)], \qquad (3.10.72)$$

where

$$N_n = \frac{h}{2} \left[1 + \frac{\sin(2\mu_n h)}{2\mu_n h} \right], \qquad (3.10.73)$$

and μ_n is the nth root of $\mu \tan(\mu h) = -\omega^2$. Equation (3.10.73) was derived assuming that all of the poles are simple; Linton and McIver discuss the case when there are higher-order poles. Note that (3.10.73) converges rapidly if t is large.

Turning to (3.10.69), we invert it term by term. This yields

$$w(z,t) = \frac{e^{-(z-\zeta)^2/(4t)}}{\sqrt{4\pi t}} + \frac{e^{-(2h+z+\zeta)^2/(4t)}}{\sqrt{4\pi t}} + \sum_{n=1}^{\infty}(-1)^n \sum_{m=1}^{4} I_n(\chi_{nm}),$$
$$(3.10.74)$$

where

$$I_n(\chi) = \sum_{j=0}^{n} \frac{n!}{j!(n-j)!}(-1)^j(2\omega^2)^{n-j}\overline{I}_{n-j}(\chi), \qquad (3.10.75)$$

and $\overline{I}_n(\chi)$ is given by the recurrence relationship

$$(n-1)\overline{I}_n(\chi) = 2t\overline{I}_{n-2}(\chi) + (\chi - 2\omega^2 t)\overline{I}_{n-1}(\chi), \qquad (3.10.76)$$

with the initial values

$$\overline{I}_0(\chi) = \frac{e^{-\chi^2/(4t)}}{\sqrt{4\pi t}}, \quad \text{and} \quad \overline{I}_1(\chi) = -\tfrac{1}{2}e^{\omega^4 t - \omega^2 \chi}\mathrm{erfc}\left(\frac{\chi}{2\sqrt{t}} - \omega^2\sqrt{t}\right).$$
$$(3.10.77)$$

Equation (3.10.74) converges rapidly for small t. It also enjoys the property that it is independent of any dispersion relationship.

If we now substitute (3.10.62) in (3.10.55), we find that it does not converge because of the $n = 0$ term in (3.10.72). Consequently, by choosing

$$\tilde{u}(\mathbf{x},t) = -u_1(x,y,t)\frac{e^{-\mu_0^2 t}}{N_0} \cos[\mu_0(z+h)] \cos[\mu_0(\zeta+h)], \quad (3.10.78)$$

the integral $\int_0^{\infty}(u + \tilde{u})\, dt$ now converges. To find $\tilde{g}(\mathbf{x})$, we must solve

$$\nabla^2 \tilde{g}(\mathbf{x}) = -\tilde{u}(\mathbf{x},0) = \frac{\delta(x-\xi)\delta(y-\eta)}{N_0} \cos[\mu_0(z+h)] \cos[\mu_0(\zeta+h)],$$
$$(3.10.79)$$

because $u_1(x, y, 0) = \delta(x - \xi)\delta(y - \eta)$. Equation (3.10.79) is a two-dimensional Poisson equation where z is a parameter. Using techniques that we will develop in Chapter 5, the solution is

$$\tilde{g}(\mathbf{x}) = \frac{K_0(\mu_0 r)}{2\pi N_0} \cos[\mu_0(z + h)] \cos[\mu_0(\zeta + h)]. \tag{3.10.80}$$

Our final task remains to evaluate $\int_0^\infty (u + \tilde{u})\, dt$. We have two choices for $w(z, t)$: there is (3.10.72) or (3.10.74). As pointed out earlier, Linton and McIver introduced the technique of breaking up the integral into two parts with the break point at $t = a$. In the first integral, we apply the representation that converges rapidly for small t, namely (3.10.74), while, in the second integral, we use the representation that converges rapidly for large t, (3.10.72). Carrying out the integrations, we find that

$$g(\mathbf{x}|\mathbf{x}_0) = \sum_{m=0}^M \frac{\Lambda_m}{N_m} \cos[\mu_m(z + h)] \cos[\mu_m(\zeta + h)]$$

$$+ \frac{1}{4\pi r} \mathrm{erfc}\left(\frac{r}{ah}\right) + \frac{1}{4\pi r'} \mathrm{erfc}\left(\frac{r'}{ah}\right) + \sum_{i=1}^4 \frac{\mathrm{erfc}[\sqrt{R^2 + \chi_{1i}^2}/(ah)]}{4\pi\sqrt{R^2 + \chi_{1i}^2}}$$

$$+ \frac{\omega^2}{2\pi} \int_0^{ah/2} e^{\omega^4 u^2 - R^2/(4u^2)} \sum_{i=1}^4 e^{-\omega^2\chi_{1i}} \mathrm{erfc}\left(\frac{\chi_{1i}}{2u} - \omega^2 u\right) \frac{du}{u}$$

$$+ \cdots, \tag{3.10.81}$$

where

$$\Lambda_0 = \frac{K_0(\mu_0 R)}{2\pi} - \int_0^{a^2 h^2/4} \frac{e^{-R^2/(4t)}}{4\pi t} e^{-\mu_0^2 t}\, dt, \tag{3.10.82}$$

and

$$\Lambda_m = \int_{a^2 h^2/4}^\infty \frac{e^{-R^2/(4t)}}{4\pi t} e^{-\mu_m^2 t}\, dt, \tag{3.10.83}$$

and M is a truncation parameter. Linton[32] used the same technique to find rapidly convergent representations for the free-space Green's function describing water waves in channels of constant depth and width.

Problems

1. By direct substitution, show[33] that

$$g(x, t|0, 0) = J_0(\sqrt{xt})H(x)H(t)$$

[32] Linton, C. M., 1999: A new representation for the free-surface channel Green's function. *Appl. Ocean Res.*, **21**, 17–25.

[33] First proven by Picard, É., 1894: Sur une équation aux dérivées partielles de la théorie de la propagation de l'électricité. *Bull. Soc. Math.*, **22**, 2–8.

is the free-space Green's function governed by

$$\frac{\partial^2 g}{\partial x \partial t} + \tfrac{1}{4}g = \delta(x)\delta(t), \qquad -\infty < x, t < \infty.$$

This Green's function is graphed in the figure captioned Problem 1.

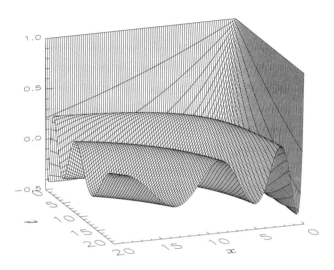

Problem 1

2. Construct the Green's function for the one-dimensional wave equation

$$\frac{\partial^2 g}{\partial t^2} - \frac{\partial^2 g}{\partial x^2} = \delta(x - \xi)\delta(t - \tau), \quad 0 < x, \xi < L, \quad 0 < t, \tau,$$

subject to the boundary conditions $g(0, t|\xi, \tau) = g_x(L, t|\xi, \tau) = 0, 0 < t$, and the initial conditions that $g(x, 0|\xi, \tau) = g_t(x, 0|\xi, \tau) = 0, 0 < x < L$. This Green's function is plotted in the figure captioned Problem 2 as functions of position x/L and time $(t - \tau)/L$ when $\xi = 0.2$.

3. Use the Green's functions from the text and the previous problem to write down the solution to the wave equation $u_{tt} = u_{xx}$ on the interval $0 < x < L$ with the initial data $u(x, 0) = \sin(x)$ and $u_t(x, 0) = 1$ and the following boundary conditions:

(a) $u(0, t) = e^{-t}$, $u(L, t) = 0$, $0 < t$,

(b) $u(0, t) = 0$, $u_x(L, t) = 1$, $0 < t$,

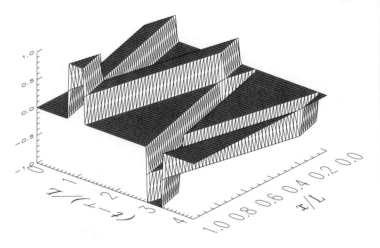

Problem 2

(c) $u_x(0, t) = \sin(t),$ $u_x(L, t) = 0,$ $0 < t.$

4. Find the Green's function[34] governed by

$$\frac{\partial^2 g}{\partial t^2} + 2\frac{\partial g}{\partial t} - \frac{\partial^2 g}{\partial x^2} = \delta(x - \xi)\delta(t - \tau), \quad 0 < x, \xi < L, \quad 0 < t, \tau,$$

subject to the boundary conditions

$$g_x(0, t|\xi, \tau) = g_x(L, t|\xi, \tau) = 0, \qquad 0 < t,$$

and the initial conditions

$$g(x, 0|\xi, \tau) = g_t(x, 0|\xi, \tau) = 0, \qquad 0 < x < L.$$

Step 1: If the Green's function can be written as the Fourier half-range cosine series

$$g(x, t|\xi, \tau) = \frac{1}{L}G_0(t|\tau) + \frac{2}{L}\sum_{n=1}^{\infty} G_n(t|\tau)\cos\left(\frac{n\pi x}{L}\right),$$

---

[34] Reprinted from *Int. J. Heat Mass Transfer*, **27**, M. N. Özişik and B. Vick, Propagation and reflection of thermal waves in a finite medium, 1845–1854, ©1984, with permission from Elsevier Science. See also Tang, D.-W., and N. Araki, 1996: Propagation of non-Fourier temperature wave in finite medium under laser-pulse heating (in Japanese). *Nihon Kikai Gakkai Rombumshu (Trans. Japan Soc. Mech. Engrs.)*, Ser. B, **62**, 1136–1141.

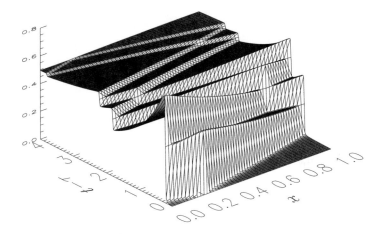

Problem 4

so that it satisfies the boundary conditions, show that $G_n(t|\tau)$ is governed by

$$G_n'' + 2G_n' + \frac{n^2\pi^2}{L^2}G_n = \cos\left(\frac{n\pi\xi}{L}\right)\delta(t-\tau), \qquad 0 \le n.$$

Step 2: Show that

$$G_0(t|\tau) = e^{-(t-\tau)}\sinh(t-\tau)H(t-\tau),$$

and

$$G_n(t|\tau) = \cos\left(\frac{n\pi\xi}{L}\right)e^{-(t-\tau)}\frac{\sin[\beta_n(t-\tau)]}{\beta_n}H(t-\tau), \quad 1 \le n,$$

where $\beta_n = \sqrt{(n\pi/L)^2 - 1}$.

Step 3: Combine the results from Steps 1 and 2 and show that the Green's function equals

$$g(x,t|\xi,\tau) = e^{-(t-\tau)}\sinh(t-\tau)H(t-\tau)/L$$
$$+ 2e^{-(t-\tau)}H(t-\tau)/L$$
$$\times \sum_{n=1}^{\infty}\frac{\sin[\beta_n(t-\tau)]}{\beta_n}\cos\left(\frac{n\pi\xi}{L}\right)\cos\left(\frac{n\pi x}{L}\right).$$

This Green's function is graphed in the figure captioned Problem 4 with $L = 1$, and $\xi = 0.2$.

5. Find the free-space Green's function[35] for the hyperbolic partial differential equation

$$\frac{\partial^2 g}{\partial y^2} - \frac{\partial^2 g}{\partial x^2} + g = \delta(x - \xi)\delta(y - \eta), \qquad -\infty < x, y, \xi, \eta < \infty,$$

subject to the boundary conditions

$$\lim_{|x| \to \infty} |g(x, y|\xi, \eta)| < \infty, \qquad -\infty < y < \infty,$$

and

$$\lim_{|y| \to \infty} |g(x, y|\xi, \eta)| < \infty, \qquad -\infty < x < \infty.$$

Step 1: Taking the Fourier transform of the partial differential equation, show that the joint transform $\mathcal{G}(k, \ell|\xi, \eta)$ is

$$\mathcal{G}(k, \ell|\xi, \eta) = \frac{e^{-ik\xi - i\ell\eta}}{k^2 - \ell^2 + 1},$$

where k and ℓ are the transform variables in the x and y directions, respectively.

Step 2: Show that the Green's function can be written as

$$\begin{aligned}
g(x, y|\xi, \eta) &= \frac{1}{4\pi^2} \int_{-\infty}^{\infty} \int_{-\infty}^{\infty} \frac{e^{ik(x-\xi)+i\ell(y-\eta)}}{k^2 - \ell^2 + 1} \, d\ell \, dk \\
&= \frac{1}{\pi^2} \int_0^{\infty} \int_0^{\infty} \frac{\cos[k(x - \xi)] \cos[\ell(y - \eta)]}{(k^2 + 1) - \ell^2} \, d\ell \, dk \\
&= \frac{1}{2\pi} \int_0^{\infty} \frac{\sin[(y - \eta)\sqrt{k^2 + 1}] \cos[k(x - \xi)]}{\sqrt{k^2 + 1}} dk \\
&= \tfrac{1}{2} J_0\left[\sqrt{(y - \eta)^2 - (x - \xi)^2}\right] H\left(|y - \eta| - |x - \xi|\right).
\end{aligned}$$

This Green's function is graphed in Figure 3.1.2 with $a = c = 1$ and y and η replacing t and τ, respectively.

6. Find the free-space Green's function[36] governed by the partial differential equation

$$\frac{\partial^2 g}{\partial t^2} + \frac{\partial^2 g}{\partial x \partial t} + g = \delta(x - \xi)\delta(t - \tau), \qquad -\infty < x, \xi < \infty, \quad 0 < t, \tau,$$

[35] Taken from Shay, L. K., R. L. Elsberry and P. G. Black, 1989: Vertical structure of the ocean current response to a hurricane. *J. Phys. Oceanogr.*, **19**, 649–669.

[36] Taken from Moore, D. W., R. C. Kloosterziel and W. S. Kessler, 1998: Evolution of mixed Rossby gravity waves. *J. Geophys. Res.*, **103**, 5331–5346. ©1998 by American Geophysical Union.

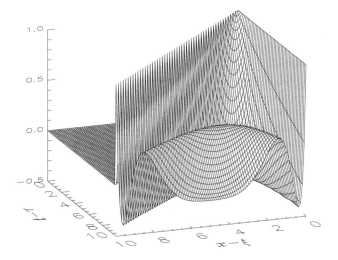

Problem 6

subject to the boundary conditions

$$\lim_{|x|\to\infty} |g(x,t|\xi,\tau)| < \infty, \qquad 0 < t,$$

and the initial conditions

$$g(x,0|\xi,\tau) = g_t(x,0|\xi,\tau) = 0, \qquad -\infty < x < \infty.$$

Step 1: Taking the Laplace transform and Fourier transform of the partial differential equation, show that the joint transform $\mathcal{G}(k,s|\xi,\tau)$ is

$$\mathcal{G}(k,s|\xi,\tau) = \frac{e^{-ik\xi-s\tau}}{iks + s^2 + 1}.$$

Step 2: By inverting the Fourier transform, show that

$$G(x,s|\xi,\tau) = \frac{e^{-(x-\xi)/s-s[\tau+(x-\xi)]}}{s}H(x-\xi).$$

Step 3: Using tables and the second shifting theorem, show that the Green's function equals

$$g(x,t|\xi,\tau) = J_0\left[2\sqrt{x-\xi}\sqrt{t-\tau-(x-\xi)}\right]H(x-\xi)H[t-\tau-(x-\xi)].$$

This Green's function is illustrated in the figure captioned Problem 6.

7. Find the free-space Green's function when the domain $x < 0$ is governed by

$$\frac{\partial^2 g_1}{\partial t^2} - \frac{\partial^2 g_1}{\partial x^2} + g_1 = \delta(t - \tau)\delta(x + \xi), \qquad \xi > 0,$$

while the domain $x > 0$ obeys

$$\frac{\partial^2 g_2}{\partial t^2} - \frac{\partial^2 g_2}{\partial x^2} = 0.$$

At the interface, we have the conditions that

$$g_1(0, t|\xi, \tau) = g_2(0, t|\xi, \tau), \quad \text{and} \quad g_{1_x}(0, t|\xi, \tau) = g_{2_x}(0, t|\xi, \tau).$$

At infinity, the Green's functions must be finite or

$$\lim_{x \to -\infty} |g_1(x, t|\xi, \tau)| < \infty, \quad \text{and} \quad \lim_{x \to \infty} |g_2(x, t|\xi, \tau)| < \infty.$$

All of the initial conditions equal zero.

Step 1: Taking the Laplace transform of the partial differential equations and boundary conditions, show that they reduce to the ordinary differential equations

$$\frac{d^2 G_1}{dx^2} - (s^2 + 1)G_1 = -\delta(x + \xi)e^{-s\tau},$$

and

$$\frac{d^2 G_2}{dx^2} - s^2 G_2 = 0,$$

with the boundary conditions

$$G_1(0, s|\xi, \tau) = G_2(0, s|\xi, \tau), \quad \text{and} \quad G_1'(0, s|\xi, \tau) = G_2'(0, s|\xi, \tau).$$

Step 2: Show that the Laplace transform of $g_1(x, t|\xi, \tau)$ and $g_2(x, t|\xi, \tau)$ are

$$G_1(x, s|\xi, \tau) = \frac{\exp(-|x + \xi|\sqrt{s^2 + 1} - s\tau)}{2\sqrt{s^2 + 1}}$$

$$+ \frac{\exp[-(\xi - x)\sqrt{s^2 + 1} - s\tau]}{2\left(s + \sqrt{s^2 + 1}\right)^2 \sqrt{s^2 + 1}} e^{-s(\xi - x + \tau)},$$

and

$$G_2(x, s|\xi, \tau) = \frac{\exp[-\xi\sqrt{s^2 + 1} - s(x + \tau)]}{s + \sqrt{s^2 + 1}}.$$

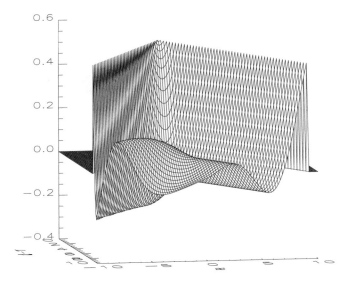

Problem 7

Step 3: Show that $g_1(x, t|\xi, \tau)$ and $g_2(x, t|\xi, \tau)$ are

$$g_1(x, t|\xi, \tau) = \tfrac{1}{2} H(t - \tau - |x + \xi|) J_0\left[\sqrt{(t-\tau)^2 - (x+\xi)^2}\right]$$
$$+ \frac{1}{2}\left[\frac{t - \tau - \xi + x}{t - \tau + \xi - x}\right] J_2\left[\sqrt{(t-\tau)^2 - (x-\xi)^2}\right]$$
$$\times H(t - \tau + x - \xi),$$

and

$$g_2(x, t|\xi, \tau) = \frac{(t - \tau - x)\, J_1\left[\sqrt{(t-\tau-x)^2 - \xi^2}\right]}{(t - \tau + \xi - x)\,\sqrt{(t-\tau-x)^2 - \xi^2}} H(t - \tau - x - \xi)$$
$$+ \frac{\xi J_1'\left[\sqrt{(t-\tau-x)^2 - \xi^2}\right]}{t - \tau + \xi - x} H(t - \tau - x - \xi).$$

This Green's function is illustrated in the figure captioned Problem 7 with $\xi = 1$.

8. Find the Green's function within a spherical cavity of radius a that is governed by the partial differential equation

$$\frac{\partial^2 g}{\partial r^2} + \frac{2}{r}\frac{\partial g}{\partial r} - \frac{1}{c^2}\frac{\partial^2 g}{\partial t^2} = \frac{1}{r}\frac{\partial^2 (rg)}{\partial r^2} - \frac{1}{c^2}\frac{\partial^2 g}{\partial t^2} = -\frac{\delta(r - \rho)\delta(t - \tau)}{4\pi r^2},$$

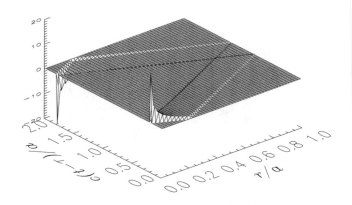

Problem 8

where $0 < r, \rho < a$, and $0 < t, \tau$, subject to the boundary conditions

$$\lim_{r \to 0} |g(r,t|\rho,\tau)| < \infty, \qquad \text{and} \qquad g(a,t|\rho,\tau) = 0, \qquad 0 < t,$$

and the initial conditions

$$g(r,0|\rho,\tau) = g_t(r,0|\rho,\tau) = 0, \qquad 0 < r < a.$$

Step 1: By setting $\varphi(r,t|\rho,\tau) = rg(r,t|\rho,\tau)$, show that the problem can be rewritten as

$$\frac{\partial^2 \varphi}{\partial r^2} - \frac{1}{c^2}\frac{\partial^2 \varphi}{\partial t^2} = -\frac{\delta(r-\rho)\delta(t-\tau)}{4\pi r}, \qquad 0 < r, \rho < a, \quad 0 < t, \tau,$$

subject to the boundary conditions

$$\varphi(0,t|\rho,\tau) = \varphi(a,t|\rho,\tau) = 0, \qquad 0 < r < a,$$

and the initial conditions

$$\varphi(r,0|\rho,\tau) = \varphi_t(r,0|\rho,\tau) = 0, \qquad 0 < t.$$

Step 2: Take the Laplace transform of the partial differential equation in Step 1 and show that it becomes the ordinary differential equation

$$\frac{d^2 \Phi}{dr^2} - \frac{s^2}{c^2}\Phi = -\frac{\delta(r-\rho)}{4\pi r}e^{-s\tau}, \qquad 0 < r, \rho < a,$$

subject to the boundary conditions $\Phi(0,s|\rho,\tau) = \Phi(a,s|\rho,\tau) = 0$.

Step 3: Show that

$$\frac{\delta(r - \rho)}{4\pi r} = \frac{1}{2\pi a \rho} \sum_{n=1}^{\infty} \sin\left(\frac{n\pi\rho}{a}\right) \sin\left(\frac{n\pi r}{a}\right).$$

Step 4: Show that the eigenfunction expansion of the ordinary differential equation in Step 2 is

$$G(r, s|\rho, \tau) = \frac{e^{-s\tau}}{2\pi a \rho r} \sum_{n=1}^{\infty} \frac{\sin(n\pi\rho/a) \sin(n\pi r/a)}{s^2/c^2 + n^2\pi^2/a^2}.$$

Step 5: Invert the Laplace transform and show that the Green's function is

$$g(r, t|\rho, \tau) = \frac{c\, H(t - \tau)}{2\pi^2 \rho r} \sum_{n=1}^{\infty} \frac{1}{n} \sin\left(\frac{n\pi\rho}{a}\right) \sin\left(\frac{n\pi r}{a}\right) \sin\left[\frac{n\pi c(t - \tau)}{a}\right].$$

This Green's function (divided by c/a^2) is plotted in the figure captioned Problem 8 for $a/100 \leq r \leq a$.

9. The Green's function for a beam of length L with internal damping is given by

$$\frac{\partial^2 g}{\partial t^2} + \frac{\partial^4 g}{\partial x^4} + a\frac{\partial^5 g}{\partial x^4 \partial t} = \delta(x - \xi)\delta(t - \tau), \quad 0 < x, \xi < L, \quad 0 < t, \tau,$$

where $a > 0$, with the boundary conditions

$$g(0, t|\xi, \tau) = g(L, t|\xi, \tau) = g_{xx}(0, t|\xi, \tau) = g_{xx}(L, t|\xi, \tau) = 0,$$

and the initial conditions

$$g(x, 0|\xi, \tau) = g_t(x, 0|\xi, \tau) = 0.$$

Find the Green's function.

Step 1: Assuming that the Green's function has the form

$$g(x, t|\xi, \tau) = \frac{2}{L} \sum_{n=1}^{\infty} G_n(t) \sin\left(\frac{n\pi\xi}{L}\right) \sin\left(\frac{n\pi x}{L}\right),$$

show that $G_n(t)$ is governed by

$$\frac{d^2 G_n}{dt^2} + a\left(\frac{n\pi}{L}\right)^4 \frac{dG_n}{dt} + \left(\frac{n\pi}{L}\right)^4 G_n = \delta(t - \tau).$$

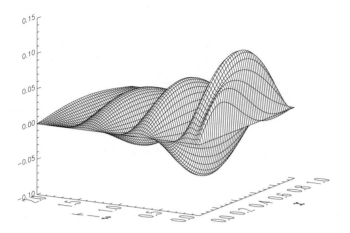

Problem 9

Why have we chosen this particular eigenfunction expansion?

Step 2: Solve the ordinary differential equation in Step 1 and show that

$$
G_n(t) = e^{-a\omega_n^2(t-\tau)/2}
\begin{cases}
\dfrac{\sin[\omega_n(t-\tau)\sqrt{1-a^2\omega_n^2/4}]}{\omega_n\sqrt{1-a^2\omega_n^2/4}}, & a\omega_n < 2, \\[12pt]
(t-\tau), & a\omega_n = 2, \\[12pt]
\dfrac{\sinh[\omega_n(t-\tau)\sqrt{a^2\omega_n^2/4-1}]}{\omega_n\sqrt{a^2\omega_n^2/4-1}}, & 2 < a\omega_n,
\end{cases}
$$

if $t > \tau$ and zero otherwise, where $\omega_n = n^2\pi^2/L^2$.

Step 3: Show that

$$
g(x,t|\xi,\tau) = \frac{2}{L}H(t-\tau)\sum_{n=1}^{\infty} e^{-a\omega_n^2(t-\tau)/2} \sin\left(\frac{n\pi\xi}{L}\right)\sin\left(\frac{n\pi x}{L}\right)
$$

$$
\times
\begin{cases}
\dfrac{\sin[\omega_n(t-\tau)\sqrt{1-a^2\omega_n^2/4}]}{\omega_n\sqrt{1-a^2\omega_n^2/4}}, & a\omega_n < 2, \\[12pt]
(t-\tau), & a\omega_n = 2, \\[12pt]
\dfrac{\sinh[\omega_n(t-\tau)\sqrt{a^2\omega_n^2/4-1}]}{\omega_n\sqrt{a^2\omega_n^2/4-1}}, & 2 < a\omega_n.
\end{cases}
$$

This Green's function is plotted in the figure captioned Problem 9 as functions of position x and time $t - \tau$ when $a = 0.2$, $L = 1$, and $\xi = 0.2$.

10. Find the free-space Green's function[37] for the system

$$\frac{\partial g_1}{\partial t} + a\frac{\partial g_1}{\partial x} - bg_2 = \delta(x - \xi)\delta(t - \tau),$$

and

$$\frac{\partial g_2}{\partial t} + cg_2 - bg_1 = 0.$$

Step 1: Taking the Laplace transform in time and the Fourier transform in space, show that the transformed equations are

$$\overline{G}_1(k, s|\xi, \tau) = \frac{(s + c)e^{-ik\xi - s\tau}}{(s + ika)(s + c) - b^2},$$

and

$$\overline{G}_2(k, s|\xi, \tau) = \frac{b\,e^{-ik\xi - s\tau}}{(s + ika)(s + c) - b^2}.$$

Step 2: Using the residue theorem, take the inverse Fourier transform and show that

$$G_1(x, s|\xi, \tau) = \frac{1}{a}\exp\left[\frac{c(x - \xi)}{a} + c\tau\right]H(x - \xi)$$
$$\times \exp\left[-\frac{(x - \xi + a\tau)(s + c)}{a} + \frac{b^2(x - \xi)}{a(s + c)}\right],$$

and

$$G_2(x, s|\xi, \tau) = \frac{b}{a}\exp\left[\frac{c(x - \xi)}{a} + c\tau\right]\frac{H(x - \xi)}{s + c}$$
$$\times \exp\left[-\frac{(x - \xi + a\tau)(s + c)}{a} + \frac{b^2(x - \xi)}{a(s + c)}\right].$$

Step 3: Using tables and the second shifting theorem, take the inverse Laplace transform and show that

---

[37] Reprinted with permission from Mounaix, Ph., D. Pesme, W. Rozmus, and M. Casanova, 1993: Space and time behavior of parametric instabilities for a finite pump wave duration in a bounded plasma. *Phys. Fluids B*, **5**, 3304–3318. ©American Institute of Physics, 1993.

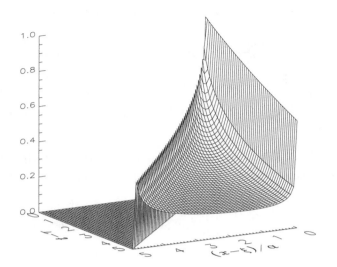

Problem 10

$$g_1(x, t|\xi, \tau) = \frac{1}{a} \exp\left[-c\left(t - \tau - \frac{x-\xi}{a}\right)\right] H(x - \xi)$$

$$\times \frac{d}{dt}\left(I_0\left\{\frac{2b}{a}\sqrt{(x-\xi)[a(t-\tau)-(x-\xi)]}\right\}\right.$$

$$\left. \times H[t - \tau - (x-\xi)/a]\right)$$

$$= \delta[a(t-\tau) - (x-\xi)]H(t-\tau)$$

$$+ \frac{b}{a}\sqrt{\frac{x-\xi}{a(t-\tau)-(x-\xi)}} \exp\left[-c\left(t-\tau-\frac{x-\xi}{a}\right)\right] H(x-\xi)$$

$$\times I_1\left\{\frac{2b}{a}\sqrt{(x-\xi)[a(t-\tau)-(x-\xi)]}\right\} H\left(t-\tau-\frac{x-\xi}{a}\right),$$

and

$$g_2(x, t|\xi, \tau) = \frac{b}{a}\exp\left[-c\left(t-\tau-\frac{x-\xi}{a}\right)\right] H(x-\xi)$$

$$\times I_0\left\{\frac{2b}{a}\sqrt{(x-\xi)[a(t-\tau)-(x-\xi)]}\right\} H\left(t-\tau-\frac{x-\xi}{a}\right).$$

The Green's function $ag_2(x, t|\xi, \tau)/b$ is plotted in the figure captioned Figure 10 as functions of distance $(x-\xi)/a$ and time $t-\tau$ when $c = 0.1$, and $b = 1$.

11. Find the free-space Green's function[38] governed by the equation

$$\left(\frac{\partial^2}{\partial r^2} + \frac{1}{r}\frac{\partial}{\partial r}\right)^2 g + \frac{1}{L^4}\frac{\partial^2 g}{\partial t^2} = \frac{\delta(r)\delta(t-\tau)}{2\pi r}, \quad 0 < r < \infty, \quad 0 < t, \tau,$$

subject to the boundary conditions

$$\lim_{r\to 0} |g(r, t|0, \tau)| < \infty, \quad \lim_{r\to\infty} |g(r, t|0, \tau)| < \infty, \quad 0 < t,$$

and the initial conditions

$$g(r, 0|0, \tau) = g_t(r, 0|0, \tau) = 0, \quad 0 < r < \infty.$$

Step 1: Let $\overline{G}(k, s|0, \tau)$ denote the joint Laplace-Hankel transform of $g(r, t|0, \tau)$, where

$$\overline{G}(k, s|0, \tau) = \int_0^\infty G(r, s|0, \tau) J_0(kr)\, r\, dr.$$

Show that the joint transform equals

$$\overline{G}(k, s|0, \tau) = \frac{e^{-s\tau}}{2\pi(k^4 + s^2/L^4)}$$

$$= \frac{L^2 e^{-s\tau}}{4\pi i s}\left(\frac{1}{k^2 + se^{-\pi i/2}/L^2} - \frac{1}{k^2 + se^{\pi i/2}/L^2}\right).$$

Step 2: Using the integral representation

$$\int_0^\infty \frac{J_0(ak)}{k^2 + z^2}\, k\, dk = K_0(az), \quad a > 0, \quad \text{Re}(z) > 0,$$

show that

$$G(r, s|0, \tau) = \frac{L^2 e^{-s\tau}}{4\pi i s}\left[K_0\left(\frac{e^{-\pi i/4} r\sqrt{s}}{L}\right) - K_0\left(\frac{e^{\pi i/4} r\sqrt{s}}{L}\right)\right].$$

Step 3: Using the second shifting theorem and the relationships that

$$\frac{F(s)}{s} = \int_0^t f(\tau)\, d\tau, \quad \text{and} \quad \mathcal{L}^{-1}\left[K_0(\sqrt{\alpha s})\right] = \frac{e^{-\alpha/(4t)}}{2t},$$

[38] Reprinted from *Int. J. Solids Struct.*, **37**, P. H. Wen, M. H. Aliabadi and A. Young, A boundary element method for dynamic plate bending problems, 5177–5188, ©2000, with permission from Elsevier Science. See also Graff, K. F., 1991: *Wave Motion in Elastic Solids*. Dover Publications, Inc., §4.2.5.

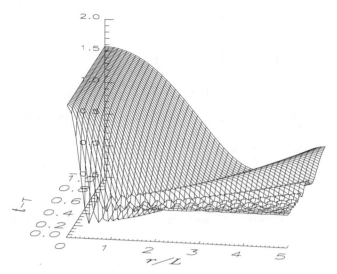

Problem 11

if $\alpha \neq 0$ and $\mathrm{Re}(\alpha) \geq 0$, show that the Green's function is

$$g(r, t|0, \tau) = \frac{H(t - \tau)L^2}{4\pi} \int_0^{t-\tau} \sin\left(\frac{r^2}{4L^2 u}\right) \frac{du}{u}$$

$$= \frac{H(t - \tau)L^2}{4\pi} \, \mathrm{si}\left[\frac{(r/L)^2}{4(t - \tau)}\right],$$

where si( ) is the sine integral

$$\mathrm{si}(u) = \int_u^\infty \frac{\sin(u)}{u} \, du.$$

The Green's function $4\pi g(r, t|0, \tau)/L^2$ is illustrated in the figure captioned Figure 11 as functions of distance r/L and time $t - \tau$.

12. Find the Green's function[39] governed by the equation

$$D_{11}\frac{\partial^4 g}{\partial x^4} + 4D_{16}\frac{\partial^4 g}{\partial x^3 \partial y} + 2(D_{12} + 2D_{66})\frac{\partial^4 g}{\partial x^2 \partial y^2}$$

$$+ 4D_{26}\frac{\partial^4 g}{\partial x \partial y^3} + D_{22}\frac{\partial^4 g}{\partial y^4} + \rho h\frac{\partial^2 g}{\partial t^2}$$

$$= \delta(x - \xi)\delta(y - \eta)\delta(t - \tau),$$

[39] Taken from Fällström, K.-E., and O. Lindblom, 1998: Transient bending wave propagation in anisotropic plates. *J. Appl. Mech.*, **65**, 930–938.

where $-\infty < x, y, \xi, \eta < \infty$, and $0 < t, \tau$. The boundary conditions are

$$\lim_{|x| \to \infty} |g(x, y, t | \xi, \eta, \tau)| < \infty, \quad \text{and} \quad \lim_{|y| \to \infty} |g(x, y, t | \xi, \eta, \tau)| < \infty.$$

The initial conditions are

$$g(x, y, 0 | \xi, \eta, \tau) = g_t(x, y, 0 | \xi, \eta, \tau) = 0.$$

Step 1: If $G(k, \ell, t | \xi, \eta, \tau)$ denotes the double Fourier transform of $g(x, y, t | \xi, \eta, \tau)$, show that

$$\rho h \frac{d^2 G}{dt^2} + \rho h \gamma^2 G = e^{-ik\xi - i\ell\eta} \delta(t - \tau),$$

where

$$\rho h \gamma^2 = D_{11} k^4 + 4 D_{16} k^3 \ell + 2(D_{12} + 2 D_{66}) k^2 \ell^2 + 4 D_{26} k \ell^3 + D_{22} \ell^4.$$

Step 2: Using Laplace transforms, show that the solution to the ordinary differential equation in Step 1 is

$$G(k, \ell, t | \xi, \eta, \tau) = \frac{e^{-ik\xi - i\ell\eta}}{\rho h} \frac{\sin[\gamma(t - \tau)]}{\gamma} H(t - \tau).$$

Step 3: Show that

$$g(x, y, t | \xi, \eta, \tau) = \frac{H(t - \tau)}{4\pi^2 \rho h} \int_{-\infty}^{\infty} \int_{-\infty}^{\infty} \frac{\sin[\gamma(t - \tau)]}{\gamma}$$
$$\times \cos[k(x - \xi) + \ell(y - \eta)] \, dk \, d\ell,$$

or

$$g(r, \theta, t | 0, 0, \tau) = \frac{H(t - \tau)}{4\pi^2 \rho h} \int_{0}^{2\pi} \int_{0}^{\infty} \frac{\sin[\gamma(t - \tau)]}{\gamma}$$
$$\times \cos[\sigma r \cos(\varphi - \theta)] \, \sigma \, d\sigma \, d\varphi,$$

where $x - \xi = r \cos(\theta)$, $y - \eta = r \sin(\theta)$, $k = \sigma \cos(\varphi)$, $\ell = \sigma \sin(\varphi)$, and

$$\frac{\rho h \gamma^2}{\sigma^4} = D_{11} \cos^4(\varphi) + 4 D_{16} \cos^3(\varphi) \sin(\varphi)$$
$$+ 2(D_{12} + 2 D_{66}) \cos^2(\varphi) \sin^2(\varphi)$$
$$+ 4 D_{26} \cos(\varphi) \sin^3(\varphi) + D_{22} \sin^4(\varphi).$$

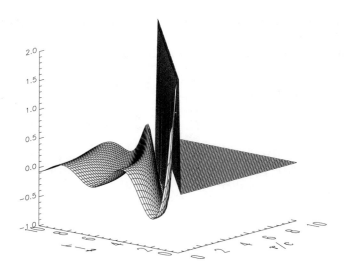

Problem 13

Step 4: If $D_{11} = D_{12} = D_{22} = D$, and $D_{16} = D_{66} = D_{26} = 0$, show that we recover $g(r, t|0, \tau)$ from the previous problem if $D/(\rho h) = L^4$.

13. Find the free-space Green's function for the two-dimensional Klein-Gordon equation:

$$\frac{\partial^2 g}{\partial x^2} + \frac{\partial^2 g}{\partial y^2} - \frac{1}{c^2}\left(\frac{\partial^2 g}{\partial t^2} + a^2 g\right) = -\delta(x - \xi)\delta(y - \eta)\delta(t - \tau),$$

where $-\infty < x, y, \xi, \eta < \infty$, and $0 < t, \tau$.

Step 1: Using Laplace transforms, show that our partial differential equation reduces to

$$\frac{\partial^2 G}{\partial x^2} + \frac{\partial^2 G}{\partial y^2} - \left(\frac{s^2 + a^2}{c^2}\right) G = -\delta(x - \xi)\delta(y - \eta)e^{-s\tau}.$$

Step 2: Using polar coordinates, show that

$$G(x, y, s|\xi, \eta, \tau) = \frac{e^{-s\tau}}{2\pi} K_0\left(r\sqrt{s^2 + a^2}/c\right).$$

Here $r = \sqrt{(x - \xi)^2 + (y - \eta)^2}$.

Step 3: Invert the Laplace transform and show that

$$g(x, y, t|\xi, \eta, \tau) = \frac{\cos\left[a\sqrt{(t - \tau)^2 - r^2/c^2}\right]}{2\pi\sqrt{(t - \tau)^2 - r^2/c^2}} H(t - \tau - r/c).$$

This Green's function (times 2π) has been graphed in the figure captioned Problem 13 as functions of position r/c and time $t - \tau$. The value of a is 1.5; no value has been allowed to be greater than two.

14. Find the Green's function[40] governed by the equation

$$\frac{\partial^2}{\partial t^2}\left(\frac{\partial^2 g}{\partial x^2} + \frac{\partial^2 g}{\partial z^2}\right) + N^2\frac{\partial^2 g}{\partial x^2} = \delta(x - \xi)\delta(z - \zeta)\delta(t - \tau),$$

where $-\infty < x, \xi < \infty$, $0 < z, \zeta < L$, and $0 < t, \tau$. The boundary conditions are

$$g(x, 0, t|\xi, \zeta, \tau) = g(x, L, t|\xi, \zeta, \tau) = 0,$$

and

$$\lim_{|x|\to\infty} |g(x, z, t|\xi, \zeta, \tau)| < \infty.$$

The initial conditions are

$$g(x, z, 0|\xi, \zeta, \tau) = g_t(x, z, 0|\xi, \zeta, \tau) = 0.$$

Step 1: If the Green's function can be written as the Fourier half-range sine expansion

$$g(x, z, t|\xi, \zeta, \tau) = \sum_{n=1}^{\infty} G_n(x, t|\xi, \tau)\sin\left(\frac{n\pi z}{L}\right),$$

which fulfills the boundary conditions in the z direction, show that $G_n(x, t|\xi, \tau)$ satisfies the equation

$$\frac{\partial^2}{\partial t^2}\left(\frac{\partial^2 G_n}{\partial x^2} - \frac{n^2\pi^2}{L^2}G_n\right) + N^2\frac{\partial^2 G_n}{\partial x^2} = \frac{2}{L}\sin\left(\frac{n\pi\zeta}{L}\right)\delta(x - \xi)\delta(t - \tau).$$

Step 2: Taking the Fourier transform of the partial differential equation in Step 1, show that the Fourier transform of $G_n(x, t|\xi, \tau)$, $\overline{G}_n(k, t|\xi, \tau)$, is governed by

$$\left(k^2 + \frac{n^2\pi^2}{L^2}\right)\overline{G}_n'' + N^2k^2\overline{G}_n = -\frac{2}{L}\sin\left(\frac{n\pi\zeta}{L}\right)e^{-ik\xi}\delta(t - \tau).$$

---

[40] Reprinted with permission from Grigor'ev, P. L., and V. A. Yakovlev, 1986: Solution of the Cauchy problem for internal waves in a medium with rigid boundaries. *Sov. Phys. Tech. Phys.*, **31**, 1254–1256. ©American Institute of Physics, 1986.

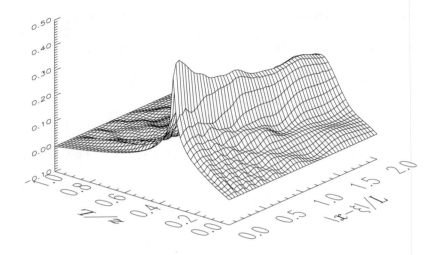

Problem 14

Step 3: Show that

$$\overline{G}_n(k,t|\xi,\tau) = -\frac{2}{NL} \sin\left(\frac{n\pi\zeta}{L}\right) e^{-ik\xi} H(t-\tau)$$
$$\times \frac{\sin[Nk(t-\tau)/\sqrt{k^2 + n^2\pi^2/L^2}\,]}{k\sqrt{k^2 + n^2\pi^2/L^2}}.$$

Step 4: Taking the inverse Fourier transform of $\overline{G}_n(k,t|\xi,\tau)$, show that

$$G_n(x,t|\xi,\tau) = -\frac{2}{NL\pi} \sin\left(\frac{n\pi\zeta}{L}\right) H(t-\tau)$$
$$\times \int_0^\infty \frac{\sin[Nk(t-\tau)/\sqrt{k^2 + n^2\pi^2/L^2}\,]}{k\sqrt{k^2 + n^2\pi^2/L^2}} \cos[k(x-\xi)]\,dk$$
$$= -\frac{2}{NL\pi} \sin\left(\frac{n\pi\zeta}{L}\right) \frac{|x-\xi|}{n} H(t-\tau)$$
$$\times \int_0^\infty \frac{\sin[N(t-\tau)\eta/\sqrt{\eta^2 + \pi^2(x-\xi)^2/L^2}\,]}{\eta\sqrt{\eta^2 + \pi^2(x-\xi)^2/L^2}} \cos(n\eta)\,d\eta.$$

Hint: Let $|x-\xi|k = n\eta$.

Step 5: Using the relationship that

$$\sum_{n=1}^\infty \frac{\cos(n\eta)}{n} \sin\left(\frac{n\pi\zeta}{L}\right) \sin\left(\frac{n\pi z}{L}\right) = \frac{1}{8} \ln\left\{ \left| \frac{\cos(\eta) - \cos[\pi(z+\zeta)/L]}{\cos(\eta) - \cos[\pi(z-\zeta)/L]} \right| \right\},$$

show that

$$g(x, z, t|\xi, \zeta, \tau) = -\frac{2|x - \xi|}{NL\pi} H(t - \tau)$$

$$\times \int_0^\infty \frac{\sin[N(t - \tau)\eta/\sqrt{\eta^2 + \pi^2(x - \xi)^2/L^2}]}{\eta\sqrt{\eta^2 + \pi^2(x - \xi)^2/L^2}}$$

$$\times \sum_{n=1}^\infty \frac{\cos(n\eta)}{n} \sin\left(\frac{n\pi\zeta}{L}\right) \sin\left(\frac{n\pi z}{L}\right) d\eta$$

$$= \frac{H(t - \tau)}{4N\pi^2} \int_0^{\pi/2} \frac{\sin[N(t - \tau)\sin(\varphi)]}{\sin(\varphi)} d\varphi$$

$$\times \ln\left\{\left|\frac{\cos(\eta) - \cos[\pi(z - \zeta)/L]}{\cos(\eta) - \cos[\pi(z + \zeta)/L]}\right|\right\},$$

where $\eta = |x - \xi|\pi\tan(\varphi)/L$. The Green's function $Ng(x, z, t|\xi, \zeta, \tau)$ is illustrated in the figure captioned Problem 14 as functions of $|x - \xi|/L$ and z/L when $\zeta/L = 1/\pi$, and $N(t - \tau) = 50$. The integration was performed using Weddle's rule with an increment of 0.001 after setting $\chi = \tan(\varphi)$.

15. Find the free-space Green's function[41] governed by the equation

$$\frac{\partial^2 g}{\partial r^2} + \frac{1 + \lambda}{r} \frac{\partial g}{\partial r} + \frac{1}{r^2}\left(\frac{\partial^2 g}{\partial \theta^2} - \frac{\partial^2 g}{\partial t^2}\right) = -\frac{\delta(r - 1)\delta(\theta - \theta')\delta(t - \tau)}{r^\lambda},$$

over the domain $0 < r < \infty$, $0 \le \theta, \theta' \le 2\pi$, and $0 < t, \tau$. The boundary conditions are

$$\lim_{r \to 0} |g(r, \theta, t|1, \theta', \tau)| < \infty, \qquad \lim_{r \to \infty} |g(r, \theta, t|1, \theta', \tau)| < \infty,$$

and $g(r, \theta, t|1, \theta', \tau) = g(r, \theta + 2n\pi, t|1, \theta', \tau)$. The initial conditions are

$$g(r, \theta, 0|1, \theta', \tau) = g_t(r, \theta, 0|1, \theta', \tau) = 0.$$

Step 1: Taking the Laplace transform of our partial differential equation, show that it becomes

$$\frac{\partial^2 G}{\partial r^2} + \frac{1 + \lambda}{r} \frac{\partial G}{\partial r} + \frac{1}{r^2}\left(\frac{\partial^2 G}{\partial \theta^2} - s^2 G\right) = -\frac{\delta(r - 1)\delta(\theta - \theta')}{r^\lambda} e^{-s\tau},$$

----

[41] Taken from Watanabe, K., 1981: Transient response of an inhomogeneous elastic solid to an impulse SH-source (variable SH-wave velocity)(in Japanese). *Nihon Kikai Gakkai Rombumshu (Trans. Japan Soc. Mech. Engrs.)*, Ser. A, **47**, 740–746.

with the boundary conditions

$$\lim_{r \to 0} |G(r,\theta,s|1,\theta',\tau)| < \infty, \qquad \lim_{r \to \infty} |G(r,\theta,s|1,\theta',\tau)| < \infty,$$

and $G(r,\theta,s|1,\theta',\tau) = G(r,\theta + 2n\pi, s|1,\theta',\tau)$.

Step 2: Introducing the complex Fourier coefficient

$$\overline{G}(r,n,s|1,\theta',\tau) = \frac{1}{2\pi} \int_0^{2\pi} G(r,\theta,s|1,\theta',\tau)\, e^{-in\theta}\, d\theta,$$

show that the partial differential equation in Step 1 reduces to the ordinary differential equation

$$\frac{d^2\overline{G}}{dr^2} + \frac{1+\lambda}{r}\frac{d\overline{G}}{dr} - \frac{(n^2+s^2)}{r^2}\overline{G} = -\frac{\delta(r-1)}{2\pi r^\lambda}e^{-in\theta'-s\tau},$$

with the boundary conditions

$$\lim_{r \to 0} |\overline{G}(r,n,s|1,\theta',\tau)| < \infty, \qquad \lim_{r \to \infty} |\overline{G}(r,n,s|1,\theta',\tau)| < \infty.$$

Step 3: Introducing the new independent variable $x = \ln(r)$, show that the ordinary differential equation in Step 2 becomes

$$\frac{d^2\overline{G}}{dx^2} + \lambda\frac{d\overline{G}}{dx} - (n^2+s^2)\overline{G} = -\frac{\delta(x)}{2\pi}e^{-in\theta'-s\tau}, \qquad -\infty < x < \infty,$$

with the boundary conditions

$$\lim_{|x| \to \infty} |\overline{G}(x,n,s|0,\theta',\tau)| < \infty.$$

Step 4: Taking the Fourier transform of $\overline{G}(x,n,s|0,\theta',\tau)$ with respect to x, show that

$$\overline{\mathcal{G}}(k,n,s|0,\theta',\tau) = \frac{e^{-in\theta'-s\tau}}{2\pi(k^2 - i\lambda k + n^2 + s^2)}.$$

Step 5: Using the residue theorem, show that

$$\overline{G}(r,n,s|1,\theta',\tau) = \frac{r^{-\lambda/2}}{4\pi}\frac{\exp[-|\ln(r)|\sqrt{s^2+\lambda_n^2}\,]}{\sqrt{s^2+\lambda_n^2}}e^{-s\tau},$$

where $\lambda_n^2 = (\lambda/2)^2 + n^2$, and $\text{Re}(\sqrt{s^2+\lambda_n^2}) \geq 0$.

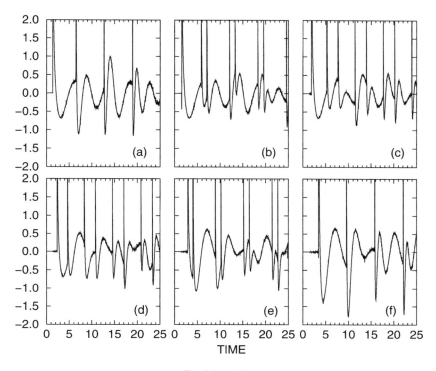

Problem 15

(a) $\theta - \theta' = 0$, (b) $\theta - \theta' = \pi/5$, (c) $\theta - \theta' = 2\pi/5$,

(d) $\theta - \theta' = 3\pi/5$, (e) $\theta - \theta' = 4\pi/5$, (f) $\theta - \theta' = \pi$

Step 6: Taking the inverse Laplace transform, show that

$$\bar{g}(r, n, t | 1, \theta', \tau) = \frac{r^{-\lambda/2}}{4\pi} H\left[t - \tau - |\ln(r)|\right] J_0\left[\lambda_n \sqrt{(t - \tau)^2 - \ln^2(r)}\right].$$

Step 7: Constructing the Fourier series, show that

$$g(r, \theta, t | 1, \theta', \tau) = \frac{r^{-\lambda/2}}{2\pi} H\left[t - \tau - |\ln(r)|\right]$$

$$\times \sum_{n=0}^{\infty} \epsilon_n J_0\left[\lambda_n \sqrt{(t - \tau)^2 - \ln^2(r)}\right] \cos[n(\theta - \theta')],$$

where $\epsilon_0 = \frac{1}{2}$, and $\epsilon_n = 1$ if $n > 0$.

The Green's function $2\pi r^{\lambda/2} g(r, \theta, t | 1, \theta', \tau)$ is illustrated in the figure captioned Problem 15 as a function of time $t - \tau$ for six values of $\theta - \theta'$, $r = 4$, and $\lambda = 2$.

16. Find the Green's function[42] governed by the equation

$$\frac{\partial^2 g}{\partial x^2} + \frac{\partial^2 g}{\partial y^2} + \frac{\partial^2 g}{\partial z^2} - \frac{1}{c^2}\frac{\partial^2 g}{\partial t^2} = -\delta(x-\xi)\delta(y-\eta)\delta(z-\zeta)\delta(t-\tau),$$

in the rectangular waveguide defined by $0 < x, \xi < a$, $0 < y, \eta < b$, $-\infty < z, \zeta < \infty$, and $0 < t, \tau$. The boundary conditions are

$$g(0, y, z, t|\xi, \eta, \zeta, \tau) = g(a, y, z, t|\xi, \eta, \zeta, \tau) = 0,$$

$$g(x, 0, z, t|\xi, \eta, \zeta, \tau) = g(x, b, z, t|\xi, \eta, \zeta, \tau) = 0,$$

and

$$\lim_{|z|\to\infty} |g(x, y, z, t|\xi, \eta, \zeta, \tau)| < \infty.$$

The initial conditions are

$$g(x, y, z, 0|\xi, \eta, \zeta, \tau) = g_t(x, y, z, 0|\xi, \eta, \zeta, \tau) = 0.$$

Step 1: If we write the Green's function as the double Fourier half-range sine expansion

$$g(x, y, z, t|\xi, \eta, \zeta, \tau) = \frac{4}{ab}\sum_{m=1}^{\infty}\sum_{n=1}^{\infty} G_{nm}(z, t|\zeta, \tau)$$

$$\times \sin\left(\frac{n\pi x}{a}\right)\sin\left(\frac{n\pi\xi}{a}\right)\sin\left(\frac{m\pi y}{b}\right)\sin\left(\frac{m\pi\eta}{b}\right),$$

show that $G_{nm}(z, t|\zeta, \tau)$ satisfies the equation

$$\frac{\partial^2 G_{nm}}{\partial z^2} - \frac{1}{c^2}\left(\frac{\partial^2 G_{nm}}{\partial t^2} + \omega_{nm}^2 G_{nm}\right) = -\delta(z-\zeta)\delta(t-\tau),$$

where

$$\omega_{nm} = c\sqrt{\left(\frac{n\pi}{a}\right)^2 + \left(\frac{m\pi}{b}\right)^2}.$$

Step 2: Show that

$$G_{nm}(z, t|\zeta, \tau) = \frac{c}{2}H[c(t-\tau)-|z-\zeta|]\,J_0\left[\omega_{nm}\sqrt{(t-\tau)^2 - \left(\frac{z-\zeta}{c}\right)^2}\right].$$

---

[42] Taken with permission from Dokuchaev, V. P., 1998: Excitation of a rectangular waveguide by pulsed electric and magnetic currents. *Radiophys. Quantum Electronics*, **41**, 301–309. Published by Plenum Publishers.

Step 3: Finish the problem and show that

$$g(x, y, z, t|\xi, \eta, \zeta, \tau) = \frac{2c}{ab} H[c(t - \tau) - |z - \zeta|]$$

$$\times \sum_{m=1}^{\infty} \sum_{n=1}^{\infty} J_0 \left[\omega_{nm} \sqrt{(t - \tau)^2 - \left(\frac{z - \zeta}{c} \right)^2} \right]$$

$$\times \sin\left(\frac{n\pi x}{a} \right) \sin\left(\frac{n\pi\xi}{a} \right) \sin\left(\frac{m\pi y}{b} \right) \sin\left(\frac{m\pi\eta}{b} \right).$$

17. Find the Green's function for

$$\frac{\partial^2 g}{\partial x^2} + \frac{\partial^2 g}{\partial y^2} + \frac{\partial^2 g}{\partial z^2} - \left(\frac{1}{c} \frac{\partial}{\partial t} + M \frac{\partial}{\partial x} \right)^2 g = \delta(x - \xi)\delta(y - \eta)\delta(z - \zeta)\delta(t - \tau),$$

where $-\infty < x, y, z, \xi, \eta, \zeta < \infty$, $0 < t, \tau$, c and M are real, positive constants, and $0 \leq M < 1$. The boundary conditions are

$$\lim_{|x| \to \infty} |g(x, y, z, t|\xi, \eta, \zeta, \tau)| < \infty,$$

$$\lim_{|y| \to \infty} |g(x, y, z, t|\xi, \eta, \zeta, \tau)| < \infty,$$

and

$$\lim_{|z| \to \infty} |g(x, y, z, t|\xi, \eta, \zeta, \tau)| < \infty.$$

The initial conditions are

$$g(x, y, z, 0|\xi, \eta, \zeta, \tau) = g_t(x, y, z, 0|\xi, \eta, \zeta, \tau) = 0.$$

Step 1: Take the Laplace transform of the partial differential equation and show that it equals

$$(1 - M^2)\frac{\partial^2 G}{\partial x^2} - \frac{2sM}{c}\frac{\partial G}{\partial x} + \frac{\partial^2 G}{\partial y^2} + \frac{\partial^2 G}{\partial z^2} - \frac{s^2}{c^2}G = \delta(x - \xi)\delta(y - \eta)\delta(z - \zeta)e^{-s\tau}$$

plus the boundary conditions

$$\lim_{|x| \to \infty} |G(x, y, z, s|\xi, \eta, \zeta, \tau)| < \infty,$$

$$\lim_{|y| \to \infty} |G(x, y, z, s|\xi, \eta, \zeta, \tau)| < \infty,$$

and

$$\lim_{|z| \to \infty} |G(x, y, z, s|\xi, \eta, \zeta, \tau)| < \infty.$$

Step 2: By introducing $x = (1 - M^2)^{1/2}x'$, $\xi = (1 - M^2)^{1/2}\xi'$, $s = (1 - M^2)^{1/2}s'$ and $G = \mathcal{G}e^{s'Mx'/c}$, show that the partial differential equation in Step 1 becomes

$$\frac{\partial^2 \mathcal{G}}{\partial x'^2} + \frac{\partial^2 \mathcal{G}}{\partial y^2} + \frac{\partial^2 \mathcal{G}}{\partial z^2} - \frac{s'^2}{c^2}\mathcal{G} = \frac{\delta(x' - \xi')\delta(y - \eta)\delta(z - \zeta)}{(1 - M^2)^{1/2}}e^{-s\tau - s'M\xi'/c}.$$

Step 3: Show that

$$G(x, y, z, s|\xi, \eta, \zeta, \tau) = \frac{e^{-s'R'/c + s'(x' - \xi')M/c - s\tau}}{4\pi(1 - M^2)^{1/2}\,R'},$$

where $R' = \sqrt{(x' - \xi')^2 + (y - \eta)^2 + (z - \zeta)^2}$.

Step 4: Show that

$$g(x, y, z, t|\xi, \eta, \zeta, \tau) = \frac{\delta(t - \tau - \tau')}{4\pi(1 - M^2)^{1/2}\,R'},$$

where $\tau' = [R' - M(x' - \xi')]/[c(1 - M^2)^{1/2}]$.

18. Show that transient solutions for the one-dimensional dielectric problem (3.9.15)–(3.9.17) are

$$g_1^{(t)}(z, t|\zeta, \tau) = \frac{c}{n^2 - 1}H\left[t - \tau + \frac{n(z - a)}{c} - \frac{\zeta - a}{c}\right]$$
$$\times \frac{\sinh(ns_n z/c)\exp[s_n(t - \tau) - s_n(\zeta - a)/c]}{(as_n/c)\sinh(ns_n a/c)},$$

and

$$g_2^{(t)}(z, t|\zeta, \tau) = \frac{c}{n^2 - 1}H\left(t - \tau - \frac{z - a}{c} - \frac{\zeta - a}{c}\right)$$
$$\times \frac{\exp[s_n(t - \tau) - s_n(\zeta - a)/c - s_n(z - a)/c]}{as_n/c},$$

where

$$\frac{as_n}{c} = \frac{1}{2n}\ln\left(\frac{1 - n}{1 + n}\right),$$

when $n < 1$. Plot this solution as a function z/a for various $c(t - \tau)/a$. Compare and contrast this solution with the solution that we found for $n > 1$.

19. Find the reflected and transmitted waves for the half-space problem solved in §3.9, (3.9.31)–(3.9.32), when $n > 1$. Do you still have head

waves? Explain your solution in terms of the physical processes that are occurring.

20. Construct the Green's function[30] governed by the two-dimensional Poisson equation

$$\frac{\partial^2 g}{\partial x^2} + \frac{\partial^2 g}{\partial z^2} = -\delta(x - \xi)\delta(z - \zeta),$$

where $-\infty < x, \xi < \infty$, $-h < z, \zeta < 0$, and $0 \le \tau < t$, with the boundary conditions

$$\frac{\partial^2 g}{\partial t^2} + \frac{\partial g}{\partial z} = 0, \quad z = 0, \quad \tau < t,$$

$$\frac{\partial g}{\partial z} = 0, \quad z = -h, \quad \tau < t,$$

and the initial conditions

$$g(x, z, \tau | \xi, \zeta, \tau) = 0, \qquad g_t(x, z, \tau | \xi, \zeta, \tau) = 0.$$

As $|x| \to \infty$, we require that g and its first two derivatives with respect to x should tend to zero in such a manner so that g possesses a Fourier transform with respect to x.

Step 1: Applying the Fourier transform to the problem, show that

$$\frac{d^2 G}{dz^2} - k^2 G = -e^{-ik\xi}\delta(z - \zeta), \qquad -h < z, \zeta < 0.$$

Step 2: Show that

$$e^{-ik\xi}G = \frac{\sinh(kz_>)\cosh[k(z_< + h)]}{k\cosh(kh)}$$

$$- \frac{\cosh[k(z + h)]\cosh[k(\zeta + h)]}{\cosh^2(kh)}\left\{\frac{1 - \cos\left[(t - \tau)\sqrt{k\tanh(kh)}\right]}{k\tanh(kh)}\right\}.$$

Step 3: By inverting the Fourier transform, show that the Green's function is

$$g(x, z, t | \xi, \zeta, \tau) = \frac{1}{2\pi}\left[\ln\left(\frac{R}{R'}\right) + 2\int_0^\infty e^{-kh}\frac{\sinh(kz)\sinh(k\zeta)}{k\cosh(kh)}\cos(kr)\,dk\right]$$

$$- \frac{1}{\pi}\int_0^\infty \frac{\cosh[k(z + h)]\cosh[k(\zeta + h)]}{\cosh^2(kh)}\cos(kr)\,dk$$

$$\times \left\{\frac{1 - \cos\left[(t - \tau)\sqrt{k\tanh(kh)}\right]}{k\tanh(kh)}\right\},$$

where $R^2 = r^2 + (z - \zeta)^2$, $R'^2 = r^2 + (z + \zeta)^2$, and $r = |x - \xi|$.

21. Find the harmonic waves in the two-dimensional case by solving

$$\frac{\partial^2 g}{\partial x^2} + \frac{\partial^2 g}{\partial z^2} = -\delta(x - \xi)\delta(z - \zeta),$$

where $-\infty < x, \xi < \infty$, and $-h < z, \zeta < 0$, while, at the free surface,

$$\frac{\partial g}{\partial z} = \omega^2 g, \qquad z = 0,$$

and at the bottom,

$$\frac{\partial g}{\partial z} = 0, \qquad z = -h.$$

Finally, we require that $\lim_{|x| \to \infty} g(x, z|\xi, \zeta) \to 0$.

Step 1: Applying the Fourier transform in the x direction, show that

$$\frac{d^2 G}{dz^2} - k^2 G = -e^{-ik\xi}\delta(z - \zeta), \qquad -h < z, \zeta < 0.$$

Step 2: Solve the ordinary differential equation in Step 1 and show that

$$G(k, z|\xi, \zeta) = \frac{e^{-ik\xi}}{k} \frac{\omega^2 \sinh(kz_>) + k\cosh(kz_>)}{k\sinh(kh) - \omega^2 \cosh(kh)} \cosh[k(z_< + h)].$$

Step 3: Taking the inverse Fourier transform, show that

$$g(x, z|\xi, \zeta) = \frac{1}{\pi} \fint_0^\infty \cosh[k(z_< + h)] \frac{\omega^2 \sinh(kz_>) + k\cosh(kz_>)}{k\sinh(kh) - \omega^2 \cosh(kh)}$$
$$\times \frac{\cos[k(x - \xi)]}{k} \, dk.$$

Our special integral sign denotes an integration along the real k axis except for passing *below* the singularity at $k = \kappa_0$, where $\kappa_0 \tanh(\kappa_0 h) = \omega^2$.

Mei[43] showed that we can rewrite this expression as

$$g(x, z|\xi, \zeta) = \frac{1}{2\pi} \ln\left(\frac{h^2}{rr'}\right) - 2\fint_0^\infty \frac{dk}{k} \left\{ \frac{\cos[k(x - \xi)]}{\omega^2 \cosh(kh) - k\sinh(kh)} \right.$$
$$\left. \times \cosh[k(z + h)] \cosh[k(\zeta + h)] - e^{-kh} \right\},$$

---

[43] Mei, C. C., 1989: *The Applied Dynamics of Ocean Surface Waves*. World Scientific, 740 pp. See pp. 379–382.

where $r^2 = (x - \xi)^2 + (z - \zeta)^2$, and $r'^2 = (x - \xi)^2 + (z + \zeta + 2h)^2$.

Step 4: Using the residue theorem,[44] show that

$$g(x, z | \xi, \zeta) = \sum_{n=0}^{\infty} \frac{\cos[\kappa_n (z + h)] \cos[\kappa_n (\zeta + h)]}{2\kappa_n N_n} e^{-\kappa_n |x - \xi|},$$

where N_n is given by (3.10.48) and κ_n is the nth simple root to $\kappa \tan(\kappa h)$ $= -\omega^2$. Note that the root κ_0 lies on the real k axis. Therefore, we pass *below* the singularity at $k = \kappa_0$ and *above* the singularity at $k = -\kappa_0$.

---

[44] See Appendix B in Dalrymple, R. A., M. A. Losada, and P. A. Martin, 1991: Reflection and transmission from porous structures under oblique wave attack. *J. Fluid Mech.*, **224**, 625–644.

Chapter 4

Green's Functions
for the Heat Equation

In this chapter, we present the Green's function[1] for the heat equation

$$\frac{\partial u}{\partial t} - a^2 \nabla^2 u = q(\mathbf{r}, t), \tag{4.0.1}$$

where ∇ is the three-dimensional gradient operator, t denotes time, $\mathbf{r}$ is the position vector, a^2 is the diffusivity, and $q(\mathbf{r}, t)$ is the source density. In addition to (4.0.1), boundary conditions must be specified to ensure the uniqueness of solution; the most common ones are Dirichlet, Neumann and Robin (a linear combination of the first two). An initial condition $u(\mathbf{r}, t = t_0)$ is also needed.

The heat equation differs in many ways from the wave equation and the Green's function must, of course, manifest these differences. The most notable one is the asymmetry of the heat equation with respect to

---

[1] See also Carslaw, H. S., and J. C. Jaeger, 1959: *Conduction of Heat in Solids.* At the Clarendon Press, Chapter 14; Beck, J. V., K. D. Cole, A. Haji-Sheikh, and B. Litkouhi, 1992: *Heat Conduction Using Green's Functions.* Hemisphere Publishing Corp., 523 pp.; Özişik, M. N., 1993: *Heat Conduction.* John Wiley & Sons, Inc., Chapter 6.

time. This merely reflects the fact that the heat equation differentiates between past and future as entropy continually increases.

The purpose of this introductory section is to prove that we can express the solution to (4.0.1) in terms of boundary conditions, the initial condition and the Green's function, which is found by solving

$$\frac{\partial g}{\partial t} - a^2 \nabla^2 g = \delta(\mathbf{r} - \mathbf{r}_0)\delta(t - \tau), \qquad (4.0.2)$$

where $\mathbf{r}_0$ denotes the position of the source. From causality[2] we know that $g(\mathbf{r}, t | \mathbf{r}_0, \tau) = 0$ if $t < \tau$. We again require that the Green's function g satisfies the homogeneous form of the boundary condition on u. For example, if u satisfies a homogeneous or inhomogeneous Dirichlet condition, then the Green's function will satisfy the corresponding *homogeneous* Dirichlet condition. Although we will focus on the mathematical aspects of the problem, (4.0.2) can be given the physical interpretation of the temperature distribution within a medium when a unit of heat is introduced at $\mathbf{r}_0$ at the time τ.

Remark. Although we will use (4.0.2) as our fundamental definition of the Green's function as it applies to the heat equation, we can also find it by solving the initial-value problem:

$$\frac{\partial u}{\partial t} - a^2 \nabla^2 u = 0, \quad t > \tau, \quad u(\mathbf{r}, \tau) = \delta(\mathbf{r} - \mathbf{r}_0). \qquad (4.0.3)$$

Then $g(\mathbf{r}, t | \mathbf{r}_0, \tau) = u(\mathbf{r}, t - \tau)H(t - \tau)$. This is most easily seen by introducing a new time variable $t' = t - \tau$ into (4.0.2)–(4.0.3) and noting that the Laplace transform of (4.0.2) and (4.0.3) are identical. Therefore, the Green's function gives the heat flow resulting a temperature distribution at $t = \tau$ which is everywhere equal to zero except at $\mathbf{r} = \mathbf{r}_0$, where it is infinite.

As in the case of the wave equation, we begin by studying *reciprocity*. We now show that $g(\mathbf{r}, t | \mathbf{r}_0, \tau) = g(\mathbf{r}_0, -\tau | \mathbf{r}, -t)$. Physically, the function $g(\mathbf{r}_0, -\tau | \mathbf{r}, -t)$ gives the effect at $\mathbf{r}_0$ and time $-\tau$ of a heat source that is introduced into the medium at $\mathbf{r}$ at a time $-t$. Because $\tau < t$, the time sequence is still properly ordered.

Another interpretation follows from considering the *adjoint* function $g^*(\mathbf{r}, t | \mathbf{r}_0, \tau)$ defined by the relationship $g(\mathbf{r}, -t | \mathbf{r}_0, -\tau) = g^*(\mathbf{r}, t | \mathbf{r}_0, \tau)$. This function g^* satisfies the time-reversed equation

$$\frac{\partial g^*}{\partial t} + a^2 \nabla^2 g^* = -\delta(\mathbf{r} - \mathbf{r}_0)\delta(t - \tau), \qquad (4.0.4)$$

---

[2] The principle stating that an event cannot precede its cause.

with $g^*(\mathbf{r}, t|\mathbf{r}_0, \tau) = 0$ if $t > \tau$. In other words, g^* gives the development backward in time of a source placed at $\mathbf{r}_0$ and at the time τ. The reciprocity condition now reads $g(\mathbf{r}, t|\mathbf{r}_0, \tau) = g^*(\mathbf{r}_0, \tau|\mathbf{r}, t)$. The function g describes the evolution as time increases, leading from the initial source to the final distribution. The function g^* describes the same process in reverse time order, beginning with the final distribution and going backward in time to the initial source.

The proof of reciprocity patterns itself after the proof given for the wave equation. The two equations are now

$$a^2 \nabla^2 g(\mathbf{r}, t|\mathbf{r}_0, \tau_0) - \frac{\partial g(\mathbf{r}, t|\mathbf{r}_0, \tau_0)}{\partial t} = -\delta(\mathbf{r} - \mathbf{r}_0)\delta(t - \tau_0), \quad (4.0.5)$$

and

$$a^2 \nabla^2 g(\mathbf{r}, -t|\mathbf{r}_1, -\tau_1) + \frac{\partial g(\mathbf{r}, -t|\mathbf{r}_1, -\tau_1)}{\partial t} = -\delta(\mathbf{r} - \mathbf{r}_1)\delta(t - \tau_1). \quad (4.0.6)$$

Multiplying (4.0.5) by $g(\mathbf{r}, -t|\mathbf{r}_1, -\tau_1)$ and (4.0.6) by $g(\mathbf{r}, t|\mathbf{r}_0, \tau_0)$, subtracting, and integrating over the volume V and over the time t from $-\infty$ to t_0^+, we eventually obtain

$$a^2 \int_{-\infty}^{t_0^+} \oiint_S \left[g(\mathbf{r}, -t|\mathbf{r}_1, -\tau_1) \, \nabla g(\mathbf{r}, t|\mathbf{r}_0, \tau_0) \right.$$

$$\left. - g(\mathbf{r}, t|\mathbf{r}_0, \tau_0) \, \nabla g(\mathbf{r}, -t|\mathbf{r}_1, -\tau_1) \right] \cdot \mathbf{n} \, dS$$

$$- \int_{-\infty}^{t_0^+} \iiint_V \left[g(\mathbf{r}, -t|\mathbf{r}_1, -\tau_1) \frac{\partial g(\mathbf{r}, t|\mathbf{r}_0, \tau_0)}{\partial t} \right.$$

$$\left. + g(\mathbf{r}, t|\mathbf{r}_0, \tau_0) \frac{\partial g(\mathbf{r}, -t|\mathbf{r}_1, -\tau_1)}{\partial t} \right] dV \, dt$$

$$= g(\mathbf{r}_1, \tau_1|\mathbf{r}_0, t_0) - g(\mathbf{r}_0, -\tau_0|\mathbf{r}_1, -\tau_1), \quad (4.0.7)$$

where S is the surface enclosing V. The first integral vanishes due to the homogeneous boundary conditions satisfied by g. In the second integral, we perform the time integration and obtain

$$g(\mathbf{r}, -t|\mathbf{r}_1, -\tau_1)g(\mathbf{r}, t|\mathbf{r}_0, t_0)\big|_{t=-\infty}^{t_0^+} .$$

At the lower limit, the second of the two factors vanishes because of the initial condition. At the upper limit, the first factor vanishes also because of the initial condition. Note that we tacitly assumed that τ_1 is within the region of integration. The reciprocity condition now follows directly.

We now establish that the solution to the nonhomogeneous heat equation can be expressed in terms of the Green's function, boundary conditions and the initial condition. We begin with the equations

$$a^2 \nabla_0^2 u(\mathbf{r}_0, t_0) - \frac{\partial u(\mathbf{r}_0, t_0)}{\partial t_0} = -q(\mathbf{r}_0, t_0), \qquad (4.0.8)$$

and

$$a^2 \nabla_0^2 g(\mathbf{r}, t | \mathbf{r}_0, t_0) + \frac{\partial g(\mathbf{r}, t | \mathbf{r}_0, t_0)}{\partial t_0} = -\delta(\mathbf{r} - \mathbf{r}_0)\delta(t - t_0). \qquad (4.0.9)$$

As we did in the previous chapter, we multiply (4.0.8) by g and (4.0.9) by u and subtract. Integrating over the volume V_0 and over t_0 from 0 to t^+, where t^+ denotes a time slightly later than t so that we avoid ending the integration exactly at the peak of the delta function, we find

$$a^2 \int_0^{t^+} \iiint_{V_0} \left[u \, \nabla_0^2 g - g \, \nabla_0^2 u \right] dV_0 \, dt_0$$

$$+ \int_0^{t^+} \iiint_{V_0} \left[u \left(\frac{\partial g}{\partial t_0} \right) + g \left(\frac{\partial u}{\partial t_0} \right) \right] dV_0 \, dt_0$$

$$= \int_0^{t^+} \iiint_{V_0} q(\mathbf{r}_0, t_0) \, g(\mathbf{r}, t | \mathbf{r}_0, t) \, dV_0 \, dt_0 - u(\mathbf{r}, t). \, (4.0.10)$$

Applying Green's second formula to the first integral of (4.0.10) and performing the time integration in the second integral, we finally obtain

$$u(\mathbf{r}, t) = \int_0^{t^+} \iiint_{V_0} q(\mathbf{r}_0, t_0) \, g(\mathbf{r}, t | \mathbf{r}_0, t_0) \, dV_0 \, dt_0$$

$$+ a^2 \int_0^{t^+} \oiint_{S_0} \left[g(\mathbf{r}, t | \mathbf{r}_0, t_0) \, \nabla_0 u(\mathbf{r}_0, t_0) \right.$$

$$\left. - u(\mathbf{r}_0, t_0) \, \nabla_0 g(\mathbf{r}, t | \mathbf{r}_0, t_0) \right] \cdot \mathbf{n} \, dS_0 \, dt_0$$

$$+ \iiint_{V_0} u(\mathbf{r}_0, 0) g(\mathbf{r}, t | \mathbf{r}_0, 0) \, dV_0, \qquad (4.0.11)$$

where we used $g(\mathbf{r}, t | \mathbf{r}_0, t^+) = 0$. The first two terms in (4.0.11) represent the familiar effects of volume sources and boundary conditions, while the third term includes the effects of the initial data.

4.1 HEAT EQUATION OVER INFINITE OR SEMI-INFINITE DOMAINS

The Green's function for the one-dimensional heat equation is governed by

$$\frac{\partial g}{\partial t} - a^2 \frac{\partial^2 g}{\partial x^2} = \delta(x - \xi)\delta(t - \tau), \quad -\infty < x, \xi < \infty, \quad 0 < t, \tau, \quad (\mathbf{4.1.1})$$

subject to the boundary conditions $\lim_{|x|\to\infty} |g(x, t|\xi, \tau)| < \infty$ and the initial condition $g(x, 0|\xi, \tau) = 0$. Let us find $g(x, t|\xi, \tau)$.

We begin by taking the Laplace transform of (4.1.1) and find that

$$\frac{d^2 G}{dx^2} - \frac{s}{a^2} G = -\frac{\delta(x - \xi)}{a^2} e^{-s\tau}. \quad (\mathbf{4.1.2})$$

Next, we take the Fourier transform of (4.1.2) so that

$$(k^2 + b^2)\overline{G}(k, s|\xi, \tau) = \frac{e^{-ik\xi} e^{-s\tau}}{a^2}, \quad (\mathbf{4.1.3})$$

where $\overline{G}(k, s|\xi, \tau)$ is the Fourier transform of $G(x, s|\xi, \tau)$ and $b^2 = s/a^2$.

To find $G(x, s|\xi, \tau)$, we use the inversion integral

$$G(x, s|\xi, \tau) = \frac{e^{-s\tau}}{2\pi a^2} \int_{-\infty}^{\infty} \frac{e^{i(x-\xi)k}}{k^2 + b^2} \, dk. \quad (\mathbf{4.1.4})$$

Transforming (4.1.4) into a closed contour via Jordan's lemma, we evaluate it by the residue theorem and find that

$$G(x, s|\xi, \tau) = \frac{e^{-|x-\xi|\sqrt{s}/a - s\tau}}{2a\sqrt{s}}. \quad (\mathbf{4.1.5})$$

From a table of Laplace transforms we finally obtain

$$g(x, t|\xi, \tau) = \frac{H(t - \tau)}{\sqrt{4\pi a^2(t - \tau)}} \exp\left[-\frac{(x - \xi)^2}{4a^2(t - \tau)}\right], \quad (\mathbf{4.1.6})$$

after applying the second shifting theorem.

In the same manner, the Green's function for

$$\frac{\partial g}{\partial t} - \frac{a^2}{r} \frac{\partial}{\partial r}\left(r\frac{\partial g}{\partial r}\right) = \frac{\delta(r - \rho)\delta(t - \tau)}{2\pi r}, \quad 0 < r, \rho < \infty, \quad 0 < t, \tau, \quad (\mathbf{4.1.7})$$

Table 4.1.1: Free-Space Green's Function in Various Coordinate Systems for the Heat Equation

Two-Dimensional

$$g(x, y, t | \xi, \eta, \tau) = \frac{H(t - \tau)}{4\pi a^2(t - \tau)} \exp\left[-\frac{(x - \xi)^2 + (y - \eta)^2}{4a^2(t - \tau)}\right]$$

$$g(r, \theta, t | \rho, \theta', \tau) = \frac{H(t - \tau)}{4\pi a^2(t - \tau)} \exp\left[-\frac{r^2 + \rho^2 - 2r\rho\cos(\theta - \theta')}{4a^2(t - \tau)}\right]$$

$$g(r, z, t | \rho, \zeta, \tau) = \frac{H(t - \tau)}{8\left[\pi a^2(t - \tau)\right]^{3/2}} I_0\left[\frac{r\rho}{2a^2(t - \tau)}\right]$$

$$\times \exp\left[-\frac{r^2 + \rho^2 + (z - \zeta)^2}{4a^2(t - \tau)}\right]$$

Three-Dimensional

$$g(x, y, z, t | \xi, \eta, \zeta, \tau) = \frac{H(t - \tau)}{\left[4\pi a^2(t - \tau)\right]^{3/2}}$$

$$\times \exp\left[-\frac{(x - \xi)^2 + (y - \eta)^2 + (z - \zeta)^2}{4a^2(t - \tau)}\right]$$

$$g(r, \theta, z, t | \rho, \theta', \zeta, \tau) = \frac{H(t - \tau)}{\left[4\pi a^2(t - \tau)\right]^{3/2}}$$

$$\times \exp\left[-\frac{r^2 + \rho^2 - 2r\rho\cos(\theta - \theta') + (z - \zeta)^2}{4a^2(t - \tau)}\right]$$

is

$$g(r, t | \rho, \tau) = \frac{H(t - \tau)}{2\pi} \int_0^\infty J_0(kr) J_0(k\rho) e^{-k^2 a^2(t - \tau)} k \, dk. \qquad \textbf{(4.1.8)}$$

The primary use of the fundamental or free-space Green's function[3] is as a *particular* solution to the Green's function problem. For this rea-

---

[3] In electromagnetic theory, a free-space Green's function is the particular solution of the differential equation valid over a domain of infinite extent, where the Green's function remains bounded as we approach infinity, or satisfies a radiation condition there.

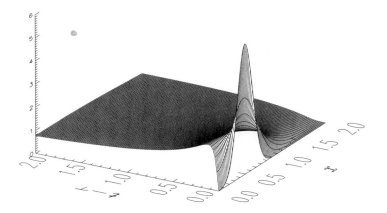

Figure 4.1.1: The Green's function (4.1.13) for the one-dimensional heat equation on the semi-infinite domain $0 < x < \infty$, and $0.01 \leq t - \tau$ when the left boundary condition is $g_x(0, t|\xi, \tau) = 0$, $\xi = 0.5$, and $a^2 = 1$.

son, it is often called the *fundamental heat conduction solution*. Consequently, we usually must find a homogeneous solution so that the sum of the free-space Green's function plus the homogeneous solution satisfies any boundary conditions. The following examples show some commonly employed techniques.

- **Example 4.1.1**

Let us find the Green's function for the following problem:

$$\frac{\partial g}{\partial t} - a^2 \frac{\partial^2 g}{\partial x^2} = \delta(x - \xi)\delta(t - \tau), \quad 0 < x, \xi < \infty, \quad 0 < t, \tau, \quad (\textbf{4.1.9})$$

subject to the boundary conditions $g(0, t|\xi, \tau) = 0$, $\lim_{x \to \infty} |g(x, t|\xi, \tau)| < \infty$, and the initial condition $g(x, 0|\xi, \tau) = 0$. From the boundary condition $g(0, t|\xi, \tau) = 0$, we deduce that $g(x, t|\xi, \tau)$ must be an odd function in x over the open interval $(-\infty, \infty)$. We find this Green's function by introducing an image source of $-\delta(x + \xi)$ and resolving (4.1.1) with the source $\delta(x - \xi)\delta(t - \tau) - \delta(x + \xi)\delta(t - \tau)$. Because (4.1.1) is linear, (4.1.6) gives the solution for each delta function and the Green's function for (4.1.9) can be written

$$g(x, t|\xi, \tau) = \frac{H(t - \tau)}{\sqrt{4\pi a^2(t - \tau)}} \left\{ \exp\left[-\frac{(x - \xi)^2}{4a^2(t - \tau)}\right] - \exp\left[-\frac{(x + \xi)^2}{4a^2(t - \tau)}\right] \right\}$$

$$(\textbf{4.1.10})$$

$$= \frac{H(t - \tau)}{\sqrt{\pi a^2(t - \tau)}} \exp\left[-\frac{x^2 + \xi^2}{4a^2(t - \tau)}\right] \sinh\left[\frac{x\xi}{2a^2(t - \tau)}\right].$$

$$(\textbf{4.1.11})$$

In a similar manner, if the boundary condition at $x = 0$ changes to $g_x(0, t|\xi, \tau) = 0$, then (4.1.10)–(4.1.11) become

$$g(x, t|\xi, \tau) = \frac{H(t - \tau)}{\sqrt{4\pi a^2(t - \tau)}} \left\{ \exp\left[-\frac{(x - \xi)^2}{4a^2(t - \tau)}\right] + \exp\left[-\frac{(x + \xi)^2}{4a^2(t - \tau)}\right] \right\}$$

$$(4.1.12)$$

$$= \frac{H(t - \tau)}{\sqrt{\pi a^2(t - \tau)}} \exp\left[-\frac{x^2 + \xi^2}{4a^2(t - \tau)}\right] \cosh\left[\frac{x\xi}{2a^2(t - \tau)}\right].$$

$$(4.1.13)$$

Equation (4.1.13) has been graphed in Figure 4.1.1 for the special case when $a^2 = 1$.

● **Example 4.1.2**

Several years ago Carslaw and Jaeger[4] showed how free-space Green's functions and Laplace transforms may be used to derive additional Green's functions. To illustrate this technique, consider the following problem:

$$\frac{\partial g}{\partial t} - a^2 \frac{\partial^2 g}{\partial x^2} = \delta(x - \xi)\delta(t - \tau), \qquad 0 < x, t, \xi, \tau, \qquad (4.1.14)$$

subject to the boundary conditions that

$$-g_x(0, t|\xi, \tau) + hg(0, t|\xi, \tau) = 0, \qquad 0 < t, \qquad (4.1.15)$$

and

$$\lim_{x \to \infty} |g(x, t|\xi, \tau)| < \infty, \qquad 0 < t, \qquad (4.1.16)$$

and the initial condition that

$$g(x, 0|\xi, \tau) = 0, \qquad 0 < x. \qquad (4.1.17)$$

Because this problem is linear, we can write the solution as

$$g(x, t|\xi, \tau) = \frac{e^{-(x-\xi)^2/[4a^2(t-\tau)]}}{2a\sqrt{\pi(t - \tau)}} H(t - \tau) + v(x, t), \qquad (4.1.18)$$

---

[4] The results in this example were first derived by Bryan, G. H., 1891: Note on a problem in the linear conduction of heat. *Proc. Camb. Phil. Soc.*, **7**, 246–248. The present derivation is due to Carslaw, H. S., and J. C. Jaeger, 1938: Some problems in the mathematical theory of the conduction of heat. *Philos. Mag.*, *Ser. 7*, **26**, 473–495. See §4. Reproduced by permission of Taylor & Francis, Ltd., http://www.tandf.co.uk/journals.

where the first term on the right side of (4.1.18) is the free-space Green's function (4.1.6). Since this Green's function is a nonhomogeneous solution of (4.1.14), $v(x,t)$ needs only satisfy the homogeneous problem

$$\frac{\partial v}{\partial t} = a^2 \frac{\partial^2 v}{\partial x^2}, \qquad 0 < x, t, \tag{4.1.19}$$

with the boundary conditions that

$$-v_x(0,t) + hv(0,t) = \frac{\partial}{\partial x}\left\{\frac{e^{-(x-\xi)^2/[4a^2(t-\tau)]}}{2a\sqrt{\pi(t-\tau)}}\right\}\Bigg|_{x=0} H(t-\tau)$$
$$- h\frac{e^{-\xi^2/[4a^2(t-\tau)]}}{2a\sqrt{\pi(t-\tau)}} H(t-\tau), \ 0 < t, \tag{4.1.20}$$

and

$$\lim_{x\to\infty} |v(x,t)| < \infty, \qquad 0 < t, \tag{4.1.21}$$

and the initial condition that

$$v(x,0) = 0, \qquad 0 < x. \tag{4.1.22}$$

To solve (4.1.19)–(4.1.22), we now employ Laplace transforms and obtain the ordinary differential equation

$$\frac{d^2 V}{dx^2} - q^2 V = 0, \qquad 0 < x, \tag{4.1.23}$$

with the boundary conditions that

$$-V'(0,s) + hV(0,s) = \frac{q-h}{2a\sqrt{s}} e^{-q\xi - s\tau}, \tag{4.1.24}$$

and

$$\lim_{x\to\infty} |V(x,s)| < \infty, \tag{4.1.25}$$

where $q^2 = s/a^2$. The solution to this ordinary differential equation is

$$V(x,s) = \frac{q-h}{q+h} \frac{e^{-q(x+\xi)-s\tau}}{2a\sqrt{s}}, \tag{4.1.26}$$

so that

$$G(x,s|\xi,\tau) = \frac{e^{-s\tau}}{2a\sqrt{s}}\left[e^{-q|x-\xi|} + e^{-q(x+\xi)} - \frac{2h}{q+h}e^{-q(x+\xi)}\right]. \tag{4.1.27}$$

Our final task is to invert (4.1.27) term by term. We find that

$$g(x,t|\xi,\tau) = \frac{H(t-\tau)}{2a\sqrt{\pi(t-\tau)}}\left\{e^{-(x-\xi)^2/[4a^2(t-\tau)]} + e^{-(x+\xi)^2/[4a^2(t-\tau)]}\right.$$

$$\left. - 2h\int_0^\infty e^{-h\eta-(x+\xi+\eta)^2/[4a^2(t-\tau)]}\,d\eta\right\}.$$

$$(4.1.28)$$

The first two terms on the right side of (4.1.28) are straightforward inversions from the tables; the third term follows from

$$\mathcal{L}^{-1}\left[\frac{he^{-q(x+\xi)}}{2a\sqrt{s}\,(q+h)}\right] = \frac{1}{2a}\int_0^\infty \left[1 - e^{-h(a\eta-x-\xi)}\right] H\,(a\eta - x - \xi)$$

$$\times \frac{\eta}{2\sqrt{\pi t^3}}e^{-\eta^2/(4t)}\,d\eta \qquad (4.1.29)$$

$$= \frac{1}{2a}\int_0^\infty (1 - e^{-h\tau})\frac{\tau+x+\xi}{2a^2t\sqrt{\pi t}}e^{-(\tau+x+\xi)^2/(4a^2t)}\,d\tau$$

$$(4.1.30)$$

$$= \frac{1}{2a\sqrt{\pi t}}\int_0^\infty \frac{\tau+x+\xi}{2a^2t}e^{-(\tau+x+\xi)^2/(4a^2t)}\,d\tau$$

$$- \frac{1}{2a\sqrt{\pi t}}\int_0^\infty \frac{\tau+x+\xi}{2a^2t}e^{-h\tau-(\tau+x+\xi)^2/(4a^2t)}\,d\tau$$

$$(4.1.31)$$

$$= \frac{h}{2a\sqrt{\pi t}}\int_0^\infty e^{-h\tau-(\tau+x+\xi)^2/(4a^2t)}\,d\tau, \qquad (4.1.32)$$

since

$$\mathcal{L}^{-1}\left[F(\sqrt{s})\right] = \int_0^\infty f(\eta)\frac{\eta}{2\sqrt{\pi t^3}}e^{-\eta^2/(4t)}\,d\eta, \qquad (4.1.33)$$

where $\mathcal{L}[f(t)] = F(s)$. Equation (4.1.28) is plotted in Figure 4.1.2 as functions of x and $t - \tau$ when $h = 1$. The effect of introducing the radiative boundary condition can be seen by comparing it with Figure 4.1.1, which gives the solution when $h = 0$.

Further insight is gained by rewriting (4.1.28) in terms of tabulated functions. After some work, it transforms into

$$g(x,t|\xi,\tau) = \frac{H(t-\tau)}{2a\sqrt{\pi(t-\tau)}}\left\{e^{-(x-\xi)^2/[4a^2(t-\tau)]} + e^{-(x+\xi)^2/[4a^2(t-\tau)]}\right\}$$

$$- |h|e^{h(x+\xi)+a^2h^2(t-\tau)}\mathrm{erfc}\left[a|h|\sqrt{t-\tau} + \frac{h(x+\xi)}{2a|h|\sqrt{t-\tau}}\right]$$

$$- 2hH(-h)e^{h(x+\xi)+a^2h^2(t-\tau)}, \qquad (4.1.34)$$

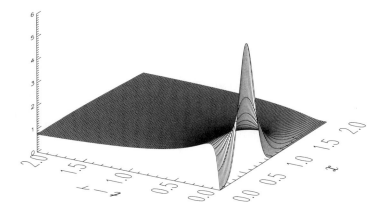

Figure 4.1.2: The Green's function for the one-dimensional heat equation on the semi-infinite domain $0 < x < \infty$, and $0.01 \leq t - \tau$ when the left boundary condition is $g_x(0, t|\xi, \tau) - hg(0, t|\xi, \tau) = 0$, $\xi = 0.5$, and $h = 1$.

if $h \neq 0$. For $h = 0$, we already have (4.1.12)–(4.1.13).

The intriguing aspect of (4.1.34) is the appearance of an extra term when $h < 0$. In this case, we have a discrete contribution to the spectrum. Bressloff[5] explained this phenomenon in terms of the time dependence of the spectral data.

• **Example 4.1.3**

Let us resolve the previous example taking a different approach. In particular, if $a^2 = h = 1$, we will find the Green's function for the heat equation

$$\frac{\partial g}{\partial t} - \frac{\partial^2 g}{\partial x^2} = \delta(x - \xi)\delta(t - \tau), \qquad 0 < x, \xi, \quad 0 < t, \tau, \qquad (4.1.35)$$

subject to the boundary conditions that

$$g_x(0, t|\xi, \tau) = g(0, t|\xi, \tau), \qquad 0 < t, \qquad (4.1.36)$$

and

$$\lim_{x \to \infty} |g(x, t|\xi, \tau)| < \infty, \qquad 0 < t, \qquad (4.1.37)$$

and the initial condition that

$$g(x, 0|\xi, \tau) = 0, \qquad 0 < x. \qquad (4.1.38)$$

[5] Bressloff, P. C., 1997: A new Green's function method for solving linear PDE's in two variables. *J. Math. Anal. Appl.*, **210**, 390–415.

We begin by considering the initial-value problem[6]

$$\frac{\partial u}{\partial t} = \frac{\partial^2 u}{\partial x^2} \qquad 0 < x, t, \tag{4.1.39}$$

subject to the boundary conditions that

$$u_x(0, t) = u(0, t), \qquad 0 < t, \tag{4.1.40}$$

and

$$\lim_{x \to \infty} |u(x, t)| < \infty, \qquad 0 < t, \tag{4.1.41}$$

and the initial condition

$$u(x, 0) = \delta(x - \xi), \qquad 0 < x. \tag{4.1.42}$$

As we showed in our introductory remarks, (4.1.39)–(4.1.42) is equivalent to the Green's function problem (4.1.35)–(4.1.38) if $\tau = 0$.

To solve (4.1.39)–(4.1.42), we introduce the new dependent variable $h(x, t) = u(x, t) - u_x(x, t)$, or

$$u(x, t) = e^x \int_x^\infty h(x', t) e^{-x'} \, dx'. \tag{4.1.43}$$

Then, (4.1.39)–(4.1.42) simplify to

$$\frac{\partial h}{\partial t} = \frac{\partial^2 h}{\partial x^2} \qquad 0 < x, t, \tag{4.1.44}$$

subject to the boundary conditions

$$h(0, t) = 0, \qquad 0 < t, \tag{4.1.45}$$

and

$$\lim_{x \to \infty} |h(x, t)| < \infty, \qquad 0 < t, \tag{4.1.46}$$

and the initial condition

$$h(x, 0) = \delta(x - \xi) - \delta'(x - \xi), \qquad 0 < x. \tag{4.1.47}$$

Why did we introduce $h(x, t)$? The main advantage is the simpler boundary condition (4.1.45) compared to (4.1.40); the resulting Green's function problem is easier to solve.

---

[6] Reprinted with permission from Nadler, W., and D. L. Stein, 1996: Reaction-diffusion description of biological transport processes in general dimension. *J. Chem. Phys.*, **104**, 1918–1936. ©American Institute of Physics, 1996.

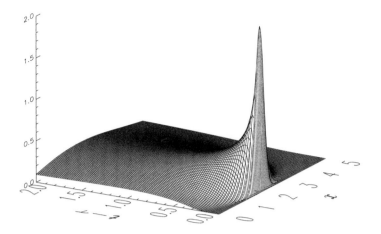

Figure 4.1.3: Same as Figure 4.1.2 except that $\xi = 2$.

We could solve (4.1.44)–(4.1.47) by Laplace transforms. As an alternative method, let us replace the right side of (4.1.47) with $\delta(x - \xi)$. We denote the Green's function to this new system by $y(x, t|\xi, 0)$. From the method of images we can immediately write down

$$y(x, t|\xi, 0) = \frac{1}{\sqrt{4\pi t}} \left[e^{-(x-\xi)^2/(4t)} - e^{-(x+\xi)^2/(4t)} \right]. \qquad (4.1.48)$$

Why did we introduce $y(x, t|\xi, 0)$? From (4.0.11), we have that

$$h(x, t) = \int_0^\infty y(x, t|\xi, 0) h(\xi, 0)\, d\xi \qquad (4.1.49)$$

$$= \int_0^\infty \delta(x - \xi) \left[y(x, t|\xi, 0) + \frac{\partial y(x, t|\xi, 0)}{\partial \xi} \right] d\xi \qquad (4.1.50)$$

$$= \frac{1}{\sqrt{4\pi t}} \left[e^{-(x-\xi)^2/(4t)} - e^{-(x+\xi)^2/(4t)} \right]$$
$$+ \frac{1}{\sqrt{\pi t}} \left[(x - \xi) e^{-(x-\xi)^2/(4t)} + (x + \xi) e^{-(x+\xi)^2/(4t)} \right]. \qquad (4.1.51)$$

After substituting (4.1.51) into (4.1.43), we obtain

$$u(x, t) = \frac{1}{\sqrt{4\pi t}} \left\{ \exp\left[-\frac{(x - \xi)^2}{4t} \right] + \exp\left[-\frac{(x + \xi)^2}{4t} \right] \right\}$$
$$- e^{x+\xi+t} \mathrm{erfc}\left[\frac{x + \xi + 2t}{2\sqrt{t}} \right], \qquad (4.1.52)$$

where erfc is the complementary error function. The point here is to show that the introduction of transformations such as (4.1.43) allows the construction of solutions to problems with complicated boundary conditions using the solutions from problems with simpler boundary conditions.

To solve our original problem (4.1.35)–(4.1.38), we merely shift the time axis and obtain

$$
g(x,t|\xi,\tau) = \frac{H(t-\tau)}{\sqrt{4\pi(t-\tau)}}\left\{\exp\left[-\frac{(x-\xi)^2}{4(t-\tau)}\right] + \exp\left[-\frac{(x+\xi)^2}{4(t-\tau)}\right]\right\}
$$
$$
- H(t-\tau)e^{x+\xi+t-\tau}\mathrm{erfc}\left[\frac{x+\xi+2(t-\tau)}{2\sqrt{t-\tau}}\right]. \qquad (4.1.53)
$$

Equation (4.1.53) has been graphed in Figure 4.1.3 with $\xi = 2$.

• Example 4.1.4: Moving boundaries

In this example, we find the Green's function for a semi-infinite domain where the left boundary moves as $x = vt$, where v is constant. Although we will only work through a problem[7] when Dirichlet conditions are present, similar considerations come into play when Neumann or Robin conditions prevail and/or the domain extends from $vt < x < L + vt$. Similar considerations[8] hold in the case when the boundaries move as $x = \alpha t \pm \beta t^2$.

Consider the Green's function problem

$$
\frac{\partial g}{\partial t} - a^2\frac{\partial^2 g}{\partial x^2} = \delta(x-\xi)\delta(t-\tau), \quad vt < x,\xi, \quad \tau < t, \qquad (4.1.54)
$$

subject to the boundary conditions that

$$
g(x,t|\xi,\tau)|_{x=vt} = 0, \qquad \tau < t, \qquad (4.1.55)
$$

and

$$
\lim_{x\to\infty}|g(x,t|\xi,\tau)| < \infty, \qquad \tau < t, \qquad (4.1.56)
$$

and the initial condition that

$$
g(x,0|\xi,\tau) = 0, \qquad 0 < x. \qquad (4.1.57)
$$

We begin by writing the Green's function as a linear combination of the free-space Green's function plus a presently unknown function

---

[7] Taken with permission from Kartashov, É. M., and G. M. Bartenev, 1969: Integral-equation construction of the Green's function for generalized boundary-value problems involving the heat-conduction equation. *Sov. Phys. J.*, **12**, 189–198. Published by Plenum Publishers.

[8] Kartashov, É. M., B. Ya. Lyubov, and G. M. Bartenev, 1970: A diffusion problem in a region with a moving boundary. *Sov. Phys. J.*, **13**, 1641–1647.

$u(x, t)$. The purpose of the free-space Green's function is to eliminate the delta functions in (4.1.54) while $u(x, t)$ is a homogeneous solution that was introduced so that $g(x, t|\xi, \tau)$ satisfies the boundary conditions. Therefore,

$$g(x, t|\xi, \tau) = \frac{e^{-(x-\xi)^2/[4a^2(t-\tau)]}}{2a\sqrt{\pi(t-\tau)}} H(t - \tau) + u(x, t). \qquad (4.1.58)$$

Substituting (4.1.58) into (4.1.54), we find that

$$\frac{\partial u}{\partial t} = a^2 \frac{\partial^2 u}{\partial x^2}, \qquad vt < x, \quad \tau < t, \qquad (4.1.59)$$

with the boundary conditions that

$$u(x, t)|_{x=vt} = -\frac{e^{-(vt-\xi)^2/[4a^2(t-\tau)]}}{2a\sqrt{\pi(t-\tau)}}, \qquad \tau < t, \qquad (4.1.60)$$

and

$$\lim_{x \to \infty} |u(x, t)| < \infty, \qquad \tau < t, \qquad (4.1.61)$$

and the initial condition that

$$u(x, 0) = 0, \qquad 0 < x. \qquad (4.1.62)$$

To eliminate the moving boundary, we introduce the new independent variables $z = x - vt$ and $t' = t - \tau$, and the dependent variable

$$u(x, t) = \exp\left[-\frac{vz}{2a^2} - \frac{v^2 t'}{4a^2}\right] w(z, t'). \qquad (4.1.63)$$

Substituting (1.4.63) into (1.4.59)–(1.4.62), we obtain

$$\frac{\partial w}{\partial t'} = a^2 \frac{\partial^2 w}{\partial z^2}, \qquad 0 < z, t', \qquad (4.1.64)$$

with the boundary conditions

$$w(0, t') = -\frac{1}{2a\sqrt{\pi t'}} \exp\left[\frac{v\xi_0}{2a^2} - \frac{\xi_0^2}{4a^2 t'}\right], \qquad 0 < t', \qquad (4.1.65)$$

and

$$\lim_{z \to \infty} |w(z, t')| < \infty, \qquad 0 < t', \qquad (4.1.66)$$

and the initial condition

$$w(z, 0) = 0, \qquad 0 < z, \qquad (4.1.67)$$

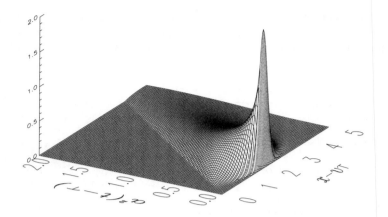

Figure 4.1.4: The Green's function (4.1.58) for the one-dimensional heat equation where a Dirichlet boundary condition applies along the left boundary which moves as $x = vt$. The parameters are $\xi - v\tau = 2$ and $a^2/v = 0.5$.

where $\xi_0 = \xi - v\tau$.

To solve (4.1.64)–(4.1.67), we take their Laplace transform and find

$$a^2 \frac{d^2 W}{dz^2} - sW = 0, \tag{4.1.68}$$

with

$$W(0,s) = -\frac{1}{2a\sqrt{s}} \exp\left(-\frac{\xi_0 \sqrt{s}}{a} + \frac{v\xi_0}{2a^2}\right), \tag{4.1.69}$$

and

$$\lim_{z \to \infty} |W(z,s)| < \infty. \tag{4.1.70}$$

The solution to (4.1.68)–(4.1.70) is

$$W(z,s) = -\frac{1}{2a\sqrt{s}} \exp\left[-\frac{(z+\xi_0)\sqrt{s}}{a} + \frac{v\xi_0}{2a^2}\right], \tag{4.1.71}$$

or

$$w(z,t') = -\frac{H(t')}{2a\sqrt{\pi t'}} \exp\left[-\frac{(z+\xi_0)^2}{4a^2 t'} + \frac{v\xi_0}{2a^2}\right], \tag{4.1.72}$$

so that

$$u(x,t) = -\frac{H(t-\tau)}{2a\sqrt{\pi(t-\tau)}} \exp\left[-\frac{(x+\xi - 2v\tau)^2}{4a^2(t-\tau)} + \frac{v(\xi - v\tau)}{a^2}\right]. \tag{4.1.73}$$

The Green's function is given by substituting (4.1.73) into (4.1.58).

• **Example 4.1.5**

Let us find the Green's function for the parabolic equation:[9]

$$\frac{\partial g}{\partial t} - \frac{\partial}{\partial x}\left(x^{2-a}\frac{\partial g}{\partial x}\right) = \delta(x-\xi)\delta(t-\tau), \quad 0 < x, \xi < \infty, \quad 0 < t, \tau,$$

$$(4.1.74)$$

where the solution remains finite over the entire interval and initially $g(x, 0|\xi, \tau) = 0$.

We begin by taking the Laplace transform of (4.1.74) or

$$\frac{d}{dx}\left(x^{2-a}\frac{dG}{dx}\right) - sG = -\delta(x-\xi)e^{-s\tau}, \quad 0 < x < \infty. \quad (4.1.75)$$

Assuming for the moment that $a \neq 0$, we solve (4.1.75) by introducing new independent variables

$$y = \frac{2\sqrt{s}}{|a|}x^{a/2}, \quad \text{and} \quad \eta = \frac{2\sqrt{s}}{|a|}\xi^{a/2}, \quad (4.1.76)$$

and the dependent variable

$$G(x, s|\xi, \tau) = \frac{2(x\xi)^{(a-1)/2}}{|a|}F(y, \eta)e^{-s\tau}. \quad (4.1.77)$$

These new variables transform (4.1.75) into

$$y^2\frac{d^2F}{dy^2} + y\frac{dF}{dy} - (\mu^2 + y^2)F = -\eta\delta(y-\eta), \quad (4.1.78)$$

where $\mu = 1 - 1/a$.

The solution to (4.1.78) is

$$F(y, \eta) = \begin{cases} AI_{|\mu|}(y), & y \leq \eta, \\ BK_{|\mu|}(y), & y \geq \eta. \end{cases} \quad (4.1.79)$$

Applying the condition that the Green's function must be continuous across $y = \eta$ and the jump condition

$$\eta^2\frac{dF}{dy}\bigg|_{y=\eta^-}^{y=\eta^+} = -\eta, \quad (4.1.80)$$

---

[9] A portion of this was first solved by Becker, P. A., 1992: First-order Fermi acceleration in spherically symmetric flows: Solutions including quadratic losses. *Astrophys. J.*, **397**, 88–116. Published by The University of Chicago Press.

we find

$$F(y, \eta) = I_{|\mu|}(y_<) K_{|\mu|}(y_>).\qquad (4.1.81)$$

Of course, (4.1.81) does not hold if $a = 0$. In that case, the solution to (4.1.75) is

$$G(x, s|\xi, \tau) = \begin{cases} A x^{(-1+\sqrt{1+4s})/2}, & x \leq \xi, \\ B x^{(-1-\sqrt{1+4s})/2}, & x \geq \xi. \end{cases}\qquad (4.1.82)$$

The values of A and B are found by again requiring that the Green's function is continuous at $x = \xi$ and satisfies the jump condition that

$$\xi^2 \frac{dG}{dx}\bigg|_{x=\xi^-}^{x=\xi^+} = -e^{-s\tau},\qquad (4.1.83)$$

or

$$G(x, s|\xi, \tau) = \frac{(x\xi)^{-1/2} e^{-s\tau}}{2\sqrt{s + \frac{1}{2}}} \left(\frac{x_<}{x_>}\right)^{\sqrt{s+\frac{1}{2}}}.\qquad (4.1.84)$$

The final step requires the inversion of (4.1.77) and (4.1.84). Applying tables,[10] we find

$$g(x, t|\xi, \tau) = \frac{(x\xi)^{(a-1)/2}}{|a|(t-\tau)} \exp\left[-\frac{x^a + \xi^a}{a^2(t-\tau)}\right] I_{|\mu|}\left[\frac{2(x\xi)^{a/2}}{a^2(t-\tau)}\right] H(t-\tau),\qquad (4.1.85)$$

if $a \neq 0$. To invert (4.1.84), we note that

$$G(x, s|\xi, \tau) = \frac{(x\xi)^{-1/2} e^{-s\tau - \ln(x_>/x_<)\sqrt{s+\frac{1}{2}}}}{2\sqrt{s + \frac{1}{2}}}.\qquad (4.1.86)$$

Employing the first and second shifting theorems,

$$g(x, t|\xi, \tau) = \frac{(x\xi)^{-1/2} H(t-\tau)}{2\sqrt{\pi(t-\tau)}} \exp\left[-\frac{\ln^2(x_>/x_<)}{4(t-\tau)} - \frac{t-\tau}{2}\right],\qquad (4.1.87)$$

if $a = 0$.

---

[10] Erdélyi, A., W. Magnus, F. Oberhettinger, and F. G. Tricomi, 1954: *Tables of Integral Transforms, Vol. I.* McGraw-Hill Co., §5.16, formula 56.

• Example 4.1.6

In his study of diffusion-controlled reactions where there is loss of population due to desorption, Agmon[11] found the Green's function for the following problem:

$$\frac{\partial g}{\partial t} - \frac{\partial^2 g}{\partial x^2} - 2c\frac{\partial g}{\partial x} = \delta(x - \xi)\delta(t - \tau), \quad 0 < x, t, \xi, \tau, \quad (4.1.88)$$

subject to the boundary conditions

$$g(0, t|\xi, \tau) = \kappa \int_0^t [g_x(0, t'|\xi, \tau) + 2c\, g(0, t'|\xi, \tau)]\, dt', \quad (4.1.89)$$

and

$$\lim_{x \to \infty} |g(x, t|\xi, \tau)| < \infty, \quad (4.1.90)$$

and the initial condition

$$g(x, 0|\xi, \tau) = 0, \quad 0 < x < \infty. \quad (4.1.91)$$

The interesting aspect of this problem is the integral condition that appears in the boundary condition at $x = 0$.

We begin by introducing the intermediate dependent variable $\varphi(x, t| \xi, \tau)$ such that $g(x, t|\xi, \tau) = e^{-c(x - \xi + ct - c\tau)}\varphi(x, t|\xi, \tau)$. Substitution into (4.1.88)–(4.1.91) results in the heat equation

$$\frac{\partial \varphi}{\partial t} - \frac{\partial^2 \varphi}{\partial x^2} = \delta(x - \xi)\delta(t - \tau), \quad 0 < x, t, \xi, \tau, \quad (4.1.92)$$

subject to the boundary conditions

$$e^{-c^2(t-\tau)}\varphi(0, t|\xi, \tau) = \kappa \int_0^t [\varphi_x(0, t'|\xi, \tau) + c\,\varphi(0, t'|\xi, \tau)]\, e^{-c^2(t'-\tau)}\, dt', \quad (4.1.93)$$

and

$$\lim_{x \to \infty} |\varphi(x, t|\xi, \tau)| < \infty, \quad (4.1.94)$$

with the initial condition

$$\varphi(x, 0|\xi, \tau) = 0, \quad 0 < x < \infty. \quad (4.1.95)$$

---

[11] Agmon, N., 1984: Diffusion with back reaction. *J. Chem. Phys.*, **81**, 2811–2817; see also Kim, H., and K. J. Shin, 1999: Exact solution of the reversible diffusion-influenced reaction for an isolated pair in three dimensions. *Phys. Rev. Lett.*, **82**, 1578–1581.

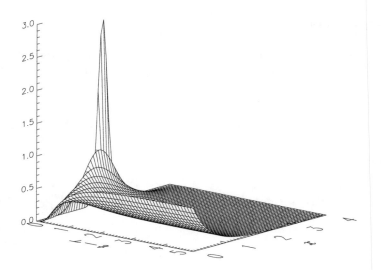

Figure 4.1.5: The Green's function (4.1.102) for the one-dimensional heat equation on the semi-infinite domain $0 < x < \infty$, and $0.01 \leq t - \tau$ when the left boundary is the integral condition (4.1.89). The parameters are $c = \kappa = 1$, and $\xi = 2$.

We rewrite (4.1.93) as

$$\varphi(0, t|\xi, \tau) = \kappa \int_0^t \left[\varphi_x(0, t'|\xi, \tau) + c\,\varphi(0, t'|\xi, \tau)\right] e^{c^2(t-t')}\, dt'. \quad (\mathbf{4.1.96})$$

To find $\varphi(x, t|\xi, \tau)$, we take the Laplace transform of (4.1.92), (4.1.94) and (4.1.96) and find that

$$\frac{d^2\Phi}{dx^2} - s\Phi = -\delta(x - \xi)e^{-s\tau}, \qquad 0 < x, \xi < \infty, \quad (\mathbf{4.1.97})$$

with

$$(s - c^2)\Phi(0, s|\xi, \tau) = \kappa\left[\Phi'(0, s|\xi, \tau) + c\,\Phi(0, s|\xi, \tau)\right], \quad (\mathbf{4.1.98})$$

and

$$\lim_{x \to \infty} |\Phi(x, s|\xi, \tau)| < \infty. \quad (\mathbf{4.1.99})$$

The general solution that satisfies (4.1.97) and (4.1.99) is

$$\Phi(x, s|\xi, \tau) = \frac{1}{2\sqrt{s}}e^{-|x-\xi|\sqrt{s}} - \frac{1}{2\sqrt{s}}e^{-(x+\xi)\sqrt{s}} + Ae^{-(x+\xi)\sqrt{s}}. \quad (\mathbf{4.1.100})$$

Substituting (4.1.100) into (4.1.98) yields an equation for A. Substituting this value into (4.1.100), we obtain

$$\begin{aligned}
\Phi(x, s|\xi, \tau) = &\frac{1}{2\sqrt{s}}e^{-|x-\xi|\sqrt{s}} - \frac{1}{2\sqrt{s}}e^{-(x+\xi)\sqrt{s}} \\
&+ \frac{\kappa e^{-(x+\xi)\sqrt{s}}}{(\kappa + 2c)(\sqrt{s} - c)} - \frac{\kappa e^{-(x+\xi)\sqrt{s}}}{(\kappa + 2c)(\sqrt{s} + c + \kappa)}. \quad (\mathbf{4.1.101})
\end{aligned}$$

Taking the inverse Laplace transform of (4.1.101), we find

$$g(x,t|\xi,\tau) = \frac{H(t-\tau)}{\sqrt{4\pi(t-\tau)}}e^{-c(x-\xi+ct-c\tau)}$$

$$\times \left\{ \exp\left[-\frac{(x-\xi)^2}{4(t-\tau)}\right] - \exp\left[-\frac{(x+\xi)^2}{4(t-\tau)}\right]\right\}$$

$$+ \frac{\kappa(\kappa+c)H(t-\tau)}{\kappa+2c}e^{[2c\xi+\kappa(x+\xi)+(\kappa^2+2c\kappa)(t-\tau)]}$$

$$\times \operatorname{erfc}\left[\frac{x+\xi+2(\kappa+c)(t-\tau)}{2\sqrt{t-\tau}}\right],$$

$$+ \frac{c\kappa H(t-\tau)}{\kappa+2c}e^{-2cx}\operatorname{erfc}\left[\frac{x+\xi-2c(t-\tau)}{2\sqrt{t-\tau}}\right], \quad (\mathbf{4.1.102})$$

where erfc is the complementary error function. This Green's function is illustrated in Figure 4.1.5.

4.2 HEAT EQUATION WITHIN A FINITE CARTESIAN DOMAIN

In this section, we find Green's functions for the heat equation within finite Cartesian domains. These solutions can be written as series involving orthonormal eigenfunctions from regular Sturm-Liouville problems.

• **Example 4.2.1**

Here we find the Green's function for the one-dimensional heat equation over the interval $0 < x < L$ associated with the problem

$$\frac{\partial u}{\partial t} - a^2\frac{\partial^2 u}{\partial x^2} = f(x,t), \quad 0 < x < L, \quad 0 < t, \quad (\mathbf{4.2.1})$$

where a^2 is the diffusivity constant.

To find the Green's function for this problem, consider the following problem:

$$\frac{\partial g}{\partial t} - a^2\frac{\partial^2 g}{\partial x^2} = \delta(x-\xi)\delta(t-\tau), \quad 0 < x,\xi < L, \quad 0 < t,\tau, \quad (\mathbf{4.2.2})$$

with the boundary conditions

$$\alpha_1 g(0,t|\xi,\tau) + \beta_1 g_x(0,t|\xi,\tau) = 0, \quad 0 < t, \quad (\mathbf{4.2.3})$$

and

$$\alpha_2 g(L,t|\xi,\tau) + \beta_2 g_x(L,t|\xi,\tau) = 0, \quad 0 < t, \quad (\mathbf{4.2.4})$$

and the initial condition

$$g(x, 0|\xi, \tau) = 0, \quad 0 < x < L. \tag{4.2.5}$$

We begin by taking the Laplace transform of (4.2.2) and find that

$$\frac{d^2G}{dx^2} - \frac{s}{a^2}G = -\frac{\delta(x - \xi)}{a^2}e^{-s\tau}, \quad 0 < x < L, \tag{4.2.6}$$

with

$$\alpha_1 G(0, s|\xi, \tau) + \beta_1 G'(0, s|\xi, \tau) = 0, \tag{4.2.7}$$

and

$$\alpha_2 G(L, s|\xi, \tau) + \beta_2 G'(L, s|\xi, \tau) = 0. \tag{4.2.8}$$

Problems similar to (4.2.6)–(4.2.8) were considered in Chapter 2. Applying this technique of eigenfunction expansions, we have

$$G(x, s|\xi, \tau) = e^{-s\tau} \sum_{n=1}^{\infty} \frac{\varphi_n(\xi)\varphi_n(x)}{s + a^2 k_n^2}, \tag{4.2.9}$$

where $\varphi_n(x)$ is the nth *orthonormal* eigenfunction to the regular Sturm-Liouville problem

$$\varphi''(x) + k^2\varphi(x) = 0, \quad 0 < x < L, \tag{4.2.10}$$

subject to the boundary conditions

$$\alpha_1\varphi(0) + \beta_1\varphi'(0) = 0, \tag{4.2.11}$$

and

$$\alpha_2\varphi(L) + \beta_2\varphi'(L) = 0. \tag{4.2.12}$$

Taking the inverse of (4.2.9), we have that the Green's function is

$$g(x, t|\xi, \tau) = \left[\sum_{n=1}^{\infty} \varphi_n(\xi)\varphi_n(x)e^{-k_n^2 a^2(t-\tau)}\right] H(t - \tau). \tag{4.2.13}$$

Let us now verify that (4.2.13) is indeed the solution to (4.2.2). We begin by computing

$$
\frac{\partial g}{\partial t} = - \left[\sum_{n=1}^{\infty} k_n^2 a^2 \varphi_n(\xi) \varphi_n(x) e^{-k_n^2 a^2 (t-\tau)} \right] H(t - \tau)
$$

$$
+ \left[\sum_{n=1}^{\infty} \varphi_n(\xi) \varphi_n(x) e^{-k_n^2 a^2 (t-\tau)} \right] \delta(t - \tau) \qquad (4.2.14)
$$

$$
= - \left[\sum_{n=1}^{\infty} k_n^2 a^2 \varphi_n(\xi) \varphi_n(x) e^{-k_n^2 a^2 (t-\tau)} \right] H(t - \tau)
$$

$$
+ \left[\sum_{n=1}^{\infty} \varphi_n(\xi) \varphi_n(x) \right] \delta(t - \tau) \qquad (4.2.15)
$$

$$
= - \left[\sum_{n=1}^{\infty} k_n^2 a^2 \varphi_n(\xi) \varphi_n(x) e^{-k_n^2 a^2 (t-\tau)} \right] H(t - \tau)
$$

$$
+ \delta(x - \xi) \delta(t - \tau), \qquad (4.2.16)
$$

because the bracketed term multiplying the delta function $\delta(t-\tau)$ equals $\delta(x - \xi)$ from (2.4.39). Therefore,

$$
\frac{\partial g}{\partial t} - a^2 \frac{\partial^2 g}{\partial x^2}
$$

$$
= \left\{ \sum_{n=1}^{\infty} \varphi_n(\xi) \left[-k_n^2 a^2 \varphi_n(x) - a^2 \varphi_n''(x) \right] e^{-k_n^2 a^2 (t-\tau)} \right\} H(t - \tau)
$$

$$
+ \delta(x - \xi) \delta(t - \tau) = \delta(x - \xi) \delta(t - \tau), \qquad (4.2.17)
$$

and (4.2.13) satisfies the differential equation (4.2.2).

• Example 4.2.2

Let us find the Green's function for the heat equation on a finite domain

$$
\frac{\partial g}{\partial t} - a^2 \frac{\partial^2 g}{\partial x^2} = \delta(x - \xi) \delta(t - \tau), \quad 0 < x, \xi < L, \quad 0 < t, \tau, \quad (4.2.18)
$$

with the boundary conditions $g(0, t|\xi, \tau) = g(L, t|\xi, \tau) = 0$, $0 < t$, and the initial condition $g(x, 0|\xi, \tau) = 0$, $0 < x < L$.

The Sturm-Liouville problem is

$$
\varphi''(x) + k^2 \varphi(x) = 0, \qquad 0 < x < L, \qquad (4.2.19)
$$

with the boundary conditions $\varphi(0) = \varphi(L) = 0$. The nth *orthonormal* eigenfunction to (4.2.19) is

$$\varphi_n(x) = \sqrt{\frac{2}{L}}\,\sin\left(\frac{n\pi x}{L}\right). \tag{4.2.20}$$

Substituting (4.2.20) into (4.2.13), we find that

$$g(x,t|\xi,\tau) = \frac{2}{L}\left\{\sum_{n=1}^{\infty}\sin\left(\frac{n\pi\xi}{L}\right)\sin\left(\frac{n\pi x}{L}\right)e^{-a^2n^2\pi^2(t-\tau)/L^2}\right\}H(t-\tau). \tag{4.2.21}$$

Let us now develop an alternative to (4.2.13) using the method of images. The free-space Green's function $g(x,t|\xi,\tau)$ for the one-dimensional heat equation was found in the previous section and equals

$$g(x,t|\xi,\tau) = \frac{H(t-\tau)}{\sqrt{4\pi a^2(t-\tau)}}\exp\left[-\frac{(x-\xi)^2}{4a^2(t-\tau)}\right]. \tag{4.2.22}$$

We will now express the Green's function to (4.2.2) in terms of a sum of (4.2.22) plus other Green's functions that we now find by the method of images.

We begin by noting that we must introduce a negative image at $x = -\xi$ so that the sum of the Green's function resulting from the source point $x = \xi$ plus the Green's function resulting from the image point $x = -\xi$ will satisfy the boundary condition at $x = 0$. These two source points will, in turn, require additional images so that the boundary condition will be satisfied at $x = L$. Eventually, we are led to an infinite sequence of source points at the points $\xi_n = \pm\xi \pm 2nL$, where $n = 0, 1, 2, \ldots$.

For this set of source points, the Green's function $g(x,t|\xi,\tau)$ will have the form

$$g(x,t|\xi,\tau) = \frac{H(t-\tau)}{\sqrt{4\pi a^2(t-\tau)}}\sum_{n=-\infty}^{\infty}\left\{\exp\left[-\frac{(x-\xi-2nL)^2}{4a^2(t-\tau)}\right]\right.$$
$$\left. -\exp\left[-\frac{(x+\xi-2nL)^2}{4a^2(t-\tau)}\right]\right\}. \tag{4.2.23}$$

The term with $n = 0$ and a positive coefficient corresponds to our free-space Green's function. Clearly (4.2.23) equals zero for $t < \tau$ and the initial condition is satisfied. A quick check shows that the series can be differentiated term by term. Because each of the terms in the series except the one corresponding to the free-space Green's function has its source point ξ_n outside the interval $0 < x < L$, (4.2.23) satisfies the differential equation.

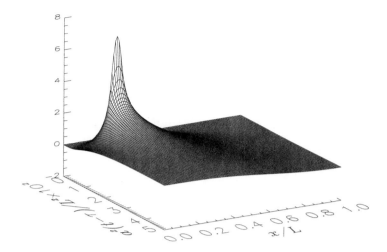

Figure 4.2.1: The Green's function (4.2.28) [times $L/2$] for the one-dimensional heat equation where the left boundary is insulated and the right boundary radiates to space when $\xi/L = 0.3$, and $hL = 1$.

Does (4.2.23) satisfy the boundary conditions? If we replace n by $-n$ in the second series in (4.2.23), it becomes identical to the first series when $x = 0$ and consequently the boundary condition is satisfied there. Similarly, if we replace n by $-n + 1$ in the second series, the two series become identical if we set $x = L$, and again, the boundary condition is satisfied.

In summary, we have shown that (4.2.23) satisfies the differential equation, the initial condition and the boundary conditions. Therefore, it is a solution to (4.2.2).

• Example 4.2.3

Let us find the Green's function for the heat equation on a finite domain

$$\frac{\partial g}{\partial t} - a^2 \frac{\partial^2 g}{\partial x^2} = \delta(x - \xi)\delta(t - \tau), \quad 0 < x, \xi < L, \quad 0 < t, \tau, \quad (\mathbf{4.2.24})$$

with the boundary conditions

$$g_x(0, t|\xi, \tau) = 0, \quad g_x(L, t|\xi, \tau) + hg(L, t|\xi, \tau) = 0, \quad 0 < t, \quad (\mathbf{4.2.25})$$

and the initial condition $g(x, 0|\xi, \tau) = 0, 0 < x < L$.

The Sturm-Liouville problem that we must now solve is

$$\varphi''(x) + \lambda\varphi(x) = 0, \quad \varphi'(0) = 0, \quad \varphi'(L) + h\varphi(L) = 0. \quad (\mathbf{4.2.26})$$

The nth *orthonormal* eigenfunction for (4.2.26) is

$$\varphi_n(x) = \sqrt{\frac{2(k_n^2 + h^2)}{L(k_n^2 + h^2) + h}}\ \cos(k_n x), \qquad (\textbf{4.2.27})$$

where k_n is the nth root of $k \tan(kL) = h$. We also used the identity that $(k_n^2 + h^2)\sin^2(k_n h) = h^2$. Substituting (4.2.27) into (4.2.13), we finally obtain

$$g(x, t|\xi, \tau) = \frac{2}{L}\left\{\sum_{n=1}^{\infty} \frac{[(k_n L)^2 + (hL)^2]\cos(k_n \xi)\cos(k_n x)}{(k_n L)^2 + (hL)^2 + hL}e^{-a^2 k_n^2(t-\tau)}\right\}$$
$$\times H(t - \tau). \qquad (\textbf{4.2.28})$$

• **Example 4.2.4**

Let us find the Green's function[12] governing the parabolic problem

$$\frac{\partial g}{\partial t} - x^{-q}\frac{\partial^2 g}{\partial x^2} = \xi^{-q}\delta(x - \xi)\delta(t - \tau), \quad 0 < x, \xi < 1, \quad 0 < t, \tau, \ (\textbf{4.2.29})$$

with the boundary conditions

$$g(0, t|\xi, \tau) = g(1, t|\xi, \tau) = 0, \quad 0 < t, \qquad (\textbf{4.2.30})$$

and the initial condition $g(x, 0|\xi, \tau) = 0$, $0 < x < 1$.

We begin by introducing the eigenfunction expansion

$$g(x, t|\xi, \tau) = \sum_{n=1}^{\infty} a_n(t)\varphi_n(x). \qquad (\textbf{4.2.31})$$

Substituting (4.2.31) into (4.2.29), we find formally that

$$\sum_{n=1}^{\infty} a_n'(t)\varphi_n(x) - x^{-q}a_n(t)\varphi_n''(x) = \xi^{-q}\delta(x - \xi)\delta(t - \tau). \qquad (\textbf{4.2.32})$$

Consider now the Sturm-Liouville problem

$$\varphi'' + \lambda x^q \varphi = 0, \qquad \varphi(0) = \varphi(1) = 0, \qquad (\textbf{4.2.33})$$

---

[12] Reprinted from *Appl. Math. Comp.*, **101**, C. Y. Chan and W. Y. Chan, Existence of classical solutions for degenerate semilinear parabolic problems, 125–149, ©1999, with permission from Elsevier Science.

where λ is the eigenvalue and x^q is the weight function. If we set $u = x^{(q+2)/2}$, then

$$\frac{d^2\varphi}{du^2} + \frac{q}{(2+q)u}\frac{d\varphi}{du} + \lambda\left(\frac{2}{2+q}\right)^2\varphi = 0, \quad \varphi(0) = \varphi(1) = 0. \quad \textbf{(4.2.34)}$$

The orthonormal eigenfunction solutions to (4.2.34) are

$$\varphi_n(x) = \sqrt{(2+q)x}\; J_{1/(2+q)}\left[\frac{2\sqrt{\lambda_n}}{2+q}x^{(2+q)/2}\right]\bigg/\left|J_{|1/(2+q)|+1}\left(\frac{2\sqrt{\lambda_n}}{2+q}\right)\right|, \quad \textbf{(4.2.35)}$$

where λ_n is the nth root of the equation $J_{1/(2+q)}\left[2\sqrt{\lambda}/(2+q)\right] = 0$. Note that the use of this particular eigenfunction in (4.2.31) ensures that $g(x,t|\xi,\tau)$ will satisfy the boundary conditions (4.2.30).

Returning to the original problem, we now use the relationship that

$$\varphi_n'' + \lambda_n x^q \varphi_n = 0 \quad \textbf{(4.2.36)}$$

to transform (4.2.32) into

$$\sum_{n=1}^{\infty}\left[a_n'(t) + \lambda_n a_n(t)\right]\varphi_n(x) = \xi^{-q}\delta(x-\xi)\delta(t-\tau), \quad \textbf{(4.2.37)}$$

or

$$a_n'(t) + \lambda_n a_n(t) = \varphi_n(\xi)\delta(t-\tau) \quad \textbf{(4.2.38)}$$

for each n. In the derivation of (4.2.38), we expanded $\delta(x-\xi)$ in terms of $\varphi_n(x)$. Integrating (4.2.38), we have that

$$a_n(t) = \varphi_n(\xi)e^{-\lambda_n(t-\tau)}H(t-\tau), \quad \textbf{(4.2.39)}$$

because $a_n(0) = 0$. Equation (4.2.39) leads directly to the Green's function

$$g(x,t|\xi,\tau) = H(t-\tau)\sum_{n=1}^{\infty}\varphi_n(\xi)\varphi_n(x)e^{-\lambda_n(t-\tau)}. \quad \textbf{(4.2.40)}$$

● **Example 4.2.5**

Let us compute the Green's function[13] governing the parabolic problem

$$ax^{\alpha}\frac{\partial g}{\partial t} - \frac{\partial}{\partial x}\left(bx^{\beta}\frac{\partial g}{\partial x}\right) = \delta(x-\xi)\delta(t-\tau), \quad 0 < x,\xi < L, \quad 0 < t,\tau,$$

$$\textbf{(4.2.41)}$$

---

[13] Reprinted from *Atmos. Environ.*, **30**, J.-S. Lin and L. M. Hildemann, Analytical solutions of the atmospheric diffusion equation with multiple sources and height-dependent wind speed and eddy diffusivities, 239–254, ©1996, with permission from Elsevier Science.

where $a, b > 0$, with the Neumann boundary conditions

$$g_x(0, t|\xi, \tau) = g_x(L, t|\xi, \tau) = 0, \quad 0 < t, \qquad (4.2.42)$$

and the initial condition $g(x, 0|\xi, \tau) = 0, 0 < x < L$.

We begin by taking the Laplace transform of (4.2.41)–(4.2.42) or

$$sG - \frac{1}{ax^\alpha} \frac{d}{dx}\left(bx^\beta \frac{dG}{dx}\right) = \frac{\delta(x - \xi)}{a\xi^\alpha} e^{-s\tau}, \qquad (4.2.43)$$

with the transformed boundary conditions

$$G'(0, s|\xi, \tau) = G'(L, s|\xi, \tau) = 0. \qquad (4.2.44)$$

Consider now the Sturm-Liouville problem

$$\frac{d^2\varphi}{dx^2} + \frac{\beta}{x}\frac{d\varphi}{dx} + \frac{ak^2 x^{\alpha-\beta}}{b}\varphi = 0, \qquad \varphi'(0) = \varphi'(L) = 0. \qquad (4.2.45)$$

If we set

$$z = x^{(\alpha-\beta+2)/2}, \qquad \text{and} \qquad \varphi(z) = z^{(1-\beta)/(\alpha-\beta+2)}\widetilde{\varphi}(z), \qquad (4.2.46)$$

(4.2.45) transforms into

$$z^2\frac{d^2\widetilde{\varphi}}{dz^2} + z\frac{d\widetilde{\varphi}}{dz} + \frac{4ak^2 z^2/b - (1-\beta)^2}{(\alpha-\beta+2)^2}\widetilde{\varphi} = 0. \qquad (4.2.47)$$

The general solution to (4.2.47) is

$$\widetilde{\varphi}(z) = AJ_\mu(\omega z) + BJ_{-\mu}(\omega z), \qquad (4.2.48)$$

where

$$\mu = \frac{1-\beta}{\alpha-\beta+2}, \qquad \text{and} \qquad \omega = \frac{2k\sqrt{a/b}}{\alpha-\beta+2}. \qquad (4.2.49)$$

assuming that $\mu > 0$. Therefore,

$$\varphi(x) = x^{(1-\beta)/2}\left\{AJ_\mu\left[\omega x^{(\alpha-\beta+2)/2}\right] + BJ_{-\mu}\left[\omega x^{(\alpha-\beta+2)/2}\right]\right\}. \qquad (4.2.50)$$

Turning to the boundary conditions and using the asymptotic formula for Bessel functions for small argument, the Neumann boundary condition at $x = 0$ leads to $A = 0$. On the other hand, the Neumann boundary condition at $x = L$ gives

$$\omega L^{(\alpha-\beta+2)/2}J'_{-\mu}\left[\omega L^{(\alpha-\beta+2)/2}\right] + \mu J_{-\mu}\left[\omega L^{(\alpha-\beta+2)/2}\right] = 0, \qquad (4.2.51)$$

or

$$J_{1-\mu}\left[\omega L^{(\alpha-\beta+2)/2}\right] = 0. \tag{4.2.52}$$

Consequently, the appropriate eigenfunctions for this problem are

$$\varphi_n(x) = x^{(1-\beta)/2} J_{-\mu}\left[\omega_n x^{(\alpha-\beta+2)/2}\right], \tag{4.2.53}$$

where ω_n is the nth root of (4.2.52).

We use these eigenfunctions to express $G(x, s|\xi, \tau)$ and $\delta(x-\xi)$. Beginning with the delta function, we assume that

$$\delta(x - \xi) = c_0 + x^{(1-\beta)/2} \sum_{n=1}^{\infty} c_n J_{-\mu}\left[\omega_n x^{(\alpha-\beta+2)/2}\right]. \tag{4.2.54}$$

This is a Dini series for $\delta(x - \xi)$ and the coefficients equal

$$c_0 = \frac{1+\alpha}{aL^{1+\alpha}}, \tag{4.2.55}$$

and

$$c_n = \frac{\alpha - \beta + 2}{aL^{\alpha-\beta+2}} \xi^{(1-\beta)/2} \frac{J_{-\mu}\left[\omega_n \xi^{(\alpha-\beta+2)/2}\right]}{J_{-\mu}^2\left[\omega_n L^{(\alpha-\beta+2)/2}\right]}. \tag{4.2.56}$$

Substituting (4.2.54) and a similar expansion for $G(x, s|\xi, \tau)$ into (4.2.43), we find that

$$G(x, s|\xi, \tau) = \frac{c_0 e^{-s\tau}}{s} + x^{(1-\beta)/2} e^{-s\tau} \sum_{n=1}^{\infty} c_n \frac{J_{-\mu}\left[\omega_n x^{(\alpha-\beta+2)/2}\right]}{s + b(\alpha - \beta + 2)^2 \omega_n^2/(4a)}. \tag{4.2.57}$$

Taking the inverse Laplace transform and applying the second shifting theorem,

$$g(x, t|\xi, \tau) = \frac{1+\alpha}{aL^{1+\alpha}} H(t - \tau) + \frac{\alpha - \beta + 2}{aL^{\alpha-\beta+2}} (x\xi)^{(1-\beta)/2} H(t - \tau)$$

$$\times \sum_{n=1}^{\infty} \frac{J_{-\mu}[\lambda_n(x/L)^{(\alpha-\beta+2)/2}] J_{-\mu}[\lambda_n(\xi/L)^{(\alpha-\beta+2)/2}]}{J_{-\mu}^2(\lambda_n)}$$

$$\times \exp\left[-\frac{b(\alpha - \beta + 2)^2 \lambda_n^2 (t - \tau)}{4aL^{\alpha-\beta+2}}\right], \tag{4.2.58}$$

where λ_n is the nth root of $J_{1-\mu}(\lambda) = 0$.

In a similar manner, when we have the Dirichlet boundary conditions

$$g(0, t|\xi, \tau) = g(L, t|\xi, \tau) = 0, \quad 0 < t, \tag{4.2.59}$$

the Green's function becomes

$$g(x,t|\xi,\tau) = \frac{\alpha - \beta + 2}{aL^{\alpha-\beta+2}} (x\xi)^{(1-\beta)/2} H(t-\tau)$$

$$\times \sum_{n=1}^{\infty} \frac{J_\mu[\lambda_n(x/L)^{(\alpha-\beta+2)/2}]J_\mu[\lambda_n(\xi/L)^{(\alpha-\beta+2)/2}]}{J_{\mu+1}^2(\lambda_n)}$$

$$\times \exp\left[-\frac{b(\alpha-\beta+2)^2\lambda_n^2(t-\tau)}{4aL^{\alpha-\beta+2}}\right], \qquad (4.2.60)$$

where λ_n is the nth root of $J_\mu(\lambda) = 0$.

Finally, for the mixed boundary conditions

$$g_x(0,t|\xi,\tau) = g(L,t|\xi,\tau) = 0, \quad 0 < t, \qquad (4.2.61)$$

and

$$g(0,t|\xi,\tau) = g_x(L,t|\xi,\tau) = 0, \quad 0 < t, \qquad (4.2.62)$$

the Green's functions are

$$g(x,t|\xi,\tau) = \frac{\alpha - \beta + 2}{aL^{\alpha-\beta+2}} (x\xi)^{(1-\beta)/2} H(t-\tau)$$

$$\times \sum_{n=1}^{\infty} \frac{J_{-\mu}[\lambda_n(x/L)^{(\alpha-\beta+2)/2}]J_{-\mu}[\lambda_n(\xi/L)^{(\alpha-\beta+2)/2}]}{J_{1-\mu}^2(\lambda_n)}$$

$$\times \exp\left[-\frac{b(\alpha-\beta+2)^2\lambda_n^2(t-\tau)}{4aL^{\alpha-\beta+2}}\right], \qquad (4.2.63)$$

where λ_n is the nth root of $J_{-\mu}(\lambda) = 0$, and

$$g(x,t|\xi,\tau) = \frac{\alpha - \beta + 2}{aL^{\alpha-\beta+2}} (x\xi)^{(1-\beta)/2} H(t-\tau)$$

$$\times \sum_{n=1}^{\infty} \frac{J_\mu[\lambda_n(x/L)^{(\alpha-\beta+2)/2}]J_\mu[\lambda_n(\xi/L)^{(\alpha-\beta+2)/2}]}{J_\mu^2(\lambda_n)}$$

$$\times \exp\left[-\frac{b(\alpha-\beta+2)^2\lambda_n^2(t-\tau)}{4aL^{\alpha-\beta+2}}\right], \qquad (4.2.64)$$

where λ_n is the nth root of $J_{\mu-1}(\lambda) = 0$, respectively.

● **Example 4.2.6**

Let us find the Green's function[14] governing the parabolic problem

$$\frac{\partial g}{\partial t} - \frac{\partial^2 g}{\partial x^2} - K_B(t)\frac{\partial^2 g}{\partial y^2} - K_H(t)\frac{\partial^2 g}{\partial z^2} + \rho(t)g = \delta(x-\xi)\delta(y-\eta)\delta(z-\zeta)\delta(t-\tau),$$

$$(4.2.65)$$

---

[14] Taken from Marinoschi, G., U. Jaekel, and H. Vereecken, 1999: Analytical solutions of three-dimensional convection-dispersion problems with time dependent coefficients. *Zeit. Angew. Math. Mech.*, **79**, 411–421.

where $-\infty < x, y, \xi, \eta < \infty$, $z_1 < z, \zeta < z_2$, and $0 < t, \tau$, with the boundary conditions

$$g(x, y, z_1, t|\xi, \eta, \zeta, \tau) = g(x, y, z_2, t|\xi, \eta, \zeta, \tau) = 0, \qquad (4.2.66)$$

and the initial condition $g(x, y, z, 0|\xi, \eta, \zeta, \tau) = 0$.

From the form of the boundary conditions, the Green's function can be written as the eigenfunction expansion

$$g(x, y, z, t|\xi, \eta, \zeta, \tau) = \sum_{n=1}^{\infty} G_n(x, y, t|\xi, \eta, \tau) \sin\left[\frac{n\pi(z - z_1)}{D}\right], \quad (4.2.67)$$

where $D = z_2 - z_1$. Substituting (4.2.67) into (4.2.65), the equation that governs $G_n(x, y, t|\xi, \eta, \tau)$ is

$$\frac{\partial G_n}{\partial t} - \frac{\partial^2 G_n}{\partial x^2} - K_B(t)\frac{\partial^2 G_n}{\partial y^2} + \left[\frac{n^2\pi^2}{D^2}K_H(t) + \rho(t)\right]G_n$$

$$= \frac{2}{D}\delta(x - \xi)\delta(y - \eta)\sin\left[\frac{n\pi(\zeta - z_1)}{D}\right]\delta(t - \tau). \quad (4.2.68)$$

To solve (4.2.68), we introduce $F_n(x, y, t|\xi, \eta, \tau)$, which is defined by

$$G_n(x, y, t|\xi, \eta, \tau) = F_n(x, y, t|\xi, \eta, \tau)$$

$$\times \exp\left\{-\int_\tau^t \left[\frac{n^2\pi^2}{D^2}K_H(\mu) + \rho(\mu)\right]d\mu\right\}, \quad (4.2.69)$$

so that the problem now becomes

$$\frac{\partial F_n}{\partial t} - \frac{\partial^2 F_n}{\partial x^2} - K_B(t)\frac{\partial^2 F_n}{\partial y^2} = \frac{2}{D}\delta(x - \xi)\delta(y - \eta)\sin\left[\frac{n\pi(\zeta - z_1)}{D}\right]\delta(t - \tau). \quad (4.2.70)$$

Finally, we must solve (4.2.70). Taking the Fourier transform of (4.2.70) in the x and y directions, we obtain the ordinary differential equation

$$\frac{d\overline{F}_n}{dt} + \left[k^2 + K_B(t)\ell^2\right]\overline{F}_n = \frac{2}{D}e^{-ik\xi - i\ell\eta}\sin\left[\frac{n\pi(\zeta - z_1)}{D}\right]\delta(t - \tau). \quad (4.2.71)$$

The solution to (4.2.71) is

$$\overline{F}_n(k, \ell, t|\xi, \eta, \tau) = \frac{2}{D}\sin\left[\frac{n\pi(\zeta - z_1)}{D}\right]e^{-ik\xi - i\ell\eta}H(t - \tau)$$

$$\times \exp\left[-k^2(t - \tau) - \ell^2\int_\tau^t K_B(\mu)\,d\mu\right]. \quad (4.2.72)$$

Applying the inverse Fourier transform,

$$F_n(x,y,t|\xi,\eta,\tau) = \frac{1}{4\pi^2} \int_{-\infty}^{\infty} \int_{-\infty}^{\infty} \overline{F}_n(k,\ell,t|\xi,\eta,\tau)e^{ikx+i\ell y} \, dk \, d\ell,$$

$$(4.2.73)$$

we obtain

$$F_n(x,y,t|\xi,\eta,\tau) = \frac{1}{2D\pi^2} \sin\left[\frac{n\pi(\zeta-z_1)}{D}\right]$$

$$\times \int_{-\infty}^{\infty} \int_{-\infty}^{\infty} e^{-s_1(k)} e^{-s_2(\ell)} \, dk \, d\ell, \quad (4.2.74)$$

where

$$s_1(k) = k^2(t-\tau) - ik(x-\xi) = \left[k\sqrt{t-\tau} - \frac{i(x-\xi)}{2\sqrt{t-\tau}}\right]^2 + \frac{(x-\xi)^2}{4(t-\tau)},$$

$$(4.2.75)$$

$$s_2(\ell) = -i\ell(y-\eta) + \ell^2 \int_{\tau}^{t} K_B(\mu) \, d\mu \quad (4.2.76)$$

$$= \left[\ell\sqrt{\int_{\tau}^{t} K_B(\mu) \, d\mu} - \frac{i(y-\eta)}{2\sqrt{\int_{\tau}^{t} K_B(\mu) \, d\mu}}\right]^2 + \frac{(y-\eta)^2}{4\int_{\tau}^{t} K_B(\mu) \, d\mu}.$$

$$(4.2.77)$$

Performing the integrals with respect to k and ℓ,

$$\int_{-\infty}^{\infty} e^{-s_1(k)} \, dk = \frac{\exp[-(x-\xi)^2/4(t-\tau)]}{\sqrt{t-\tau}} \int_{-\infty}^{\infty} e^{-\mu^2} \, d\mu \quad (4.2.78)$$

$$= \frac{\sqrt{\pi}}{\sqrt{t-\tau}} \exp\left[-\frac{(x-\xi)^2}{4(t-\tau)}\right], \quad (4.2.79)$$

and

$$\int_{-\infty}^{\infty} e^{-s_2(\ell)} \, d\ell = \frac{\sqrt{\pi}}{\sqrt{\int_{\tau}^{t} K_B(\mu) \, d\mu}} \exp\left[-\frac{(y-\eta)^2}{4\int_{\tau}^{t} K_B(\mu) \, d\mu}\right], \quad (4.2.80)$$

so that we finally obtain

$$G_n(x,y,t|\xi,\eta,\tau) = \frac{2}{D}\sin\left[\frac{n\pi(z-z_1)}{D}\right] \frac{\exp\left[-\dfrac{(x-\xi)^2}{4(t-\tau)}\right]}{\sqrt{4\pi(t-\tau)}}$$

$$\times \frac{\exp\left[-\dfrac{(y-\eta)^2}{4\int_{\tau}^{t} K_B(\mu) \, d\mu}\right]}{\sqrt{4\pi \int_{\tau}^{t} K_B(\mu) \, d\mu}} H(t-\tau)$$

$$\times \exp\left\{-\int_{\tau}^{t} \left[\frac{n^2\pi^2}{D^2}K_H(\mu) + \rho(\mu)\right] d\mu\right\}. \quad (4.2.81)$$

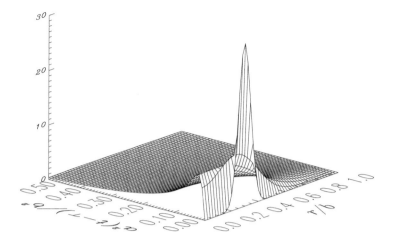

Figure 4.3.1: The Green's function (4.3.7) for the axisymmetric heat equation (4.3.1) with a Dirichlet boundary condition at $r = b$. Here $\rho/b = 0.3$ and the graph starts at $a^2(t - \tau)/b^2 = 0.001$ to avoid the delta function at $t - \tau = 0$.

The complete solution is given by substituting (4.2.81) into (4.2.67).

4.3 HEAT EQUATION WITHIN A CYLINDER

In this section, we turn our attention to cylindrical domains. The techniques used here have much in common with those used in the previous section. However, in place of sines and cosines from a Sturm-Liouville problem, we will encounter Bessel functions.

• Example 4.3.1

In this example, we find the Green's function for the heat equation in cylindrical coordinates

$$\frac{\partial g}{\partial t} - \frac{a^2}{r} \frac{\partial}{\partial r} \left(r \frac{\partial g}{\partial r} \right) = \frac{\delta(r - \rho)\delta(t - \tau)}{2\pi r}, \quad 0 < r, \rho < b, \quad 0 < t, \tau,$$

(**4.3.1**)

subject to the boundary conditions $\lim_{r \to 0} |g(r, t|\rho, \tau)| < \infty$, $g(b, t|\rho, \tau) = 0$, and the initial condition $g(r, 0|\rho, \tau) = 0$.

As usual, we begin by taking the Laplace transform (4.3.1) or

$$\frac{1}{r} \frac{d}{dr} \left(r \frac{dG}{dr} \right) - \frac{s}{a^2} G = -\frac{e^{-s\tau}}{2\pi a^2 r} \delta(r - \rho).$$

(**4.3.2**)

Next we re-express $\delta(r - \rho)/r$ as the Fourier-Bessel expansion

$$\frac{\delta(r - \rho)}{2\pi r} = \sum_{n=1}^{\infty} A_n J_0(k_n r/b), \tag{4.3.3}$$

where k_n is the nth root of $J_0(k) = 0$, and

$$A_n = \frac{2}{b^2 J_1^2(k_n)} \int_0^b \frac{\delta(r - \rho)}{2\pi r} J_0(k_n r/b) \, r \, dr = \frac{J_0(k_n \rho/b)}{\pi b^2 J_1^2(k_n)} \tag{4.3.4}$$

so that

$$\frac{1}{r} \frac{d}{dr} \left(r \frac{dG}{dr} \right) - \frac{s}{a^2} G = -\frac{e^{-s\tau}}{\pi a^2 b^2} \sum_{n=1}^{\infty} \frac{J_0(k_n \rho/b) J_0(k_n r/b)}{J_1^2(k_n)}. \tag{4.3.5}$$

The solution to (4.3.5) is

$$G(r, s|\rho, \tau) = \frac{e^{-s\tau}}{\pi} \sum_{n=1}^{\infty} \frac{J_0(k_n \rho/b) J_0(k_n r/b)}{(sb^2 + a^2 k_n^2) J_1^2(k_n)}. \tag{4.3.6}$$

Taking the inverse of (4.3.6) and applying the second shifting theorem,

$$g(r, t|\rho, \tau) = \frac{H(t - \tau)}{\pi b^2} \sum_{n=1}^{\infty} \frac{J_0(k_n \rho/b) J_0(k_n r/b)}{J_1^2(k_n)} e^{-a^2 k_n^2 (t-\tau)/b^2}. \tag{4.3.7}$$

If we modify the boundary condition at $r = b$ so that it now reads

$$g_r(b, t|\rho, \tau) + h g(b, t|\rho, \tau) = 0, \tag{4.3.8}$$

where $h \geq 0$, our analysis now leads to

$$g(r, t|\rho, \tau) = \frac{H(t - \tau)}{\pi b^2} \sum_{n=1}^{\infty} \frac{J_0(k_n \rho/b) J_0(k_n r/b)}{J_0^2(k_n) + J_1^2(k_n)} e^{-a^2 k_n^2 (t-\tau)/b^2}, \tag{4.3.9}$$

where k_n are the positive roots of $k J_1(k) - h b J_0(k) = 0$. If $h = 0$, we must add $1/(\pi b^2)$ to (4.3.9).

• Example 4.3.2

An example of great practical importance is finding the Green's function for the heat equation in cylindrical coordinates

$$\frac{\partial g}{\partial t} - \frac{a^2}{r} \frac{\partial}{\partial r} \left(r \frac{\partial g}{\partial r} \right) = \frac{\delta(r - \rho)\delta(t - \tau)}{2\pi r}, \qquad \alpha < r, \rho < \beta, \quad 0 < t, \tau,$$

$$\tag{4.3.10}$$

subject to the boundary conditions that

$$g(\alpha, t|\rho, \tau) = g(\beta, t|\rho, \tau) = 0, \tag{4.3.11}$$

and the initial condition $g(r, 0|\rho, \tau) = 0$.

Once again, we begin by taking the Laplace transform (4.3.10) and obtain

$$\frac{1}{r}\frac{d}{dr}\left(r\frac{dG}{dr}\right) - \frac{s}{a^2}G = -\frac{e^{-s\tau}}{2\pi a^2 r}\delta(r - \rho). \tag{4.3.12}$$

Next, we express $\delta(r - \rho)/r$ as a Fourier-Bessel expansion by considering the regular Sturm-Liouville problem

$$\frac{1}{r}\frac{d}{dr}\left(r\frac{d\varphi}{dr}\right) + k^2\varphi = 0, \qquad \varphi(\alpha) = \varphi(\beta) = 0. \tag{4.3.13}$$

The eigenfunctions that satisfy (4.3.13) are

$$\varphi_n(r) = Y_0(k_n\alpha)J_0(k_nr) - J_0(k_n\alpha)Y_0(k_nr), \tag{4.3.14}$$

provided k_n is the nth zero of $J_0(k\alpha)Y_0(k\beta) - J_0(k\beta)Y_0(k\alpha) = 0$. Therefore, the expansion for the delta function in terms of $\varphi_n(r)$ is

$$\frac{\delta(r - \rho)}{r} = \sum_{n=1}^{\infty} A_n\varphi_n(r), \tag{4.3.15}$$

where

$$A_n = \frac{\int_\alpha^\beta \delta(r - \rho)\varphi_n(r)\,dr}{\int_\alpha^\beta \varphi_n^2(r)\,r\,dr}. \tag{4.3.16}$$

Using the orthogonality conditions[15] that

$$\int_\alpha^\beta J_0^2(k_nr)\,r\,dr = \tfrac{1}{2}r^2\left[J_0^2(k_nr) + J_1^2(k_nr)\right]\Big|_\alpha^\beta, \tag{4.3.17}$$

$$\int_\alpha^\beta J_0(k_nr)Y_0(k_nr)\,r\,dr = \tfrac{1}{2}r^2\left[J_0(k_nr)Y_0(k_nr) + J_1(k_nr)Y_1(k_nr)\right]\Big|_\alpha^\beta, \tag{4.3.18}$$

and

$$\int_\alpha^\beta Y_0^2(k_nr)\,r\,dr = \tfrac{1}{2}r^2\left[Y_0^2(k_nr) + Y_1^2(k_nr)\right]\Big|_\alpha^\beta, \tag{4.3.19}$$

---

[15] Watson, G. N., 1966: *A Treatise on the Theory of Bessel Functions.* Cambridge University Press, §5.12, equation 2.

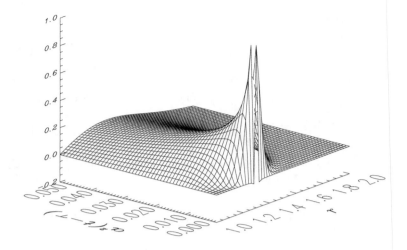

Figure 4.3.2: The Green's function (4.3.23) for the axisymmetric heat equation (4.3.10) in an annulus with Dirichlet boundary conditions at $r = \alpha$ and $r = \beta$ when $\alpha = 1$, $\beta = 2$, and $\rho = 1.3$. The plot begins at $a^2(t - \tau) = 10^{-5}$ to avoid the delta function at $t - \tau = 0$. Values greater than one are not plotted.

and the Wronskian relationship[16]

$$J_0(z)Y_1(z) - J_1(z)Y_0(z) = -\frac{2}{\pi z}, \qquad (4.3.20)$$

we find that

$$\int_\alpha^\beta \varphi_n^2(r)\, r\, dr = \frac{2}{\pi^2 k_n^2}\left[\frac{J_0^2(k_n\alpha)}{J_0^2(k_n\beta)} - 1\right]. \qquad (4.3.21)$$

Therefore, $G(r, s|\rho, \tau)$ is

$$G(r, s|\rho, \tau) = \frac{\pi}{4}e^{-s\tau}\sum_{n=1}^\infty \frac{k_n^2 J_0^2(k_n\beta)\varphi_n(\rho)\varphi_n(r)}{[J_0^2(k_n\alpha) - J_0^2(k_n\beta)]\,(s + a^2 k_n^2)}. \qquad (4.3.22)$$

Inverting the Laplace transform, we obtain the final answer

$$g(r, t|\rho, \tau) = \frac{\pi}{4}H(t - \tau)\sum_{n=1}^\infty \frac{k_n^2 J_0^2(k_n\beta)\varphi_n(\rho)\varphi_n(r)}{J_0^2(k_n\alpha) - J_0^2(k_n\beta)}e^{-a^2 k_n^2 (t-\tau)}.$$
$$(4.3.23)$$

In a similar manner, we can solve the general problem[17] of (4.3.10) with Robin boundary conditions

$$a_1 g_r(\alpha, t|\rho, \tau) - a_2 g(\alpha, t|\rho, \tau) = 0, \qquad (4.3.24)$$

---

[16] *Ibid.*, §3.6, equation 12.

[17] Naji *et al.* [Naji, M., M. Al-Nimr, and S. Masoud, 2000: Transient thermal behavior of a cylindrical brake system. *Heat Mass Transfer*, **36**, 45–49.] have found the Green's function for a two-layer ring with radiative boundary conditions.

and

$$b_1 g_r(\beta, t|\rho, \tau) + b_2 g(\beta, t|\rho, \tau) = 0, \qquad (\mathbf{4.3.25})$$

where $a_1, a_2, b_1, b_2 \geq 0$. The solution for this problem is

$$g(r, t|\rho, \tau) = \frac{\pi}{4} H(t - \tau) \sum_{n=1}^{\infty} \frac{k_n^2}{F(k_n)} \left[b_1 k_n J_1(k_n\beta) - b_2 J_0(k_n\beta) \right]^2$$

$$\times \varphi_n(\rho)\varphi_n(r) e^{-a^2 k_n^2 (t-\tau)}, \qquad (\mathbf{4.3.26})$$

where

$$\varphi_n(r) = J_0(k_n r) \left[a_1 k_n Y_1(k_n\alpha) + a_2 Y_0(k_n\alpha) \right]$$
$$- Y_0(k_n r) \left[a_1 k_n J_1(k_n\alpha) + a_2 J_0(k_n\alpha) \right], \qquad (\mathbf{4.3.27})$$

and

$$F(k_n) = (b_1^2 k_n^2 + b_2^2) \left[a_1 k_n J_1(k_n\alpha) + a_2 J_0(k_n\alpha) \right]^2$$
$$- (a_1^2 k_n^2 + a_2^2) \left[b_1 k_n J_1(k_n\beta) - b_2 J_0(k_n\beta) \right]^2. \qquad (\mathbf{4.3.28})$$

If $a_2 = b_2 = 0$, we must add the extra term $1/[\pi(\beta^2 - \alpha^2)]$ to (4.3.26).

● **Example 4.3.3**

In Example 4.1.2, we utilized the one-dimensional, free-space Green's function to develop additional Green's functions by finding the homogeneous solution via Laplace transforms. Similar considerations hold in the two-dimensional case. For example, let us find the Green's function for

$$\frac{\partial g}{\partial t} - a^2 \left(\frac{\partial^2 g}{\partial r^2} + \frac{1}{r} \frac{\partial g}{\partial r} + \frac{1}{r^2} \frac{\partial^2 g}{\partial \theta^2} \right) = \frac{\delta(r - \rho)\delta(\theta - \theta')\delta(t - \tau)}{2\pi r},$$
$$(\mathbf{4.3.29})$$

where $\alpha < r, \rho < \beta$, $0 \leq \theta, \theta' \leq 2\pi$, and $0 < t, \tau$. The boundary conditions are

$$g(\alpha, \theta, t|\rho, \theta', \tau) = g(\beta, \theta, t|\rho, \theta', \tau) = 0, \quad 0 \leq \theta \leq 2\pi, \quad 0 < t,$$
$$(\mathbf{4.3.30})$$

with the initial condition that

$$g(r, \theta, 0|\rho, \theta', \tau) = 0, \quad \alpha < r < \beta, \quad 0 \leq \theta \leq 2\pi. \qquad (\mathbf{4.3.31})$$

Obviously, the solution must be periodic in θ.

Because the differential equation is linear, we can write

$$g(r, \theta, t|\rho, \theta', \tau) = \frac{e^{-R^2/[4a^2(t-\tau)]}}{4\pi a^2 (t - \tau)} H(t - \tau) + v(r, \theta, t), \qquad (\mathbf{4.3.32})$$

Table 4.3.1: Green's Functions for the Two-Dimensional Heat Conduction Equation

$$\frac{\partial g}{\partial t} - \kappa \left(\frac{\partial^2 g}{\partial r^2} + \frac{1}{r} \frac{\partial g}{\partial r} + \frac{1}{r^2} \frac{\partial^2 g}{\partial \theta^2} \right) = \frac{\delta(r - \rho)\delta(\theta - \theta')\delta(t)}{2\pi r}$$

for Various Annuli with Dirichlet Boundary Conditions

---

1. $0 \le r < a$

$$g = \frac{1}{\pi a^2} \sum_{m=-\infty}^{\infty} \sum_{n=1}^{\infty} \frac{J_m(\alpha_n r) J_m(\alpha_n \rho)}{[J_m'(\alpha_n a)]^2} \cos[m(\theta - \theta')] e^{-\kappa \alpha_n^2 t},$$

where α_n is the nth positive root of $J_m(\alpha a) = 0$.

2. $a < r < \infty$

$$g = \frac{1}{2\pi} \sum_{n=-\infty}^{\infty} \cos[n(\theta - \theta')] \int_0^{\infty} e^{-\kappa \alpha^2 t} \frac{U_n(\alpha r) U_n(\alpha \rho)}{J_n^2(\alpha a) + Y_n^2(\alpha a)} \, \alpha \, d\alpha,$$

where $U_n(\alpha r) = J_n(\alpha r) Y_n(\alpha a) - J_n(\alpha a) Y_n(\alpha r)$.

3. $a < r < b$

$$g = \frac{\pi}{4} \sum_{m=-\infty}^{\infty} \sum_{n=1}^{\infty} \alpha_n^2 \frac{J_m^2(\alpha_n b) U_m(\alpha_n r) U_m(\alpha_n \rho)}{J_m^2(\alpha_n a) - J_m^2(\alpha_n b)} \cos[m(\theta - \theta')] e^{-\kappa \alpha_n^2 t},$$

where α_n is the nth positive root of $U_m(\alpha b) = 0$ with $U_m(\alpha r) = J_m(\alpha r) Y_m(\alpha a) - J_m(\alpha a) Y_m(\alpha r)$.

---

where $R^2 = r^2 + \rho^2 - 2\rho r \cos(\theta - \theta')$. Consequently, we must find the homogeneous solution $v(r, \theta, t)$ that satisfies

$$\frac{\partial v}{\partial t} = a^2 \left(\frac{\partial^2 v}{\partial r^2} + \frac{1}{r} \frac{\partial v}{\partial r} + \frac{1}{r^2} \frac{\partial^2 v}{\partial \theta^2} \right), \tag{4.3.33}$$

subject to the boundary conditions

$$v(\alpha, \theta, t) = -\frac{H(t - \tau)}{4\pi a^2(t - \tau)} \exp\left[-\frac{\alpha^2 + \rho^2 - 2\alpha\rho\cos(\theta - \theta')}{4a^2(t - \tau)} \right], \tag{4.3.34}$$

and

$$v(\beta, \theta, t) = -\frac{H(t - \tau)}{4\pi a^2(t - \tau)} \exp\left[-\frac{\beta^2 + \rho^2 - 2\beta\rho\cos(\theta - \theta')}{4a^2(t - \tau)} \right]. \tag{4.3.35}$$

Table 4.3.2: Green's Functions for the Two-Dimensional Heat Conduction Equation

$$\frac{\partial g}{\partial t} - \kappa \left(\frac{\partial^2 g}{\partial r^2} + \frac{1}{r} \frac{\partial g}{\partial r} + \frac{1}{r^2} \frac{\partial^2 g}{\partial \theta^2} \right) = \frac{\delta(r - \rho)\delta(\theta - \theta')\delta(t)}{2\pi r}$$

for the Sector $0 < \theta, \theta' < \theta_0$ and Dirichlet Boundary Conditions

1. $0 \leq r < \infty$

$$g = \frac{2}{\theta_0} \sum_{m=1}^{\infty} \sin(s\theta) \sin(s\theta') \int_0^{\infty} \alpha J_s(\alpha\rho) J_s(\alpha r) e^{-\kappa \alpha^2 t} \, d\alpha,$$

where $s = m\pi/\theta_0$.

2. $0 \leq r < a$

$$g = \frac{4}{a^2 \theta_0} \sum_{m=1}^{\infty} \sin(s\theta) \sin(s\theta') \sum_{n=1}^{\infty} \frac{J_s(\alpha_n r) J_s(\alpha_n \rho)}{[J_s'(\alpha_n a)]^2} e^{-\kappa \alpha_n^2 t},$$

where α_n is the nth positive root of $J_s(\alpha a) = 0$, and $s = m\pi/\theta_0$.

3. $a < r < \infty$

$$g = \frac{2}{\theta_0} \sum_{m=1}^{\infty} \sin(s\theta) \sin(s\theta') \int_0^{\infty} \alpha \frac{U_s(\alpha r) U_s(\alpha\rho)}{J_s^2(\alpha a) + Y_s^2(\alpha a)} e^{-\kappa \alpha^2 t} \, d\alpha,$$

where $U_s(\alpha r) = J_s(\alpha r) Y_s(\alpha a) - J_s(\alpha a) Y_s(\alpha r)$, and $s = m\pi/\theta_0$.

4. $a < r < b$

$$g = \frac{\pi^2}{\theta_0} \sum_{m=1}^{\infty} \sin(s\theta) \sin(s\theta') \sum_{n=1}^{\infty} \alpha_n^2 \frac{J_s^2(\alpha_n b) U_s(\alpha_n r) U_s(\alpha_n \rho)}{J_s^2(\alpha_n a) - J_s^2(\alpha_n b)} e^{-\kappa \alpha_n^2 t},$$

where $U_s(\alpha r) = J_s(\alpha r) Y_s(\alpha a) - J_s(\alpha a) Y_s(\alpha r)$, α_n is the nth positive root of $U_s(\alpha b) = 0$, and $s = m\pi/\theta_0$.

and the initial condition that

$$v(r, \theta, 0) = 0, \quad \alpha < r < \beta, \quad 0 \leq \theta \leq 2\pi. \tag{4.3.36}$$

Next, we take the Laplace transform of (4.3.33)–(4.3.35),

$$\frac{\partial^2 V}{\partial r^2} + \frac{1}{r} \frac{\partial V}{\partial r} + \frac{1}{r^2} \frac{\partial^2 V}{\partial \theta^2} - q^2 V = 0, \tag{4.3.37}$$

where $q^2 = s/a^2$. The boundary conditions become

$$V(\alpha, \theta, s) = -\frac{e^{-s\tau}}{2\pi a^2} K_0 \left[q\sqrt{\alpha^2 + \rho^2 - 2\alpha\rho\cos(\theta - \theta')} \right], \quad \textbf{(4.3.38)}$$

and

$$V(\beta, \theta, s) = -\frac{e^{-s\tau}}{2\pi a^2} K_0 \left[q\sqrt{\beta^2 + \rho^2 - 2\beta\rho\cos(\theta - \theta')} \right]. \quad \textbf{(4.3.39)}$$

Using separation of variables to solve (4.3.37), we find that

$$V(r, \theta, s) = \sum_{n=-\infty}^{\infty} [A_n I_n(qr) + B_n K_n(qr)] \cos[n(\theta - \theta')]. \quad \textbf{(4.3.40)}$$

To compute A_n and B_n, we take advantage of the property[18] that

$$K_0(qR) = \sum_{n=-\infty}^{\infty} \cos[n(\theta - \theta')] I_n(qr_<) K_n(qr_>). \quad \textbf{(4.3.41)}$$

We then find that

$$V(r, \theta, s) = -\frac{e^{-s\tau}}{2\pi a^2} \sum_{n=-\infty}^{\infty} \frac{\cos[n(\theta - \theta')]}{I_n(q\alpha)K_n(q\beta) - K_n(q\alpha)I_n(q\beta)}$$

$$\times \left\{ K_n(q\beta)I_n(qr)[I_n(q\alpha)K_n(q\rho) - K_n(q\alpha)I_n(q\rho)] \right.$$

$$\left. - I_n(q\alpha)K_n(qr)[I_n(q\beta)K_n(q\rho) - K_n(q\beta)I_n(q\rho)] \right\}.$$

$$\textbf{(4.3.42)}$$

Therefore,

$$G(r, \theta, s|\xi, \theta', \tau) = \frac{e^{-s\tau}}{2\pi a^2} \sum_{n=-\infty}^{\infty} \frac{\cos[n(\theta - \theta')]}{I_n(q\alpha)K_n(q\beta) - K_n(q\alpha)I_n(q\beta)}$$

$$\times [I_n(q\alpha)K_n(qr_<) - K_n(q\alpha)I_n(qr_<)]$$

$$\times [I_n(q\beta)K_n(qr_>) - K_n(q\beta)I_n(qr_>)].$$

$$\textbf{(4.3.43)}$$

Upon applying the residue theorem to (4.3.43), we obtain

$$g(r, \theta, t|\xi, \theta', \tau) = \frac{\pi}{4} H(t - \tau) \sum_{m=-\infty}^{\infty} \cos[m(\theta - \theta')]$$

$$\times \left\{ \sum_{n=1}^{\infty} k_n^2 \frac{J_m^2(k_n\beta)U_m(k_n\rho)U_m(k_nr)}{J_m^2(k_n\alpha) - J_m^2(k_n\beta)} e^{-a^2 k_n^2(t-\tau)} \right\},$$

$$\textbf{(4.3.44)}$$

---

[18] Gray, A., and G. B. Mathews, 1966: *Treatise on Bessel Functions and Their Applications to Physics*. Dover Publications, p. 74.

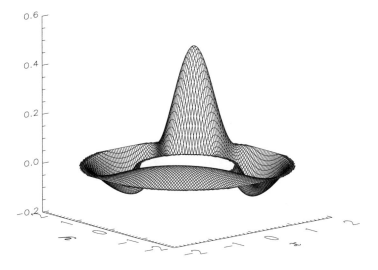

Figure 4.3.3: The Green's function (4.3.44) for the two-dimensional heat equation (4.3.29) in an annulus with Dirichlet boundary conditions at $r = \alpha$ and $r = \beta$ when $t - \tau = 0.1$, $\alpha = 1$, $\beta = 2$, $\rho = \sqrt{2}$, and $\theta' = \pi/4$.

where

$$U_m(kr) = J_m(kr)Y_m(k\alpha) - J_m(k\alpha)Y_m(kr), \qquad (4.3.45)$$

and k_n is the nth positive root of $U_m(k\beta) = 0$.

Green's functions for a sector $0 < \theta, \theta' < \theta_0$ of the annulus $\alpha < r, \rho < \beta$ were found by Carslaw and Jaeger[19] using the same technique and are presented in Tables 4.3.1 and 4.3.2.

• Example 4.3.4

In this example, we find the Green's function for the three-dimensional equation

$$\frac{\partial g}{\partial t} - a^2 \left(\frac{\partial^2 g}{\partial r^2} + \frac{1}{r} \frac{\partial g}{\partial r} + \frac{1}{r^2} \frac{\partial^2 g}{\partial \theta^2} + \frac{\partial^2 g}{\partial z^2} \right)$$
$$= \frac{\delta(r - \rho)\delta(\theta - \theta')\delta(z - \zeta)\delta(t - \tau)}{2\pi r}, \quad (4.3.46)$$

---

[19] Carslaw, H. S., and J. C. Jaeger, 1941: The determination of Green's function for line sources for the equation of conduction of heat in cylindrical coordinates by the Laplace transformation. *Philos. Mag., Ser. 7*, **31**, 204–208. Reproduced by permission of Taylor & Francis, Ltd., http://www.tandf.co.uk/journals.

Table 4.3.3: Green's Functions for the Three-Dimensional Heat Conduction Equation

$$\frac{\partial g}{\partial t} - \kappa \left(\frac{\partial^2 g}{\partial r^2} + \frac{1}{r}\frac{\partial g}{\partial r} + \frac{1}{r^2}\frac{\partial^2 g}{\partial \theta^2} + \frac{\partial^2 g}{\partial z^2} \right) = \frac{\delta(r-\rho)\delta(\theta-\theta')\delta(z)\delta(t)}{2\pi r}$$

for Infinitely Long Cylinders with Dirichlet Boundary Conditions

---

1. $0 \le r < a$

$$g = \frac{e^{-z^2/4\kappa t}}{2\pi a^2 \sqrt{\pi \kappa t}} \sum_{m=-\infty}^{\infty} \sum_{n=1}^{\infty} \frac{J_m(\alpha_n r) J_m(\alpha_n \rho)}{[J_m'(\alpha_n a)]^2} \cos[m(\theta - \theta')] e^{-\kappa \alpha_n^2 t},$$

where α_n is the nth positive root of $J_m(\alpha a) = 0$.

2. $a < r < \infty$

$$g = \frac{e^{-z^2/4\kappa t}}{4\pi\sqrt{\pi \kappa t}} \sum_{n=-\infty}^{\infty} \cos[n(\theta - \theta')] \int_0^\infty e^{-\kappa \alpha^2 t} \frac{U_n(\alpha r) U_n(\alpha \rho)}{J_n^2(\alpha a) + Y_n^2(\alpha a)}\, \alpha\, d\alpha,$$

where $U_n(\alpha r) = J_n(\alpha r) Y_n(\alpha a) - J_n(\alpha a) Y_n(\alpha r)$.

3. $a < r < b$

$$g = \frac{\sqrt{\pi}}{8\sqrt{\kappa t}} e^{-z^2/4\kappa t} \sum_{m=-\infty}^{\infty} \sum_{n=1}^{\infty} \alpha_n^2 \frac{J_m^2(\alpha_n b) U_m(\alpha_n r) U_m(\alpha_n \rho)}{J_m^2(\alpha_n a) - J_m^2(\alpha_n b)}$$

$$\times \cos[m(\theta - \theta')] e^{-\kappa \alpha_n^2 t},$$

where α_n is the nth positive root of $U_m(\alpha b) = 0$ with $U_m(\alpha r) = J_m(\alpha r) Y_m(\alpha a) - J_m(\alpha a) Y_m(\alpha r)$.

---

where $0 < r, \rho < \infty$, $0 \le \theta, \theta' \le 2\pi$, $0 < z, \zeta < L$, and $0 < t, \tau$. The boundary conditions are

$$\lim_{r \to 0} |g(r,\theta,z,t|\rho,\theta',\zeta,\tau)| < \infty, \qquad \lim_{r \to \infty} |g(r,\theta,z,t|\rho,\theta',\zeta,\tau)| < \infty,$$

$$(4.3.47)$$

if $0 \le \theta \le 2\pi$, $0 < z < L$, and $0 < t$, and

$$g(r,\theta,0,t|\rho,\theta',\zeta,\tau) = g(r,\theta,L,t|\rho,\theta',\zeta,\tau) = 0, \qquad (4.3.48)$$

if $0 < r < \infty$, $0 \le \theta \le 2\pi$, and $0 < t$. The initial condition is

$$g(r,\theta,z,0|\rho,\theta',\zeta,\tau) = 0, \quad 0 < r < \infty, \quad 0 \le \theta \le 2\pi, \quad 0 < z < L.$$

$$(4.3.49)$$

Table 4.3.4: Green's Functions for the Three-Dimensional Heat Conduction Equation

$$\frac{\partial g}{\partial t} - \kappa \left(\frac{\partial^2 g}{\partial r^2} + \frac{1}{r}\frac{\partial g}{\partial r} + \frac{1}{r^2}\frac{\partial^2 g}{\partial \theta^2} + \frac{\partial^2 g}{\partial z^2} \right) = \frac{\delta(r-\rho)\delta(\theta-\theta')\delta(z-\zeta)\delta(t)}{2\pi r}$$

for Right Cylinders between $0 < z, \zeta < L$ with Dirichlet Boundary Conditions

---

1. $0 \le r < a$

$$g = \frac{2}{\pi a^2 L} \sum_{k=1}^{\infty} e^{-\kappa k^2 \pi^2 t/L^2} \sin\left(\frac{k\pi\zeta}{L}\right) \sin\left(\frac{k\pi z}{L}\right)$$

$$\times \left\{ \sum_{m=-\infty}^{\infty} \sum_{n=1}^{\infty} \frac{J_m(\alpha_n r) J_m(\alpha_n \rho)}{[J'_m(\alpha_n a)]^2} \cos[m(\theta - \theta')] e^{-\kappa \alpha_n^2 t} \right\},$$

where α_n is the nth positive root of $J_m(\alpha a) = 0$.

2. $a < r < \infty$

$$g = \frac{1}{\pi L} \sum_{k=1}^{\infty} e^{-\kappa k^2 \pi^2 t/L^2} \sin\left(\frac{k\pi\zeta}{L}\right) \sin\left(\frac{k\pi z}{L}\right)$$

$$\times \left\{ \sum_{n=-\infty}^{\infty} \cos[n(\theta - \theta')] \int_0^{\infty} e^{-\kappa \alpha^2 t} \frac{U_n(\alpha r) U_n(\alpha \rho)}{J_n^2(\alpha a) + Y_n^2(\alpha a)} \alpha\, d\alpha \right\},$$

where $U_n(\alpha r) = J_n(\alpha r) Y_n(\alpha a) - J_n(\alpha a) Y_n(\alpha r)$.

3. $a < r < b$

$$g = \frac{\pi}{2L} \sum_{k=1}^{\infty} e^{-\kappa k^2 \pi^2 t/L^2} \sin\left(\frac{k\pi\zeta}{L}\right) \sin\left(\frac{k\pi z}{L}\right)$$

$$\times \left\{ \sum_{m=-\infty}^{\infty} \sum_{n=1}^{\infty} \alpha_n^2 \frac{J_m^2(\alpha_n b) U_m(\alpha_n r) U_m(\alpha_n \rho)}{J_m^2(\alpha_n a) - J_m^2(\alpha_n b)} \cos[m(\theta - \theta')] e^{-\kappa \alpha_n^2 t} \right\},$$

where α_n is the nth positive root of $U_m(\alpha b) = 0$, with $U_m(\alpha r) = J_m(\alpha r) Y_m(\alpha a) - J_m(\alpha a) Y_m(\alpha r)$.

Obviously, the solution must be periodic in θ.

Because the differential equation is linear, we can write

$$g(r, \theta, z, t | \rho, \theta', \zeta, \tau) = \frac{e^{-[R^2 + (z-\zeta)^2]/[4a^2(t-\tau)]}}{8a^3[\pi(t-\tau)]^{3/2}} H(t-\tau) + v(r, \theta, z, t),$$

$$(4.3.50)$$

Table 4.3.5: Green's Functions for the Three-Dimensional Heat Conduction Equation

$$\frac{\partial g}{\partial t} - \kappa \left(\frac{\partial^2 g}{\partial r^2} + \frac{1}{r}\frac{\partial g}{\partial r} + \frac{1}{r^2}\frac{\partial^2 g}{\partial \theta^2} + \frac{\partial^2 g}{\partial z^2} \right) = \frac{\delta(r-\rho)\delta(\theta-\theta')\delta(z)\delta(t)}{2\pi r}$$

for the Sector $0 < \theta, \theta' < \theta_0$ Cut from an Infinitely Long Cylinder with Dirichlet Boundary Conditions

---

1. $0 \le r < a$

$$g = \frac{2e^{-z^2/4\kappa t}}{a^2\theta_0\sqrt{\pi\kappa t}} \sum_{m=1}^{\infty} \sin(s\theta)\sin(s\theta') \sum_{n=1}^{\infty} \frac{J_s(\alpha_n r)J_s(\alpha_n\rho)}{[J_s'(\alpha_n a)]^2} e^{-\kappa\alpha_n^2 t},$$

where α_n is the nth positive root of $J_s(\alpha a) = 0$, and $s = m\pi/\theta_0$.

2. $a < r < \infty$

$$g = \frac{e^{-z^2/4\kappa t}}{\theta_0\sqrt{\pi\kappa t}} \sum_{m=1}^{\infty} \sin(s\theta)\sin(s\theta') \int_0^{\infty} e^{-\kappa\alpha^2 t} \frac{U_s(\alpha r)U_s(\alpha\rho)}{J_s^2(\alpha a) + Y_s^2(\alpha a)} \alpha\, d\alpha,$$

where $U_s(\alpha r) = J_s(\alpha r)Y_s(\alpha a) - J_s(\alpha a)Y_s(\alpha r)$, and $s = m\pi/\theta_0$.

3. $a < r < b$

$$g = \frac{\pi^2 e^{-z^2/4\kappa t}}{2\theta_0\sqrt{\pi\kappa t}} \sum_{m=1}^{\infty} \sin(s\theta)\sin(s\theta')$$

$$\times \left\{ \sum_{n=1}^{\infty} \alpha_n^2 \frac{J_s^2(\alpha_n b)U_s(\alpha_n r)U_s(\alpha_n\rho)}{J_s^2(\alpha_n a) - J_s^2(\alpha_n b)} e^{-\kappa\alpha_n^2 t} \right\},$$

where $U_s(\alpha r) = J_s(\alpha r)Y_s(\alpha a) - J_s(\alpha a)Y_s(\alpha r)$, α_n is the nth positive root of $U_s(\alpha b) = 0$, and $s = m\pi/\theta_0$.

---

where $R^2 = r^2 + \rho^2 - 2\rho r \cos(\theta - \theta')$. The first term in (4.3.50) is the three-dimensional, free-space Green's function in cylindrical coordinates; it is the particular solution. On the other hand, $v(r, \theta, z, t)$ is the homogeneous solution that satisfies

$$\frac{\partial v}{\partial t} = a^2 \left(\frac{\partial^2 v}{\partial r^2} + \frac{1}{r}\frac{\partial v}{\partial r} + \frac{1}{r^2}\frac{\partial^2 v}{\partial \theta^2} + \frac{\partial^2 v}{\partial z^2} \right), \tag{4.3.51}$$

subject to the boundary conditions

$$\lim_{r\to 0} |v(r, \theta, z, t)| < \infty, \qquad \lim_{r\to\infty} |v(r, \theta, z, t)| < \infty, \tag{4.3.52}$$

$$v(r, \theta, 0, t) = -\frac{e^{-(R^2+\zeta^2)/[4a^2(t-\tau)]}}{8a^3[\pi(t-\tau)]^{3/2}} H(t-\tau), \qquad (4.3.53)$$

and

$$v(r, \theta, L, t) = -\frac{e^{-[R^2+(L-\zeta)^2]/[4a^2(t-\tau)]}}{8a^3[\pi(t-\tau)]^{3/2}} H(t-\tau), \qquad (4.3.54)$$

and the initial condition

$$v(r, \theta, z, 0) = 0, \qquad 0 < r < \infty, \quad 0 \le \theta \le 2\pi, \quad 0 < z < L. \quad (4.3.55)$$

Next, we take the Laplace transform of (4.3.51)–(4.3.54),

$$\frac{\partial^2 V}{\partial r^2} + \frac{1}{r}\frac{\partial V}{\partial r} + \frac{1}{r^2}\frac{\partial^2 V}{\partial \theta^2} + \frac{\partial^2 V}{\partial z^2} - q^2 V = 0, \qquad (4.3.56)$$

where $q^2 = s/a^2$. The boundary conditions are

$$\lim_{r \to 0} |V(r, \theta, z, s)| < \infty, \qquad \lim_{r \to \infty} |V(r, \theta, z, s)| < \infty, \qquad (4.3.57)$$

$$V(r, \theta, 0, s) = -\frac{e^{-s\tau - q\sqrt{R^2+\zeta^2}}}{4\pi a^2 \sqrt{R^2 + \zeta^2}} \qquad (4.3.58)$$

$$= -\frac{e^{-s\tau}}{4\pi a^2} \int_0^\infty e^{-\zeta\sqrt{\xi^2+q^2}} \frac{J_0(\xi R)}{\sqrt{\xi^2 + q^2}} \xi \, d\xi, \quad (4.3.59)$$

and

$$V(r, \theta, L, s) = -\frac{e^{-s\tau - q\sqrt{R^2+(L-\zeta)^2}}}{2\pi a^2 \sqrt{R^2 + (L-\zeta)^2}} \qquad (4.3.60)$$

$$= -\frac{e^{-s\tau}}{2\pi a^2} \int_0^\infty e^{-(L-\zeta)\sqrt{\xi^2+q^2}} \frac{J_0(\xi R)}{\sqrt{\xi^2 + q^2}} \xi \, d\xi. \quad (4.3.61)$$

We will shortly show why we introduced these integral representations.[20]
Using Hankel transforms, the solution to (4.3.56)–(4.3.61) can be expressed

$$V(r, \theta, z, s) = -\frac{e^{-s\tau}}{4\pi a^2} \int_0^\infty \frac{J_0(\xi R)}{\sqrt{\xi^2 + q^2}} \xi \, d\xi$$

$$\times \left\{ \frac{e^{-\zeta\sqrt{\xi^2+q^2}} \sinh[(L-z)\sqrt{\xi^2 + q^2}]}{\sinh(L\sqrt{\xi^2 + q^2})} \right.$$

$$\left. + \frac{e^{-(L-\zeta)\sqrt{\xi^2+q^2}} \sinh[z\sqrt{\xi^2 + q^2}]}{\sinh(L\sqrt{\xi^2 + q^2})} \right\} \qquad (4.3.62)$$

---

[20] Watson, G. N., 1966: *A Treatise on the Theory of Bessel Functions.* Cambridge University Press, §13.47, equation 2.

for $0 < z < L$.

Consequently,

$$G(r, \theta, z, s|\rho, \theta', \zeta, \tau)$$

$$= V(r, \theta, z, s) + \frac{e^{-s\tau}}{2\pi a^2} \int_0^\infty e^{-|z-\zeta|\sqrt{\xi^2+q^2}} \frac{J_0(\xi R)}{\sqrt{\xi^2 + q^2}} \xi \, d\xi$$

(4.3.63)

$$= \frac{e^{-s\tau}}{2\pi a^2} \int_0^\infty \frac{\sinh[(L-\zeta)\sqrt{\xi^2+q^2}\,]\sinh(z\sqrt{\xi^2+q^2}\,)}{\sqrt{\xi^2+q^2}\,\sinh(L\sqrt{\xi^2+q^2}\,)} J_0(\xi R)\,\xi \, d\xi$$

(4.3.64)

$$= \frac{e^{-s\tau}}{2\pi^2 i a^2} \int_{-\infty i}^{\infty i} \frac{\sinh[(L-\zeta)\sqrt{q^2-\xi^2}\,]\sinh(z\sqrt{q^2-\xi^2}\,)}{\sqrt{q^2-\xi^2}\,\sinh(L\sqrt{q^2-\xi^2}\,)} K_0(\xi R)\,\xi \, d\xi$$

(4.3.65)

for $0 < z < L$. Evaluating (4.3.65) by the residue theorem where simple poles occur at $\xi_m = \sqrt{q^2 + n^2\pi^2/L^2}$, we find that

$$G(r, \theta, z, s|\rho, \theta', \zeta, \tau)$$

$$= \frac{e^{-s\tau}}{\pi a^2 L} \sum_{n=1}^\infty \sin\left(\frac{n\pi\zeta}{L}\right) \sin\left(\frac{n\pi z}{L}\right) K_0\left(R\sqrt{\frac{s}{a^2} + \frac{n^2\pi^2}{L^2}}\right). \quad (4.3.66)$$

Using tables, we invert (4.3.66) and obtain

$$g(r, \theta, z, t|\rho, \theta', \zeta, \tau) = \frac{H(t-\tau)}{2\pi a^2 L} \sum_{n=1}^\infty \sin\left(\frac{n\pi\zeta}{L}\right) \sin\left(\frac{n\pi z}{L}\right)$$

$$\times \frac{e^{-R^2/[4a^2(t-\tau)]-a^2n^2\pi^2(t-\tau)/L^2}}{t-\tau}.$$

(4.3.67)

In Tables 4.3.3–4.3.6, Green's functions[21] are listed for the heat equation involving various three-dimensional cylindrical geometries.

4.4 HEAT EQUATION WITHIN A SPHERE

Let us find the Green's function for the radially symmetric heat equation within a sphere of radius b. Mathematically, we must solve

$$\frac{\partial g}{\partial t} - \frac{a^2}{r}\frac{\partial^2(rg)}{\partial r^2} = \frac{\delta(r-\rho)\delta(t-\tau)}{4\pi r^2}, \quad 0 < r, \rho < b, \quad 0 < t, \tau, \quad (4.4.1)$$

---

[21] Given in Carslaw, H. S., and J. C. Jaeger, 1940: The determination of Green's function for the equation of conduction of heat in cylindrical coordinates by the Laplace transformation. *J. London Math. Soc.*, **15**, 273–281.

Table 4.3.6: Green's Functions for the Three-Dimensional Heat Conduction Equation

$$\frac{\partial g}{\partial t} - \kappa\left(\frac{\partial^2 g}{\partial r^2} + \frac{1}{r}\frac{\partial g}{\partial r} + \frac{1}{r^2}\frac{\partial^2 g}{\partial \theta^2} + \frac{\partial^2 g}{\partial z^2}\right) = \frac{\delta(r-\rho)\delta(\theta-\theta')\delta(z-\zeta)\delta(t)}{2\pi r}$$

within the Wedge $0 < \theta, \theta' < \theta_0$, $0 < z, \zeta < L$ with Dirichlet Boundary Conditions

---

1. $0 \le r < \infty$

$$g = \frac{4}{L\theta_0}\sum_{k=1}^{\infty} e^{-\kappa k^2 \pi^2 t/L^2}\sin\left(\frac{k\pi z}{L}\right)\sin\left(\frac{k\pi\zeta}{L}\right)$$

$$\times\left\{\sum_{m=1}^{\infty}\sin(s\theta)\sin(s\theta')\int_0^\infty \alpha J_s(\alpha\rho)J_s(\alpha r)e^{-\kappa\alpha^2 t}\,d\alpha\right\},$$

where $s = m\pi/\theta_0$.

2. $0 \le r < a$

$$g = \frac{8}{a^2 L\theta_0}\sum_{k=1}^{\infty} e^{-\kappa k^2 \pi^2 t/L^2}\sin\left(\frac{k\pi z}{L}\right)\sin\left(\frac{k\pi\zeta}{L}\right)$$

$$\times\left\{\sum_{m=1}^{\infty}\sin(s\theta)\sin(s\theta')\sum_{n=1}^{\infty}\frac{J_s(\alpha_n r)J_s(\alpha_n\rho)}{[J_s'(\alpha_n a)]^2}e^{-\kappa\alpha_n^2 t}\right\},$$

where α_n is the nth positive root of $J_s(\alpha a) = 0$, and $s = m\pi/\theta_0$.

3. $a < r < \infty$

$$g = \frac{4}{L\theta_0}\sum_{k=1}^{\infty} e^{-\kappa k^2 \pi^2 t/L^2}\sin\left(\frac{k\pi z}{L}\right)\sin\left(\frac{k\pi\zeta}{L}\right)$$

$$\times\left\{\sum_{m=1}^{\infty}\sin(s\theta)\sin(s\theta')\int_0^\infty \alpha\frac{U_s(\alpha r)U_s(\alpha\rho)}{J_s^2(\alpha a) + Y_s^2(\alpha a)}e^{-\kappa\alpha^2 t}\,d\alpha\right\},$$

where $U_s(\alpha r) = J_s(\alpha r)Y_s(\alpha a) - J_s(\alpha a)Y_s(\alpha r)$, and $s = m\pi/\theta_0$.

4. $a < r < b$

$$g = \frac{2\pi^2}{L\theta_0}\sum_{k=1}^{\infty} e^{-\kappa k^2 \pi^2 t/L^2}\sin\left(\frac{k\pi z}{L}\right)\sin\left(\frac{k\pi\zeta}{L}\right)$$

$$\times\left\{\sum_{m=1}^{\infty}\sin(s\theta)\sin(s\theta')\sum_{n=1}^{\infty}\alpha_n^2\frac{J_s^2(\alpha_n b)U_s(\alpha_n r)U_s(\alpha_n\rho)}{J_s^2(\alpha_n a) - J_s^2(\alpha_n b)}e^{-\kappa\alpha_n^2 t}\right\},$$

where $U_s(\alpha r) = J_s(\alpha r)Y_s(\alpha a) - J_s(\alpha a)Y_s(\alpha r)$, α_n is the nth positive root of $U_s(\alpha b) = 0$, and $s = m\pi/\theta_0$.

---

with the boundary conditions $\lim_{r \to 0} |g(r, t|\rho, \tau)| < \infty$, $g(b, t|\rho, \tau) = 0$, and the initial condition $g(r, 0|\rho, \tau) = 0$.

We begin by introducing the dependent variable $u(r, t|\rho, \tau) = rg(r, t |\rho, \tau)$ so that (4.4.1) becomes

$$\frac{\partial u}{\partial t} - a^2 \frac{\partial^2 u}{\partial r^2} = \frac{\delta(r - \rho)\delta(t - \tau)}{4\pi r}, \quad 0 < r, \rho < b, \quad 0 < t, \tau, \quad \textbf{(4.4.2)}$$

with the boundary conditions $u(0, t|\rho, \tau) = u(b, t|\rho, \tau) = 0$, and the initial condition $u(r, 0|\rho, \tau) = 0$. Taking the Laplace transform of (4.4.2), we obtain

$$\frac{d^2 U}{dr^2} - \frac{s}{a^2} U = -\frac{e^{-s\tau}}{4\pi a^2 r} \delta(r - \rho). \quad \textbf{(4.4.3)}$$

Let us expand $\delta(r - \rho)/r$ in the Fourier sine series

$$\frac{\delta(r - \rho)}{r} = \sum_{n=1}^{\infty} B_n \sin\left(\frac{n\pi r}{b}\right), \quad \textbf{(4.4.4)}$$

where

$$B_n = \frac{2}{b} \int_0^b \frac{\delta(r - \rho)}{r} \sin\left(\frac{n\pi r}{b}\right) dr = \frac{2}{b\rho} \sin\left(\frac{n\pi\rho}{b}\right). \quad \textbf{(4.4.5)}$$

Therefore, we rewrite (4.4.3) as

$$\frac{d^2 U}{dr^2} - \frac{s}{a^2} U = -\frac{e^{-s\tau}}{2\pi a^2 b\rho} \sum_{n=1}^{\infty} \sin\left(\frac{n\pi\rho}{b}\right) \sin\left(\frac{n\pi r}{b}\right), \quad \textbf{(4.4.6)}$$

and

$$U(r, s|\rho, \tau) = \frac{e^{-s\tau}}{2\pi b\rho} \sum_{n=1}^{\infty} \frac{1}{s + a^2 n^2 \pi^2 / b^2} \sin\left(\frac{n\pi\rho}{b}\right) \sin\left(\frac{n\pi r}{b}\right). \quad \textbf{(4.4.7)}$$

Because this particular solution also satisfies the boundary conditions, we do not require any homogeneous solutions so that the sum of the particular and homogeneous solutions satisfies the boundary conditions. Indeed, this is why we choose to expand the delta function in terms of the eigenfunction $\sin(n\pi r/b)$. Taking the inverse of (4.4.7), using the second shifting theorem and substituting for $u(r, t|\rho, \tau)$, we finally obtain

$$g(r, t|\rho, \tau) = \frac{H(t - \tau)}{2\pi b r \rho} \sum_{n=1}^{\infty} \sin\left(\frac{n\pi\rho}{b}\right) \sin\left(\frac{n\pi r}{b}\right) e^{-a^2 n^2 \pi^2 (t - \tau)/b^2}.$$

$$\textbf{(4.4.8)}$$

For the case of a hollow sphere $\alpha < r < \beta$, the Green's function can be found by introducing the new independent variable $x = r - \alpha$ into the governing partial differential equation

$$\frac{\partial g}{\partial t} - \frac{a^2}{r}\frac{\partial^2(rg)}{\partial r^2} = \frac{\delta(r - \rho)\delta(t - \tau)}{4\pi r^2}, \quad \alpha < r, \rho < \beta, \quad 0 < t, \tau, \quad (\textbf{4.4.9})$$

with the boundary conditions $g(\alpha, t|\rho, \tau) = g(\beta, t|\rho, \tau) = 0$, and the initial condition $g(r, 0|\rho, \tau) = 0$. Therefore, the Green's function in this particular case is

$$g(r, t|\rho, \tau) = \frac{H(t - \tau)}{2\pi(\beta - \alpha)r\rho} \sum_{n=1}^{\infty} \sin\left[\frac{n\pi(\rho - \alpha)}{\beta - \alpha}\right] \sin\left[\frac{n\pi(r - \alpha)}{\beta - \alpha}\right]$$

$$\times \exp\left[-\frac{a^2 n^2 \pi^2 (t - \tau)}{(\beta - \alpha)^2}\right]. \quad (\textbf{4.4.10})$$

Finally we find the Green's function for a sphere that varies in the two spatial dimensions r and $\mu = \cos(\theta)$. The governing equation is

$$\frac{\partial g}{\partial t} - a^2\left(\frac{\partial^2 g}{\partial r^2} + \frac{2}{r}\frac{\partial g}{\partial r}\right) - \frac{a^2}{r^2}\frac{\partial}{\partial \mu}\left[(1 - \mu^2)\frac{\partial g}{\partial \mu}\right]$$

$$= \frac{\delta(r - \rho)\delta(\mu - \mu')\delta(t - \tau)}{2\pi r^2 (1 - \mu^2)^{1/2}}, \quad (\textbf{4.4.11})$$

where $0 < r, \rho < b$, $\mu' = \cos(\theta')$, and $0 < t, \tau$, with the boundary conditions $\lim_{r \to 0} |g(r, \mu, t|\rho, \mu', \tau)| < \infty$, $g(b, \mu, t|\rho, \mu', \tau) = 0$, and the initial condition $g(r, \mu, 0|\rho, \mu', \tau) = 0$.

Our analysis begins by taking the Laplace transform of (4.4.11). This yields

$$\frac{\partial^2 G}{\partial r^2} + \frac{2}{r}\frac{\partial G}{\partial r} + \frac{1}{r^2}\frac{\partial}{\partial \mu}\left[(1 - \mu^2)\frac{\partial G}{\partial \mu}\right] - s'G = -\frac{\delta(r - \rho)\delta(\mu - \mu')}{2\pi r^2 (1 - \mu^2)^{1/2}}e^{-s\tau},$$
$$(\textbf{4.4.12})$$

where $s' = s/a^2$. Let $U(r, \mu, s|\rho, \mu', \tau) = \sqrt{r}\, G(r, \mu, s|\rho, \mu', \tau)$. Then

$$\frac{\partial^2 U}{\partial r^2} + \frac{1}{r}\frac{\partial U}{\partial r} - \frac{U}{4r^2} + \frac{1}{r^2}\frac{\partial}{\partial \mu}\left[(1 - \mu^2)\frac{\partial U}{\partial \mu}\right] - s'U$$

$$= -\frac{\delta(r - \rho)\delta(\mu - \mu')}{2\pi r^{3/2} (1 - \mu'^2)^{1/2}}e^{-s\tau}. \quad (\textbf{4.4.13})$$

Since

$$\frac{\delta(\mu - \mu')}{\sqrt{1 - \mu'^2}} = \sum_{n=0}^{\infty} \left(n + \tfrac{1}{2}\right) P_n(\mu)P_n(\mu'), \quad (\textbf{4.4.14})$$

where $P_n(\ )$ is the Legendre polynomial of order n, we have

$$U(r, \mu, s|\rho, \mu', \tau) = e^{-s\tau} \sum_{n=0}^{\infty} \left(n + \tfrac{1}{2}\right) U_n(r|\rho) P_n(\mu) P_n(\mu'). \quad \textbf{(4.4.15)}$$

Substituting (4.4.14) and (4.4.15) into (4.4.13),

$$\frac{d^2 U_n}{dr^2} + \frac{1}{r} \frac{dU_n}{dr} - \frac{\left(n + \tfrac{1}{2}\right)^2}{r^2} U_n - s' U_n = -\frac{\delta(r - \rho)}{2\pi \rho^{3/2}} \quad \textbf{(4.4.16)}$$

with $U_n(b|\rho) = 0$. Solving (4.4.16), we find that

$$U_n(r|\rho) = \frac{1}{\pi b^2 \rho^{1/2}} \sum_{m=1}^{\infty} \frac{J_{n+\frac{1}{2}}(\lambda_{mn} r) J_{n+\frac{1}{2}}(\lambda_{mn} \rho)}{\left[J'_{n+\frac{1}{2}}(\lambda_{mn} b)\right]^2 \left[s' + \lambda_{mn}^2\right]}, \quad \textbf{(4.4.17)}$$

where λ_{mn} is the mth root of $J_{n+\frac{1}{2}}(\lambda b) = 0$. Substituting (4.4.17) into (4.4.15) and inverting the Laplace transform, we obtain

$$g(r, \theta, t|\rho, \theta', \tau) = \frac{a^2 H(t - \tau)}{\pi b^2 \sqrt{r\rho}} \sum_{n=0}^{\infty} \left(n + \tfrac{1}{2}\right) P_n(\mu) P_n(\mu')$$

$$\times \left\{ \sum_{m=1}^{\infty} \frac{J_{n+\frac{1}{2}}(\lambda_{mn} r) J_{n+\frac{1}{2}}(\lambda_{mn} \rho)}{\left[J'_{n+\frac{1}{2}}(\lambda_{mn} b)\right]^2} e^{-a^2 \lambda_{mn}^2 (t-\tau)} \right\}, \quad \textbf{(4.4.18)}$$

where $\mu = \cos(\theta)$, and $\mu' = \cos(\theta')$.

4.5 PRODUCT SOLUTION

So far, we used integral transforms, eigenfunction expansions and the method of images to find Green's functions for the heat equation. We conclude this chapter by showing how products of Green's functions from one-dimensional problems can be multiplied together to give Green's functions for two and three dimensions in rectangular and cylindrical coordinate systems. The method fails in spherical coordinates.

Consider the Green's function in a two-dimensional rectangular coordinate system

$$\frac{\partial g}{\partial t} - \frac{\partial^2 g}{\partial x^2} - \frac{\partial^2 g}{\partial y^2} = \delta(x - \xi) \delta(y - \eta) \delta(t - \tau) \quad \textbf{(4.5.1)}$$

over some domain with linear boundary conditions and $g(x, y, 0|\xi, \eta, \tau) = 0$. We showed in the introductory remarks that we can find the Green's function by solving the initial-value problem

$$\frac{\partial u}{\partial t} - \frac{\partial^2 u}{\partial x^2} - \frac{\partial^2 u}{\partial y^2} = 0 \quad \textbf{(4.5.2)}$$

over the same domain with $u(x, y, \tau) = \delta(x - \xi)\delta(y - \eta)$ and the same boundary conditions.

We solve (4.5.2) by assuming that $u(x, y, t) = u_1(x, t)u_2(y, t)$. By direct substitution,

$$u_2 \left(\frac{\partial u_1}{\partial t} - \frac{\partial^2 u_1}{\partial x^2} \right) + u_1 \left(\frac{\partial u_2}{\partial t} - \frac{\partial^2 u_2}{\partial y^2} \right) = 0. \qquad (4.5.3)$$

Equation (4.5.3) will be satisfied if

$$\frac{\partial u_1}{\partial t} - \frac{\partial^2 u_1}{\partial x^2} = 0, \qquad (4.5.4)$$

and

$$\frac{\partial u_2}{\partial t} - \frac{\partial^2 u_2}{\partial y^2} = 0. \qquad (4.5.5)$$

Next, consider the initial conditions. Direct substitution yields

$$u(x, y, \tau) = u_1(x, \tau)u_2(y, \tau) = \delta(x - \xi)\delta(y - \eta). \qquad (4.5.6)$$

Equation (4.5.6) will be satisfied if we choose $u_1(x, \tau) = \delta(x - \xi)$, and $u_2(y, \tau) = \delta(y - \eta)$.

Finally, we must examine the boundary conditions. For our problem, there are four boundaries: $x = x_i$, and $y = y_i$, where $i = 1, 2$. For the case when x is held constant, the most general boundary condition is

$$k_i \frac{\partial u}{\partial x}\bigg|_{x=x_i} + h_i u\bigg|_{x=x_i} = 0, \qquad (4.5.7)$$

or

$$u_2 \left(k_i \frac{\partial u_1}{\partial x}\bigg|_{x=x_i} + h_i u_1\bigg|_{x=x_i} \right) = 0, \qquad (4.5.8)$$

or

$$k_i \frac{\partial u_1}{\partial x}\bigg|_{x=x_i} + h_i u\bigg|_{x=x_i} = 0, \qquad (4.5.9)$$

where we have assumed that h_i is constant. Similarly, at the other two boundaries,

$$k_j \frac{\partial u_2}{\partial y}\bigg|_{y=y_j} + h_j u_2\bigg|_{y=y_j} = 0. \qquad (4.5.10)$$

Consequently, we may find $u(x, y, t)$ by solving (4.5.4) and (4.5.9) with $u_1(x, \tau) = \delta(x - \xi)$, as well as (4.5.5) and (4.5.10) with $u_2(y, \tau) = \delta(y - \eta)$. This corresponds to finding the Green's function for the problems

$$\frac{\partial g_1}{\partial t} - \frac{\partial^2 g_1}{\partial x^2} = \delta(x - \xi)\delta(t - \tau), \qquad (4.5.11)$$

with the boundary conditions

$$k_i \frac{\partial g_1}{\partial x}\bigg|_{x=x_i} + h_i g_1 \bigg|_{x=x_i} = 0, \qquad (4.5.12)$$

and

$$\frac{\partial g_2}{\partial t} - \frac{\partial^2 g_2}{\partial y^2} = \delta(y - \eta)\delta(t - \tau), \qquad (4.5.13)$$

with the boundary conditions

$$k_j \frac{\partial g_2}{\partial y}\bigg|_{y=y_j} + h_j g_2 \bigg|_{y=y_j} = 0. \qquad (4.5.14)$$

Therefore, $g(x, y, t|\xi, \eta, \tau) = u(x, y, t)H(t - \tau) = g_1(x, t|\xi, \tau)g_2(y, t|\eta, \tau)$. In general, one-dimensional Green's functions multiply in rectangular coordinates to give multidimensional Green's functions regardless of whether the domain is unlimited or we have Dirichlet, Neumann or Robin boundary conditions.

Does this hold for cylindrical coordinates? It does, but in a more limited sense. Let us denote the one-dimensional Green's functions in each direction by g_r, g_φ and g_z. Then the multi-dimensional Green's function g_{rz}, $g_{\varphi z}$ and $g_{r\varphi z}$ are given by $g_{rz} = g_r g_z$, $g_{\varphi z} = g_\varphi g_z$, and $g_{r\varphi z} = g_{r\varphi} g_z$, if we are dealing with free-space Green's functions or have Dirichlet or Neumann boundary conditions. Note that $g_{r\varphi}$ *cannot* be found by product solution. We can also include Robin boundary conditions if we require that h does not vary along any of the boundaries.

• Example 4.5.1

Let us find the Green's function for

$$\frac{\partial g}{\partial t} - a^2 \left(\frac{\partial^2 g}{\partial x^2} + \frac{\partial^2 g}{\partial y^2} \right) = \delta(x - \xi)\delta(y - \eta)\delta(t - \tau), \qquad (4.5.15)$$

where $-\infty < x, \xi < \infty$, $0 < y, \eta < \infty$, and $0 < t, \tau$, subject to the boundary conditions

$$\lim_{|x|\to\infty} |g(x, y, t|\xi, \eta, \tau)| < \infty, \quad \lim_{y\to\infty} |g(x, y, t|\xi, \eta, \tau)| < \infty, \quad (4.5.16)$$

and

$$g_y(x, 0, t|\xi, \eta, \tau) - hg(x, 0, t|\xi, \eta, \tau) = 0, \qquad (4.5.17)$$

with the initial condition

$$g(x, y, 0|\xi, \eta, \tau) = 0. \qquad (4.5.18)$$

Using the method of product solutions, we express the Green's function as a product of two Green's function

$$g(x, y, t|\xi, \eta, \tau) = g_1(x, t|\xi, \tau)g_2(y, t|\eta, \tau), \tag{4.5.19}$$

where $g_1(x, t|\xi, \tau)$ is given by the one-dimensional problem

$$\frac{\partial g_1}{\partial t} - a^2 \frac{\partial^2 g_1}{\partial x^2} = \delta(x-\xi)\delta(t-\tau), \quad -\infty < x, \xi < \infty, \quad 0 < t, \tau, \tag{4.5.20}$$

with the boundary conditions

$$\lim_{|x| \to \infty} |g_1(x, t|\xi, \tau)| < \infty, \tag{4.5.21}$$

and the initial condition

$$g_1(x, 0|\xi, \tau) = 0, \tag{4.5.22}$$

while $g_2(y, t|\eta, \tau)$ is given by the one-dimensional problem

$$\frac{\partial g_2}{\partial t} - a^2 \frac{\partial^2 g_2}{\partial y^2} = \delta(y - \eta)\delta(t - \tau), \quad 0 < y, \eta < \infty, \quad 0 < t, \tau, \tag{4.5.23}$$

with the boundary conditions

$$\lim_{y \to \infty} |g_2(y, t|\eta, \tau)| < \infty, \quad g_{2y}(0, t|\eta, \tau) - hg_2(0, t|\eta, \tau) = 0, \tag{4.5.24}$$

and the initial condition

$$g_2(x, 0|\eta, \tau) = 0. \tag{4.5.25}$$

The solution to (4.5.20)–(4.5.22) is the free-space Green's function given by (4.1.6). We found the Green's function $g_2(y, t|\eta, \tau)$ in Example 4.1.2. Therefore,

$$g(x, y, t|\xi, \eta, \tau) = \frac{H(t - \tau)}{4\pi a^2(t - \tau)} \left\{ \exp\left[-\frac{(x - \xi)^2 + (y + \eta)^2}{4a^2(t - \tau)} \right] \right.$$
$$+ \exp\left[-\frac{(x - \xi)^2 + (y - \eta)^2}{4a^2(t - \tau)} \right]$$
$$- 2h \exp\left[-\frac{(x - \xi)^2}{4a^2(t - \tau)} \right]$$
$$\left. \times \int_0^\infty \exp\left[-h\chi - \frac{(y + \eta + \chi)^2}{4a^2(t - \tau)} \right] d\chi \right\}. \tag{4.5.26}$$

- **Example 4.5.2**

Using the product solution method, we find in this example the Green's function[22] for

$$\frac{\partial g}{\partial t} - a^2 \left[\frac{1}{r}\frac{\partial}{\partial r}\left(r\frac{\partial g}{\partial r} \right) + \frac{\partial^2 g}{\partial z^2} \right] = \frac{\delta(r-\rho)\delta(z-\zeta)\delta(t-\tau)}{2\pi r}, \quad (4.5.27)$$

where $0 < r, \rho < \infty$, $0 < z, \zeta < \infty$, and $0 < t, \tau$. The boundary conditions are

$$\lim_{r\to 0} |g(r,z,t|\rho,\zeta,\tau)| < \infty, \qquad \lim_{r\to\infty} |g(r,z,t|\rho,\zeta,\tau)| < \infty, \quad (4.5.28)$$

$$g_z(r,0,t|\rho,\zeta,\tau) = 0, \quad \text{and} \quad \lim_{z\to\infty} |g(r,z,t|\rho,\zeta,\tau)| < \infty. \quad (4.5.29)$$

We begin by finding a product solution of the form

$$g(r,z,t|\rho,\zeta,\tau) = g_1(r,t|\rho,\tau)g_2(z,t|\zeta,\tau), \quad (4.5.30)$$

where

$$\frac{\partial g_1}{\partial t} - \frac{a^2}{r}\frac{\partial}{\partial r}\left(r\frac{\partial g_1}{\partial r} \right) = \frac{\delta(r-\rho)\delta(t-\tau)}{2\pi r}, \quad (4.5.31)$$

with

$$\lim_{r\to 0} |g_1(r,t|\rho,\tau)| < \infty, \quad \text{and} \quad \lim_{r\to\infty} |g_1(r,t|\rho,\tau)| < \infty, \quad (4.5.32)$$

and

$$\frac{\partial g_2}{\partial t} - a^2\frac{\partial^2 g_2}{\partial z^2} = \delta(z-\zeta)\delta(t-\tau), \quad (4.5.33)$$

with

$$\frac{\partial g_2(0,t|\zeta,\tau)}{\partial z} = 0, \quad \text{and} \quad \lim_{z\to\infty} |g_2(z,t|\zeta,\tau)| < \infty. \quad (4.5.34)$$

As you will show in Problem 8,

$$g_1(r,t|\rho,\tau) = \frac{H(t-\tau)}{4\pi a^2(t-\tau)} \exp\left[-\frac{r^2+\rho^2}{4a^2(t-\tau)} \right] I_0\left[\frac{r\rho}{2a^2(t-\tau)} \right]. \quad (4.5.35)$$

Using the method of images, we have that

$$g_2(z,t|\zeta,\tau) = \frac{H(t-\tau)}{\sqrt{4\pi a^2(t-\tau)}}\left\{ \exp\left[-\frac{(z-\zeta)^2}{4a^2(t-\tau)} \right] + \exp\left[-\frac{(z+\zeta)^2}{4a^2(t-\tau)} \right] \right\}. \quad (4.5.36)$$

---

[22] For an example of its use, see Nelson, D. J., and B. Vick, 1995: Thermal limitations in optical recording. *IEEE Trans. Comp., Packag., and Manufact. Technol.—Part A*, **18**, 521–526.

Therefore,

$$g(r, z, t | \rho, \zeta, \tau) = \frac{H(t - \tau)}{[4\pi a^2 (t - \tau)]^{3/2}} \exp\left[-\frac{r^2 + \rho^2}{4a^2(t - \tau)}\right] I_0\left[\frac{r\rho}{2a^2(t - \tau)}\right]$$
$$\times \left\{\exp\left[-\frac{(z - \zeta)^2}{4a^2(t - \tau)}\right] + \exp\left[-\frac{(z + \zeta)^2}{4a^2(t - \tau)}\right]\right\}. \quad (4.5.37)$$

4.6 ABSOLUTE AND CONVECTIVE INSTABILITY

One of the striking features of dynamical flows is the propensity of some of them to be unstable to small perturbations. This is especially true of fluids where the study of hydrodynamic instabilities dates back to the nineteenth century and involves such luminaries as Helmholtz, Kelvin, Rayleigh and G. I. Taylor.

The analysis begins by deriving a set of linearized equations that describe the response of a simple basic state that satisfies the governing equations to small perturbations. The partial differential equation

$$\frac{\partial u}{\partial t} = \frac{\partial^2 u}{\partial x^2} + a\frac{\partial u}{\partial x} + u, \quad -\infty < x < \infty, \quad 0 < t, \quad (4.6.1)$$

where a is real, is prototypical[23] of the perturbation equations that exhibit instability. Because of its homogeneity in space and time, solutions to (4.6.1) can be expressed as a superposition of normal modes, namely

$$u(x, t) = Ae^{i(kx - \lambda t)}. \quad (4.6.2)$$

Upon substituting (4.6.2) into (4.6.1), we find that the relationship between the frequency λ and wavenumber k is governed by

$$\lambda = -ak + i(1 - k^2). \quad (4.6.3)$$

In classical stability calculations, instability occurs if for some *real* wavenumber k a *complex* λ with positive $\text{Im}(\lambda)$ is obtained from the dispersion equation and we have temporal growth of spatial periodic disturbances of infinite extent. For our particular example, instability occurs when $-1 < k < 1$, and $\text{Im}(\lambda) = 1 - k^2 > 0$.

During the 1950s, plasma physicists further refined this concept of stability. This was necessary because most initial perturbations are

---

[23] This is a simplified version of the linearized Ginzburg-Landau equation. In the case of fluids, the Ginzburg-Landau equation governs the amplitude of perturbations in a Stewartson-Stuart [Stewartson, K., and J. T. Stuart, 1971: A non-linear instability theory for a wave system in plane Poiseuille flow. *J. Fluid Mech.*, **48**, 529–545. See equation (4.9).] stability calculation of a fluid flow.

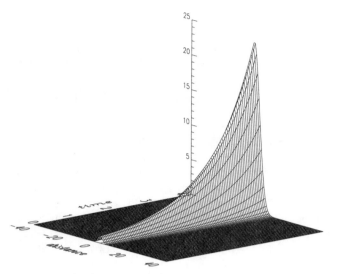

Figure 4.6.1: Plot of $u(x,t)$ given by (4.6.5) as a function of distance x and time t when $a = \frac{1}{2}$ and $b = 1$. This is an example of an absolutely unstable solution.

not monochromatic and their presence, even if the normal modes were unstable, would not ensure instability, because they might simply propagate away, leaving nothing. In this sense, the system would be stable in the sense that the amplitude in a fixed, local region *eventually* decays. To illustrate this, consider an initial pulse that consists of the Gaussian distribution

$$u(x,0) = e^{-bx^2/4}, \qquad b > 0. \tag{4.6.4}$$

The exact solution of (4.6.1) in this case is

$$u(x,t) = \frac{e^t}{\sqrt{1+bt}} \exp\left[-\frac{b(x+at)^2}{4(1+bt)}\right]. \tag{4.6.5}$$

What happens as $t \to \infty$? At $x = 0$,

$$u(0,t) = \frac{e^t}{\sqrt{1+bt}} \exp\left[-\frac{ba^2t^2}{4(1+bt)}\right] \tag{4.6.6}$$

$$= \frac{1}{\sqrt{1+bt}} \exp\left\{\frac{t}{1+bt}\left[1 + bt\left(1 - \tfrac{1}{4}a^2\right)\right]\right\}. \tag{4.6.7}$$

In the limit $t \to \infty$, $u(0,t)$ will grow in time without limit if $|a| < 2$. This is also true of any other *fixed* point in the absolute frame of reference. Such an instability is called an *absolute* or *globalized* instability. See Figure 4.6.1. This type of instability is generally catastrophic, growing exponentially in time at all points in space leading to immediate and

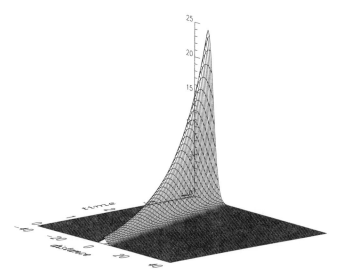

Figure 4.6.2: Same as Figure 4.6.1 except that $a = 4$. This is an example of a convectively unstable solution.

dramatic changes in the flow. On the other hand, if $|a| > 2$, $u(0, t)$ decays exponentially to zero and we have absolute stability.

If we now introduce a moving reference frame with velocity $-a$, then (4.6.5) grows exponentially in time for any real a. Consequently, if the convection speed $|a|$ is large enough, the instability in the absolute reference frame is localized in space around the moving point $x = -at$. In this case, the group velocity of the disturbance is greater than the speed at which it spreads. This type of instability is called a *convective* or *localized* instability.[24] See Figure 4.6.2. Because a convective or localized instability can convect away from a region of interest, it may be possible to control.

The fact that a flow may not be absolutely unstable but convectively unstable has had a profound effect on our concept of what stability and instability are. The application of Green's function to this problem yields a clean and concise understanding of this phenomenon because of the simple nature of the forcing.

Consider the Green's function associated with the partial differential equation (4.6.1)

$$\frac{\partial g}{\partial t} - \frac{\partial^2 g}{\partial x^2} - a\frac{\partial g}{\partial x} - g = \delta(x - \xi)\delta(t - \tau), \quad -\infty < x, \xi < \infty, \quad 0 < t, \tau. \tag{4.6.8}$$

---

[24] Some authors prefer *spatial* instability because it frees the phenomenon from its association with just fluids.

Our study of instability will focus on this two-dimensional partial differential equation, which is invariant with respect to the time displacement and uniform in the spatial dimension.[25] Upon applying Laplace transforms in time and Fourier transforms in space, the joint transform $\mathcal{G}(k, s | \xi, \tau)$ is

$$(s + k^2 - iak - 1)\mathcal{G}(k, s | \xi, \tau) = e^{-ik\xi - s\tau}, \qquad (4.6.9)$$

and

$$g(x, t | \xi, \tau) = \frac{1}{4\pi^2 i} \int_{c-\infty i}^{c+\infty i} \int_{-\infty}^{\infty} \frac{e^{ik(x-\xi)+s(t-\tau)}}{s + k^2 - iak - 1} \, dk \, ds. \qquad (4.6.10)$$

To compute $g(x, t | \xi, \tau)$, we first invert the Laplace transform in (4.6.10), which yields

$$g(x, t | \xi, \tau) = \frac{H(t - \tau)}{2\pi} \int_{-\infty}^{\infty} e^{ik(x-\xi)-(k^2 - iak - 1)(t-\tau)} \, dk. \qquad (4.6.11)$$

Performing the k-integration, we obtain

$$g(x, t | \xi, \tau) = \frac{e^{(t-\tau) - a^2(t-\tau)/4 - a(x-\xi)/2}}{2\sqrt{\pi(t - \tau)}} \exp\left[-\frac{(x - \xi)^2}{4(t - \tau)}\right] H(t - \tau).$$
$$(4.6.12)$$

Once again, we see that if $|a| > 2$, the solution will decay to zero for any fixed value of x as $t \to \infty$. Conversely, if $|a| < 2$, there is exponential growth for any fixed value of x and we have absolute instability.

So far, we have shown that the Green's function formulation provides a straightforward method for determining the stability of (4.6.1). At this point, we would like to extend this method so that it applies to the general problem when we cannot write $g(x, t | \xi, \tau)$ in an analytic form, as we just did. The remaining portions of this section is devoted to developing analytic tools for the treatment of this general case.

We begin by noting that any linear, constant coefficient partial differential equation that varies in x and t has the Green's function

$$g(x, t | \xi, \tau) = \frac{1}{4\pi^2 i} \int_{c-\infty i}^{c+\infty i} \int_{-\infty}^{\infty} \frac{e^{ik(x-\xi)+s(t-\tau)}}{D(k, s)} \, dk \, ds, \qquad (4.6.13)$$

or

$$g(x, t | \xi, \tau) = \frac{1}{4\pi^2} \int_{-\infty+i\sigma}^{\infty+i\sigma} \int_{-\infty}^{\infty} \frac{e^{ik(x-\xi)-i\omega(t-\tau)}}{D(k, \omega)} \, dk \, d\omega, \qquad (4.6.14)$$

---

[25] Equations that are nonuniform in space have been explored by Brevdo, L., and T. J. Bridges, 1996: Absolute and convective instabilities of spatially periodic flows. *Phil. Trans. R. Soc. Lond.*, **A354**, 1027–1064; Huerre, P., and P. A. Monkewitz, 1990: Local and global instabilities in spatial developing flows. *Ann. Rev. Fluid Mech.*, **22**, 473–537.

where we have rewritten the Laplace inversion in (4.6.14) as an inversion of a Fourier transform by letting $s = -\omega i$. The contour in the complex frequency plane, commonly called the *Bromwich contour*, is a straight horizontal line located above all of the singularities of the integrand as dictated by causality, which requires that $g(x,t|\xi,\tau) = 0$ for all x when $t < \tau$. Hence, the σ in (4.6.14) is greater than the imaginary part of any ω given by $D(k,\omega) = 0$. The path in the complex wavenumber plane is initially taken along the real axis.

For simplicity, let us assume that (4.6.14) has a single discrete temporal mode $\omega(k)$. Then the Green's function $g(x,t|\xi,\tau)$ can be formally obtained from a residue calculation in the ω-plane at $\omega = \omega(k)$. Carrying out the calculation, we have that

$$g(x,t|\xi,\tau) = -\frac{i}{2\pi}H(t-\tau)\int_{-\infty}^{\infty} \frac{e^{i[k(x-\xi)-\omega(k)(t-\tau)]}}{D_\omega[k,\omega(k)]}\, dk. \qquad (4.6.15)$$

We can quickly check (4.6.15) by using our earlier example. In this case, $D_\omega[k,\omega(k)] = -i$, and $\omega(k) = -ak + i(1-k^2)$. Substituting into (4.6.15), we recover (4.6.11).

Ideally, we would like to evaluate (4.6.15) exactly as we did (4.6.11) Unfortunately, most of the time we cannot and must pursue another avenue of attack. Presently, we choose to compute the Fourier integral by the method of steepest descent[26] for large t and fixed $(x-\xi)/(t-\tau)$. Although the exact steepest descent path depends upon the particular form of $\omega(k)$, we presently assume that there is only a single stationary point k_* for the phase of the integrand where $\partial\omega(k_*)/\partial k = (x-\xi)/(t-\tau)$. Under this condition, the original contour of integration along the real k-axis can be deformed into a steepest descent path passing through the saddle point k_*. Performing the calculation,

$$g(x,t|\xi,\tau) \sim -\frac{e^{\pi i/4}}{\sqrt{2\pi}}\frac{e^{i[k_*(x-\xi)-\omega(k_*)(t-\tau)]}}{D_\omega[k_*,\omega(k_*)]\sqrt{\omega''(k_*)(t-\tau)}}. \qquad (4.6.16)$$

For instance, in our present example, $\omega''(k_*) = -2i$,

$$k_* = \frac{ai}{2} + \frac{i}{2}\left(\frac{x-\xi}{t-\tau}\right), \qquad (4.6.17)$$

---

[26] Recently, Lingwood [Lingwood, R. J., 1997: On the application of the Briggs' and steepest-descent methods to a boundary-layer flow. *Stud. Appl. Math.*, **98**, 213–254.] provided examples of saddle points through which the steepest descent path cannot be made to pass. For this reason, care must be taken to establish the topography of the phase function. Otherwise, calculations of the wave packet evolution can be seriously flawed and incorrect conclusions can be drawn about the convective or absolute instability of the flow.

and

$$\omega(k_*) = i - \frac{a^2 i}{4} + \frac{i}{4}\left(\frac{x - \xi}{t - \tau}\right)^2. \qquad (4.6.18)$$

Upon substituting these results into (4.6.16), the asymptotic evaluation of (4.6.15) equals the exact solution (4.6.12); this is generally not true.

Although (4.6.16) is only the asymptotic evaluation of (4.6.15), that is sufficient for our stability analysis. Examining (4.6.16) more closely, we see that the Green's function behaves like a wave packet in the (x, t)-plane. Along each ray $(x - \xi)/(t - \tau)$ within the packet, the response is dominated by a specific complex wavenumber k_* such that its real group velocity satisfies $\omega_k(k_*) = (x - \xi)/(t - \tau)$. Along each ray, the temporal growth rate is $\mathrm{Im}\left[\omega(k_*) - (x - \xi)k_*/(t - \tau)\right]$.

Let us return to (4.6.15). As k varies from $-\infty$ to ∞, $\mathrm{Im}[\omega(k)]$ also varies. Let us denote by $k_{\max}$ the wavenumber where $\mathrm{Im}[\omega(k)]$ has its maximum value, $\omega_{i,\max} = \mathrm{Im}[\omega(k_{\max})]$. Clearly, for the system to be unstable, $\mathrm{Im}[\omega(k)] > 0$ at $k = k_{\max}$ and we arrive at the following criteria for linear instability:

- If $\omega_{i,\max} > 0$, then the system is linearly unstable;

- If $\omega_{i,\max} < 0$, then the system is linearly stable.

Although we now have criteria for instability, how do we discern an absolute instability from a convective instability? Recall that the difference between them involves the behavior of a *fixed* point $x - \xi$ as $t - \tau \to \infty$: absolute instability occurs when the solution grows with time, while convective instability occurs when the solution damps. Consequently, if we examine (4.6.16) along the ray $(x - \xi)/(t - \tau) = 0$ as $t - \tau \to \infty$, we can develop a criterion.

Let us denote the k_* for the $(x - \xi)/(t - \tau) = 0$ ray by k_0 and the corresponding $\omega(k_*)$ by $\omega_0 = \omega(k_0)$. Then the *absolute growth rate* is $\omega_{0,i} = \mathrm{Im}[\omega_0(k_0)]$. From (4.6.16) it follows that

- If $\omega_{0,i} > 0$, the system is absolutely unstable;

- if $\omega_{0,i} < 0$, the system is convectively unstable.

We may illustrate these concepts with our earlier example involving (4.6.8). Because $\omega(k) = -ak + i(1 - k^2)$, $k_{\max} = 0$ and $\omega_{i,\max} = 1$. The system is unstable for any value of a. Turning to the criteria for convective/absolute instability, a quick calculation shows that $k_0 = ai/2$ from (4.6.17) since $(x - \xi)/(t - \tau) = 0$; $\omega_0 = \omega(k_0) = i(1 - a^2/4)$ and

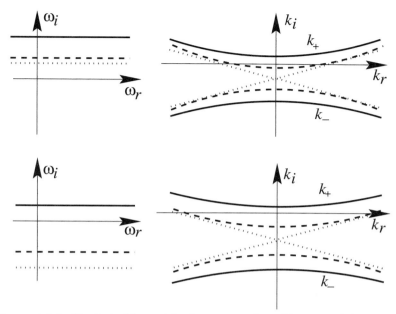

Figure 4.6.3: The loci of k_+ and k_- for our sample problem as $\operatorname{Re}(\omega)$ varies for various fixed $\operatorname{Im}(\omega)$. As $\operatorname{Im}(\omega)$ decreases (first the solid line, then the dashed line, and finally the dotted line), k_+ and k_- move toward each other, finally pinching at a single point. In the top frame, the case for absolute instability, $a = -1$. In the bottom frame, the case for convective instability, $a = -4$.

$\omega_{0,i} = 1 - a^2/4$ from (4.6.18). Therefore, if $|a| > 2$, we have convective instability; if $|a| < 2$, we have absolute instability.

This method of determining absolute/convective instability from $\omega_{0,i}$ may not be precise enough when $\omega(k)$ is particularly complicated. For example, $\omega(k)$ may be multivalued so that we must consider branch cuts and Riemann surfaces. To formulate alternative criteria, consider the case where the temporal mode $\omega(k)$ exhibits a single second-order algebraic branch point ω_0 with only two spatial branches $k_+(\omega)$ and $k_-(\omega)$. The example that we have been using is just such a case because $\omega - \omega_0 = -i(k - k_0)^2$ and ω_0, the absolute frequency, is an algebraic branch point of the function $k(\omega)$ on the ω-plane; k_0 is a saddle point of $\omega(k)$.

Let us now follow the loci of $k_+(\omega)$ and $k_-(\omega)$ as ω travels along the Bromwich contour and we lower the contour to the real axis of the ω-plane. We do this so that we can discover the asymptotic response of the inverse Laplace transform; the response in the limit $t \to \infty$ is governed by the *highest* singularity in $D(k,\omega)$ in the ω-plane. When the Bromwich contour is high enough, the spatial branches k_+ and k_- are located on opposite sides of the real k-axis. See the solid lines in Figure 4.6.3. As the Bromwich integral is displaced downward, k_+

and k_- move toward each other and the original path along the real k-axis in the Fourier inversion must be deformed to remain between them. This simultaneous deformation of Bromwich and Fourier contours cannot proceed further if the Bromwich contour touches ω_0 and Fourier contour becomes pinched between the branches $k_+(\omega)$ and $k_-(\omega)$. See the dotted lines in Figure 4.6.3. Pinching occurs precisely at the point k_0 where the group velocity equals zero. If the corresponding ω_0 is located in the upper half of the ω-plane, as is the case for the top frame shown in Figure 4.6.3, the instability is absolute; otherwise, it is convective. This graphical method of determining absolute or convective instability is commonly called the Briggs' criterion.[27]

• Example 4.6.1: The Ginzburg-Landau equation

The time-dependent, generalized Ginzburg-Landau equation is a simplified version of the dynamical equations for fluid mechanical systems such as Rayleigh-Bénard convection, Taylor-Couette flow, plane Poiseuille flow and wind-generated water waves. If a, b, and c denote complex constants with $\operatorname{Re}(b) \geq 0$, the Ginzburg-Landau equation is

$$\frac{\partial u}{\partial t} + v_g \frac{\partial u}{\partial x} - au - b\frac{\partial^2 u}{\partial x^2} + c|u|^2 u = 0, \qquad (4.6.19)$$

where v_g denotes the group velocity.

Consider the free-space Green's function for the linearized form of (4.6.19):

$$\frac{\partial g}{\partial t} + v_g \frac{\partial g}{\partial x} - ag - b\frac{\partial^2 g}{\partial x^2} = \delta(x - \xi)\delta(t - \tau). \qquad (4.6.20)$$

The last term in (4.6.19) disappears because we linearized about $u = 0$. The Green's function is

$$g(x,t|\xi,\tau) = \frac{e^{a(t-\tau)} H(t - \tau)}{2\sqrt{\pi b(t - \tau)}} \exp\left\{-\frac{[x - \xi - v_g(t - \tau)]^2}{4b(t - \tau)}\right\}. \qquad (4.6.21)$$

In the development of our criteria for absolute and convective instability, we focused on the dispersion relationship. We will do so momentarily. Before that, we will first examine an alternative definition of these instabilities given by Bers.[28] They are based more on physical intuition.

---

[27] Briggs, R. J., 1964: *Electron-Stream Interaction With Plasmas*. MIT Press, chapter 2.

[28] Bers, A., 1983: Space-time evolution of plasma instabilities—absolute and convective. *Basic Plasma Physics, Vol. I*, A. A. Galeev and R. N. Sudan, Eds., North-Holland, 451–517.

Bers states that in plasma problems there are three different types of stability possible. The first type is defined by

$$\lim_{t \to \infty} |g(x, t | \xi, \tau)| \to \infty \qquad (4.6.22)$$

for an arbitrary, fixed value of x. This condition corresponds to *absolute instability*. Checking (4.6.21), (4.6.22) will be satisfied if

$$\text{Re}\left(a - \frac{v_g^2}{4b}\right) > 0, \qquad \text{or} \qquad \text{Re}(a) - \frac{v_g^2 \, \text{Re}(b)}{4|b|^2} > 0. \qquad (4.6.23)$$

Because $\text{Re}(b) > 0$, absolute instability occurs here if $\text{Re}(a)$ is sufficiently large.

The second type of stability is defined by

$$\lim_{t \to \infty} |g(\Xi + vt, t | \xi, \tau)| \to 0 \qquad (4.6.24)$$

for some v and for arbitrary fixed value of Ξ. This condition corresponds to *absolute stability* and disturbances will die out in any frame of reference. From (4.6.21), this will occur when

$$\text{Re}\left[a - \frac{(v - v_g)^2}{4b}\right] < 0. \qquad (4.6.25)$$

Therefore, regardless of the values of v and v_g, the flow will be absolutely stable if $\text{Re}(a) < 0$.

Finally, the third type of stability is dictated by

$$\lim_{t \to \infty} |g(x, t | \xi, \tau)| \to 0, \quad \text{and} \quad \lim_{t \to \infty} |g(\Xi + vt, t | \xi, \tau)| \to \infty \qquad (4.6.26)$$

for some v and for arbitrary fixed values of x and Ξ, respectively. For the present problem, this occurs for $0 < \text{Re}(a) < v_g^2 \text{Re}(b)/(4|b|^2)$. This condition corresponds to *convective* or *spatial instability*. In this case, the disturbance damps as $t \to \infty$ at *any* given stationary point but a moving frame of reference may be found in which the disturbance grows. As the disturbance convects away, the system returns to its original, undisturbed state.

Let us now determine the stability criterion from our pulse analysis. The temporal growth rate along any ray $(x - \xi)/(t - \tau)$ is

$$\text{Im}\left[\omega(k_*) - (x - \xi)k_*/(t - \tau)\right], \qquad (4.6.27)$$

where k_* satisfies $\omega_k(k_*) = (x - \xi)/(t - \tau)$. In our current problem,

$$k_* = -\frac{v_g i}{2b} + \frac{i}{2b}\left(\frac{x - \xi}{t - \tau}\right), \qquad (4.6.28)$$

and

$$\text{Im}\left[\omega(k_*) - \frac{x - \xi}{t - \tau} k_*\right]$$

$$= \text{Re}(a) + \left[\frac{v_g}{2}\left(\frac{x - \xi}{t - \tau}\right) - \frac{v_g^2}{4} - \frac{1}{4}\left(\frac{x - \xi}{t - \tau}\right)^2\right]\frac{\text{Re}(b)}{|b|^2}. \quad (4.6.29)$$

Maximizing (4.6.29) with respect to $(x-\xi)/(t-\tau)$ gives $(x-\xi)/(t-\tau) = v_g$. Along this ray $[(x-\xi)/(t-\tau)]_{max}$ there is maximum temporal growth that equals $\text{Re}(a)$. Thus, the system becomes convectively unstable whenever $\text{Re}(a) > 0$, the same result that we obtained earlier. In this case, there is a reference frame moving at the group velocity in which a localized disturbance will appear to grow although in the stationary frame $(x - \xi)/(t - \tau) = 0$, it decays to zero.

For $\text{Re}(a) > 0$, there exists a pair of values, $[(x-\xi)/(t-\tau)]_{slow}$ and $[(x-\xi)/(t-\tau)]_{fast}$, on either side of $[(x-\xi)/(t-\tau)]_{max}$ for which (4.6.29) returns to zero. These values of $(x - \xi)/(t - \tau)$ represent *fronts* of the disturbance along which the system is marginally stable. Eventually, for some value of a, $[(x - \xi)/(t - \tau)]_{slow} = 0$ and the system becomes absolutely unstable. From (4.6.29) this occurs when

$$\text{Re}(a) - \frac{v_g^2 \text{Re}(b)}{4|b|^2} = 0. \quad (4.6.30)$$

When (4.6.30) is satisfied, the growth rate is simply $\text{Im}[\omega(k_*)]$ for any fixed $x - \xi$ as $t \to \infty$. Any localized disturbance will grow exponentially with time at its point of origin even though the disturbance itself may propagate away as a wave packet. On the other hand, for $0 < \text{Re}(a) < v_g^2 \text{Re}(b)/(4|b|^2)$, the disturbance propagates away quickly enough relative to its growth and spreading so that the point of origin settles back to its undisturbed state.

In summary, we see that as the value of $\text{Re}(a)$ changes, so does the behavior of the solution. For $\text{Re}(a) < 0$, the flow is absolutely stable. As soon as $\text{Re}(a) > 0$, the flow becomes convectively or spatially unstable. Eventually, for large enough and positive $\text{Re}(a)$, the flow is absolutely unstable.

Problems

Unlimited Domain

1. Find the free-space Green's function for

$$\frac{\partial g}{\partial t} - a^2\left(\frac{\partial^2 g}{\partial x^2} + \frac{\partial^2 g}{\partial y^2} - b\frac{\partial g}{\partial x}\right) = \delta(x - \xi)\delta(y - \eta)\delta(t - \tau),$$

subject to the boundary conditions

$$\lim_{|x|,|y|\to\infty} |g(x,y,t|\xi,\eta,\tau)| < \infty, \quad 0 < t,$$

and the initial condition

$$g(x,y,0|\xi,\eta,\tau) = 0, \quad -\infty < x,y < \infty.$$

Step 1: Apply Laplace transforms to the partial differential equation and show that it becomes

$$\frac{\partial^2 G}{\partial x^2} + \frac{\partial^2 G}{\partial y^2} - b\frac{\partial G}{\partial x} - \frac{s}{a^2}G = -\frac{\delta(x-\xi)\delta(y-\eta)}{a^2}e^{-s\tau},$$

with

$$\lim_{|x|,|y|\to\infty} |G(x,y,s|\xi,\eta,\tau)| < \infty.$$

Step 2: Taking the Fourier transform in *both* the x and y directions, show that the partial differential equation in Step 1 reduces to

$$\overline{G}(k,\ell,s|\xi,\eta,\tau) = \frac{e^{-ik\xi-i\ell\eta-s\tau}}{a^2(k^2+\ell^2+ibk)+s},$$

where $\overline{G}(k,\ell,s|\xi,\eta,\tau)$ denotes the double Fourier transform of $G(x,y,s|\xi,\eta,\tau)$.

Step 3: Following the example of Problem 7 in Chapter 5, show that

$$G(x,y,s|\xi,\eta,\tau) = \frac{e^{b(x-\xi)/2-s\tau}}{2\pi a^2}K_0\left[\sqrt{(x-\xi)^2+(y-\eta)^2}\sqrt{\frac{b^2}{4}+\frac{s}{a^2}}\right].$$

Step 4: Using tables, show that

$$g(x,y,t|\xi,\eta,\tau) = \frac{H(t-\tau)}{4\pi a^2(t-\tau)}\exp\left\{-\frac{[(x-\xi)-a^2b(t-\tau)]^2+(y-\eta)^2}{4a^2(t-\tau)}\right\}.$$

2. Find the free-space Green's function for the three-dimensional heat equation[29]

$$\frac{\partial g}{\partial t} - a^2\left(\frac{\partial^2 g}{\partial x^2} + \frac{\partial^2 g}{\partial y^2} + \frac{\partial^2 g}{\partial z^2}\right) = a^2\delta(x-\xi)\delta(y-\eta)\delta(z-\zeta)\delta(t-\tau).$$

---

[29] Reprinted with permission from Bowler, J. R., 1999: Time domain half-space dyadic Green's functions for eddy-current calculations. *J. Appl. Phys.*, **86**, 6494–6500. ©American Institute of Physics, 1999.

Step 1: Apply Laplace transforms to the partial differential equation and show that it becomes

$$\frac{\partial^2 G}{\partial x^2} + \frac{\partial^2 G}{\partial y^2} + \frac{\partial^2 G}{\partial z^2} - \frac{s}{a^2} G = -\delta(x - \xi)\delta(y - \eta)\delta(z - \zeta)e^{-s\tau}.$$

Step 2: Taking the Fourier transform in *both* the x and y directions, show that the partial differential equation in Step 1 reduces to the ordinary differential equation

$$\frac{d^2 \overline{G}}{dz^2} - (\kappa^2 + s/a^2)\overline{G} = -\delta(z - \zeta)e^{-ik\xi - i\ell\eta - s\tau},$$

where k and ℓ are the Fourier transform parameters in the x and y directions, respectively, $\kappa^2 = k^2 + \ell^2$, and $\overline{G}(k, \ell, z, s|\xi, \eta, \zeta, \tau)$ denotes the double Fourier transform of $G(x, y, z, s|\xi, \eta, \zeta, \tau)$.

Step 3: Show that the solution to Step 2 is

$$\overline{G}(k, \ell, z, s|\xi, \eta, \zeta, \tau) = \frac{e^{-\gamma|z - \zeta| - ik\xi - i\ell\eta - s\tau}}{2\gamma},$$

where $\gamma = \sqrt{\kappa^2 + s/a^2}$.

Step 4: Show that

$$G(x, y, z, s|\xi, \eta, \zeta, \tau) = \frac{e^{-s\tau}}{8\pi^2} \int_{-\infty}^{\infty} \int_{-\infty}^{\infty} \frac{e^{ik(x-\xi) + i\ell(y-\eta) - |z-\zeta|\gamma}}{\gamma} \, dk \, d\ell$$

$$= \frac{e^{-s\tau}}{4\pi} \int_0^{\infty} \frac{e^{-|z-\zeta|\sqrt{\kappa^2 + s/a^2}}}{\sqrt{\kappa^2 + s/a^2}} J_0(\kappa r) \, \kappa \, d\kappa$$

$$= \frac{\exp(-R\sqrt{s}/a - s\tau)}{4\pi R},$$

where $k = \kappa\cos(\theta)$, $\ell = \kappa\sin(\theta)$, $x - \xi = r\cos(\beta)$, $y - \eta = r\sin(\beta)$, and $R^2 = (x - \xi)^2 + (y - \eta)^2 + (z - \zeta)^2$. Hint: Use integral tables.

Step 5: Complete the problem by using transform tables and show that

$$g(x, y, z, t|\xi, \eta, \zeta, \tau) = \frac{H(t - \tau)}{[4\pi a^2(t - \tau)]^{3/2}}$$

$$\times \exp\left[-\frac{(x - \xi)^2 + (y - \eta)^2 + (z - \zeta)^2}{4a^2(t - \tau)}\right].$$

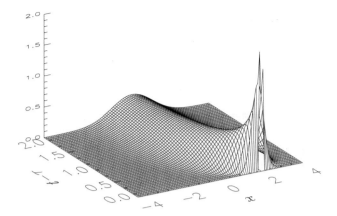

Problem 4

3. Find the free-space Green's function for the linearized Ginzburg-Landau equation[30]

$$\frac{\partial g}{\partial t} + v\frac{\partial g}{\partial x} - ag - b\frac{\partial^2 g}{\partial x^2} = \delta(x - \xi)\delta(t - \tau), \quad -\infty < x, \xi < \infty, \ 0 < t, \tau,$$

with $b > 0$.

Step 1: Taking the Laplace transform of the partial differential equation, show that it reduces to the ordinary differential equation

$$b\frac{d^2 G}{dx^2} - v\frac{dG}{dx} + aG - sG = -\delta(x - \xi)e^{-s\tau}.$$

Step 2: Using Fourier transforms, show that

$$G(x, s|\xi, \tau) = \frac{e^{-s\tau}}{2\pi} \int_{-\infty}^{\infty} \frac{e^{ik(x-\xi)}}{s + ikv + bk^2 - a} \, dk,$$

or

$$g(x, t|\xi, \tau) = \frac{e^{a(t-\tau)}}{\pi} H(t - \tau) \int_{0}^{\infty} e^{-b(t-\tau)k^2} \cos[x - \xi - v(t - \tau)] \, dk.$$

Step 3: Evaluate the second integral and show that

$$g(x, t|\xi, \tau) = \frac{e^{a(t-\tau)} H(t - \tau)}{2\sqrt{\pi b(t - \tau)}} \exp\left\{-\frac{[x - \xi - v(t - \tau)]^2}{4b(t - \tau)}\right\}.$$

---

[30] See Deissler, R. J., 1985: Noise-sustained structure, intermittency, and the Ginzburg-Landau equation. *J. Stat. Phys.*, **40**, 371–395.

4. Show that the free-space Green's function[31] governed by the equation

$$\frac{\partial g}{\partial t} - \frac{\partial^2 g}{\partial x^2} - \frac{\partial(xg)}{\partial x} = \delta(x - \xi)\delta(t - \tau)$$

is given by

$$g(x, t|\xi, \tau) = \frac{H(t - \tau)}{\sqrt{2\pi\{1 - \exp[-2(t - \tau)]\}}} \exp\left(-\frac{\{x - \xi\exp[-(t - \tau)]\}^2}{2\{1 - \exp[-2(t - \tau)]\}}\right).$$

This Green's function is illustrated in the figure captioned Problem 4 as functions of x and $t - \tau$ with $\xi = 2$.

5. Find the free-space Green's function[32] governed by

$$\frac{\partial g}{\partial t} - a^2\frac{\partial^2 g}{\partial x^2} + b\sin(\omega t + \varphi)\frac{\partial g}{\partial x} = \delta(x-\xi)\delta(t-\tau), \ -\infty < x, \xi < \infty, \ 0 < t, \tau,$$

subject to the boundary conditions

$$\lim_{|x|\to\infty} |g(x, t|\xi, \tau)| < \infty, \qquad 0 < t,$$

and the initial condition

$$g(x, 0|\xi, \tau) = 0, \qquad -\infty < x < \infty.$$

Step 1: Taking the Fourier transform of the partial differential equation, show that the Fourier transform of $g(x, t|\xi, \tau)$ is governed by the ordinary differential equation

$$\frac{dG}{dt} + a^2 k^2 G + ikb\sin(\omega t + \varphi)G = e^{-ik\xi}\delta(t - \tau).$$

Step 2: Show that

$$G(k, t|\xi, \tau) = H(t - \tau)\exp\bigg\{-ik\xi - a^2 k^2(t - \tau)$$

$$+ \frac{ikb}{\omega}[\cos(\omega t + \varphi) - \cos(\omega \tau + \varphi)]\bigg\}.$$

---

[31] Reprinted with permission from Mahajan, S. M., P. M. Valanju, and W. L. Rowan, 1992: Closed, form invariant solutions of convective-diffusive systems with applications to impurity transport. *Phys. Fluids B*, **4**, 2495–2498. ©American Institute of Physics, 1992.

[32] Taken with permission from Zon, B. A., 1974: Forced diffusion in an alternating external field. *Sov. Phys. J.*, **17**, 1594–1595. Published by Plenum Publishers.

Step 3: Show that $g(x, t|\xi, \tau)$ equals

$$g(x, t|\xi, \tau) = \frac{H(t - \tau)}{2\pi} \int_{-\infty}^{\infty} \exp\left\{ ik(x - \xi) - a^2 k^2(t - \tau) \right.$$

$$\left. + \frac{ikb}{\omega} [\cos(\omega t + \varphi) - \cos(\omega\tau + \varphi)] \right\} dk$$

$$= \frac{H(t - \tau)}{\sqrt{4\pi a^2(t - \tau)}}$$

$$\times \exp\left[-\frac{\{x - \xi + b[\cos(\omega t + \varphi) - \cos(\omega\tau + \varphi)]/\omega\}^2}{4a^2(t - \tau)} \right].$$

6. Show that the free-space Green's function[31] governed by the equation

$$\frac{\partial g}{\partial t} - \frac{\partial}{\partial x}\left(x\frac{\partial g}{\partial x} \right) - \frac{\partial(xg)}{\partial x} = \delta(x - \xi)\delta(t - \tau), \qquad 0 < x, \xi, t, \tau,$$

is given by

$$g(x, t|\xi, \tau) = e^{-x} H(t - \tau) \sum_{n=0}^{\infty} e^{-n(t-\tau)} L_n(\xi) L_n(x)$$

$$= \frac{H(t - \tau)}{1 - \exp[-(t - \tau)]} \exp\left\{ -\frac{x + \xi \exp[-(t - \tau)]}{1 - \exp[-(t - \tau)]} \right\}$$

$$\times I_0 \left\{ 2\frac{\sqrt{x\xi \exp[-(t - \tau)]}}{1 - \exp[-(t - \tau)]} \right\},$$

where $L_n(\ )$ denotes a Laguerre polynomial of order n. This Green's function is graphed in the figure captioned Problem 6 as functions of x and $t - \tau$ with $\xi = 2$.

7. Find the free-space Green's function[33] governed by

$$\frac{\partial^2 g}{\partial x \partial t} + U(t)\frac{\partial^2 g}{\partial x^2} + a\frac{\partial g}{\partial x} + bg = \delta(x - \xi)\delta(t - \tau), \quad -\infty < x, \xi < \infty, \ 0 < t, \tau,$$

with $b > 0$, subject to the boundary conditions

$$\lim_{|x| \to 0} g(x, t|\xi, \tau) \to 0, \qquad 0 < t,$$

---

[33] Taken from Kozlov, V. F., 1980: Formation of a Rossby wave under the action of disturbances in a nonstationary barotropic oceanic flow. *Izv. Atmos. Oceanic Phys.*, **16**, 275–279. ©1980 by the American Geophysical Union.

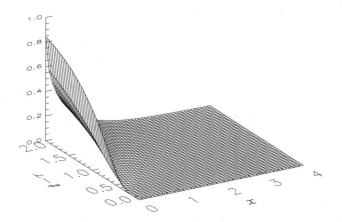

Problem 6

and the initial condition

$$g(x,0|\xi,\tau) = 0, \qquad -\infty < x < \infty.$$

Step 1: Taking the Fourier transform of the governing equation, show that it becomes

$$ik\frac{dG}{dt} - k^2 U(t)G + iakG + bG = e^{-ik\xi}\delta(t-\tau),$$

with $G(k,0|\xi,\tau) = 0$.

Step 2: Defining $s(t) = \int_0^t U(t')\,dt'$, show that

$$G(k,t|\xi,\tau) = \frac{e^{-ik\xi}}{ik}H(t-\tau)\exp\{-ik[s(t)-s(\tau)]-a(t-\tau)-b(t-\tau)/(ik)\}.$$

Step 3: Taking the inverse Fourier transform, show that

$$g(x,t|\xi,\tau) = \frac{e^{-a(t-\tau)}}{\pi}H(t-\tau)\int_0^\infty \frac{\sin[\eta k + b(t-\tau)/k]}{k}\,dk$$
$$= e^{-a(t-\tau)}H(t-\tau)H(\eta)J_0[2\sqrt{\eta b(t-\tau)}\,],$$

where $\eta = x - \xi - s(t) + s(\tau)$.

8. Find the free-space Green's function[34] governed by

$$\frac{\partial g}{\partial t} - a^2 \left(\frac{\partial^2 g}{\partial r^2} + \frac{1}{r} \frac{\partial g}{\partial r} \right) = \frac{\delta(r-\rho)\delta(t-\tau)}{r}, \quad 0 < r, \rho < \infty, \quad 0 < t, \tau,$$

subject to the boundary conditions

$$\lim_{r \to 0} |g(r, t|\rho, \tau)| < \infty, \quad 0 < t,$$

and

$$\lim_{r \to \infty} |g(r, t|\rho, \tau)| < \infty, \quad 0 < t,$$

and the initial condition

$$g(r, 0|\rho, \tau) = 0, \quad 0 < r < \infty.$$

Step 1: Taking the Hankel transform of the governing equation, show that it becomes

$$\frac{dG}{dt} + a^2 k^2 G = J_0(k\rho)\delta(t-\tau),$$

with $G(k, 0|\rho, \tau) = 0$, where

$$G(k, t|\rho, \tau) = \int_0^\infty g(r, t|\rho, \tau) J_0(kr) \, r \, dr.$$

Step 2: Use Laplace transforms and show that

$$G(k, t|\rho, \tau) = J_0(k\rho) e^{-a^2 k^2 (t-\tau)} H(t-\tau).$$

Step 3: Take the inverse of the Hankel transform and show that

$$g(r, t|\rho, \tau) = H(t-\tau) \int_0^\infty J_0(k\rho) J_0(kr) e^{-k^2(t-\tau)} \, k \, dk$$

$$= \frac{H(t-\tau)}{2a^2(t-\tau)} \exp\left[-\frac{r^2+\rho^2}{4a^2(t-\tau)} \right] I_0\left[\frac{r\rho}{2a^2(t-\tau)} \right].$$

---

[34] Taken from Golitsyn, G. S., and N. N. Romanova, 1970: Heat conduction in a rarefied and inhomogeneous atmosphere. *Geomag. Aeronomy*, **10**, 80–85. ©1970 by the American Geophysical Union.

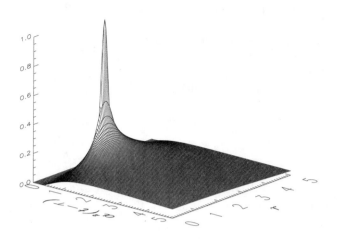

Problem 8

This Green's function is illustrated in the figure captioned Problem 8 as functions of r and $a^2(t - \tau)$ with $\rho = 0.3$.

9. Find the free-space Green's function[35] governed by

$$\frac{1}{r}\frac{\partial g}{\partial t} = \frac{1}{r^2}\frac{\partial}{\partial r}\left(r^{2+\beta}\frac{\partial g}{\partial r}\right) + \delta(r - 1)\delta(t), \quad 0 < r < \infty, \quad 0 < t,$$

where the solution remains finite over the entire interval and initially $g(r, 0|1, 0) = 0$.

Step 1: Take the Laplace transform of the partial differential equation and show that it equals

$$\frac{1}{r^2}\frac{d}{dr}\left(r^{2+\beta}\frac{dG}{dr}\right) - \frac{s}{r}G = -\delta(r - 1), \quad 0 < r < \infty,$$

with $\lim_{r \to 0}|G(r, s|1, 0)| < \infty$, and $\lim_{r \to \infty}|G(r, s|1, 0)| < \infty$.

Step 2: Show that Step 1 may be rewritten

$$\frac{1}{r^2}\frac{d}{dr}\left(r^{2+\beta}\frac{dG_i}{dr}\right) - \frac{s}{r}G_i = 0, \quad i = 1, 2, \quad 0 < r < \infty,$$

[35] Toptygin, I. N., 1973: Direct and inverse problem on cosmic-ray propagation in interplanetary space. *Geomagnet. Aeronomy*, **13**, 181–186. ©1973 by the American Geophysical Union.

with $\lim_{r\to 0}|G_1(r,s|1,0)| < \infty$, and $\lim_{r\to\infty}|G_2(r,s|1,0)| < \infty$, where $G_1(r,s|1,0)$ is the Laplace transform of the Green's function in the domain $0 < r < 1$ while $G_2(r,s|1,0)$ is the Laplace transform of the Green's function in the domain $1 < r < \infty$.

Step 3: Show that the solution in each domain is

$$G_1(r,s|1,0) = Ae^{-(1+\beta)r/2}I_\nu\left[\frac{2}{1-\beta}\sqrt{s}\ r^{(1-\beta)/2}\right], \qquad 0 < r < 1,$$

and

$$G_2(r,s|1,0) = Be^{-(1+\beta)r/2}K_\nu\left[\frac{2}{1-\beta}\sqrt{s}\ r^{(1-\beta)/2}\right], \qquad 1 < r < \infty,$$

if $\beta \neq 1$; otherwise,

$$G_1(r,s|1,0) = C\frac{r^{\sqrt{s+1}}}{r}, \qquad 0 < r < 1,$$

and

$$G_2(r,s|1,0) = D\frac{r^{-\sqrt{s+1}}}{r}, \qquad 1 < r < \infty,$$

where $\nu = (1+\beta)/(1-\beta)$.

Step 4: Show that $G_1(1,s|1,0) = G_2(1,s|1,0)$.

Step 5: By integrating the Laplace transformed version of the governing equation from 1^- to 1^+, where $r = 1^-$ and $r = 1^+$ are points just below and above $r = 1$, respectively, obtain the condition that

$$G_2'(1^+,s|1,0) - G_1'(1^-,s|1,0) = -1.$$

Step 6: Show that the solutions that satisfy the transformed partial differential equation and the boundary conditions are

$$G_1(r,s|1,0) = \frac{2}{1-\beta}r^{-(1+\beta)/2}K_\nu\left(\frac{2}{1-\beta}\sqrt{s}\right)I_\nu\left[\frac{2}{1-\beta}r^{(1-\beta)/2}\sqrt{s}\right],$$

for $0 < r < 1$, and

$$G_2(r,s|1,0) = \frac{2}{1-\beta}r^{-(1+\beta)/2}I_\nu\left(\frac{2}{1-\beta}\sqrt{s}\right)K_\nu\left[\frac{2}{1-\beta}r^{(1-\beta)/2}\sqrt{s}\right],$$

for $1 < r < \infty$, if $\beta \neq 1$;

$$G_1(r, s|1, 0) = \frac{1}{2\sqrt{s+1}} \frac{r^{\sqrt{s+1}}}{r}, \quad 0 < r < 1,$$

and

$$G_2(r, s|1, 0) = \frac{1}{2\sqrt{s+1}} \frac{r^{-\sqrt{s+1}}}{r}, \quad 1 < r < \infty,$$

if $\beta = 1$.

Step 7: Using a very good table[10] of Laplace transforms, show that the Green's function is

$$g(r, t|1, 0) = \frac{t^{-1}}{1 - \beta} r^{-(1+\beta)/2} I_\nu \left[\frac{2r^{(1-\beta)/2}}{(1-\beta)^2 t} \right] \exp\left[-\frac{1 + r^{1-\beta}}{(1-\beta)^2 t} \right],$$

if $\beta \neq 1$; otherwise,

$$g(r, t|1, 0) = \frac{1}{2r\sqrt{\pi t}} \exp\left[-t - \frac{\ln^2(r)}{4t} \right],$$

if $\beta = 1$.

10. Find the free-space Green's function[34] governed by

$$\frac{\partial g}{\partial t} - x^m \frac{\partial^2 g}{\partial x^2} = \delta(x - \xi)\delta(t - \tau), \quad 0 < x, \xi < \infty, \quad 0 < t, \tau,$$

subject to the boundary conditions

$$\lim_{x \to 0} |g(x, t|\xi, \tau)| < \infty, \quad 0 < t,$$

and

$$\lim_{x \to \infty} |g(x, t|\xi, \tau)| < \infty, \quad 0 < t,$$

and the initial condition

$$g(x, 0|\xi, \tau) = 0, \quad 0 < x < \infty,$$

with $m > 0$.

Step 1: For the moment, let us assume that $m \neq 2$; the case of $m = 2$ is treated later. Show that the original differential equation can be rewritten as

$$\frac{\partial R}{\partial t} - \frac{1}{4}\left(\frac{\partial^2 R}{\partial r^2} + \frac{1}{r}\frac{\partial R}{\partial r} - \frac{R}{a^2 r^2} \right) = \frac{\delta(r - \rho)\delta(t - \tau)}{(a\rho)^{1/a}},$$

where $0 < r, \rho < \infty$, $0 < t, \tau$, $R(x, t|\xi, \tau) = g(x, t|\xi, \tau)/\sqrt{x}$, $r = x^{a/2}/a$, $\rho = \xi^{a/2}/a$, and $a = |2 - m|$.

Step 2: Taking the Hankel transform of the equation in Step 1, show that it becomes

$$\frac{d\mathcal{R}}{dt} + \frac{k^2}{4}\mathcal{R} = \frac{\rho J_{1/a}(k\rho)\delta(t - \tau)}{(a\rho)^{1/a}}, \qquad 0 < t,$$

with $\mathcal{R}(k, 0|\rho, \tau) = 0$, where

$$\mathcal{R}(k, t|\rho, \tau) = \int_0^\infty R(r, t|\rho, \tau) J_{1/a}(kr) r \, dr.$$

Step 3: Use Laplace transforms and show that

$$\mathcal{R}(k, t|\rho, \tau) = \frac{\rho \, J_{1/a}(k\rho)}{(a\rho)^{1/a}} e^{-k^2(t-\tau)/4} H(t - \tau).$$

Step 4: Take the inverse of the Hankel transform and show that

$$g(x, t|\xi, \tau) = \frac{\xi^{a/2}\sqrt{x}}{a\sqrt{\xi}} H(t - \tau) \int_0^\infty J_{1/a}\left(\frac{k\xi^{a/2}}{a}\right) J_{1/a}\left(\frac{kx^{a/2}}{a}\right)$$

$$\times e^{-k^2(t-\tau)/4} \, k \, dk$$

$$= \frac{4\xi^{(a-1)/2}\sqrt{x}}{a(t - \tau)} \exp\left[-\frac{\xi^a + x^a}{a^2(t - \tau)}\right] I_{1/a}\left[\frac{2(a\xi)^{a/2}}{a^2(t - \tau)}\right] H(t - \tau).$$

Step 5: Consider now the case $m = 0$. Show that the original differential equation can be rewritten as

$$\frac{\partial R}{\partial t} - \frac{\partial^2 R}{\partial r^2} + \frac{R}{4} = e^{-\rho/2}\delta(r - \rho)\delta(t - \tau),$$

where $-\infty < r, \rho < \infty$, $0 < t, \tau$, $R(x, t|\xi, \tau) = g(x, t|\xi, \tau)/\sqrt{x}$, $r = \ln(x)$, and $\rho = \ln(\xi)$.

Step 6: Taking the Fourier transform of the equation in Step 5, show that it becomes

$$\frac{d\mathcal{R}}{dt} + \left(k^2 + \tfrac{1}{4}\right)\mathcal{R} = e^{-\rho/2}e^{-ik\rho}\delta(t - \tau), \qquad 0 < t,$$

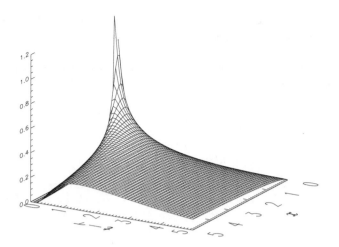

Problem 10

with $\mathcal{R}(k, 0|\rho, \tau) = 0$, where

$$R(r, t|\rho, \tau) = \frac{1}{2\pi} \int_{-\infty}^{\infty} \mathcal{R}(k, t|\rho, \tau) \, e^{ikr} \, dk.$$

Step 7: Use Laplace transforms and show that

$$\mathcal{R}(k, t|\rho, \tau) = e^{-\rho/2} e^{-ik\rho} e^{-(k^2 + \frac{1}{4})(t-\tau)} H(t - \tau).$$

Step 8: Take the inverse of the Fourier transform and show that

$$g(x, t|\xi, \tau) = \frac{H(t-\tau)}{\pi} \sqrt{\frac{x}{\xi}} \int_0^{\infty} \cos\{k [\ln(x) - \ln(\xi)]\} \, e^{-(k^2 + \frac{1}{4})(t-\tau)} \, dk$$

$$= \frac{H(t-\tau)}{2\sqrt{\pi(t-\tau)}} \exp\left[-\frac{\ln^2(x/\xi)}{4(t-\tau)} - \frac{\ln(\xi/x)}{2} - \frac{t-\tau}{4}\right]$$

$$= \frac{H(t-\tau)}{2\sqrt{\pi(t-\tau)}} \exp\left\{-\frac{\ln^2\left[\xi e^{(t-\tau)}/x\right]}{4(t-\tau)}\right\},$$

in agreement with Zhukova and Saichev.[36] This Green's function is plotted in the figure captioned Problem 10 as functions of x and $t - \tau$ when $\xi = 0.5$.

---

[36] Zhukova, I. S., and A. I. Saichev, 1997: Two-point statistical properties of a passive tracer in a turbulent compressible medium. *Radiophys. Quantum Electron.*, **40**, 682–693.

11. With the advent of financial instruments known as *derivatives*, partial differential equations entered into finance in a big way. During a study involving options on actively managed funds, Hyer *et al.*[37] found the free-space Green's function governed by

$$\frac{\partial g}{\partial t} - \frac{\partial^2 g}{\partial x^2} - \delta(x)g = \delta(x - \xi), \quad -\infty < x, \xi < \infty, \quad 0 < t,$$

subject to the boundary condition $\lim_{|x| \to \infty} |g(x, t|\xi)| < \infty$, $0 < t$, and the initial condition $g(x, 0|\xi) = 0$, $-\infty < x < \infty$. Let us retrace their analysis.

Step 1: The interesting aspect of this problem is the presence of the delta function as a coefficient in the partial differential equation. Noting that the partial differential equation can be written

$$\frac{\partial g}{\partial t} - \frac{\partial^2 g}{\partial x^2} = f(\xi, t)\delta(x) + \delta(x - \xi) = g(0, t|\xi)\delta(x) + \delta(x - \xi),$$

show that

$$g(x, t|\xi) = \frac{e^{-(x-\xi)^2/(4t)}}{2\sqrt{\pi t}} + \int_0^t \frac{e^{-x^2/[4(t-\tau)]}}{2\sqrt{\pi(t-\tau)}} f(\xi, \tau) \, d\tau,$$

using joint Fourier and Laplace transforms.

Step 2: Setting $x = 0$, show that the solution in Step 1 reduces to the integral equation

$$f(\xi, t) = \frac{e^{-\xi^2/(4t)}}{2\sqrt{\pi t}} + \int_0^t \frac{f(\xi, \tau)}{2\sqrt{\pi(t-\tau)}} \, d\tau.$$

Step 3: Use Laplace transforms and show that

$$f(\xi, t) = \frac{\exp[-\xi^2/(4t)]}{2\sqrt{\pi t}} + \frac{1}{4}e^{t/4 - |\xi|/2} \operatorname{erfc}\left(\frac{|\xi|}{2\sqrt{t}} - \frac{\sqrt{t}}{2}\right),$$

and

$$g(x, t|\xi) = \frac{e^{-(x-\xi)^2/(4t)}}{2\sqrt{\pi t}} + \frac{1}{4}e^{t/4 - (|x|+|\xi|)/2} \operatorname{erfc}\left(\frac{|x| + |\xi|}{2\sqrt{t}} - \frac{\sqrt{t}}{2}\right),$$

[37] Hyer, T., A. Lipton-Lifschifz, and D. Pugachevsky, 1997: Passport to success. *Risk*, **10(9)**, 127–131.

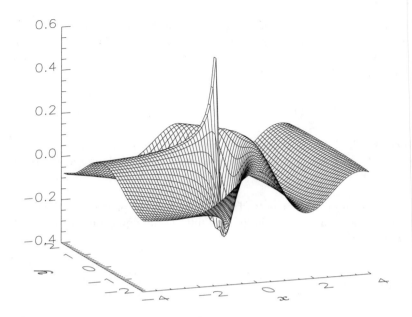

Problem 12

where erfc( ) is the complementary error function. Hint: Don't evaluate the convolution integral directly but use the convolution theorem and tables.

12. Find the free-space Green's function[38] governed by

$$\frac{\partial}{\partial t}\left(\frac{\partial^2 g}{\partial x^2} + \frac{\partial^2 g}{\partial y^2}\right) + \beta\frac{\partial g}{\partial x} = \delta(x - \xi)\delta(y - \eta)\delta(t - \tau),$$

with $-\infty < x, y, \xi, \eta < \infty$, and $0 < t, \tau$, where the boundary conditions are

$$\lim_{|x|\to\infty} |g(x, y, t|\xi, \eta, \tau)| < \infty, \qquad \lim_{|y|\to\infty} |g(x, y, t|\xi, \eta, \tau)| < \infty, \qquad 0 < t,$$

and the initial condition is

$$g(x, y, 0|\xi, \eta, \tau) = 0, \qquad -\infty < x, y < \infty.$$

Step 1: Take the Laplace transform of the partial differential equation and show that it equals

$$s\left(\frac{\partial^2 G}{\partial x^2} + \frac{\partial^2 G}{\partial y^2}\right) + \beta\frac{\partial G}{\partial x} = \delta(x - \xi)\delta(y - \eta)e^{-s\tau}.$$

---

[38] Taken from Veronis, G., 1958: On the transient response of a β-plane ocean. *J. Oceanogr. Soc. Japan*, **14**, 1–5.

Step 2: Setting $G(x, y, s|\xi, \eta, \tau) = e^{-\beta x/(2s)} U(x, y, s|\xi, \eta, \tau)$, show that the partial differential equation in Step 1 reduces to

$$\frac{\partial^2 U}{\partial x^2} + \frac{\partial^2 U}{\partial y^2} - \frac{\beta^2}{4s^2} U = e^{\beta \xi/(2s) - s\tau} \delta(x - \xi)\delta(y - \eta).$$

Step 3: Show that

$$G(x, y, s|\xi, \eta, \tau) = -\frac{e^{-(x-\xi)\beta/(2s) - s\tau}}{2s} K_0\left(\frac{\beta r}{2s}\right),$$

where $r = \sqrt{(x - \xi)^2 + (y - \eta)^2}$. Hint: See problem 7 in the next chapter.

Step 4: Use the convolution theorem and a good set of tables to show that

$$g(x, y, t|\xi, \eta, \tau) = -\int_\tau^t \frac{K_0[2\sqrt{\beta r(t - u)}]\cos[\sqrt{2(x + r - \xi)\beta(u - \tau)}]}{\pi \sqrt{(t - u)(u - \tau)}}\, du,$$

if $t > \tau$, and $g(x, y, t|\xi, \eta, \tau) = 0$, otherwise.

Step 5: Let $u - \tau = (t - \tau)\sin^2(\varphi)$, $\alpha^2 = \beta r(t - \tau)$, $\gamma = \cos(\theta/2)$, and $x - \xi = r\cos(\theta)$. Using these variables and the integrals

$$\int_0^{\pi/2} \cos[a\,\sin(x)]\cos[b\,\cos(x)]\, dx = \frac{\pi}{2} J_0\left(\sqrt{a^2 + b^2}\right),$$

and

$$K_0(\alpha) = \int_0^\infty \frac{\cos(\alpha z)}{\sqrt{z^2 + 1}}\, dz,$$

show that $g(x, y, t|\xi, \eta, \tau)$ can be rewritten as

$$g(x, y, t|\xi, \eta, \tau) = -H(t - \tau)\int_0^\infty \frac{J_0\left(2\alpha\sqrt{z^2 + \gamma^2}\right)}{\sqrt{z^2 + 1}}\, dz.$$

The integral portion of $g(x, y, t|\xi, \eta, \tau)$ is illustrated for $\beta(t - \tau) = 10$ in the figure captioned Problem 12.

Semi-Infinite Domain

13. Use Green's functions to show that the solution[39] to

$$\frac{\partial u}{\partial t} = a^2 \frac{\partial^2 u}{\partial x^2}, \qquad 0 < x, t,$$

---

[39] Reprinted with permission from Gilev, S. D., and T. Yu. Mikhaĭlova, 1996: Current wave in shock compression of matter in a magnetic field. *Tech. Phys.*, **41**, 407–411. ©American Institute of Physics, 1996.

subject to the boundary conditions

$$u(0,t) = 0, \quad \lim_{x \to \infty} |u(x,t)| < \infty, \qquad 0 < t,$$

and the initial condition

$$u(x,0) = f(x), \qquad 0 < x < \infty,$$

is

$$u(x,t) = \frac{e^{-x^2/(4a^2t)}}{a\sqrt{\pi t}} \int_0^\infty f(\tau) \sinh\left(\frac{x\tau}{2a^2t}\right) e^{-\tau^2/(4a^2t)} \, d\tau.$$

14. Find the Green's function[40] governed by

$$\frac{\partial g}{\partial t} - a^2 \frac{\partial^2 g}{\partial x^2} + \beta g = \delta(x - \xi)\delta(t - \tau), \quad 0 < x, t, \xi, \tau,$$

subject to the boundary conditions

$$g(0,t|\xi,\tau) = 0, \quad \lim_{x \to \infty} |g(x,t|\xi,\tau)| < \infty,$$

and the initial condition

$$g(x,0|\xi,\tau) = 0, \qquad 0 < x < \infty.$$

Step 1: Taking the Laplace transform of the governing equation and boundary conditions, show that it becomes the ordinary differential equation

$$\frac{d^2G}{dx^2} - \frac{s+\beta}{a^2} G = -\frac{\delta(x-\xi)}{a^2} e^{-s\tau}, \qquad 0 < x, \xi < \infty,$$

with

$$G(0,s|\xi,\tau) = 0, \quad \lim_{x \to \infty} |G(x,s|\xi,\tau)| < \infty.$$

Step 2: Consider the similar problem of

$$\frac{d^2G}{dx^2} - \frac{s+\beta}{a^2} G = -\frac{\delta(x-\xi)}{a^2} e^{-s\tau}, \qquad -\infty < x, \xi < \infty,$$

---

[40] Taken from Khantadze, A. G., and B. Ya. Chekhoskvili, 1975: Effect of plasma flow on the nighttime F region. *Geomagnet. Aeron.*, **15**, 440–441.

with

$$\lim_{|x|\to\infty} |G(x,s|\xi,\tau)| < \infty.$$

Show that the solution to this problem is

$$G(x,s|\xi,\tau) = \frac{1}{2a\sqrt{s'}} e^{-s\tau - |x-\xi|\sqrt{s'}/a},$$

where $s' = s + \beta$.

Step 3: Using the method of images and the results from Step 2, show that the solution to Step 1 is

$$G(x,s|\xi,\tau) = \frac{e^{-s\tau}}{2a\sqrt{s'}} \left[e^{-|x-\xi|\sqrt{s'}/a} - e^{-(x+\xi)\sqrt{s'}/a} \right].$$

Step 4: Using the shifting theorems and tables, show that the Green's function is

$$g(x,t|\xi,\tau) = \frac{e^{-\beta(t-\tau)} H(t-\tau)}{2a\sqrt{\pi(t-\tau)}}$$
$$\times \left\{ e^{-(x-\xi)^2/[4a^2(t-\tau)]} - e^{-(x+\xi)^2/[4a^2(t-\tau)]} \right\}.$$

15. Find the Green's function[41] for

$$\frac{\partial g}{\partial t} - \frac{\partial^2 g}{\partial x^2} = \delta(x-\xi)\delta(t-\tau), \quad 0 < x, t, \xi, \tau,$$

subject to the boundary conditions

$$g(0,t|\xi,\tau) = \kappa \int_0^t g_x(0,t'|\xi,\tau)\, dt', \quad \text{and} \quad \lim_{x\to\infty} |g(x,t|\xi,\tau)| < \infty,$$

and the initial condition

$$g(x,0|\xi,\tau) = 0, \quad 0 < x < \infty.$$

The interesting aspect of this problem is the presence of the integral condition in the boundary condition.

---

[41] Reprinted with permission from Agmon, N., 1984: Diffusion with back reaction. *J. Chem. Phys.*, **81**, 2811–2817. ©American Institute of Physics, 1984.

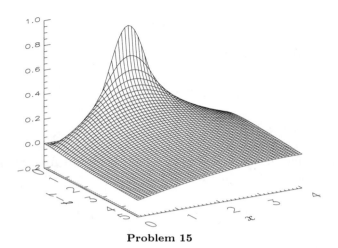

Problem 15

Step 1: Taking the Laplace transform of the governing equation and the boundary conditions, show that it becomes the ordinary differential equation

$$\frac{d^2 G}{dx^2} - sG = -\delta(x - \xi)e^{-s\tau}, \qquad 0 < x, \xi < \infty,$$

with

$$G(0, s|\xi, \tau) = \kappa G'(0, s|\xi, \tau)/s, \quad \text{and} \quad \lim_{x \to \infty} |G(x, s|\xi, \tau)| < \infty.$$

Step 2: Show that the solution to the ordinary differential equation in Step 1 is

$$G(x, s|\xi, \tau) = \frac{e^{-|x - \xi|\sqrt{s} - s\tau}}{2\sqrt{s}} - \frac{e^{-(x+\xi)\sqrt{s} - s\tau}}{2\sqrt{s}} + \frac{\kappa \, e^{-(x+\xi)\sqrt{s} - s\tau}}{\sqrt{s}(\kappa + \sqrt{s})}.$$

Step 3: Taking the inverse Laplace transform, show that

$$g(x, t|\xi, \tau) = \frac{H(t - \tau)}{\sqrt{4\pi(t - \tau)}} \left\{ \exp\left[-\frac{(x - \xi)^2}{4(t - \tau)}\right] - \exp\left[-\frac{(x + \xi)^2}{4(t - \tau)}\right] \right\}$$

$$+ \kappa H(t - \tau) e^{\kappa[x + \xi + \kappa(t - \tau)]} \text{erfc}\left[\frac{x + \xi + 2\kappa(t - \tau)}{2\sqrt{t - \tau}}\right],$$

where erfc is the complementary error function. This Green's function is plotted in the figure captioned Problem 15 when $\kappa = -0.1$, and $\xi = 2$.

16. Find the Green's function for the linearized Ginzburg-Landau equation

$$\frac{\partial g}{\partial t} + v\frac{\partial g}{\partial x} - ag - b\frac{\partial^2 g}{\partial x^2} = \delta(x - \xi)\delta(t - \tau), \quad 0 < x, \xi < \infty, \quad 0 < t, \tau,$$

with $b > 0$, subject to the boundary conditions

$$g(0, t|\xi, \tau) = 0, \quad \text{and} \quad \lim_{x \to \infty} |g(x, t|\xi, \tau)| < \infty,$$

and the initial condition

$$g(x, 0|\xi, \tau) = 0, \quad 0 < x < \infty.$$

Step 1: Taking the Laplace transform of the partial differential equation, show that it reduces to

$$b\frac{d^2 G}{dx^2} - v\frac{dG}{dx} + aG - sG = -\delta(x - \xi)e^{-s\tau},$$

with $G(0, s|\xi, \tau) = 0$ and $\lim_{x \to \infty} |G(x, s|\xi, \tau)| < \infty$.

Step 2: Setting $G(x, s|\xi, \tau) = e^{vx/(2b)}U(x, s|\xi, \tau)$, show that the ordinary differential equation in Step 1 becomes

$$b\frac{d^2 U}{dx^2} - \frac{v^2}{4b}U + aU - sU = -\delta(x - \xi)e^{-v\xi/(2b)-s\tau},$$

with $U(0, s|\xi, \tau) = 0$ and $\lim_{x \to \infty} U(x, s|\xi, \tau) \to 0$.

Step 3: Using Fourier sine transforms, show that

$$G(x, s|\xi, \tau) = \frac{2}{\pi}\exp\left[\frac{v(x - \xi)}{2b} - s\tau\right]\int_0^\infty \frac{\sin(k\xi)\sin(kx)}{s + bk^2 + v^2/(4b) - a}\,dk.$$

Step 4: Invert the Laplace transform and show that

$$g(x, t|\xi, \tau) = \frac{2}{\pi}\exp\left[\frac{v(x - \xi)}{2b}\right]H(t - \tau)e^{a(t-\tau)-v^2(t-\tau)/(4b)}$$

$$\times \int_0^\infty e^{-b(t-\tau)k^2}\sin(k\xi)\sin(kx)\,dk$$

$$= \frac{e^{a(t-\tau)-v^2(t-\tau)/(4b)}}{2\sqrt{\pi b(t - \tau)}}\exp\left[\frac{v(x - \xi)}{2b}\right]H(t - \tau)$$

$$\times \left\{\exp\left[-\frac{(x - \xi)^2}{4b(t - \tau)}\right] - \exp\left[-\frac{(x + \xi)^2}{4b(t - \tau)}\right]\right\}.$$

17. Find the Green's function[42] governed by

$$\frac{\partial g}{\partial t} - \frac{\partial^2 g}{\partial x^2} - \frac{\partial g}{\partial x} = \delta(x - \xi)\delta(t - \tau), \quad 0 < x, t, \xi, \tau,$$

subject to the boundary conditions

$$g_x(0, t|\xi, \tau) + (1 - h)g(0, t|\xi, \tau) = 0, \quad \lim_{x \to \infty} |g(x, t|\xi, \tau)| < \infty,$$

and the initial condition

$$g(x, 0|\xi, \tau) = 0, \quad 0 < x < \infty.$$

Step 1: Taking the Laplace transform of the governing equation and boundary conditions, show that it becomes the ordinary differential equation

$$\frac{d^2 G}{dx^2} + \frac{dG}{dx} - sG = -\delta(x - \xi)e^{-s\tau}, \quad 0 < x, \xi < \infty,$$

with

$$G'(0, s|\xi, \tau) + (1 - h)G(0, s|\xi, \tau) = 0, \quad \lim_{x \to \infty} |G(x, s|\xi, \tau)| < \infty.$$

Step 2: Show that the solution to the ordinary differential equation in Step 1 is

$$G(x, s|\xi, \tau) = \frac{e^{-(x-\xi)/2 - |x-\xi|\sqrt{s+\frac{1}{4}} - s\tau}}{2\sqrt{s + \frac{1}{4}}} - \frac{e^{-(x-\xi)/2 - (x+\xi)\sqrt{s+\frac{1}{4}} - s\tau}}{2\sqrt{s + \frac{1}{4}}}$$

$$+ \frac{e^{-(x-\xi)/2 - (x+\xi)\sqrt{s+\frac{1}{4}} - s\tau}}{h - \frac{1}{2} + \sqrt{s + \frac{1}{4}}}.$$

Step 3: Taking the inverse Laplace transform, show that

$$g(x, t|\xi, \tau) = H(t - \tau)\left(\tfrac{1}{2} - h\right)e^{-(1-h)[x + h(t-\tau)] + h\xi}$$

$$\times \operatorname{erfc}\left[\frac{x + \xi}{2\sqrt{t - \tau}} - \left(\tfrac{1}{2} - h\right)\sqrt{t - \tau}\right]$$

$$+ \frac{H(t - \tau)}{2\sqrt{\pi(t - \tau)}}e^{-(x-\xi)/2 - (t-\tau)/4}$$

$$\times \left\{e^{-(x+\xi)^2/[4(t-\tau)]} + e^{-(x-\xi)^2/[4(t-\tau)]}\right\},$$

---

[42] Reprinted from *J. Cryst. Growth*, **183**, S.-H. Kim, S. A. Korpela, and A. Chait, Axial segregation in unsteady diffusion-dominated solidification of a binary alloy, 490–496, ©1998, with permission from Elsevier Science.

where erfc is the complementary error function.

18. Find the Green's function[43] for

$$x^\alpha \frac{\partial g}{\partial t} - \frac{\partial}{\partial x}\left(x^{1-\alpha}\frac{\partial g}{\partial x}\right) = x^\alpha \delta(x-\xi)\delta(t-\tau), \quad 0 < x, t, \xi, \tau,$$

where $0 \le \alpha \le 1$, subject to the boundary conditions

$$g_x(0, t|\xi, \tau) = 0, \qquad \lim_{x\to\infty} |g(x, t|\xi, \tau)| < \infty,$$

and the initial condition

$$g(x, 0|\xi, \tau) = 0, \qquad 0 < x < \infty.$$

Step 1: Taking the Laplace transform of the governing equation and boundary conditions, show that it becomes the ordinary differential equation

$$\frac{d^2 G}{dx^2} + \frac{1-\alpha}{x}\frac{dG}{dx} - sx^{2\alpha-1}G = -x^{2\alpha-1}\delta(x-\xi)e^{-s\tau}, \quad 0 < x, \xi < \infty,$$

with

$$G'(0, s|\xi, \tau) = 0, \qquad \lim_{x\to\infty} |G(x, s|\xi, \tau)| < \infty.$$

Step 2: Show that the solution to the ordinary differential equation in Step 1 is

$$G(x, s|\xi, \tau) = \frac{2}{1+2\alpha}(x\xi)^{\alpha/2}e^{-s\tau}I_{-\alpha/(1+2\alpha)}\left[\frac{2}{1+2\alpha}\sqrt{s}\,x_<^{(1+2\alpha)/2}\right]$$

$$\times K_{\alpha/(1+2\alpha)}\left[\frac{2}{1+2\alpha}\sqrt{s}\,x_>^{(1+2\alpha)/2}\right].$$

Step 3: Find the inverse Laplace transform by applying Bromwich's integral with the branch cut for $\sqrt{s}$ taken along the negative axis of the s-plane. Show that

$$g(x, t|\xi, \tau) = \frac{1+2\alpha}{2}(x\xi)^{\alpha/2}H(t-\tau)$$

$$\times \int_0^\infty J_{-\alpha/(1+2\alpha)}\left[\eta\xi^{(1+2\alpha)/2}\right]J_{-\alpha/(1+2\alpha)}\left[\eta x^{(1+2\alpha)/2}\right]$$

$$\times \exp\left[-(1+2\alpha)^2(t-\tau)\eta^2/4\right]\eta\,d\eta.$$

---

[43] Similar to a problem solved by Smith, F. B., 1957: The diffusion of smoke from a continuous elevated point-source into a turbulent atmosphere. *J. Fluid Mech.*, **2**, 49–76. Reprinted with the permission of Cambridge University Press.

Step 4: Using integral tables, evaluate the integral and show that

$$g(x,t|\xi,\tau) = \frac{(x\xi)^{\alpha/2} H(t-\tau)}{(1+2\alpha)(t-\tau)} \exp\left[-\frac{x^{1+2\alpha} + \xi^{1+2\alpha}}{(1+2\alpha)^2(t-\tau)}\right]$$

$$\times I_{-\alpha/(1+2\alpha)}\left[\frac{2(x\xi)^{(1+2\alpha)/2}}{(1+2\alpha)^2(t-\tau)}\right].$$

19. Find the Green's function governed by

$$\frac{\partial g}{\partial t} - \frac{1}{r}\frac{\partial}{\partial r}\left(r\frac{\partial g}{\partial r}\right) - \frac{\partial^2 g}{\partial z^2} + 2\frac{\partial g}{\partial z} = \frac{\delta(r-\rho)\delta(z-\zeta)\delta(t-\tau)}{2\pi r},$$

where $0 < r, \rho < 1$, $|z - \zeta| < \infty$, and $0 < t, \tau$, with the boundary conditions

$$g_r(0,z,t|\rho,\zeta,\tau) = g_r(1,z,t|\rho,\zeta,\tau) = 0,$$

and

$$\lim_{|z|\to\infty} |g(r,z,t|\rho,\zeta,\tau)| < \infty,$$

and the initial condition that $g(r,z,0|\rho,\zeta,\tau) = 0$.

Step 1: Taking the Laplace transform of the partial differential equation, show that it becomes

$$sG - \frac{1}{r}\frac{\partial}{\partial r}\left(r\frac{\partial G}{\partial r}\right) - \frac{\partial^2 G}{\partial z^2} + 2\frac{\partial G}{\partial z} = \frac{\delta(r-\rho)\delta(z-\zeta)}{2\pi r}e^{-s\tau},$$

with the boundary conditions

$$G_r(0,z,s|\rho,\zeta,\tau) = G_r(1,z,s|\rho,\zeta,\tau) = 0,$$

and

$$\lim_{|z|\to\infty} |G(r,z,s|\rho,\zeta,\tau)| < \infty.$$

Step 2: Assuming that $G(r,z,s|\rho,\zeta,\tau)$ can be expressed in terms of the expansion

$$G(r,z,s|\rho,\zeta,\tau) = \sum_{n=1}^{\infty} Z_n(z) J_0(k_n r),$$

where k_n is the nth zero of $J_1(k) = 0$, show that $Z_n(z)$ is governed by

$$\frac{d^2 Z_n}{dz^2} - 2\frac{dZ_n}{dz} - \left(s + k_n^2\right) Z_n = -\frac{J_0(k_n\rho)}{\pi J_0^2(k_n)}\delta(z-\zeta)e^{-s\tau}.$$

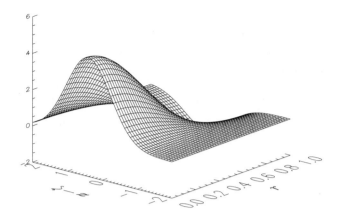

Problem 19

Why did we choose this particular expansion?

Step 3: Show that

$$Z_n(z) = \frac{e^{z-\zeta-s\tau} J_0(k_n\rho)}{2\pi J_0^2(k_n)\sqrt{s+k_n^2+1}} e^{-|z-\zeta|\sqrt{s+k_n^2+1}}.$$

Step 4: Finish up the problem by inverting the Laplace transform and show that

$$g(r,z,t|\rho,\zeta,\tau) = \frac{e^{z-\zeta-(z-\zeta)^2/[4(t-\tau)]}}{2\pi\sqrt{\pi(t-\tau)}} H(t-\tau)$$
$$\times \sum_{n=1}^{\infty} \frac{J_0(k_n\rho)J_0(k_nr)}{J_0^2(k_n)} e^{-(1+k_n^2)(t-\tau)}.$$

This Green's function (multiplied by 100) is illustrated in the figure captioned Problem 19 as functions of r and $z - \zeta$ when $t - \tau = 0.2$, and $\rho = 0.3$.

Finite Domain

Construct the Green's function for the one-dimensional heat equation $g_t - g_{xx} = \delta(x - \xi)\delta(t - \tau)$, on the interval $0 < x < L$, with the initial condition that $g(x, 0|\xi, \tau) = 0$ and the following boundary conditions:

20. $\qquad g(0, t|\xi, \tau) = g(L, t|\xi, \tau) = 0, \qquad\qquad 0 < t,$

21. $g(0, t|\xi, \tau) = g_x(L, t|\xi, \tau) = 0,$ $0 < t,$

22. $g_x(0, t|\xi, \tau) = g_x(L, t|\xi, \tau) = 0,$ $0 < t.$

Use the Green's function from the previous problem to write down the
solution to the heat equation $u_t = u_{xx}$ on the interval $0 < x < L$ with
the initial data $u(x, 0) = 1$ and the boundary conditions:

23. $u(0, t) = e^{-t},$ $u(L, t) = 0,$ $0 < t,$

24. $u(0, t) = \sin(t),$ $u_x(L, t) = 0,$ $0 < t,$

25. $u_x(0, t) = 1,$ $u_x(L, t) = 0,$ $0 < t.$

26. Find the Green's function for

$$\frac{\partial g}{\partial t} - a^2 \frac{\partial^2 g}{\partial x^2} + a^2 k^2 g = \delta(x - \xi)\delta(t - \tau), \quad 0 < x, \xi < L, \quad 0 < t, \tau,$$

subject to the boundary conditions

$$g(0, t|\xi, \tau) = g_x(L, t|\xi, \tau) = 0, \qquad 0 < t,$$

and the initial condition

$$g(x, 0|\xi, \tau) = 0, \qquad 0 < x < L,$$

where a and k are real constants.

27. Find the Green's function[44] for

$$\frac{\partial g}{\partial t} - a^2 \frac{\partial^2 g}{\partial x^2} = \delta(x - \xi)\delta(t - \tau), \quad 0 < x, \xi < L, \quad 0 < t, \tau,$$

subject to the boundary conditions

$$a^2 g_x(0, t|\xi, \tau) - hg(0, t|\xi, \tau) = -\alpha\varphi(t),$$

and

$$a^2 g_x(L, t|\xi, \tau) + hg(L, t|\xi, \tau) = \alpha\psi(t),$$

where

$$\varphi'(t) + \alpha\varphi(t) = hg(0, t|\xi, \tau), \qquad \varphi(0) = 0,$$

---

[44] Reprinted with permission from Kim, H., and K. J. Shin, 2000: On the diffusion-influenced reversible trapping problem in one dimension. *J. Chem. Phys.*, **112**,
8312–8317. ©American Institute of Physics, 2000.

and

$$\psi'(t) + \alpha\psi(t) = hg(L, t|\xi, \tau), \qquad \psi(0) = 0.$$

Step 1: Take the Laplace transform of the partial differential equation and boundary conditions and show that we obtain the transformed set

$$sG - a^2 \frac{d^2 G}{dx^2} = \delta(x - \xi)e^{-s\tau},$$

with

$$a^2 G'(0, s|\xi, \tau) = a^2 q X(q) G(0, s|\xi, \tau),$$

and

$$a^2 G'(L, s|\xi, \tau) = -a^2 q X(q) G(L, s|\xi, \tau),$$

where

$$q^2 = s/a^2, \qquad \text{and} \qquad X(q) = \frac{qh}{a^2 q^2 + \alpha}.$$

Step 2: Show that

$$G(x, s|\xi, \tau) = \frac{e^{-s\tau}}{a^2 q Y(q, L)} [\cosh(qx_<) + X(q) \sinh(qx_<)]$$
$$\times \{\cosh[a(L - x_>)] + X(q) \sinh[q(L - x_>)]\},$$

where

$$Y(q, L) = [1 + X^2(q)] \sinh(qL) + 2X(q) \cosh(qL).$$

Step 3: Show that $G(x, s|\xi, \tau)$ has singularities at $q = 0$ and q_n, where q_n is the nth root of $Y(q, L) = 0$ and $n = 1, 2, 3, \ldots$

Step 4: Conclude the problem by showing that

$$g(x, t|\xi, \tau) = \frac{H(t - \tau)}{L + 2h/\alpha}$$
$$+ 2H(t - \tau) \sum_{n=1}^{\infty} \frac{W(q_n, x)W(q_n, \xi)}{Z(q_n, L)} \exp\left[-\frac{a^2 \beta_n^2 (t - \tau)}{L^2}\right],$$

where $W(q, x) = \cosh(qx) + X(q) \sinh(qx)$, $Z(q, L) = L[1 - X^2(q)] + 2X'(q)$, and $q_n = \beta_n i/L$.

28. Find the Green's function[45] governed by

$$\frac{\partial g}{\partial t} - \frac{\partial^2 g}{\partial x^2} - \frac{b}{x} \frac{\partial g}{\partial x} = \delta(x - \xi)\delta(t - \tau), \quad b < 1, \quad 0 < x, \xi < 1, \quad 0 < t, \tau,$$

---

[45] Taken from Chan, C. Y., and B. M. Wong, 1995: Existence of classical solutions for singular parabolic problems. *Q. Appl. Math.*, **53**, 201–213.

subject to the boundary conditions

$$g(0, t | \xi, \tau) = g(1, t | \xi, \tau) = 0, \qquad 0 < t,$$

and the initial condition

$$g(x, 0 | \xi, \tau) = 0, \qquad 0 < x < 1.$$

Step 1: Consider the Sturm-Liouville problem

$$\frac{d}{dx}\left(x^b \frac{d\varphi}{dx}\right) + \lambda x^b \varphi = 0, \qquad \varphi(0) = \varphi(1) = 0.$$

Show that the orthonormal eigenfunctions to this Sturm-Liouville problem are

$$\varphi_n(x) = \sqrt{2}\, x^\nu J_\nu(x\sqrt{\lambda_n}\,)/|J_{\nu+1}(\sqrt{\lambda_n}\,)|,$$

where $\nu = (1 - b)/2$, and λ_n is the nth root of $J_\nu(\sqrt{\lambda}\,) = 0$.

Step 2: Assuming that $g(x, t | \xi, \tau)$ can be written as

$$g(x, t | \xi, \tau) = \sum_{n=1}^{\infty} a_n(t) \varphi_n(x),$$

show that $a_n(t)$ is governed by

$$a_n'(t) + \lambda_n a_n(t) = \xi^b \varphi_n(\xi) \delta(t - \tau).$$

Step 3: Show that

$$a_n(t) = \xi^b \varphi_n(\xi) e^{-\lambda_n(t-\tau)} H(t - \tau).$$

Step 4: Show that

$$g(x, t | \xi, \tau) = H(t - \tau) \sum_{n=1}^{\infty} \xi^b \varphi_n(\xi) \varphi_n(x) e^{-\lambda_n(t-\tau)}.$$

29. Find the Green's function[46] governed by

$$\frac{\partial g}{\partial t} - a^2 \left(\frac{\partial^2 g}{\partial x^2} + \frac{\partial^2 g}{\partial y^2}\right) = \delta(x - \xi)\delta(y - \eta)\delta(t - \tau),$$

---

[46] Taken from Kidawa-Kukla, J., 1997: Vibration of a beam induced by harmonic motion of a heat source. *J. Sound Vibr.*, **205**, 213–222. Published by Academic Press Ltd., London, U.K.

where $0 < x, \xi < L$, and $0 < y, \eta < h$, with the boundary conditions

$$g(0, y, t|\xi, \eta, \tau) = g(L, y, t|\xi, \eta, \tau) = 0,$$

$$g_y(x, 0, t|\xi, \eta, \tau) - \alpha_0 g(x, 0, t|\xi, \eta, \tau) = 0,$$

and

$$g_y(x, h, t|\xi, \eta, \tau) + \alpha_1 g(x, h, t|\xi, \eta, \tau) = 0,$$

and the initial condition that $g(x, y, 0|\xi, \eta, \tau) = 0$.

Step 1: Taking the Laplace transform of the partial differential equation, show that it becomes

$$sG - a^2 \left(\frac{\partial^2 G}{\partial x^2} + \frac{\partial^2 G}{\partial y^2} \right) = \delta(x - \xi)\delta(y - \eta)e^{-s\tau}.$$

Step 2: Assuming that $G(x, y, s|\xi, \eta, \tau)$ can be expressed in terms of the expansion

$$G(x, y, s|\xi, \eta, \tau) = \frac{2}{L} \sum_{m=1}^{\infty} G_m(y|\eta) \sin\left(\frac{m\pi\xi}{L}\right) \sin\left(\frac{m\pi x}{L}\right),$$

show that $G_m(y|\eta)$ is governed by

$$\frac{d^2 G_m}{dy^2} - \left(\frac{s}{a^2} + \frac{m^2 \pi^2}{L^2} \right) G_m = -\frac{\delta(y - \eta)}{a^2} e^{-s\tau},$$

with

$$G'_m(0|\eta) - \alpha_0 G_m(0|\eta) = 0,$$

and

$$G'_m(h|\eta) + \alpha_1 G_m(h|\eta) = 0.$$

Why have we chosen this particular eigenfunction expansion?

Step 3: Consider the eigenfunctions

$$\varphi_n(y) = \beta_n \cos(\beta_n y) + \alpha_0 \sin(\beta_n y), \qquad n = 1, 2, 3, \ldots$$

Show that they satisfy the differential equation $\varphi_n'' + \beta_n^2 \varphi_n = 0$, with the boundary conditions $\varphi'(0) - \alpha_0 \varphi_n(0) = 0$ and $\varphi'(h) + \alpha_1 \varphi_n(h) = 0$, if β_n is chosen so that $(\alpha_0 \alpha_1 - \beta_n^2) + (\alpha_0 + \alpha_1)\beta_n \cot(\beta_n h) = 0$.

Step 4: Show that the solution to the ordinary differential equation in Step 2 is

$$G_m(y|\eta) = \frac{2}{h} e^{-s\tau} \sum_{n=1}^{\infty} \frac{\varphi_n(\eta)\varphi_n(y)}{q_n^2 (s + a^2 \beta_n^2 + a^2 m^2 \pi^2 / L^2)},$$

where

$$q_n^2 = \alpha_0^2 + \beta_n^2 + \frac{(\alpha_0 + \alpha_1)(\alpha_0\alpha_1 + \beta_n^2)}{h(\alpha_1^2 + \beta_n^2)}.$$

Hint: You will need to show that

$$[(\alpha_0\alpha_1 - \beta_n^2)^2 + (\alpha_0 + \alpha_1)^2\beta_n^2]\sin^2(\beta_n h) = (\alpha_0 + \alpha_1)^2\beta_n^2.$$

Step 5: Taking the inverse Laplace transform,

$$g(x, y, t|\xi, \eta, \tau) = \frac{4}{hL}H(t - \tau)\sum_{m=1}^{\infty}\sum_{n=1}^{\infty}\frac{\varphi_n(\eta)\varphi_n(y)}{q_n^2}\sin\left(\frac{m\pi\xi}{L}\right)$$
$$\times \sin\left(\frac{m\pi x}{L}\right)\exp\left[-a^2\left(\frac{m^2\pi^2}{L^2} + \beta_n^2\right)(t - \tau)\right].$$

Chapter 5
Green's Functions
for the Helmholtz Equation

In the previous chapters, we sought solutions to the heat and wave equations via Green's function. In this chapter, we turn to the reduced wave equation

$$\nabla^2 u + \lambda u = -f(\mathbf{r}). \tag{5.0.1}$$

Equation (5.0.1), generally known as *Helmholtz's equation*, includes the special case of *Poisson's equation* when $\lambda = 0$. Poisson's equation has a special place in the theory of Green's functions because George Green invented his technique for its solution.

The reduced wave equation arises during the solution of the harmonically forced wave equation[1] by separation of variables. In one spatial dimension, the problem is

$$\frac{\partial^2 u}{\partial x^2} - \frac{1}{c^2}\frac{\partial^2 u}{\partial t^2} = -f(x)e^{-i\omega t}. \tag{5.0.2}$$

Equation (5.0.2) occurs, for example, in the mathematical analysis of a stretched string over some interval subject to an external, harmonic

---

[1] See, for example, Graff, K. F., 1991: *Wave Motion in Elastic Solids*. Dover Publications, Inc., §1.4.

forcing. Assuming that $u(x,t)$ is bounded everywhere, we seek solutions of the form $u(x,t) = y(x)e^{-i\omega t}$. Upon substituting this solution into (5.0.2), we obtain the ordinary differential equation

$$y'' + k_0^2 y = -f(x), \qquad (5.0.3)$$

where $k_0^2 = \omega^2/c^2$. This is an example of the one-dimensional Helmholtz equation.

Similar considerations hold as we include more spatial dimensions. For example, in three spatial dimensions we have

$$\frac{\partial^2 g}{\partial x^2} + \frac{\partial^2 g}{\partial y^2} + \frac{\partial^2 g}{\partial z^2} + \lambda g = -\delta(x - \xi)\delta(y - \eta)\delta(z - \zeta), \qquad (5.0.4)$$

or in vectorial form

$$\nabla^2 g(\mathbf{r}|\mathbf{r}_0) + \lambda g(\mathbf{r}|\mathbf{r}_0) = -\delta(\mathbf{r} - \mathbf{r}_0). \qquad (5.0.5)$$

In this notation $\mathbf{r}_0$, the *source variable*, denotes the position of the point source (ξ, η, ζ) and appears in the partial differential equation as a parameter. On the other hand, $\mathbf{r}$, the *field variable*, is the variable with respect to which differentiation is carried out.

Consider now the equations for the Green's function for two different source points $\mathbf{r}_1$ and $\mathbf{r}_2$:

$$\nabla^2 g(\mathbf{r}|\mathbf{r}_1) + \lambda g(\mathbf{r}|\mathbf{r}_1) = -\delta(\mathbf{r} - \mathbf{r}_1), \qquad (5.0.6)$$

and

$$\nabla^2 g(\mathbf{r}|\mathbf{r}_2) + \lambda g(\mathbf{r}|\mathbf{r}_2) = -\delta(\mathbf{r} - \mathbf{r}_2). \qquad (5.0.7)$$

Multiplying (5.0.6) by $g(\mathbf{r}|\mathbf{r}_2)$ and (5.0.7) by $g(\mathbf{r}|\mathbf{r}_1)$, subtracting and employing Green's second formula, we find

$$\oiint_S [g(\mathbf{r}|\mathbf{r}_2)\nabla g(\mathbf{r}|\mathbf{r}_1) - g(\mathbf{r}|\mathbf{r}_1)\nabla g(\mathbf{r}|\mathbf{r}_2)] \cdot \mathbf{n}\, dS = g(\mathbf{r}_2|\mathbf{r}_1) - g(\mathbf{r}_1|\mathbf{r}_2),$$
$$(5.0.8)$$

where the closed surface S is a piece-wise smooth surface that includes the points $\mathbf{r}_1$ and $\mathbf{r}_2$. To proceed further, we need to discuss the boundary conditions. The most common ones are

• the *Dirichlet boundary condition*, where g vanishes on the boundary,

• the *Neumann boundary condition*, where the normal gradient of g vanishes on the boundary $\nabla g \cdot \mathbf{n} = 0$, and

• the *Robin boundary condition*, which is the linear combination of the Dirichlet and Neumann conditions.

When any of these conditions hold, the surface integral vanishes and we obtain $g(\mathbf{r}_2|\mathbf{r}_1) = g(\mathbf{r}_1|\mathbf{r}_2)$. Because $\mathbf{r}_1$ and $\mathbf{r}_2$ are two arbitrary points inside the region, we obtain our first important result that

$$g(\mathbf{r}|\mathbf{r}_0) = g(\mathbf{r}_0|\mathbf{r}), \qquad (5.0.9)$$

which establishes the symmetry or *reciprocity* of the Green's function under the interchange of field and source variables.

Let us now use Green's functions to solve the Helmholtz equation

$$\nabla^2 u(\mathbf{r}) + \lambda u(\mathbf{r}) = -f(\mathbf{r}), \qquad (5.0.10)$$

where $f(\mathbf{r})$ is a specified scalar function of the field variables. If we multiply (5.0.10) by $g(\mathbf{r}|\mathbf{r}_0)$ and (5.0.5) by $u(\mathbf{r})$, subtract and integrate, we find that

$$u(\mathbf{r}_0) = \iiint_V \left[g(\mathbf{r}|\mathbf{r}_0)\nabla^2 u(\mathbf{r}) - u(\mathbf{r})\nabla^2 g(\mathbf{r}|\mathbf{r}_0) \right] dV$$
$$+ \iiint_V f(\mathbf{r})g(\mathbf{r}|\mathbf{r}_0) \, dV. \qquad (5.0.11)$$

Applying Green's second formula to the first volume integral, equation (5.0.11) becomes

$$u(\mathbf{r}_0) = \iiint_V f(\mathbf{r})g(\mathbf{r}|\mathbf{r}_0) \, dV$$
$$- \oiint_S \left[u(\mathbf{r})\nabla g(\mathbf{r}|\mathbf{r}_0) - g(\mathbf{r}|\mathbf{r}_0)\nabla u(\mathbf{r}) \right] \cdot \mathbf{n} \, dS. \qquad (5.0.12)$$

Since $\mathbf{r}_0$ is an arbitrary point inside of volume V, we denote it in general by $\mathbf{r}$. Furthermore, the variable $\mathbf{r}$ is now merely a dummy integration variable that we now denote by $\mathbf{r}_0$. Finally, to emphasize the fact that the gradient operation is occurring with the variable $\mathbf{r}_0$, we write the gradient operator as ∇_0. Upon making these substitutions and using the symmetry condition, we have that

$$u(\mathbf{r}) = \iiint_{V_0} f(\mathbf{r}_0)g(\mathbf{r}|\mathbf{r}_0) \, dV_0 - \oiint_{S_0} u(\mathbf{r}_0) \left[\nabla_0 g(\mathbf{r}_0|\mathbf{r}) \cdot \mathbf{n} \right] dS_0$$
$$+ \oiint_{S_0} g(\mathbf{r}|\mathbf{r}_0) \left[\nabla_0 u(\mathbf{r}_0) \cdot \mathbf{n} \right] dS_0. \qquad (5.0.13)$$

Equation (5.0.13) shows that the solution of Helmholtz's equation depends upon the sources inside the volume V and values of $u(\mathbf{r})$ and $\nabla u(\mathbf{r})$ on the enclosing surface. On the other hand, we must still find the particular Green's function for a given problem; this Green's function depends directly upon the boundary conditions. At this point, we work out several special cases.

1. *Nonhomogeneous Helmholtz equation and homogeneous Dirichlet boundary conditions*

In this case, let us assume that we can find a Green's function that also satisfies the same Dirichlet boundary conditions as $u(\mathbf{r})$. Once the Green's function is found, then (5.0.13) reduces to

$$u(\mathbf{r}) = \iiint_{V_0} f(\mathbf{r}_0) g(\mathbf{r}|\mathbf{r}_0) \, dV_0. \qquad (5.0.14)$$

A possible source of difficulty would be the nonexistence of the Green's function. From our experience in Chapter 2, we know that this will occur if λ equals one of the eigenvalues of the corresponding homogeneous problem. An example of this occurs in acoustics when the Green's function for the Helmholtz equation does not exist at *resonance*.

2. *Homogeneous Helmholtz equation and nonhomogeneous Dirichlet boundary conditions*

In this particular case $f(\mathbf{r}) = 0$. For convenience, let us use the Green's function from the previous example so that $g(\mathbf{r}|\mathbf{r}_0)$ equals zero along the boundary. Under these conditions, (5.0.13) becomes

$$u(\mathbf{r}) = \oiint_{S_0} u(\mathbf{r}_0) \frac{\partial g(\mathbf{r}_0|\mathbf{r})}{\partial n_0} \, dS_0, \qquad (5.0.15)$$

where n_0 is the unit, inwardly pointing normal to the boundary. Consequently, the solution is determined once we compute the normal gradient of the Green's function along the boundary.

3. *Nonhomogeneous Helmholtz equation and homogeneous Neumann boundary conditions*

If we require that $u(\mathbf{r})$ satisfies the nonhomogeneous Helmholtz equation with homogeneous Neumann boundary conditions, then the governing equations are

$$\nabla^2 u(\mathbf{r}) + \lambda u(\mathbf{r}) = -f(\mathbf{r}), \qquad \text{and} \qquad \frac{\partial u}{\partial n} = 0, \quad \text{on } S. \qquad (5.0.16)$$

Integrating (5.0.16) and using Gauss's divergence theorem, we obtain

$$\oiint_S \nabla u \cdot \mathbf{n} \, dS + \lambda \iiint_V u(\mathbf{r}) \, dV = - \iiint_V f(\mathbf{r}) \, dV. \qquad (5.0.17)$$

Because the first integral in (5.0.17) must vanish in the case of homogeneous Neumann boundary conditions, equation (5.0.17) cannot be satisfied if $\lambda = 0$ unless

$$\iiint_V f(\mathbf{r}) \, dV = 0. \tag{5.0.18}$$

A physical interpretation of (5.0.18) is as follows: Consider the physical process of steady-state heat conduction within a finite region V. The temperature u is given by Poisson's equation $\nabla^2 u = -f$, where f is proportional to the density of the heat sources and sinks. The boundary condition $\partial u/\partial n = 0$ implies that there is no heat exchange across the boundary. Consequently, no steady-state temperature distribution can exist unless the heat sources are balanced by heat sinks. This balance of heat sources and sinks is given by (5.0.18).

This same problem can be solved using Green's functions. Equation (5.0.13) suggests that it would be convenient to choose a Green's function that also satisfies the homogeneous boundary condition $\partial g/\partial n = 0$ on S. However, the differential equation governing the Green's function is $\nabla^2 g + \lambda g = -\delta(\mathbf{r} - \mathbf{r}_0)$. Integrating this equation over the volume leads to $\lambda \iiint_V g(\mathbf{r}|\mathbf{r}_0) \, dV = -1$. This cannot be true if $\lambda = 0$. On the other hand, we just showed that solutions to Poisson's equation with homogeneous Neumann boundary conditions exist if $\iiint_V f(\mathbf{r}) \, dV = 0$. This contradiction forces us to abandon the homogeneous Neumann boundary condition for another. Let us try $\partial g/\partial n = C$, a constant. To find C, we integrate $\nabla^2 g = -\delta(\mathbf{r} - \mathbf{r}_0)$ and find that $C \oiint_S dS = -1$ or $C = -1/A_s$, where A_s is the area of the surface S.

A disadvantage of using this nonhomogeneous Neumann boundary condition is the loss of the symmetry condition $g(\mathbf{r}|\mathbf{r}_0) = g(\mathbf{r}_0|\mathbf{r})$. However, retracing our steps in the symmetry proof, we see that if we impose the additional constraint that $\oiint_S g(\mathbf{r}|\mathbf{r}_0) \, dS = 0$, then symmetry still holds. Equation (5.0.13) still applies and yields

$$u(\mathbf{r}) = \iiint_{V_0} f(\mathbf{r}_0) g(\mathbf{r}|\mathbf{r}_0) \, dV_0 + \frac{1}{A_s} \oiint_{S_0} u(\mathbf{r}_0) \, dS_0. \tag{5.0.19}$$

While the last term cannot be evaluated unless $u(\mathbf{r})$ is known, it is a constant. Therefore, (5.0.19) yields $u(\mathbf{r})$ up to an arbitrary constant. For most physical problems, this is adequate. We omit the proof that a Green's function $g(\mathbf{r}|\mathbf{r}_0)$ that satisfies all of our conditions can be found.

5.1 FREE-SPACE GREEN'S FUNCTION FOR HELMHOLTZ'S AND POISSON'S EQUATIONS

Before we plunge into the solution of the various shades of the Helmholtz equation, let us focus on the simple problem of finding the

Green's function for the Helmholtz equation

$$\nabla^2 g(\mathbf{r}|\mathbf{r}_0) + k_0^2 g(\mathbf{r}|\mathbf{r}_0) = -\delta(\mathbf{r} - \mathbf{r}_0) \qquad (\mathbf{5.1.1})$$

when there are no boundaries present. The primary use of this *free-space Green's function*[2] is as a *particular* solution to the Green's function problem.

• *One-dimensional Helmholtz equation*

Clearly, the simplest problem is the one-dimensional case. Let us find the Green's function for

$$g'' + k_0^2 g = -\delta(x - \xi), \qquad -\infty < x, \xi < \infty. \qquad (\mathbf{5.1.2})$$

If we solve (5.1.2) by piecing together homogeneous solutions, then

$$g(x|\xi) = Ae^{-ik_0(x-\xi)} + Be^{ik_0(x-\xi)}, \qquad (\mathbf{5.1.3})$$

for $x < \xi$ while

$$g(x|\xi) = Ce^{-ik_0(x-\xi)} + De^{ik_0(x-\xi)}, \qquad (\mathbf{5.1.4})$$

for $\xi < x$.

Let us examine the solution (5.1.3) more closely. The solution represents two propagating waves. Because $x < \xi$, the first term is a wave propagating out to infinity, while the second term gives a wave propagating in from infinity. This is seen most clearly by including the $e^{-i\omega t}$ term into (5.1.3) or

$$g(x|\xi)e^{-i\omega t} = Ae^{-ik_0(x-\xi)-i\omega t} + Be^{ik_0(x-\xi)-i\omega t}. \qquad (\mathbf{5.1.5})$$

Because we have a source only at $x = \xi$, solutions that represent waves originating at infinity are nonphysical and we must discard them. This requirement that there are only outwardly propagating wave solutions is commonly called *Sommerfeld's radiation condition*.[3] Similar considerations hold for (5.1.4) and we must take $C = 0$.

To evaluate A and D, we use the continuity conditions on the Green's function:

$$g(\xi^+|\xi) = g(\xi^-|\xi), \quad \text{and} \quad g'(\xi^+|\xi) - g'(\xi^-|\xi) = -1, \qquad (\mathbf{5.1.6})$$

---

[2] In electromagnetic theory, a free-space Green's function is the particular solution of the differential equation valid over a domain of infinite extent, where the Green's function remains bounded as we approach infinity, or satisfies a radiation condition there.

[3] Sommerfeld, A., 1912: Die Greensche Funktion der Schwingungsgleichung. *Jahresber. Deutschen Math.-Vereinung*, **21**, 309–353.

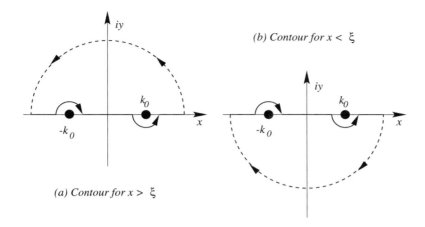

(b) Contour for x < ξ

(a) Contour for x > ξ

Figure 5.1.1: Contour used to evaluate (5.1.10).

or

$$A = D, \quad \text{and} \quad ik_0 D + ik_0 A = -1. \tag{5.1.7}$$

Therefore,

$$g(x|\xi) = \frac{i}{2k_0} e^{ik_0|x-\xi|}. \tag{5.1.8}$$

We can also solve (5.1.2) by Fourier transforms. Assuming that the Fourier transform of $g(x|\xi)$ exists and denoting it by $G(k|\xi)$, we find that

$$G(k|\xi) = \frac{e^{-ik\xi}}{k^2 - k_0^2}, \tag{5.1.9}$$

and

$$g(x|\xi) = \frac{1}{2\pi} \int_{-\infty}^{\infty} \frac{e^{ik(x-\xi)}}{k^2 - k_0^2} \, dk. \tag{5.1.10}$$

Immediately we see that there is a problem with the singularities lying on the path of integration at $k = \pm k_0$. How do we avoid them?

There are four possible ways that we might circumvent the singularities. One of them is shown in Figure 5.1.1. Applying Jordan's lemma to close the line integral along the real axis (as shown in Figure 5.1.1),

$$g(x|\xi) = \frac{1}{2\pi} \oint_C \frac{e^{iz(x-\xi)}}{z^2 - k_0^2} \, dz. \tag{5.1.11}$$

Table 5.1.1: Free-Space Green's Function for the Poisson and Helmholtz Equations

Dimension	Poisson Eq.	Helmholtz Eq.
One	no solution	$g(x\|\xi) = \dfrac{i}{2k_0}e^{ik_0\|x-\xi\|}$
Two	$g(x,y\|\xi,\eta) = -\dfrac{\ln(r)}{2\pi}$	$g(x,y\|\xi,\eta) = \dfrac{i}{4}H_0^{(1)}(k_0 r)$
Three	$g(x,y,z\|\xi,\eta,\zeta) = \dfrac{1}{4\pi R}$	$g(x,y,z\|\xi,\eta,\zeta) = \dfrac{e^{ik_0 R}}{4\pi R}$

$$r = \sqrt{(x-\xi)^2 + (y-\eta)^2}, \quad R = \sqrt{(x-\xi)^2 + (y-\eta)^2 + (z-\zeta)^2}$$

Note: For the Helmholtz equation, we have taken the temporal forcing to be $e^{-i\omega t}$ and $k_0 = \omega/c$.

For $x < \xi$,

$$g(x|\xi) = -i \operatorname{Res}\left[\frac{e^{iz(x-\xi)}}{z^2 - k_0^2}; -k_0\right] = \frac{i}{2k_0}e^{-ik_0(x-\xi)}, \qquad (5.1.12)$$

while

$$g(x|\xi) = i \operatorname{Res}\left[\frac{e^{iz(x-\xi)}}{z^2 - k_0^2}; k_0\right] = \frac{i}{2k_0}e^{ik_0(x-\xi)}, \qquad (5.1.13)$$

for $x > \xi$. A quick check shows that these solutions agree with (5.1.8). If we try the three other possible paths around the singularities, we obtain incorrect solutions.

• *One-dimensional Poisson equation*

There is no solution for the domain $-\infty < x < \infty$.

• *Two-dimensional Helmholtz equation*

At this point, we have found two forms of the free-space Green's function for the one-dimensional Helmholtz equation. The first form is the analytic solution (5.1.8), while the second is the integral representation (5.1.10) where the line integration along the real axis is shown in Figure 5.1.1.

In the case of two dimensions, the Green's function[4] for the Helmholtz equation symmetric about the point (ξ, η) is the solution of the equation

$$\frac{d^2 g}{dr^2} + \frac{1}{r}\frac{dg}{dr} + k_0^2 g = -\frac{\delta(r)}{2\pi r}, \qquad (5.1.14)$$

where $r = \sqrt{(x - \xi)^2 + (y - \eta)^2}$. The homogeneous form of (5.1.14) is Bessel's differential equation of order zero. Consequently, the general solution in terms of Hankel functions[5] is

$$g(\mathbf{r}|\mathbf{r}_0) = A\, H_0^{(1)}(k_0 r) + B\, H_0^{(2)}(k_0 r). \qquad (5.1.15)$$

Why have we chosen to use Hankel functions rather than $J_0(\ )$ and $Y_0(\ )$? As we argued earlier, solutions to the Helmholtz equation must represent *outwardly* propagating waves (the Sommerfeld radiation condition). If we again assume that the temporal behavior is $e^{-i\omega t}$ and use the asymptotic expressions for Hankel functions given in §C.2, we see that $H_0^{(1)}(k_0 r)$ represents outwardly propagating waves and $B = 0$.

What is the value of A? Integrating (5.1.14) over a small circle around the point $r = 0$ and taking the limit as the radius of the circle vanishes, $A = i/4$ and

$$g(\mathbf{r}|\mathbf{r}_0) = \frac{i}{4} H_0^{(1)}(k_0 r). \qquad (5.1.16)$$

If a real function is needed, then the free-space Green's function equals the Neumann function $Y_0(k_0 r)$ divided by -4.

As we did in the one-dimensional case, let us solve the two-dimensional Helmholtz equation by Fourier transforms.[6] The governing equation is

$$\frac{\partial^2 g}{\partial x^2} + \frac{\partial^2 g}{\partial y^2} + k_0^2 g = -\delta(x - \xi)\delta(y - \eta), \qquad -\infty < x, y, \xi, \eta < \infty. \quad (5.1.17)$$

---

[4] For an alternative derivation, see Graff, K. F., 1991: *Wave Motion in Elastic Solids*. Dover Publications, Inc., pp. 284–285.

[5] For the reader who is unfamiliar with Hankel functions, a short summary is given in §C.2.

[6] Reprinted with permission from Gunda, R., S. M. Vijayakar, R. Singh, and J. E. Farstad, 1998: Harmonic Green's functions of a semi-infinite plate with clamped or free edges. *J. Acoust. Soc. Am.*, **103**, 888–899. ©Acoustical Society of America, 1998. See §I.D.

Taking the Fourier transform of (5.1.17), we find that

$$\frac{d^2 G}{dy^2} + \ell^2 G = -\delta(y - \eta)e^{-ik\xi}, \tag{5.1.18}$$

where $\ell^2 = k_0^2 - k^2$, and k is the transform variable in the x direction. Using the results from the one-dimensional Helmholtz equation,

$$G(k, y | \xi, \eta) = \frac{i}{2\ell} e^{i\ell |y-\eta| - ik\xi}, \tag{5.1.19}$$

with $\text{Re}(\ell)$ and $\text{Im}(\ell) \geq 0$. These choices ensure outwardly radiating waves if the temporal behavior is $e^{-i\omega t}$. Taking the inverse of (5.1.19), we obtain

$$g(x, y | \xi, \eta) = \frac{i}{4\pi} \int_{-\infty}^{\infty} \frac{1}{\ell} e^{i\ell |y-\eta| + ik(x-\xi)} \, dk \tag{5.1.20}$$

$$= \frac{i}{2\pi} \int_{0}^{\infty} \frac{\cos[k(x-\xi)]e^{i\ell |y-\eta|}}{\ell} \, dk. \tag{5.1.21}$$

As an interesting aside, we can use (5.1.16) and (5.1.21) to obtain an integral property involving Hankel functions. Because there is only one unique free-space Green's function for the two-dimensional Helmholtz equation, (5.1.16) and (5.1.21) must be equivalent. Setting these two representations equal to each other, we obtain the Cartesian decomposition of the zero order Hankel function of the first kind

$$H_0^{(1)}(k_0 r) = \frac{1}{\pi} \int_{-\infty}^{\infty} \frac{e^{ik(x-\xi) - i\sqrt{k_0^2 - k^2} \, |y-\eta|}}{\sqrt{k_0^2 - k^2}} \, dk, \tag{5.1.22}$$

where the imaginary part of the radical must be zero or negative and $\text{Im}(k_0) \geq 0$.

In §3.5 we showed how the free-space Green's function for the three-dimensional Helmholtz equation can be written as a superposition of cylindrical waves (3.5.29). Here we have shown that the free-space Green's function for the two-dimensional Helmholtz equation can be written as a superposition of plane waves,[7] (5.1.21). Why are such decompositions useful? Imagine a situation where we wish to use the free-space Green's function but we need to evaluate it along some straight edge. In that case, (5.1.21) would be crucial to our analysis. This ability to re-express the solutions of Helmholtz's equation in one coordinate

---

[7] This result has been generalized by Cincotti, G., F. Gori, M. Santarsiero, F. Frezza, F. Furnò, and G. Schettini, 1993: Plane wave expansion of cylindrical functions. *Optics Comm.*, **95**, 192–198.

system in terms of another coordinate system is further examined in Appendix D.

We can also solve the planar Helmholtz equation in terms of polar coordinates

$$\frac{1}{r}\frac{\partial}{\partial r}\left(r\frac{\partial g}{\partial r}\right) + \frac{1}{r^2}\frac{\partial^2 g}{\partial \theta^2} + k_0^2 g = -\frac{\delta(r-\rho)\delta(\theta-\theta')}{r}, \qquad (5.1.23)$$

where $0 < r, \rho < \infty$, and $0 \leq \theta, \theta' \leq 2\pi$. Because

$$\delta(\theta-\theta') = \frac{1}{2\pi} + \frac{1}{\pi}\sum_{n=1}^{\infty}\cos[n(\theta-\theta')] = \frac{1}{2\pi}\sum_{n=-\infty}^{\infty}\cos[n(\theta-\theta')], \quad (5.1.24)$$

we assume a solution of the form

$$g(r,\theta|\rho,\theta') = \sum_{n=-\infty}^{\infty} g_n(r|\rho)\cos[n(\theta-\theta')]. \qquad (5.1.25)$$

Substituting (5.1.24)–(5.1.25) into (5.1.23) and simplifying, we find that

$$\frac{1}{r}\frac{d}{dr}\left(r\frac{dg_n}{dr}\right) - \frac{n^2}{r^2}g_n + k_0^2 g_n = -\frac{\delta(r-\rho)}{2\pi r}. \qquad (5.1.26)$$

The homogeneous solution of (5.1.26) is

$$g_n(r|\rho) = \begin{cases} AJ_n(k_0 r), & 0 \leq r \leq \rho, \\ BH_n^{(1)}(k_0 r), & \rho \leq r < \infty. \end{cases} \qquad (5.1.27)$$

This solution possesses the properties of remaining finite as $r \to 0$ and corresponding to a radiating wave solution in the limit of $r \to \infty$. If we now use the requirements that the solution is continuous at $r = \rho$, and that

$$\rho\left.\frac{dg_n(r|\rho)}{dr}\right|_{r=\rho^-}^{r=\rho^+} = -\frac{1}{2\pi}, \qquad (5.1.28)$$

where ρ^+ and ρ^- denote points slightly greater and less than ρ, then

$$g_n(r|\rho) = \frac{i}{4}H_n^{(1)}(k_0 r_>)J_n(k_0 r_<), \qquad (5.1.29)$$

so that

$$g(r,\theta|\rho,\theta') = \frac{i}{4}\sum_{n=-\infty}^{\infty}\cos[n(\theta-\theta')]H_n^{(1)}(k_0 r_>)J_n(k_0 r_<). \qquad (5.1.30)$$

- **Example 5.1.1**

Let us find the free-space Green's function for

$$\frac{1}{r}\frac{\partial}{\partial r}\left(r\frac{\partial g}{\partial r}\right) + \frac{\partial^2 g}{\partial z^2} + k_0^2 g = -\frac{\delta(r-\rho)\delta(z-\zeta)}{r}, \tag{5.1.31}$$

where $0 < r, \rho < \infty$, and $-\infty < z, \zeta < \infty$.

Taking the Fourier transform of (5.1.31) with respect to z, we find that

$$\frac{1}{r}\frac{d}{dr}\left(r\frac{dG}{dr}\right) + \kappa^2 G = -\frac{\delta(r-\rho)}{r}e^{-ik\zeta}, \tag{5.1.32}$$

where $\kappa = \sqrt{k_0^2 - k^2}$ with the $\mathrm{Im}(\kappa) \geq 0$. The homogeneous solution of (5.1.32) is

$$G(r, k|\rho, \zeta) = \begin{cases} AJ_0(\kappa r), & 0 \leq r \leq \rho, \\ BH_0^{(1)}(\kappa r), & \rho \leq r < \infty. \end{cases} \tag{5.1.33}$$

This solution possesses the properties of remaining finite as $r \to 0$ and corresponding to a radiating wave solution in the limit of $r \to \infty$. If we now use the requirements that the solution is continuous at $r = \rho$ so that $G(\rho^-, k|\rho, \zeta) = G(\rho^+, k|\rho, \zeta)$ and that

$$\left.\frac{dG(r, k|\rho, \zeta)}{dr}\right|_{r=\rho^-}^{r=\rho^+} = -\frac{1}{\rho}, \tag{5.1.34}$$

where ρ^+ and ρ^- denote points slightly greater and less than ρ, the coefficients A and B are determined by

$$\begin{pmatrix} J_0(\kappa\rho) & -H_0^{(1)}(\kappa\rho) \\ \kappa J_0'(\kappa\rho) & -\kappa H'_0^{(1)}(\kappa\rho) \end{pmatrix} \begin{pmatrix} A \\ B \end{pmatrix} = \begin{pmatrix} 0 \\ 1/\rho \end{pmatrix}. \tag{5.1.35}$$

Because the Wronskian determinant is

$$W\left[J_0(\kappa\rho), H_0^{(1)}(\kappa\rho)\right] = \frac{2i}{\pi\kappa\rho}, \tag{5.1.36}$$

$$A = \frac{\pi i}{2}H_0^{(1)}(\kappa\rho), \quad \text{and} \quad B = \frac{\pi i}{2}J_0(\kappa\rho). \tag{5.1.37}$$

Therefore,

$$G(r, k|\rho, \zeta) = \frac{\pi i}{2}J_0(\kappa r_<)H_0^{(1)}(\kappa r_>), \tag{5.1.38}$$

and

$$g(r, z|\rho, \zeta) = \frac{i}{4}\int_{-\infty}^{\infty} J_0(\kappa r_<)H_0^{(1)}(\kappa r_>)e^{ik(z-\zeta)}\, dk. \tag{5.1.39}$$

• *Two-dimensional Laplace equation*

In this subsection, we find the free-space Green's function for Poisson's equation in two dimensions. This Green's function is governed by

$$\frac{1}{r}\frac{\partial}{\partial r}\left(r\frac{\partial g}{\partial r}\right) + \frac{1}{r^2}\frac{\partial^2 g}{\partial \theta^2} = -\frac{\delta(r-\rho)\delta(\theta-\theta')}{r}. \tag{5.1.40}$$

If we now choose our coordinate system so that the origin is located at the point source, $r = \sqrt{(x-\xi)^2 + (y-\eta)^2}$ and $\rho = 0$. Multiplying both sides of this simplified (5.1.40) by $r\,dr\,d\theta$ and integrating over a circle of radius ϵ, we obtain -1 on the right side from the surface integration over the delta functions. On the left side,

$$\int_0^{2\pi} r\frac{\partial g}{\partial r}\bigg|_{r=\epsilon} d\theta = -1. \tag{5.1.41}$$

The Green's function $g(r,\theta|0,\theta') = -\ln(r)/(2\pi)$ satisfies (5.1.41).

To find an alternative form of the free-space Green's function when the point of excitation and the origin of the coordinate system do not coincide, we first note that

$$\delta(\theta-\theta') = \frac{1}{2\pi}\sum_{n=-\infty}^{\infty} e^{in(\theta-\theta')}. \tag{5.1.42}$$

This suggests that the Green's function should be of the form

$$g(r,\theta|\rho,\theta') = \sum_{n=-\infty}^{\infty} g_n(r|\rho)e^{in(\theta-\theta')}. \tag{5.1.43}$$

Substituting (5.1.42) and (5.1.43) into (5.1.40), we obtain the ordinary differential equation

$$\frac{1}{r}\frac{d}{dr}\left(r\frac{dg_n}{dr}\right) - \frac{n^2}{r^2}g_n = -\frac{\delta(r-\rho)}{2\pi r}. \tag{5.1.44}$$

The homogeneous solution to (5.1.44) is

$$g_0(r|\rho) = \begin{cases} a, & 0 \le r \le \rho, \\ b\ln(r), & \rho \le r < \infty. \end{cases} \tag{5.1.45}$$

and

$$g_n(r|\rho) = \begin{cases} c(r/\rho)^n, & 0 \le r \le \rho, \\ d(\rho/r)^n, & \rho \le r < \infty, \end{cases} \tag{5.1.46}$$

if $n \ne 0$.

At $r = \rho$, the g_n's must be continuous, in which case,

$$a = b\ln(\rho), \qquad \text{and} \qquad c = d. \qquad (5.1.47)$$

On the other hand,

$$\rho\frac{dg_n}{dr}\Big|_{r=\rho^-}^{r=\rho^+} = -\frac{1}{2\pi}, \qquad (5.1.48)$$

or

$$a = -\frac{\ln(\rho)}{2\pi}, \quad b = -\frac{1}{2\pi}, \quad \text{and} \quad c = d = \frac{1}{4\pi n}. \qquad (5.1.49)$$

Therefore,

$$g(r,\theta|\rho,\theta') = -\frac{\ln(r_>)}{2\pi} + \frac{1}{2\pi}\sum_{n=1}^{\infty}\frac{1}{n}\left(\frac{r_<}{r_>}\right)^n\cos[n(\theta-\theta')]. \qquad (5.1.50)$$

We can simplify (5.1.50) by noting that

$$\ln\left[1 + \rho^2 - 2\rho\cos(\theta-\theta')\right] = -2\sum_{n=1}^{\infty}\frac{\rho^n\cos[n(\theta-\theta')]}{n}, \qquad (5.1.51)$$

if $|\rho| < 1$. Applying this relationship to (5.1.50), we find that

$$g(r,\theta|\rho,\theta') = -\frac{1}{4\pi}\ln\left[r^2 + \rho^2 - 2r\rho\cos(\theta-\theta')\right]. \qquad (5.1.52)$$

Note that when $\rho = 0$ we recover $g(r,\theta|0,\theta') = -\ln(r)/(2\pi)$.

- *Three-dimensional Helmholtz equation*

Finally, we treat the three-dimensional Helmholtz equation. In particular, we now prove that

$$g(\mathbf{r}|\mathbf{r}_0) = \frac{e^{ik_0 R}}{4\pi R}, \qquad (5.1.53)$$

where $R = |\mathbf{r} - \mathbf{r}_0|$. This particular solution is the *three-dimensional free-space Green's function*.

We begin by noting that (5.1.53) satisfies $\nabla^2 g(\mathbf{r}|\mathbf{r}_0) + k_0^2\, g(\mathbf{r}|\mathbf{r}_0) = 0$ everywhere except at the point $\mathbf{r}_0$. Next, we integrate both sides of this equation over a sphere of very small radius ϵ, centered at $\mathbf{r}_0$. On the right side, we obtain -1 from the volume integration over the delta functions. The integration over the left side consists of two integrals. The integral involving $k_0^2 g(\mathbf{r}|\mathbf{r}_0)$ vanishes because it tends to zero as ϵ^3 as $\epsilon \to 0$. The volume integral of $\nabla^2 g(\mathbf{r}|\mathbf{r}_0)$ can be replaced by a closed surface integral of $\nabla g(\mathbf{r}|\mathbf{r}_0) \cdot \mathbf{n}$ by Gauss's divergence theorem. This radial gradient, centered at $\mathbf{r}_0$, equals $\partial g(\mathbf{r}|\mathbf{r}_0)/\partial r = [-1/(4\pi R^2) + ik_0/(4\pi R)]e^{ik_0 R}$, which is constant over the spherical surface $R = \epsilon$. Thus, the closed surface integral equals $\partial g(\mathbf{r}|\mathbf{r}_0)/\partial r$ times the surface area $4\pi\epsilon^2$ of the sphere. As $\epsilon \to 0$, this integral approaches the value -1, which also equals the right side, verifying our solution for $g(\mathbf{r}|\mathbf{r}_0)$. We chose $e^{ik_0 R}/(4\pi R)$ rather than $e^{-ik_0 R}/(4\pi R)$ for the Green's function because the former represents an *outwardly* radiating solution.

- *Three-dimensional Poisson equation*

Because Helmholtz's equation becomes Poisson's equation if $k_0 = 0$, $1/(4\pi R)$ is the free-space Green's function for the three-dimensional Poisson equation.

- **Example 5.1.2**

Let us show that the three-dimensional, free-space Green's function for Poisson's equation in cylindrical coordinates is given by

$$g(r,\theta,z|\rho,\theta',\zeta) = \frac{1}{4\pi} \int_0^\infty e^{-k|z-\zeta|} J_0(kR)\,dk \qquad (5.1.54)$$

$$= \frac{1}{4\pi} \int_0^\infty e^{-k|z-\zeta|} \sum_{n=-\infty}^\infty J_n(k\rho)J_n(kr)\cos[n(\theta-\theta')]\,dk, \qquad (5.1.55)$$

where $R = \sqrt{r^2 + \rho^2 - 2r\rho\cos(\theta-\theta')}$.

To prove these relationships, we begin with

$$\frac{1}{r}\frac{\partial}{\partial r}\left(r\frac{\partial g}{\partial r}\right) + \frac{\partial^2 g}{\partial z^2} + \frac{1}{r^2}\frac{\partial^2 g}{\partial \theta^2} = -\frac{\delta(r-\rho)\delta(z-\zeta)\delta(\theta-\theta')}{r}. \qquad (5.1.56)$$

Because $\delta(\theta-\theta')$ can be re-expressed as (5.1.24), we write the Green's function as

$$g(r,\theta,z|\rho,\theta',\zeta) = \sum_{n=-\infty}^\infty g_n(r,z|\rho,\zeta)\cos[n(\theta-\theta')]. \qquad (5.1.57)$$

Substituting (5.1.24) and (5.1.57) into (5.1.56) and simplifying, we find that

$$\frac{1}{r}\frac{\partial}{\partial r}\left(r\frac{\partial g_n}{\partial r}\right) + \frac{\partial^2 g_n}{\partial z^2} - \frac{n^2}{r^2}g_n = -\frac{\delta(r-\rho)\delta(z-\zeta)}{2\pi r}. \qquad (5.1.58)$$

Taking the Hankel transform of (5.1.58), we obtain

$$\frac{d^2 G_n}{dz^2} - k^2 G_n = -\frac{J_n(k\rho)\delta(z - \zeta)}{2\pi}. \tag{5.1.59}$$

The solution to (5.1.59) is

$$G_n(k, z | \rho, \zeta) = \frac{J_n(k\rho) \exp(-k|z - \zeta|)}{4\pi k}. \tag{5.1.60}$$

Taking the inverse of (5.1.60) and substituting the inverse into (5.1.57), we obtain (5.1.55).

To obtain (5.1.54), we recall the addition theorem for Bessel functions:

$$J_0\left[\sqrt{\zeta^2 + z^2 - 2\zeta z \cos(\varphi)}\right] = \sum_{n=-\infty}^{\infty} J_n(\zeta) J_n(z) \cos(n\varphi). \tag{5.1.61}$$

Setting $\zeta = k\rho$, $z = kr$, and $\varphi = \theta - \theta'$ in (5.1.61) and then substituting this modified version of (5.1.61) into (5.1.55), we obtain (5.1.54).

An alternative to (5.1.55) can be derived starting with (5.1.58). However, at this point, we take the Fourier cosine transform with respect to z:

$$f(z) = \frac{1}{\pi} \int_0^\infty F(k) \cos[k(z - \zeta)]\, dk, \tag{5.1.62}$$

so that the transformed (5.1.58) is

$$\frac{1}{r}\frac{d}{dr}\left(r\frac{dG_n}{dr}\right) - \left(k^2 + \frac{n^2}{r^2}\right) G_n = -\frac{\delta(r - \rho)}{2\pi r}. \tag{5.1.63}$$

Applying the techniques of §2.4, we solve (5.1.63) and find that

$$G_n(r|\rho) = \frac{I_n(kr_<) K_n(kr_>)}{4\pi}. \tag{5.1.64}$$

Therefore,

$$g(r, \theta, z | \rho, \theta', \zeta) = \frac{1}{2\pi^2} \sum_{n=-\infty}^{\infty} \int_0^\infty e^{in(\theta - \theta')} \cos[k(z - \zeta)]$$

$$\times I_n(kr_<) K_n(kr_>)\, dk \tag{5.1.65}$$

$$= \frac{1}{4\pi^2} \sum_{n=-\infty}^{\infty} \int_{-\infty}^{\infty} e^{in(\theta - \theta') + ik(z - \zeta)} I_n(|k|r_<) K_n(|k|r_>)\, dk.$$

$$\tag{5.1.66}$$

- **Example 5.1.3**

As an example of how we can re-express a Green's function in one coordinate system (spherical coordinates) in terms of another one (cylindrical coordinates), let us show that

$$\frac{e^{ik_0 R}}{R} = \sum_{n=-\infty}^{\infty} \cos[n(\theta - \theta')] \left[\int_0^{k_0} \frac{i\xi J_n(r\xi) J_n(\rho\xi)}{\sqrt{k_0^2 - \xi^2}} e^{i\sqrt{k_0^2 - \xi^2}\,|z - \varsigma|}\,d\xi \right.$$

$$\left. + \int_{k_0}^{\infty} \frac{\xi J_n(r\xi) J_n(\rho\xi)}{\sqrt{\xi^2 - k_0^2}} e^{-\sqrt{\xi^2 - k_0^2}\,|z - \varsigma|}\,d\xi \right].$$

$$(\mathbf{5.1.67})$$

Consider the modified Helmholtz equation

$$\frac{1}{r}\frac{\partial}{\partial r}\left(r\frac{\partial g}{\partial r}\right) + \frac{\partial^2 g}{\partial z^2} + \frac{1}{r^2}\frac{\partial^2 g}{\partial \theta^2} - k_0^2 g = -\frac{\delta(r - \rho)\delta(z - \varsigma)\delta(\theta - \theta')}{r}.$$

$$(\mathbf{5.1.68})$$

Because $\delta(\theta - \theta')$ is given by (5.1.24), we assume a solution of the form

$$g(r, \theta, z | \rho, \theta', \varsigma) = \sum_{n=-\infty}^{\infty} g_n(r, z | \rho, \varsigma) \cos[n(\theta - \theta')]. \qquad (\mathbf{5.1.69})$$

Substituting (5.1.24) and (5.1.69) into (5.1.68) and simplifying, we find that

$$\frac{1}{r}\frac{\partial}{\partial r}\left(r\frac{\partial g_n}{\partial r}\right) + \frac{\partial^2 g_n}{\partial z^2} - \frac{n^2}{r^2} g_n - k_0^2 g_n = -\frac{\delta(r - \rho)\delta(z - \varsigma)}{2\pi r}. \qquad (\mathbf{5.1.70})$$

Next, we take the Hankel transform

$$G_n(k, z | \rho, \varsigma) = \int_0^{\infty} g_n(r, z | \rho, \varsigma) J_n(kr)\, r\, dr \qquad (\mathbf{5.1.71})$$

of (5.1.70) and obtain

$$\frac{d^2 G_n}{dz^2} - \left(k_0^2 + k^2\right) G_n = -\frac{J_n(k\rho)\delta(z - \varsigma)}{2\pi}. \qquad (\mathbf{5.1.72})$$

The solution to (5.1.72) is

$$G_n(k, z | \rho, \varsigma) = \frac{J_n(k\rho)\exp(-|z - \varsigma|\sqrt{k_0^2 + k^2})}{4\pi\sqrt{k_0^2 + k^2}}. \qquad (\mathbf{5.1.73})$$

Table 5.1.2: Free-Space Green's Function for the Axisymmetric Modified Helmholtz Equation

$$\frac{1}{r}\frac{\partial}{\partial r}\left(r\frac{\partial g}{\partial r}\right) + \frac{\partial^2 g}{\partial z^2} - k_0^2 g - \frac{n^2}{r^2}g = -\frac{\delta(r-\rho)\delta(z-\zeta)}{2\pi r}$$

k_0	Free-Space Green's Function
0	$g = \dfrac{1}{4\pi}\displaystyle\int_0^\infty J_n(r\xi)J_n(\rho\xi)e^{-\xi|z-\zeta|}\,d\xi$
k_0	$g = \dfrac{1}{4\pi}\displaystyle\int_0^\infty \frac{\xi}{\sqrt{\xi^2+k_0^2}}J_n(r\xi)J_n(\rho\xi)e^{-\sqrt{\xi^2+k_0^2}\,|z-\zeta|}\,d\xi$
si	$g = \dfrac{1}{4\pi}\displaystyle\int_0^s \frac{i\xi J_n(r\xi)J_n(\rho\xi)}{\sqrt{s^2-\xi^2}}e^{i\sqrt{s^2-\xi^2}\,|z-\zeta|}\,d\xi$ $\quad + \dfrac{1}{4\pi}\displaystyle\int_s^\infty \frac{\xi J_n(r\xi)J_n(\rho\xi)}{\sqrt{\xi^2-s^2}}e^{-\sqrt{\xi^2-s^2}\,|z-\zeta|}\,d\xi$

Taking the inverse of (5.1.73), we obtain

$$g_n(r,z|\rho,\zeta) = \frac{1}{4\pi}\int_0^\infty \frac{\xi}{\sqrt{k_0^2+\xi^2}}J_n(r\xi)J_n(\rho\xi)e^{-|z-\zeta|\sqrt{k_0^2+\xi^2}}\,d\xi.$$

$$(5.1.74)$$

Equation (5.1.74) serves as the free-space Green's function for the axisymmetric modified Helmholtz equation (5.1.70). Additional variations are listed in Table 5.1.2. Lu[8] developed equivalent expressions when $n = 0$ and 1 in terms of (1) a power series of spherical Hankel functions and (2) a sum of a singular term involving a Legendre function of degree $\frac{1}{2}$ of the second kind and a regular term.

The final result follows from the substitution of (5.1.74) into (5.1.69) or

$$g(r,\theta,z|\rho,\theta',\zeta)$$
$$= \frac{1}{4\pi}\sum_{n=-\infty}^{\infty}\cos[n(\theta-\theta')]\left[\int_0^\infty \frac{\xi J_n(r\xi)J_n(\rho\xi)}{\sqrt{k_0^2+\xi^2}}e^{-|z-\zeta|\sqrt{k_0^2+\xi^2}}\,d\xi\right].$$

$$(5.1.75)$$

---

[8] Lu, Y., 1998: The elastodynamic Green's function for a torsional ring source. *J. Appl. Mech.*, **65**, 566–568 and Lu, Y., 1999: The full-space elastodynamic Green's function for time-harmonic radial and axial ring sources in a homogeneous isotropic medium. *J. Appl. Mech.*, **66**, 639–645.

Because $e^{-k_0 R}/(4\pi R)$, where $R^2 = r^2 + \rho^2 - 2r\rho\cos(\theta - \theta') + (z - \zeta)^2$, is also a free-space Green's function, we find that

$$\frac{e^{-k_0 R}}{R} = \sum_{n=-\infty}^{\infty} \cos[n(\theta - \theta')] \left[\int_0^\infty \frac{\xi J_n(r\xi) J_n(\rho\xi)}{\sqrt{k_0^2 + \xi^2}} e^{-|z-\zeta|\sqrt{k_0^2 + \xi^2}} \, d\xi \right].$$

(5.1.76)

Equation (5.1.67) follows by replacing k_0 with $-ik_0$.

In the case when there is no θ dependence and $\rho = 0$, a similar derivation leads to the famous Lamb or Sommerfeld integral

$$\frac{e^{ik_0 R}}{R} = \int_0^{k_0} \frac{i\xi J_n(r\xi) J_n(\rho\xi)}{\sqrt{k_0^2 - \xi^2}} e^{i\sqrt{k_0^2 - \xi^2}\,|z-\zeta|} \, d\xi$$
$$+ \int_{k_0}^\infty \frac{\xi J_n(r\xi) J_n(\rho\xi)}{\sqrt{\xi^2 - k_0^2}} e^{-\sqrt{\xi^2 - k_0^2}\,|z-\zeta|} \, d\xi, \qquad (5.1.77)$$

where R^2 now equals $r^2 + (z - \zeta)^2$.

Finally, for the special case $k_0 = 0$,

$$\frac{1}{R} = \sum_{n=-\infty}^{\infty} \cos[n(\theta - \theta')] \int_0^\infty J_n(r\xi) J_n(\rho\xi) e^{-|z-\zeta|\xi} \, d\xi. \qquad (5.1.78)$$

Since the integral[9]

$$\int_0^\infty e^{-at} J_m(bt) J_m(ct) \, dt = \frac{1}{\pi\sqrt{bc}} Q_{m-\frac{1}{2}}\left(\frac{a^2 + b^2 + c^2}{2bc} \right), \qquad (5.1.79)$$

where $Q_{m-\frac{1}{2}}(\ )$ is the half-integer degree Legendre function of the second kind, Cohl and Tohline[10] found that

$$\frac{1}{R} = \frac{1}{\pi\sqrt{\rho r}} \sum_{n=-\infty}^{\infty} \cos[n(\theta - \theta')] Q_{n-\frac{1}{2}} \left[\frac{r^2 + \rho^2 + (z - \zeta)^2}{2r\rho} \right] \qquad (5.1.80)$$

$$= \frac{1}{\pi\sqrt{\rho r}} \sum_{n=0}^{\infty} \epsilon_n \cos[n(\theta - \theta')] Q_{n-\frac{1}{2}} \left[\frac{r^2 + \rho^2 + (z - \zeta)^2}{2r\rho} \right], \qquad (5.1.81)$$

since $Q_{n-\frac{1}{2}}(\chi) = Q_{-n-\frac{1}{2}}(\chi)$, where $\epsilon_0 = 1$ and $\epsilon_n = 2$ if $n > 0$. Equations (5.1.80)–(5.1.81) are significantly more compact and easier to numerically evaluate than (5.1.78).

---

[9] Gradshteyn, I. S., and I. M. Ryzhik, 1965: *Tables of Integrals, Series, and Products*. Academic Press, formula 6.612, number 3.

[10] Cohl, H. S., and J. E. Tohline, 1999: A compact cylindrical Green's function expansion for the solution of potential problems. *Astrophys. J.*, **527**, 86–101. Published by The University of Chicago Press.

Table 5.1.3: Free-Space Green's Function for Some Axisymmetric Helmholtz-like Equations: LHS $= -\delta(r - \rho)\delta(z - \zeta)$

LHS	Free-Space Green's Function
$\dfrac{1}{r}\dfrac{\partial}{\partial r}\left(r\dfrac{\partial g}{\partial r}\right) + \dfrac{\partial^2 g}{\partial z^2}$	$g = \dfrac{\rho}{4\pi}\displaystyle\int_0^{2\pi}\dfrac{1}{R}\,d\theta$
$\dfrac{1}{r}\dfrac{\partial}{\partial r}\left(r\dfrac{\partial g}{\partial r}\right) + \dfrac{\partial^2 g}{\partial z^2} - k_0^2 g$	$g = \dfrac{\rho}{4\pi}\displaystyle\int_0^{2\pi}\dfrac{e^{-k_0 R}}{R}\,d\theta$
$\dfrac{1}{r}\dfrac{\partial}{\partial r}\left(r\dfrac{\partial g}{\partial r}\right) + \dfrac{\partial^2 g}{\partial z^2} - \dfrac{n^2 g}{r^2}$	$g = \dfrac{\rho}{4\pi}\displaystyle\int_0^{2\pi}\dfrac{\cos(n\theta)}{R}\,d\theta$
$\dfrac{1}{r}\dfrac{\partial}{\partial r}\left(r\dfrac{\partial g}{\partial r}\right) + \dfrac{\partial^2 g}{\partial z^2} - \dfrac{g}{r^2} + k_0^2 g$	$g = \dfrac{\rho}{4\pi}\displaystyle\int_0^{2\pi}\dfrac{\cos(\theta)}{R}e^{-ik_0 R}\,d\theta$
$\dfrac{1}{r}\dfrac{\partial}{\partial r}\left(r\dfrac{\partial g}{\partial r}\right) + \dfrac{\partial^2 g}{\partial z^2} - \dfrac{g}{r^2} - k_0^2 g$	$g = \dfrac{\rho}{4\pi}\displaystyle\int_0^{2\pi}\dfrac{\cos(\theta)}{R}e^{-k_0 R}\,d\theta$

$$R^2 = r^2 + \rho^2 - 2r\rho\cos(\theta) + (z - \zeta)^2$$

Reprinted from *Appl. Math. Modelling*, **14**, M. Tsuchimoto, K. Miya, T. Honma, and H. Igarashi, Fundamental solutions of the axisymmetric Helmholtz-type equations, 605–611, ©1990, with permission from Elsevier Science.

• **Example 5.1.4**

We can use the free-space Green's function $e^{-k_0 R}/(4\pi R)$, where $R^2 = r^2 + \rho^2 - 2r\rho\cos(\theta - \theta') + (z - \zeta)^2$, for (5.1.68) to generate additional axisymmetric Green's functions by simply eliminating the θ dependence by integration. We have listed several solutions in Table 5.1.3. Tezer-Sezgin and Dost[11] give additional Green's functions for Helmholtz-type equations.

• **Example 5.1.5: Green's function for a quarter plane**

Just as with the heat equation, the method of images can be used to find Green's functions when boundaries are present. As a simple example, let us find the Green's function for the planar Poisson equation

$$\frac{\partial^2 g}{\partial x^2} + \frac{\partial^2 g}{\partial y^2} = -\delta(x - \xi)\delta(y - \eta), \tag{5.1.82}$$

[11] Tezer-Sezgin, M., and S. Dost, 1993: On the fundamental solutions of the axisymmetric Helmholtz-type equations. *Appl. Math. Modelling*, **17**, 47–51.

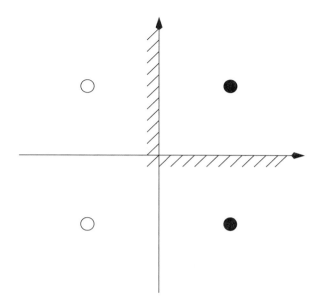

Figure 5.1.2: Configuration of images to find the Green's function for the planar Poisson equation with the boundary conditions $g(0, y|\xi, \eta) = g_y(x, 0|\xi, \eta) = 0$. The positive images are filled circles while the negative images are open circles.

for the quarter space $0 < x, y$, with the boundary conditions

$$g(0, y|\xi, \eta) = g_y(x, 0|\xi, \eta) = 0, \qquad (5.1.83)$$

and

$$\lim_{x, y \to \infty} |g(x, y|\xi, \eta)| < \infty. \qquad (5.1.84)$$

One method for solving (5.1.82)–(5.1.84) is to use Fourier sine transforms in the x direction. Taking the transform

$$G(k, y|\xi, \eta) = \int_0^\infty g(x, y|\xi, \eta) \, \sin(kx) \, dx \qquad (5.1.85)$$

of (5.1.82)–(5.1.84), they become the ordinary differential equation

$$\frac{d^2 G}{dy^2} - k^2 G = -\sin(k\xi)\delta(y - \eta), \qquad (5.1.86)$$

with $G'(k, 0|\xi, \eta) = 0$, and $\lim_{y \to \infty} |G(k, y|\xi, \eta)| < \infty$. The solution to (5.1.86) is

$$G(k, y|\xi, \eta) = \frac{\sin(k\xi)}{2k} \left[e^{-k|y-\eta|} + e^{-k(y+\eta)} \right]. \qquad (5.1.87)$$

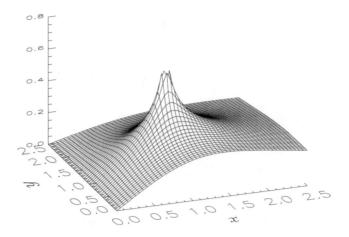

Figure 5.1.3: The Green's function for the planar Poisson equation over the quarter plane $0 < x, y$, with the boundary conditions $g(0, y|\xi, \eta) = g_y(x, 0|\xi, \eta) = 0$ when $\xi = \eta = 1$.

Therefore,

$$g(x, y|\xi, \eta) = \frac{1}{\pi} \int_0^\infty \left[e^{-k|y-\eta|} + e^{-k(y+\eta)} \right] \sin(k\xi)\, \sin(kx)\, \frac{dk}{k}.$$

$$(5.1.88)$$

Performing the integration and simplifying,

$$g(x, y|\xi, \eta) = \frac{1}{4\pi} \ln\left\{ \frac{[(x+\xi)^2 + (y-\eta)^2][(x+\xi)^2 + (y+\eta)^2]}{[(x-\xi)^2 + (y-\eta)^2][(x-\xi)^2 + (y+\eta)^2]} \right\}.$$

$$(5.1.89)$$

Now let us do this problem by the method of images. To satisfy the boundary conditions, we must introduce three images, which are shown in Figure 5.1.2. The filled circles denote positive images while the open circles are negative images. Using the free-space Green's function for each image and simplifying, we obtain (5.1.89). This solution is shown in Figure 5.1.3. Note that the point $x = y = 1$ has *not* been plotted.

- **Example 5.1.6: Free-space Green's function for the biharmonic equation**

An equation similar to Laplace's equation is the biharmonic equation

$$\nabla^4 g = \frac{\partial^4 g}{\partial x^4} + 2\frac{\partial^4 g}{\partial x^2 \partial y^2} + \frac{\partial^4 g}{\partial y^4} = \delta(x - \xi)\delta(y - \eta).$$

$$(5.1.90)$$

Let us find its free-space Green's function. We start by taking the Fourier transform in both directions. If we denote the transform variables in the x and y directions by k and ℓ, respectively, we find that

$$g(x, y|\xi, \eta) = \frac{1}{4\pi^2} \int_{-\infty}^{\infty} \int_{-\infty}^{\infty} \frac{e^{ik(x-\xi) + i\ell(y-\eta)}}{(k^2 + \ell^2)^2}\, dk\, d\ell.$$

$$(5.1.91)$$

If we try to evaluate (5.1.91) directly, we fail because the integral diverges. To avoid this difficulty, we first compute

$$g_{xy}(x, y|\xi, \eta) = -\frac{1}{4\pi^2} \int_{-\infty}^{\infty} \int_{-\infty}^{\infty} \frac{k\ell \, e^{ik(x-\xi)+i\ell(y-\eta)}}{(k^2 + \ell^2)^2} \, dk \, d\ell. \quad (5.1.92)$$

Then

$$g_{xy}(x, y|\xi, \eta) = \frac{1}{2\pi^2 i} \int_{0}^{\infty} \left\{ \int_{-\infty}^{\infty} \frac{k \, e^{ik(x-\xi)}}{(k^2 + \ell^2)^2} \, dk \right\} \ell \sin[\ell(y - \eta)] \, d\ell$$
$$(5.1.93)$$

$$= \frac{(x - \xi)}{4\pi} \int_{0}^{\infty} \sin[\ell(y - \eta)] \, e^{-|x-\xi|\ell} \, d\ell \quad (5.1.94)$$

$$= \frac{1}{4\pi} \frac{(x - \xi)(y - \eta)}{(x - \xi)^2 + (y - \eta)^2}. \quad (5.1.95)$$

Integrating twice yields

$$g(x, y|\xi, \eta) = \frac{1}{8\pi} \left[(x - \xi)^2 + (y - \eta)^2 \right] \left\{ \ln \left[\sqrt{(x - \xi)^2 + (y - \eta)^2} \right] - 1 \right\}$$
$$+ f(x - \xi) + g(y - \eta), \quad (5.1.96)$$

where $f(\ )$ and $g(\ )$ are arbitrary, differentiable functions. Actually, only that portion of (5.1.96) that contains the logarithm is the free-space Green's function because it contains the singular nature of the function. Hayes *et al.*[12] have found the free-space Green's function for the slightly different equation

$$\frac{\partial g}{\partial x} + \tau \left(\frac{\partial^4 g}{\partial x^4} + 2\frac{\partial^4 g}{\partial x^2 \partial y^2} + \frac{\partial^4 g}{\partial y^4} \right) = \delta(x - \xi)\delta(y - \eta). \quad (5.1.97)$$

5.2 TWO-DIMENSIONAL POISSON'S EQUATION OVER RECTANGULAR AND CIRCULAR DOMAINS

Consider the two-dimensional Poisson equation

$$\frac{\partial^2 u}{\partial x^2} + \frac{\partial^2 u}{\partial y^2} = -f(x, y). \quad (5.2.1)$$

This equation arises in equilibrium problems, such as the static deflection of a rectangular membrane. In that case, $f(x, y)$ represents the

---

[12] Hayes, M., S. B. G. O'Brien, and J. H. Lammers, 2000: Green's function for steady flow over a small two-dimensional topography. *Phys. Fluids*, **12**, 2845–2858.

external load per unit area, divided by the tension in the membrane. The solution $u(x,y)$ must satisfy certain boundary conditions. For the present, let us choose $u(0,y) = u(a,y) = 0$, and $u(x,0) = u(x,b) = 0$.

• *Rectangular Area*

To find the Green's function for (5.2.1) we must solve the partial differential equation

$$\frac{\partial^2 g}{\partial x^2} + \frac{\partial^2 g}{\partial y^2} = -\delta(x-\xi)\delta(y-\eta), \quad 0 < x,\xi < a, \quad 0 < y,\eta < b, \quad \textbf{(5.2.2)}$$

subject to the boundary conditions

$$g(0,y|\xi,\eta) = g(a,y|\xi,\eta) = g(x,0|\xi,\eta) = g(x,b|\xi,\eta) = 0. \qquad \textbf{(5.2.3)}$$

From (5.0.14),

$$u(x,y) = \int_0^a \int_0^b g(x,y|\xi,\eta) f(\xi,\eta) \, d\eta \, d\xi. \qquad \textbf{(5.2.4)}$$

One approach to finding the Green's function is to expand it in terms of the eigenfunctions $\varphi(x,y)$ of the differential equation

$$\frac{\partial^2 \varphi}{\partial x^2} + \frac{\partial^2 \varphi}{\partial y^2} = -\lambda\varphi, \qquad \textbf{(5.2.5)}$$

and the boundary conditions (5.2.3). The eigenvalues are

$$\lambda_{nm} = \frac{n^2\pi^2}{a^2} + \frac{m^2\pi^2}{b^2}, \qquad \textbf{(5.2.6)}$$

where $n = 1,2,3,\ldots$, $m = 1,2,3,\ldots$, and the corresponding eigenfunctions are

$$\varphi_{nm}(x,y) = \sin\left(\frac{n\pi x}{a}\right) \sin\left(\frac{m\pi y}{b}\right). \qquad \textbf{(5.2.7)}$$

Therefore, we seek $g(x,y|\xi,\eta)$ in the form

$$g(x,y|\xi,\eta) = \sum_{n=1}^{\infty} \sum_{m=1}^{\infty} A_{nm} \sin\left(\frac{n\pi x}{a}\right) \sin\left(\frac{m\pi y}{b}\right). \qquad \textbf{(5.2.8)}$$

Because the delta functions can be written

$$\delta(x-\xi)\delta(y-\eta) = \frac{4}{ab} \sum_{n=1}^{\infty} \sum_{m=1}^{\infty} \sin\left(\frac{n\pi\xi}{a}\right) \sin\left(\frac{m\pi\eta}{b}\right) \sin\left(\frac{n\pi x}{a}\right) \sin\left(\frac{m\pi y}{b}\right),$$

$$\textbf{(5.2.9)}$$

we find that

$$\left(\frac{n^2\pi^2}{a^2} + \frac{m^2\pi^2}{b^2}\right) A_{nm} = \frac{4}{ab} \sin\left(\frac{n\pi\xi}{a}\right) \sin\left(\frac{m\pi\eta}{b}\right), \qquad (\mathbf{5.2.10})$$

after substituting (5.2.8) and (5.2.9) into the partial differential equation (5.2.2) and setting the corresponding harmonics equal to each other. Therefore, the *bilinear formula* for the Green's function of Poisson's equation is

$$g(x,y|\xi,\eta) = \frac{4}{ab} \sum_{n=1}^{\infty} \sum_{m=1}^{\infty} \frac{\sin\left(\frac{n\pi x}{a}\right) \sin\left(\frac{n\pi\xi}{a}\right) \sin\left(\frac{m\pi y}{b}\right) \sin\left(\frac{m\pi\eta}{b}\right)}{n^2\pi^2/a^2 + m^2\pi^2/b^2}.$$

$$(\mathbf{5.2.11})$$

Thus, solutions to Poisson's equation can be now written as

$$u(x,y) = \sum_{n=1}^{\infty} \sum_{m=1}^{\infty} \frac{a_{nm}}{n^2\pi^2/a^2 + m^2\pi^2/b^2} \sin\left(\frac{n\pi x}{a}\right) \sin\left(\frac{m\pi y}{b}\right),$$

$$(\mathbf{5.2.12})$$

where a_{nm} are the Fourier coefficients for the function $f(x,y)$:

$$a_{nm} = \frac{4}{ab} \int_0^a \int_0^b f(x,y) \sin\left(\frac{n\pi x}{a}\right) \sin\left(\frac{m\pi y}{b}\right) dy\, dx. \qquad (\mathbf{5.2.13})$$

We save as problem 10 the derivation of the Green's function with Neumann boundary conditions.

Another form of the Green's function can be obtained by considering each direction separately. To satisfy the boundary conditions along the edges $y = 0$ and $y = b$, we write the Green's function as the Fourier series

$$g(x,y|\xi,\eta) = \sum_{m=1}^{\infty} G_m(x|\xi) \sin\left(\frac{m\pi y}{b}\right), \qquad (\mathbf{5.2.14})$$

where the coefficients $G_m(x|\xi)$ are left as undetermined functions of x, ξ, and η. Substituting this series into the partial differential equation for g, multiplying by $2\sin(n\pi y/b)/b$, and integrating over y, we find that

$$\frac{d^2 G_n}{dx^2} - \frac{n^2\pi^2}{b^2} G_n = -\frac{2}{b} \sin\left(\frac{n\pi\eta}{b}\right) \delta(x - \xi). \qquad (\mathbf{5.2.15})$$

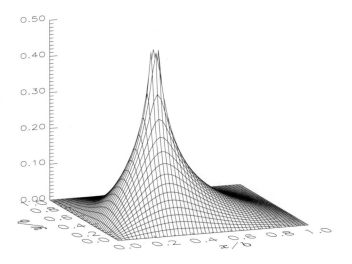

Figure 5.2.1: The Green's function (5.2.22) or (5.2.23) for the planar Poisson equation over a rectangular area with Dirichlet boundary conditions on all sides when $a = b$, and $\xi/b = \eta/b = 0.3$.

This differential equation shows that the expansion coefficients $G_n(x|\xi)$ are one-dimensional Green's functions; we can find them, as we did in Chapter 2, by piecing together homogeneous solutions to (5.2.15) that are valid over various intervals. For the region $0 \leq x \leq \xi$, the solution to (5.2.15) that vanishes at $x = 0$ is

$$G_n(x|\xi) = A_n \sinh\left(\frac{n\pi x}{b}\right), \tag{5.2.16}$$

where A_n is presently arbitrary. The corresponding solution for $\xi \leq x \leq a$ is

$$G_n(x|\xi) = B_n \sinh\left[\frac{n\pi(a - x)}{b}\right]. \tag{5.2.17}$$

Note that this solution vanishes at $x = a$. Because the Green's function must be continuous at $x = \xi$,

$$A_n \sinh\left(\frac{n\pi\xi}{b}\right) = B_n \sinh\left[\frac{n\pi(a - \xi)}{b}\right]. \tag{5.2.18}$$

On the other hand, the appropriate jump discontinuity of $G_n'(x|\xi)$ yields

$$-\frac{n\pi}{b} B_n \cosh\left[\frac{n\pi(a - \xi)}{b}\right] - \frac{n\pi}{b} A_n \cosh\left(\frac{n\pi\xi}{b}\right) = -\frac{2}{b} \sin\left(\frac{n\pi\eta}{b}\right). \tag{5.2.19}$$

Solving for A_n and B_n,

$$A_n = \frac{2}{n\pi} \sin\left(\frac{n\pi\eta}{b}\right) \frac{\sinh[n\pi(a-\xi)/b]}{\sinh(n\pi a/b)}, \tag{5.2.20}$$

and

$$B_n = \frac{2}{n\pi} \sin\left(\frac{n\pi\eta}{b}\right) \frac{\sinh(n\pi\xi/b)}{\sinh(n\pi a/b)}. \tag{5.2.21}$$

This yields the Green's function

$$g(x,y|\xi,\eta) = \frac{2}{\pi} \sum_{n=1}^{\infty} \frac{\sinh[n\pi(a-x_>)/b]\sinh(n\pi x_</b)}{n\,\sinh(n\pi a/b)}$$
$$\times \sin\left(\frac{n\pi\eta}{b}\right) \sin\left(\frac{n\pi y}{b}\right). \tag{5.2.22}$$

Equation (5.2.22) is illustrated in Figure 5.2.1 in the case of a square domain with $\xi/b = \eta/b = 0.3$.

If we began with a Fourier expansion in the y direction, we would have obtained

$$g(x,y|\xi,\eta) = \frac{2}{\pi} \sum_{m=1}^{\infty} \frac{\sinh[m\pi(b-y_>)/a]\sinh(m\pi y_</a)}{m\,\sinh(m\pi b/a)}$$
$$\times \sin\left(\frac{m\pi\xi}{a}\right) \sin\left(\frac{m\pi x}{a}\right). \tag{5.2.23}$$

• *Circular Disk*

Let us now find the Green's function for Poisson's equation for the circular domain $r \leq 1$ with Dirichlet boundary conditions. We begin by writing the Green's function as a sum of the free-space Green's function g_p plus a homogeneous solution g_H such that $\nabla^2 g_H = 0$. From separation of variables it is easily shown that

$$g_H = \frac{a_0}{2} + \sum_{n=1}^{\infty} r^n \left[a_n \cos(n\theta) + b_n \sin(n\theta)\right], \tag{5.2.24}$$

while the free-space Green's function (5.1.52) is

$$g_p = -\frac{1}{4\pi} \ln\left[r^2 + \rho^2 - 2r\rho\cos(\theta - \theta')\right]. \tag{5.2.25}$$

Along $r = 1$, it follows that

$$g_H = -g_p = \frac{1}{4\pi} \ln\left[1 + \rho^2 - 2\rho\cos(\theta - \theta')\right]. \tag{5.2.26}$$

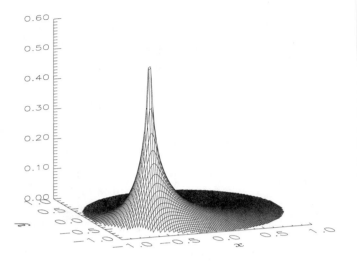

Figure 5.2.2: The Green's function (5.2.36) for the planar Poisson equation over a unit disk with a Dirichlet boundary condition when $\rho = 0.55\sqrt{2}$, and $\theta' = 5\pi/4$.

Next, we note that

$$\ln\left[1 + \rho^2 - 2\rho\cos(\theta - \theta')\right] = \ln\left\{1 + \rho^2 - \rho\left[e^{i(\theta-\theta')} + e^{-i(\theta-\theta')}\right]\right\} \tag{5.2.27}$$

$$= \ln\left\{\left[1 - \rho e^{i(\theta-\theta')}\right]\left[1 - \rho e^{-i(\theta-\theta')}\right]\right\} \tag{5.2.28}$$

$$= \ln\left[1 - \rho e^{i(\theta-\theta')}\right] + \ln\left[1 - \rho e^{-i(\theta-\theta')}\right] \tag{5.2.29}$$

$$= -\left[\rho e^{i(\theta-\theta')} + \tfrac{1}{2}\rho^2 e^{2i(\theta-\theta')} + \cdots\right]$$
$$- \left[\rho e^{-i(\theta-\theta')} + \tfrac{1}{2}\rho^2 e^{-2i(\theta-\theta')} + \cdots\right] \tag{5.2.30}$$

$$= -2\sum_{n=1}^{\infty} \frac{\rho^n \cos[n(\theta - \theta')]}{n}, \tag{5.2.31}$$

provided $|\rho| < 1$. Therefore, when $r = 1$, we have

$$-2\sum_{n=1}^{\infty} \frac{\rho^n \cos[n(\theta - \theta')]}{n} = \frac{a_0}{2} + \sum_{n=1}^{\infty} a_n \cos(n\theta) + b_n \sin(n\theta) \tag{5.2.32}$$

so that

$$a_0 = 0, \quad a_n = \frac{\rho^n}{2\pi n} \cos(n\theta'), \quad b_n = \frac{\rho^n}{2\pi n} \sin(n\theta'). \tag{5.2.33}$$

Consequently,

$$g_H = \frac{1}{2\pi} \sum_{n=1}^{\infty} \frac{(r\rho)^n}{n} \cos[n(\theta - \theta')] \tag{5.2.34}$$

$$= -\frac{1}{4\pi} \ln \left[1 + (r\rho)^2 - 2r\rho \cos(\theta - \theta') \right], \tag{5.2.35}$$

and

$$g(r, \theta | \rho, \theta') = -\frac{1}{4\pi} \ln \left[r^2 + \rho^2 - 2r\rho \cos(\theta - \theta') \right]$$
$$+ \frac{1}{4\pi} \ln \left[1 + (r\rho)^2 - 2r\rho \cos(\theta - \theta') \right] \tag{5.2.36}$$

$$= -\frac{1}{4\pi} \ln \left[r^2 + \rho^2 - 2r\rho \cos(\theta - \theta') \right]$$
$$+ \frac{1}{4\pi} \ln \left[r^2 + (1/\rho)^2 - (2r/\rho) \cos(\theta - \theta') \right] + \frac{1}{4\pi} \ln(\rho^2). \tag{5.2.37}$$

Equation (5.2.37) shows that the homogeneous portion of the Green's function behaves (within an additive constant) in the same manner as the free-space Green's function except that its source is located at $(1/\rho, \theta')$. Because $1/\rho$ is the inverse of ρ, the relationship between (ρ, θ') and $(1/\rho, \theta')$ is called inversion with respect to the circle. We will see a similar inversion pattern when we find the Green's function for Poisson's equation with a Dirichlet boundary condition on a sphere of radius a.

We can apply (5.2.36) to find the solution to Laplace's equation with the Dirichlet boundary condition $u(1, \theta)$. Using (5.0.13) with $f(\mathbf{r}_0) = 0$ and

$$\nabla_0 g(\mathbf{r}_0 | \mathbf{r}) = \left. \frac{\partial g}{\partial \rho} \right|_{\rho=1} \hat{\mathbf{r}} = -\frac{1}{2\pi} \frac{1 - r^2}{1 + r^2 - 2r \cos(\theta - \theta')} \hat{\mathbf{r}}, \tag{5.2.38}$$

we obtain Poisson's integral formula

$$u(r, \theta) = \frac{1}{2\pi} \int_0^{2\pi} \frac{1 - r^2}{1 + r^2 - 2r \cos(\theta - \theta')} u(1, \theta') \, d\theta'. \tag{5.2.39}$$

An alternative to (5.2.36) can be found by solving

$$\frac{1}{r} \frac{\partial}{\partial r} \left(r \frac{\partial g}{\partial r} \right) + \frac{1}{r^2} \frac{\partial^2 g}{\partial \theta^2} = -\frac{\delta(r - \rho)\delta(\theta - \theta')}{r}, \tag{5.2.40}$$

where $0 < r, \rho < a$, and $0 \leq \theta, \theta' \leq 2\pi$, with the boundary conditions

$$\lim_{r \to 0} |g(r, \theta|\rho, \theta')| < \infty, \qquad g(a, \theta|\rho, \theta') = 0, \qquad 0 \leq \theta, \theta' \leq 2\pi. \tag{5.2.41}$$

The Green's function must be periodic in θ.

We begin by expressing $\delta(\theta - \theta')$ by the Fourier cosine series

$$\delta(\theta - \theta') = \frac{1}{2\pi} + \frac{1}{\pi} \sum_{n=1}^{\infty} \cos[n(\theta - \theta')] = \frac{1}{2\pi} \sum_{n=-\infty}^{\infty} \cos[n(\theta - \theta')]. \tag{5.2.42}$$

Thus, the form of the Green's function is

$$g(r, \theta|\rho, \theta') = \sum_{n=-\infty}^{\infty} g_n(r|\rho) \cos[n(\theta - \theta')]. \tag{5.2.43}$$

Substituting (5.2.42) and (5.2.43) into (5.2.40) and simplifying, we find that

$$\frac{1}{r} \frac{d}{dr} \left(r \frac{dg_n}{dr} \right) - \frac{n^2}{r^2} g_n = -\frac{\delta(r - \rho)}{2\pi r}. \tag{5.2.44}$$

Because

$$\frac{\delta(r - \rho)}{2\pi r} = \frac{1}{\pi a^2} \sum_{m=1}^{\infty} \frac{J_n(k_{nm}\rho) J_n(k_{nm}r)}{J'^2_n(k_{nm}a)}, \tag{5.2.45}$$

where k_{nm} is the mth root of $J_n(k_{nm}a) = 0$, we take

$$g_n(r|\rho) = \sum_{m=1}^{\infty} A_{nm} J_n(k_{nm}r). \tag{5.2.46}$$

Substituting (5.2.45) and (5.2.46) into (5.2.44), we obtain

$$k^2_{nm} A_{nm} = \frac{1}{\pi a^2} \frac{J_n(k_{nm}\rho)}{J'^2_n(k_{nm}a)}. \tag{5.2.47}$$

after using the differential equation that governs Bessel functions of order n and the first kind. The Green's function follows by substituting (5.2.46) and (5.2.47) into (5.2.43) or

$$g(r, \theta|\rho, \theta') = \frac{1}{\pi a^2} \sum_{n=-\infty}^{\infty} \sum_{m=1}^{\infty} \frac{J_n(k_{nm}\rho) J_n(k_{nm}r)}{k^2_{nm} J'^2_n(k_{nm}a)} \cos[n(\theta - \theta')].$$

$$(5.2.48)$$

• *Infinitely long, axisymmetric cylinder*

Let us find the Green's function for the exterior of an infinitely long, axisymmetric cylinder,[13] where the Green's function equals zero along the boundary $r = a$. This problem is governed by the partial differential equation

$$\frac{\partial^2 g}{\partial r^2} + \frac{1}{r}\frac{\partial g}{\partial r} + \frac{\partial^2 g}{\partial z^2} = -\frac{\delta(r - \rho)\delta(z - \zeta)}{2\pi r}, \qquad (5.2.49)$$

where $a < r, \rho < \infty$, and $-\infty < z, \zeta < \infty$, with the boundary conditions

$$\lim_{|z|\to\infty} |g(r, z|\rho, \zeta)| < \infty, \qquad a < r < \infty, \qquad (5.2.50)$$

and

$$g(a, L|\rho, \zeta) = 0, \quad \lim_{r\to\infty} |g(r, z|\rho, \zeta)| < \infty, \quad |z| < \infty. \qquad (5.2.51)$$

We begin by taking the Fourier transform of (5.2.49), which yields

$$\frac{d^2 G}{dr^2} + \frac{1}{r}\frac{dG}{dr} - k^2 G = -\frac{\delta(r - \rho)}{2\pi r}e^{-ik\zeta}, \qquad a < r < \infty, \qquad (5.2.52)$$

with $G(a|\rho) = 0$, and $\lim_{r\to\infty} |G(r|\rho)| < \infty$. Because the homogeneous solutions of (5.2.52) are $I_0(kr)$ and $K_0(kr)$, a Green's function that satisfies (5.2.52) and the boundary conditions is

$$G(r|\rho) = A[I_0(ka)K_0(|k|r_<) - K_0(|k|a)I_0(kr_<)]K_0(|k|r_>). \qquad (5.2.53)$$

Equation (5.2.53) enjoys the additional property of being continuous at the point $r = \rho$.

To compute A, we use the condition that the Green's function must satisfy the jump condition

$$\frac{dG}{dr}\bigg|_{r=\rho^-}^{r=\rho^+} = -\frac{e^{-ik\zeta}}{2\pi\rho}. \qquad (5.2.54)$$

---

[13] Reprinted with permission from Marr-Lyon, M. J., D. B. Thiessen, F. J. Blonigen, and P. L. Marston, 2000: Stabilization of electrically conducting capillary bridges using feedback control of radial electrostatic stresses and the shapes of extended bridges. *Phys. Fluids*, **12**, 986–995. ©American Institute of Physics, 2000.

Upon substituting (5.2.53) into (5.2.54) and using the Wronskian relationship, we find that

$$2\pi A K_0(|k|a) = -e^{-ik\zeta}. \tag{5.2.55}$$

Therefore,

$$G(r,k|\rho,\zeta) = \frac{[K_0(|k|a)I_0(kr_<) - I_0(ka)K_0(|k|r_<)]K_0(|k|r_>)}{2\pi K_0(|k|a)},$$
$$\tag{5.2.56}$$

and

$$g(r,z|\rho,\zeta) = \frac{1}{4\pi^2} \int_{-\infty}^{\infty} \frac{[K_0(|k|a)I_0(kr_<) - I_0(ka)K_0(|k|r_<)]K_0(|k|r_>)}{K_0(|k|a)}$$
$$\times e^{ik(z-\zeta)}\, dk. \tag{5.2.57}$$

A problem that is very similar to (5.2.49)–(5.2.51) involves solving (5.2.49) in the interior of a cylinder of radius a with the boundary conditions

$$\lim_{|z|\to\infty} |g(r,z|\rho,\zeta)| < \infty, \qquad 0 < r < a, \tag{5.2.58}$$

and

$$\lim_{r\to 0} |g(r,z|\rho,\zeta)| < \infty, \quad g(a,z|\rho,\zeta) = 0, \quad |z| < \infty. \tag{5.2.59}$$

We begin as before and take the Fourier transform of (5.2.49) with respect to z. This leads to (5.2.52). The solution to (5.2.52) in the present case is

$$G(r|\rho) = A[I_0(ka)K_0(|k|r_>) - K_0(|k|a)I_0(kr_>)]I_0(kr_<) \tag{5.2.60}$$

with

$$2\pi A I_0(ka) = e^{-ik\zeta}. \tag{5.2.61}$$

Therefore,

$$g(r,z|\rho,\zeta) = \frac{1}{4\pi^2} \int_{-\infty}^{\infty} \frac{[I_0(ka)K_0(|k|r_>) - K_0(|k|a)I_0(kr_>)]I_0(kr_<)}{I_0(ka)}$$
$$\times e^{ik(z-\zeta)}\, dk. \tag{5.2.62}$$

An alternative to (5.2.62) can be found[14] by closing the line integration (5.2.62) with a semicircle of infinite radius as dictated by Jordan's lemma and applying the residue theorem. This yields

$$g(r,z|\rho,\zeta) = \frac{1}{2\pi a} \sum_{n=1}^{\infty} \frac{J_0(k_n\rho/a)J_0(k_n r/a)}{k_n J_1^2(k_n)} e^{-k_n|z-\zeta|/a}, \tag{5.2.63}$$

---

[14] See Allen, C. K., N. Brown, and M. Reiser, 1994: Image effects for bunched beams in axisymmetric systems. *Particle Accel.*, **45**, 149–165.

where k_n is the nth root of $J_0(k) = 0$.

- *Annulus of finite height*

In the same vein let us find the Green's function for an annular region $r_1 < r < r_2$ of finite depth L where the Green's function equals zero along the boundary. Mathematically,[15] this is equivalent to solving

$$\frac{\partial^2 g}{\partial r^2} + \frac{1}{r}\frac{\partial g}{\partial r} + \frac{\partial^2 g}{\partial z^2} = -\frac{\delta(r - \rho)\delta(z - \zeta)}{2\pi r}, \qquad (5.2.64)$$

where $r_1 < r, \rho < r_2$, and $0 < z, \zeta < L$, with the boundary conditions

$$g(r_1, z|\rho, \zeta) = g(r_2, z|\rho, \zeta) = 0, \quad 0 < z < L, \qquad (5.2.65)$$

and

$$g(r, 0|\rho, \zeta) = g(r, L|\rho, \zeta) = 0, \quad r_1 < r < r_2. \qquad (5.2.66)$$

We begin by noting that since

$$\delta(z - \zeta) = \frac{2}{L}\sum_{n=1}^{\infty}\sin\left(\frac{n\pi\zeta}{L}\right)\sin\left(\frac{n\pi z}{L}\right), \qquad (5.2.67)$$

we can express the Green's function via the expansion

$$g(r, z|\rho, \zeta) = \sum_{n=1}^{\infty} G_n(r|\rho)\sin\left(\frac{n\pi\zeta}{L}\right)\sin\left(\frac{n\pi z}{L}\right). \qquad (5.2.68)$$

Substituting (5.2.67) and (5.2.68) into (5.2.64), we obtain the ordinary differential equation

$$\frac{d^2 G_n}{dr^2} + \frac{1}{r}\frac{dG_n}{dr} - \frac{n^2\pi^2}{L^2}G_n = -\frac{\delta(r - \rho)}{L\pi r}, \qquad r_1 < r < r_2, \quad (5.2.69)$$

for $G_n(r|\rho)$. Because the homogeneous solutions of (5.2.69) are $I_0(n\pi r/L)$ and $K_0(n\pi r/L)$, solutions to (5.2.69) that satisfy the boundary conditions are

$$G_n(r|\rho) = \begin{cases} A[K_0(n\pi r/L)I_0(n\pi r_1/L) - K_0(n\pi r_1/L)I_0(n\pi r/L)], \\ \qquad\qquad r_1 \le r \le \rho \\ B[K_0(n\pi r/L)I_0(n\pi r_2/L) - K_0(n\pi r_2/L)I_0(n\pi r/L)], \\ \qquad\qquad \rho \le r \le r_2. \end{cases}$$
$$(5.2.70)$$

---

[15] A generalization of a problem from Oliveira, J. C., and C. H. Amon, 1998: Pressure-based semi-analytical solution of radial Stokes flows. *J. Math. Anal. Appl.*, **217**, 95–114.

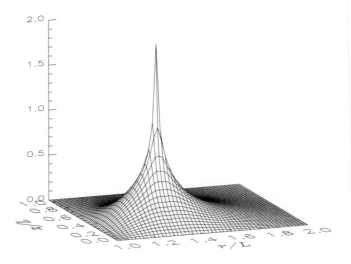

Figure 5.2.3: The Green's function (5.2.74) [times $\rho\pi^2$] for the planar Poisson equation within an annulus of finite height with Dirichlet boundary conditions on all sides when $r_1/L = 1$, $r_2/L = 2$, $\zeta/L = 1/\pi$, and $\rho/L = 1 + 1/\pi$.

To compute A and B, we first use the condition that the Green's function must be continuous from $r = \rho^-$ to $r = \rho^+$ or

$$A[K_0(n\pi\rho/L)I_0(n\pi r_1/L) - K_0(n\pi r_1/L)I_0(n\pi\rho/L)]$$
$$= B[K_0(n\pi\rho/L)I_0(n\pi r_2/L) - K_0(n\pi r_2/L)I_0(n\pi\rho/L)]. \quad (5.2.71)$$

On the other hand, integrating (5.2.69) from $r = \rho^-$ to $r = \rho^+$ yields the jump condition

$$\left.\frac{dG_n}{dr}\right|_{r=\rho^-}^{r=\rho^+} = -\frac{1}{\pi\rho L}, \quad (5.2.72)$$

or

$$A[K_1(n\pi\rho/L)I_0(n\pi r_1/L) + K_0(n\pi r_1/L)I_1(n\pi\rho/L)]$$
$$- B[K_1(n\pi\rho/L)I_0(n\pi r_2/L) + K_0(n\pi r_2/L)I_1(n\pi\rho/L)] = -\frac{1}{n\pi^2\rho}. \quad (5.2.73)$$

Solving for A and B and substituting into (5.2.70), and then (5.2.68), we obtain

$$g(r, z|\rho, \zeta) = -\sum_{n=1}^{\infty} \sin\left(\frac{n\pi\zeta}{L}\right)\sin\left(\frac{n\pi z}{L}\right)\frac{R_1(r_<)R_2(r_>)}{n\pi^2\rho C_n}, \quad (5.2.74)$$

where

$$R_m(r) = K_0(n\pi r/L)I_0(n\pi r_m/L) - K_0(n\pi r_m/L)I_0(n\pi r/L), \quad \textbf{(5.2.75)}$$

$$C_n = [K_0(n\pi\rho/L)I_1(n\pi\rho/L) + I_0(n\pi\rho/L)K_1(n\pi\rho/L)]$$
$$\times [K_0(n\pi r_1/L)I_0(n\pi r_2/L) - I_0(n\pi r_1/L)K_0(n\pi r_2/L)].$$
$$\textbf{(5.2.76)}$$

Equation (5.2.74) is illustrated in Figure 5.2.3.

• *Green's functions via complex variables*

It is well known that the real and imaginary parts of any analytic function satisfy not only the Cauchy-Riemann but also Laplace's equation. Can we use this fact to find the Green's function for Laplace equation? For the biharmonic equation, Dean[16] applied complex variables to find the Green's functions when there is an elliptic hole in the plane.

Consider the complex function

$$\log\left[\frac{1}{f(z;z_0)}\right] = -\ln|f(z;z_0)| + i\arg\left[\frac{1}{f(z;z_0)}\right] \quad \textbf{(5.2.77)}$$

$$= g(x,y|\xi,\eta) - i\theta(x,y|\xi,\eta), \quad \textbf{(5.2.78)}$$

where f is an analytic function within the domain D surrounded by the boundary C with $f(z_0;z_0) = 0$, $z = x + iy$, and $z_0 = \xi + i\eta$. Of course, the point z_0 lies within D. If we define $g(\xi,\eta|\xi,\eta)$ as the limit of $g(x,y|\xi,\eta)$ as $z \to z_0$, we have that $g(\xi,\eta|\xi,\eta) = \infty$. Next, we note that any point in D other than ξ,η, $\nabla^2 g = 0$ because g is the real part of a harmonic function. Furthermore, $g(x,y|\xi,\eta) = g(\xi,\eta|x,y)$. Finally, if $g(x,y|\xi,\eta)$ satisfies certain homogeneous boundary conditions, then g satisfies all of the conditions required of a two-dimensional Green's function for Poisson's equation in the closed domain D.

Because $g(x,y|\xi,\eta)$ is analytic at each point on the boundary C, we have from the Cauchy-Riemann relationships that

$$\frac{\partial g}{\partial x} = -\frac{\partial\theta}{\partial y}, \quad \text{and} \quad \frac{\partial g}{\partial y} = \frac{\partial\theta}{\partial x}. \quad \textbf{(5.2.79)}$$

We next observe that

$$\frac{\partial g}{\partial s} = \frac{\partial g}{\partial y}\cos(n,y) + \frac{\partial g}{\partial x}\cos(n,x) = \frac{\partial g}{\partial n}, \quad \textbf{(5.2.80)}$$

---

[16] Dean, W. R., 1954: Note on the Green's function of an elastic plate. *Proc. Camb. Phil. Soc.*, **50**, 623–627.

where s and n denote any pair of mutually perpendicular directions. For example, $s = -y$ and $n = x$, or $s = x$ and $n = y$. The cosine terms are the directional cosines. From Green's identity, we then have that

$$u(z_0) = \frac{1}{2\pi} \oint_C u(\zeta) \frac{\partial g(\zeta|z_0)}{\partial n} \, ds, \qquad (5.2.81)$$

or

$$u(z) = \frac{1}{2\pi} \oint_C \tilde{u}(\zeta) \frac{\partial g(\zeta|z_0)}{\partial n} \, ds, \qquad (5.2.82)$$

where $\tilde{u}(\zeta)$ denotes any real function defined and continuous on the boundary C of D and we have written $g(x, y|\xi, \eta)$ using complex variables as $g(z|z_0)$. Equations (5.2.81) and (5.2.82) are the complex variable form of *Green's formula*. It expresses the value of a harmonic function u at any point z_0 in the domain D in terms of the values taken by this function on the boundary of D.

• Example 5.2.1: Poisson's integral formula

Consider the function

$$f(z; z_0) = \frac{R(z - z_0)}{R^2 - zz_0^*}, \qquad |z| < R. \qquad (5.2.83)$$

Then

$$g(z|z_0) = -\ln \left| \frac{R(z - z_0)}{R^2 - zz_0^*} \right|. \qquad (5.2.84)$$

Let $z_0 = re^{i\varphi}$ and $\zeta = Re^{i\psi}$, a point on the boundary $|z| = R$. Then $ds = R \, d\psi$ and

$$\frac{\partial g}{\partial n} = \frac{\partial \arg[f(\zeta; z_0)]}{\partial n} = \frac{1}{R} \frac{\partial \arg[f(\zeta; z_0)]}{\partial \psi}. \qquad (5.2.85)$$

Here arg denotes a branch of the argument that is continuous in a neighborhood surrounding the boundary point ζ. If the Green's function satisfies a Dirichlet boundary condition, $g(\zeta|z_0) = 0$ on C and we have that

$$\frac{\partial \arg[f(\zeta; z_0)]}{\partial \psi} = -i \frac{\partial \log[f(\zeta; z_0)]}{\partial \psi}, \qquad (5.2.86)$$

and

$$\frac{\partial g}{\partial n} = -i \frac{\partial \log[f(\zeta; z_0)]}{\partial \psi} \qquad (5.2.87)$$

$$= -\frac{i}{R} \frac{\partial}{\partial \psi} \left\{ \log \left[\frac{Re^{i\psi} - re^{i\varphi}}{R - re^{i(\psi - \varphi)}} \right] \right\} \qquad (5.2.88)$$

$$= \frac{1}{R} \frac{R^2 - r^2}{R^2 - 2Rr\cos(\psi - \varphi) + r^2}. \qquad (5.2.89)$$

Substituting this result in (5.2.82) and writing $u(Re^{i\psi})$ for $u(\zeta)$, we obtain

$$u(re^{i\varphi}) = \frac{1}{2\pi} \int_0^{2\pi} u(Re^{i\psi}) \frac{R^2 - r^2}{R^2 - 2Rr\cos(\psi - \varphi) + r^2} \, d\psi. \qquad (\mathbf{5.2.90})$$

This result is known as *Poisson's integral formula*; it gives the solution to Dirichlet's problem for a disc.

A similar formula can be obtained for the upper half-plane $y > 0$. Using

$$f(z; z_0) = \frac{z - z_0}{z - z_0^*}, \qquad \text{Im}(z) > 0, \qquad (\mathbf{5.2.91})$$

in place of (5.2.83) and proceeding as before, we obtain *Poisson's integral formula for the upper half-plane*,

$$u(x, y) = \frac{1}{\pi} \int_{-\infty}^{\infty} u(\xi) \frac{y}{(x - \xi)^2 + y^2} \, d\xi. \qquad (\mathbf{5.2.92})$$

Here $u(\xi)$ denotes the prescribed function giving the boundary values of the harmonic function on the real axis; it is assumed that $u(\xi)$ is such that it has only a finite number of finite discontinuities along the x axis and the integral in (5.2.92) converges.

5.3 TWO-DIMENSIONAL HELMHOLTZ EQUATION OVER RECTANGULAR AND CIRCULAR DOMAINS

One of the classic problems of mathematical physics is the forced vibrations of a membrane over a rectangular region. Because the edges are usually clamped, the solution equals zero there. After the time dependence is removed, we are left with the two-dimensional Helmholtz equation

$$\frac{\partial^2 u}{\partial x^2} + \frac{\partial^2 u}{\partial y^2} + k_0^2 u = -f(x, y). \qquad (\mathbf{5.3.1})$$

In this section, we obtain the Green's function for (5.3.1).

• *Rectangular Area*

The problem to be solved is

$$\frac{\partial^2 g}{\partial x^2} + \frac{\partial^2 g}{\partial y^2} + k_0^2 g = -\delta(x - \xi)\delta(y - \eta), \qquad (\mathbf{5.3.2})$$

where $0 < x, \xi < a$, and $0 < y, \eta < b$, subject to the boundary conditions that

$$g(0, y|\xi, \eta) = g(a, y|\xi, \eta) = g(x, 0|\xi, \eta) = g(x, b|\xi, \eta) = 0. \qquad (\mathbf{5.3.3})$$

We use the same technique to solve (5.3.2) as we did in §5.2 by assuming that the Green's function has the form

$$g(x,y|\xi,\eta) = \sum_{m=1}^{\infty} G_m(x|\xi) \sin\left(\frac{m\pi y}{b}\right), \qquad (5.3.4)$$

where the coefficients $G_m(x|\xi)$ are undetermined functions of x, ξ, and η. Substituting this series into (5.3.2), multiplying by $2\sin(n\pi y/b)/b$, and integrating over y, we find that

$$\frac{d^2 G_n}{dx^2} - \left(\frac{n^2\pi^2}{b^2} - k_0^2\right) G_n = -\frac{2}{b}\sin\left(\frac{n\pi\eta}{b}\right)\delta(x-\xi). \qquad (5.3.5)$$

The first method for solving (5.3.5) involves writing

$$\delta(x-\xi) = \frac{2}{a}\sum_{m=1}^{\infty} \sin\left(\frac{m\pi\xi}{a}\right)\sin\left(\frac{m\pi x}{a}\right), \qquad (5.3.6)$$

and

$$G_n(x|\xi) = \frac{2}{a}\sum_{m=1}^{\infty} a_{nm}\sin\left(\frac{m\pi x}{a}\right). \qquad (5.3.7)$$

Upon substituting (5.3.6) and (5.3.7) into (5.3.5), we obtain

$$\sum_{m=1}^{\infty}\left(k_0^2 - \frac{m^2\pi^2}{a^2} - \frac{n^2\pi^2}{b^2}\right) a_{nm}\sin\left(\frac{m\pi x}{a}\right)$$

$$= -\frac{4}{ab}\sum_{m=1}^{\infty}\sin\left(\frac{n\pi\eta}{b}\right)\sin\left(\frac{m\pi\xi}{a}\right)\sin\left(\frac{m\pi x}{a}\right). \qquad (5.3.8)$$

Matching similar harmonics,

$$a_{nm} = \frac{4\sin(m\pi\xi/a)\sin(n\pi\eta/b)}{ab(m^2\pi^2/a^2 + n^2\pi^2/b^2 - k_0^2)}, \qquad (5.3.9)$$

and the *bilinear form of the Green's function* is

$$g(x,y|\xi,\eta) = \frac{4}{ab}\sum_{n=1}^{\infty}\sum_{m=1}^{\infty} \frac{\sin(m\pi\xi/a)\sin(n\pi\eta/b)\sin(m\pi x/a)\sin(n\pi y/b)}{m^2\pi^2/a^2 + n^2\pi^2/b^2 - k_0^2}.$$

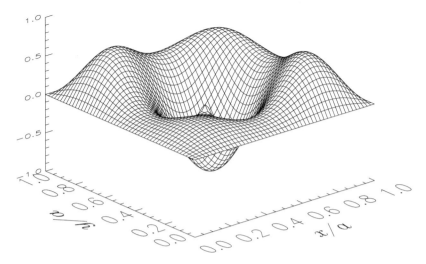

Figure 5.3.1: The Green's function (5.3.10) for Helmholtz's equation over a rectangular region with a Dirichlet boundary condition on the sides when $a = b$, $k_0 a = 10$, and $\xi/a = \eta/a = 0.35$.

$$(5.3.10)$$

The bilinear form of the Green's function for the two-dimensional Helmholtz equation with Neumann boundary conditions and the three-dimensional Helmholtz equation with Dirichlet and Neumann boundary conditions are left as problems 21, 31, and 32, respectively.

As in the previous section, we can construct an alternative to the bilinear form of the Green's function (5.3.10) by writing (5.3.5) as

$$\frac{d^2 G_n}{dx^2} - k_n^2 G_n = -\frac{2}{b} \sin\left(\frac{n\pi\eta}{b}\right) \delta(x - \xi), \qquad (5.3.11)$$

where $k_n^2 = n^2\pi^2/b^2 - k_0^2$. The homogeneous solution to (5.3.11) is now

$$G_n(x|\xi) = \begin{cases} A_n \sinh(k_n x), & 0 \le x \le \xi, \\ B_n \sinh[k_n(a - x)], & \xi \le x \le a. \end{cases} \qquad (5.3.12)$$

This solution satisfies the boundary conditions at both end points.

Because $G_n(x|\xi)$ must be continuous at $x = \xi$,

$$A_n \sinh(k_n \xi) = B_n \sinh[k_n(a - \xi)]. \qquad (5.3.13)$$

On the other hand, the jump discontinuity involving $G'_n(x|\xi)$ yields

$$-k_n B_n \cosh[k_n(a - \xi)] - k_n A_n \cosh(k_n \xi) = -\frac{2}{b} \sin\left(\frac{n\pi\eta}{b}\right). \qquad (5.3.14)$$

Solving for A_n and B_n,

$$A_n = \frac{2}{bk_n} \sin\left(\frac{n\pi\eta}{b}\right) \frac{\sinh[k_n(a-\xi)]}{\sinh(k_n a)}, \qquad (5.3.15)$$

and

$$B_n = \frac{2}{bk_n} \sin\left(\frac{n\pi\eta}{b}\right) \frac{\sinh(k_n \xi)}{\sinh(k_n a)}. \qquad (5.3.16)$$

This yields the Green's function

$$
g(x,y|\xi,\eta) = \frac{2}{b} \sum_{n=1}^{N} \frac{\sin[\kappa_n(a-x_>)]\sin(\kappa_n x_<)}{\kappa_n \sin(\kappa_n a)} \sin\left(\frac{n\pi\eta}{b}\right) \sin\left(\frac{n\pi y}{b}\right)
$$
$$
+ \frac{2}{b} \sum_{n=N+1}^{\infty} \frac{\sinh[k_n(a-x_>)]\sinh(k_n x_<)}{k_n \sinh(k_n a)} \sin\left(\frac{n\pi\eta}{b}\right) \sin\left(\frac{n\pi y}{b}\right).
$$
$$(5.3.17)$$

Here N denotes the largest value of n such that $k_n^2 < 0$, and $\kappa_n^2 = k_0^2 - n^2\pi^2/b^2$. If we began with a Fourier expansion in the y direction, we would have obtained

$$
g(x,y|\xi,\eta) = \frac{2}{a} \sum_{m=1}^{M} \frac{\sin[\kappa_m(b-y_>)]\sin(\kappa_m y_<)}{\kappa_m \sin(\kappa_m b)} \sin\left(\frac{m\pi\xi}{a}\right) \sin\left(\frac{m\pi x}{a}\right)
$$
$$
+ \frac{2}{a} \sum_{m=M+1}^{\infty} \frac{\sinh[k_m(b-y_>)]\sinh(k_m y_<)}{k_m \sinh(k_m b)} \sin\left(\frac{m\pi\xi}{a}\right) \sin\left(\frac{m\pi x}{a}\right),
$$
$$(5.3.18)$$

where M denotes the largest value of m such that $k_m^2 < 0$, $k_m^2 = m^2\pi^2/a^2 - k_0^2$, and $\kappa_m^2 = k_0^2 - m^2\pi^2/a^2$.

• *Circular Disk*

In this subsection, we find the Green's function for the Helmholtz equation when the domain consists of the circular region $0 < r < a$. The Green's function is governed by the partial differential equation

$$\frac{1}{r}\frac{\partial}{\partial r}\left(r\frac{\partial g}{\partial r}\right) + \frac{1}{r^2}\frac{\partial^2 g}{\partial\theta^2} + k_0^2 g = -\frac{\delta(r-\rho)\delta(\theta-\theta')}{r}, \qquad (5.3.19)$$

where $0 < r, \rho < a$, and $0 \le \theta, \theta' \le 2\pi$, with the boundary conditions

$$\lim_{r\to 0} |g(r,\theta|\rho,\theta')| < \infty, \qquad g(a,\theta|\rho,\theta') = 0, \qquad 0 \le \theta, \theta' \le 2\pi.$$
$$(5.3.20)$$

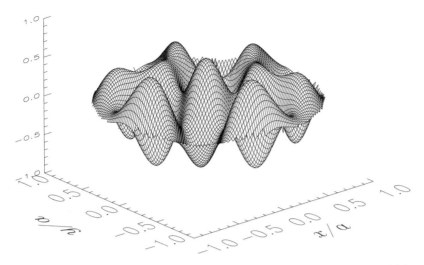

Figure 5.3.2: The Green's function (5.3.27) for Helmholtz's equation within a circular disk with a Dirichlet boundary condition on the rim when $k_0 a = 10$, $\rho/a = 0.35\sqrt{2}$, and $\theta' = \pi/4$.

The Green's function must be periodic in θ.

We begin by noting that

$$\delta(\theta - \theta') = \frac{1}{2\pi} + \frac{1}{\pi} \sum_{n=1}^{\infty} \cos[n(\theta - \theta')] = \frac{1}{2\pi} \sum_{n=-\infty}^{\infty} \cos[n(\theta - \theta')]. \quad (\mathbf{5.3.21})$$

Therefore, the solution has the form

$$g(r, \theta | \rho, \theta') = \sum_{n=-\infty}^{\infty} g_n(r|\rho) \cos[n(\theta - \theta')]. \quad (\mathbf{5.3.22})$$

Substituting (5.3.21) and (5.3.22) into (5.3.19) and simplifying, we find that

$$\frac{1}{r} \frac{d}{dr} \left(r \frac{dg_n}{dr} \right) - \frac{n^2}{r^2} g_n + k_0^2 g_n = -\frac{\delta(r - \rho)}{2\pi r}. \quad (\mathbf{5.3.23})$$

The solution to (5.3.23) is the Fourier-Bessel series

$$g_n(r|\rho) = \sum_{m=1}^{\infty} A_{nm} J_n\left(\frac{k_{nm} r}{a} \right), \quad (\mathbf{5.3.24})$$

where k_{nm} is the mth root of $J_n(k) = 0$. Upon substituting (5.3.24) into (5.3.23) and solving for A_{nm}, we have that

$$(k_0^2 - k_{nm}^2/a^2) A_{nm} = -\frac{1}{\pi a^2 J_n'^2(k_{nm})} \int_0^a \delta(r - \rho) J_n\left(\frac{k_{nm} r}{a} \right) dr, \quad (\mathbf{5.3.25})$$

or

$$A_{nm} = \frac{J_n(k_{nm}\rho/a)}{\pi(k_{nm}^2 - k_0^2 a^2) J'^2_n(k_{nm})}. \qquad (\textbf{5.3.26})$$

Thus, the Green's function[17] is

$$g(r,\theta|\rho,\theta') = \frac{1}{\pi} \sum_{n=-\infty}^{\infty} \sum_{m=1}^{\infty} \frac{J_n(k_{nm}\rho/a) J_n(k_{nm}r/a)}{(k_{nm}^2 - k_0^2 a^2) J'^2_n(k_{nm})} \cos[n(\theta - \theta')].$$

$$(\textbf{5.3.27})$$

An alternative to (5.3.27) can be obtained as follows: As we noted earlier, a particular solution to (5.3.19) is $-Y_0(k_0 R)/4$, where $R^2 = r^2 + \rho^2 - 2r\rho\cos(\theta - \theta')$. Consequently, the most general solution to (5.3.19) consists of this particular solution plus a homogeneous solution:

$$g(r,\theta|\rho,\theta') = -\tfrac{1}{4}Y_0(k_0 R) + \sum_{n=-\infty}^{\infty} A_n J_n(k_0 r) \cos[n(\theta - \theta')]. \quad (\textbf{5.3.28})$$

Along $r = a$, $g(a,\theta|\rho,\theta') = 0$, and

$$\tfrac{1}{4}Y_0\left[k_0\sqrt{a^2 + \rho^2 - 2a\rho\cos(\theta - \theta')}\right] = \sum_{n=-\infty}^{\infty} A_n J_n(k_0 a)\cos[n(\theta - \theta')].$$

$$(\textbf{5.3.29})$$

However, from the addition theorem of Bessel functions,

$$Y_0\left[k_0\sqrt{a^2 + \rho^2 - 2a\rho\cos(\theta - \theta')}\right]$$

$$= \sum_{n=-\infty}^{\infty} J_n(k_0\rho) Y_n(k_0 a)\cos[n(\theta - \theta')]. \qquad (\textbf{5.3.30})$$

Therefore,

$$A_n = \frac{J_n(k_0\rho) Y_n(k_0 a)}{4 J_n(k_0 a)} \qquad (\textbf{5.3.31})$$

---

[17] For an example of its use, see Zhang, D. R., and C. F. Foo, 1999: Fields analysis in a solid magnetic toroidal core with circular cross section based on Green's function. *IEEE Trans. Magnetics*, **35**, 3760–3762.

so that

$$g(r, \theta | \rho, \theta') = -\tfrac{1}{4} Y_0 \left[k_0 \sqrt{r^2 + \rho^2 - 2r\rho \cos(\theta - \theta')} \right]$$

$$+ \tfrac{1}{4} \sum_{n=-\infty}^{\infty} \frac{J_n(k_0\rho) Y_n(k_0 a)}{J_n(k_0 a)} J_n(k_0 r) \cos[n(\theta - \theta')]. \quad (\mathbf{5.3.32})$$

Again, we can use the addition theorem and find that

$$g(r, \theta | \rho, \theta') = -\tfrac{1}{4} \sum_{n=-\infty}^{\infty} \cos[n(\theta - \theta')]$$

$$\times J_n(k_0 r_<) \left[Y_n(k_0 r_>) - \frac{Y_n(k_0 a)}{J_n(k_0 a)} J_n(k_0 r_>) \right]. \quad (\mathbf{5.3.33})$$

Let us now turn to the case of Neumann boundary conditions. In a manner[18] similar to that used to derive (5.3.27), we find that the Green's function for Helmholtz's equation with Neumann boundary conditions is

$$g(r, \theta | \rho, \theta') = -\frac{1}{\pi a^2 k_0^2}$$

$$+ \frac{2}{\pi} \sum_{n=0}^{\infty} \sum_{m=1}^{\infty} \frac{J_n(k_{nm}\rho) J_n(k_{nm} r) \cos[n(\theta - \theta')]}{\epsilon_n (a^2 - n^2/k_{nm}^2)(k_{nm}^2 - k_0^2) J_n^2(k_{nm} a)}, \quad (\mathbf{5.3.34})$$

where k_{nm} in the mth root of $J_n'(k_{nm} a) = 0$, $\epsilon_0 = 2$, and $\epsilon_n = 1$ for $n > 0$.

Equation (5.3.34) has two disadvantages. First, we must compute two power series as well as the Bessel functions. Second, we must find the eigenvalue k_{nm}. Alhargan and Judah[19] showed how both of these difficulties can be eliminated by summing the inner power series exactly using Mittag-Leffler's expansion theory. This yields

$$g(r, \theta | \rho, \theta') = \frac{1}{2} \sum_{n=0}^{\infty} \frac{J_n(k_0 r_<) F_n(k_0 r_>)}{\epsilon_n J_n'(k_0 a)} \cos[n(\theta - \theta')] \quad (\mathbf{5.3.35})$$

$$= -\tfrac{1}{4} Y_0 \left[k_0 \sqrt{r^2 + \rho^2 - 2r\rho \cos(\theta - \theta')} \right]$$

$$+ \tfrac{1}{2} \sum_{n=-\infty}^{\infty} \frac{J_n(k_0\rho) Y_n'(k_0 a)}{\epsilon_n J_n'(k_0 a)} J_n(k_0 r) \cos[n(\theta - \theta')], \quad (\mathbf{5.3.36})$$

---

[18] See Okoshi, T., T. Takeuchi, and J.-P. Hsu, 1975: Planar 3-dB hybrid circuit. *Electron. Commun. Japan*, **58-B(8)**, 80–90.

[19] Alhargan, F. A., and S. R. Judah, 1991: Reduced form of the Green's function for disks and annular rings. *IEEE Trans. Microwave Theory Tech.*, **MMT-39**, 601–604. ©1991 IEEE

where
$$F_n(kr) = Y_n'(ka)J_n(kr) - J_n'(ka)Y_n(kr). \tag{5.3.37}$$

McIver[20] applied (5.3.36) to understand how water waves propagate through an array of cylindrical structures.

• *Circular Sector*

As the previous examples showed, Green's functions for various two-dimensional domains can be expressed in terms of the orthonormal eigenfunction expansion

$$g(\mathbf{r}|\mathbf{r}_0) = \sum_n \frac{u_n(\mathbf{r})u_n(\mathbf{r}_0)}{k_n^2 - k_0^2}, \tag{5.3.38}$$

where the eigenfunction u_n satisfies the differential equation and homogeneous boundary conditions and k_n is the eigenvalue, respectively, of the Helmholtz equation

$$\nabla^2 u_n + k_n^2 u_n = 0. \tag{5.3.39}$$

Similar considerations were developed in §2.4.

In this subsection, we use (5.3.38) to immediately find the Green's function for various circular sectors.[21] We begin by computing the Green's function for a circular sector defined over the region $0 < r < a$ and $0 < \theta < \alpha$. We will find the Green's function with Dirichlet conditions along the boundary.

Consider the product solution:

$$u_{m\gamma}(r,\theta) = J_\gamma(k_{m\gamma}r)\sin(\gamma\theta), \tag{5.3.40}$$

where $\gamma = n\pi/\alpha$, and n is any positive integer. A quick check shows that (5.3.40) satisfies the homogeneous Helmholtz equation and boundary conditions if $J_\gamma(k_{m\gamma}a) = 0$. Here the term $k_{m\gamma}a$ is the mth zero of the Bessel function $J_\gamma(\ )$.

Equation (5.3.40) does not form an orthogonal set unless γ is an integer or $\alpha = \pi/\ell$, where ℓ is a positive integer. In that case,

$$\int_0^a J_\gamma^2(k_{m\gamma}r)\, r\, dr = \tfrac{1}{2}a^2 J_{\gamma+1}^2(k_{m\gamma}a). \tag{5.3.41}$$

[20] McIver, P., 2000: Water-wave propagation through an infinite array of cylindrical structures. *J. Fluid Mech.*, **424**, 101–125.

[21] A summary is given in Lozyanoy, V. I., I. V. Petrusenko, I. G. Prokhoda, and V. P. Prudkiy, 1984: Green's function of simple regions in a polar coordinate system for solving electromagnetic problems. *Radio Engng. Electron. Phys.*, **29(6)**, 8–13.

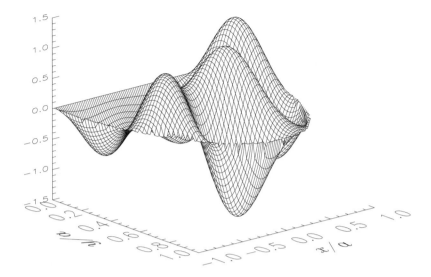

Figure 5.3.3: The Green's function (5.3.42) for Helmholtz's equation within a circular sector with a Dirichlet boundary condition on the rim when $k_0 a = 10$, $\alpha = \pi$, $\rho/a = 0.35\sqrt{2}$, and $\theta' = \pi/4$.

Employing (5.3.41) to normalize the eigenfunctions and then using (5.3.38), we immediately obtain the Green's function

$$g(r, \theta | \rho, \theta') = \frac{4\ell}{\pi a^2} \sum_{n=1}^{\infty} \sum_{m=1}^{\infty} \frac{J_\gamma(k_{m\gamma}\rho) J_\gamma(k_{m\gamma}r) \sin(\gamma\theta) \sin(\gamma\theta')}{\epsilon_n (k_{m\gamma}^2 - k_0^2) J_{\gamma+1}^2(k_{m\gamma}a)}.$$

$$(\mathbf{5.3.42})$$

Similar considerations hold if we wish to find the Green's function for Helmholtz's equation with a Neumann condition. The product solution[22] in this case is

$$u_{m\gamma}(r, \theta) = J_\gamma(k_{m\gamma}r) \cos(\gamma\theta), \qquad (\mathbf{5.3.43})$$

where $n = 0, 1, 2, \ldots$ and $k_{m\gamma}a$ is the mth root of $J_\gamma'(\ )$. In addition to these roots, which we will number $m = 1, 2, \ldots$, we have the additional solution $k_{00} = 0$ when $n = 0$. The orthogonality condition here is

$$\int_0^a J_\nu^2(k_{m\nu}r) \, r \, dr = \begin{cases} \frac{1}{2}a^2, & m = \nu = 0, \\[2mm] \frac{1}{2}\left[a^2 - \nu^2/k_{m\nu}^2\right] J_\nu^2(k_{m\nu}a), & \text{otherwise,} \end{cases}$$

$$(\mathbf{5.3.44})$$

---

[22] Several useful product solutions that satisfy Neumann boundary conditions are summarized in Lo, Y. T., D. Solomon, and W. F. Richards, 1979: Theory and experiment on microstrip antennas. *IEEE Trans. Antennas Propagat.*, **AP-27**, 137–145.

where $\nu = n\ell$. Note that ν is an nonnegative integer. Performing the same analysis[23] as above, we now have that

$$g(r, \theta | \rho, \theta') = -\frac{2\ell}{\pi a^2 k_0^2}$$

$$+ \frac{4\ell}{\pi} \sum_{n=0}^{\infty} \sum_{m=1}^{\infty} \frac{J_\nu(k_{m\nu}\rho) J_\nu(k_{m\nu}r) \cos(\nu\theta) \cos(\nu\theta')}{\epsilon_\nu (a^2 - n^2\ell^2/k_{m\nu}^2)(k_{m\nu}^2 - k_0^2) J_\nu^2(k_{m\nu}a)}.$$

$$(5.3.45)$$

• *Annular rings*

In the case of an annular ring defined by $a < r, \rho < b$, the mutually orthogonal eigenfunctions that satisfy Dirichlet conditions along $r = a$ and $r = b$ are

$$u_{mn}(r, \theta) = [Y_n(k_{mn}a) J_n(k_{mn}r) - J_n(k_{mn}a) Y_n(k_{mn}r)] \begin{cases} \cos(n\theta) \\ \sin(n\theta), \end{cases}$$

$$(5.3.46)$$

where $m = 1, 2, 3, \ldots$ and $n = 0, 1, 2, \ldots$ The eigenvalues k_{mn} are computed from

$$\frac{J_n(k_{mn}a)}{Y_n(k_{mn}a)} = \frac{J_n(k_{mn}b)}{Y_n(k_{mn}b)}. \qquad (5.3.47)$$

Because

$$\int_a^b [Y_n(k_{mn}a) J_n(k_{mn}r) - J_n(k_{mn}a) Y_n(k_{mn}r)]^2 \, r \, dr$$

$$= \frac{1}{2} \left\{ b^2 [Y_n(k_{mn}a) J_n'(k_{mn}b) - J_n(k_{mn}a) Y_n'(k_{mn}b)]^2 \right.$$

$$\left. - a^2 [Y_n(k_{mn}a) J_n'(k_{mn}a) - J_n(k_{mn}a) Y_n'(k_{mn}b)]^2 \right\}$$

$$(5.3.48)$$

and using the normalizing factors for $\cos(n\theta)$ and $\sin(n\theta)$, we obtain the Green's function

$$g(r, \theta | \rho, \theta') = \frac{2}{\pi} \sum_{n=0}^{\infty} \sum_{m=1}^{\infty} \frac{F_{mn}(\rho) F_{mn}(r) \cos[n(\theta - \theta')]}{\epsilon_n [b^2 F_{mn}'^2(b) - a^2 F_{mn}'^2(a)] (k_{nm}^2 - k_0^2)},$$

$$(5.3.49)$$

---

[23] See Okoshi, T., and T. Miyoshi, 1972: The planar circuit—An approach to microwave integrated circuitry. *IEEE Trans. Microwave Theory Tech.*, **MTT-20**, 245–252. ©1972 IEEE; Chadha, R., and K. C. Gupta, 1981: Green's functions for circular sectors, annular rings, and annular sectors in planar microwave circuits. *IEEE Trans. Microwave Theory Tech.*, **MTT-29**, 68–71. ©1981 IEEE

where

$$F_{mn}(r) = Y_n(k_{mn}a)J_n(k_{mn}r) - J_n(k_{mn}a)Y_n(k_{mn}r), \qquad (5.3.50)$$

$\epsilon_0 = 2$, and $\epsilon_n = 1$ for $n > 0$.

When there are Neumann boundary conditions, the eigenfunctions are now

$$u_{mn}(r,\theta) = [Y_n'(k_{mn}a)J_n(k_{mn}r) - J_n'(k_{mn}a)Y_n(k_{mn}r)] \begin{cases} \cos(n\theta), \\ \sin(n\theta). \end{cases}$$
$$(5.3.51)$$

The eigenvalues k_{mn} are given by the solution of

$$\frac{J_n'(k_{mn}a)}{Y_n'(k_{mn}a)} = \frac{J_n'(k_{mn}b)}{Y_n'(k_{mn}b)}, \qquad (5.3.52)$$

and the orthogonality condition becomes

$$\int_a^b [Y_n'(k_{mn}a)J_n(k_{mn}r) - J_n'(k_{mn}a)Y_n(k_{mn}r)]^2 \, r \, dr$$

$$= \tfrac{1}{2} \Big\{ \left(b^2 - n^2/k_{mn}^2\right) [Y_n'(k_{mn}a)J_n(k_{mn}b) - J_n'(k_{mn}a)Y_n(k_{mn}b)]^2$$

$$- \left(a^2 - n^2/k_{mn}^2\right) [Y_n'(k_{mn}a)J_n(k_{mn}a) - J_n'(k_{mn}a)Y_n(k_{mn}a)]^2 \Big\}.$$
$$(5.3.53)$$

We have the addition eigenvalue $k_{00} = 0$ when $n = 0$.

Proceeding as before, we obtain the Green's function[24]

$$g(r,\theta|\rho,\theta') = -\frac{1}{\pi(b^2 - a^2)k_0^2}$$

$$+ \frac{2}{\pi} \sum_{n=0}^{\infty} \sum_{m=1}^{\infty} \frac{F_{mn}(\rho)F_{mn}(r)}{\epsilon_n \left[(b^2 - \frac{n^2}{k_{nm}^2})F_{mn}^2(b) - (a^2 - \frac{n^2}{k_{nm}^2})F_{mn}^2(a)\right]}$$

$$\times \frac{\cos[n(\theta - \theta')]}{k_{mn}^2 - k_0^2}, \qquad (5.3.54)$$

where

$$F_{mn}(r) = Y_n'(k_{mn}a)J_n(k_{mn}r) - J_n'(k_{mn}a)Y_n(k_{mn}r). \qquad (5.3.55)$$

---

[24] As given in Chadha, R., and K. C. Gupta, 1981: Green's functions for circular sectors, annular rings, and annular sectors in planar microwave circuits. *IEEE Trans. Microwave Theory Tech.*, **MMT-29**, 68–71. ©1981 IEEE

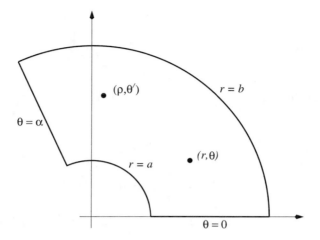

Figure 5.3.4: The geometry of the annular sector.

One of the disadvantages of (5.3.54) is the evaluation of a double series as well as the Bessel functions. We can reduce this double Fourier series to a single one using Mittag-Leffler's expansion theorem.[19] This yields

$$g(r, \theta | \rho, \theta') = \frac{1}{2} \sum_{n=0}^{\infty} \frac{F_n(r_<, a) F_n(r_>, b)}{\epsilon_n F_n'(b, a)} \cos[n(\theta - \theta')], \qquad (5.3.56)$$

where

$$F_n(r, a) = Y_n'(k_0 a) J_n(k_0 r) - J_n'(k_0 a) Y_n(k_0 r), \qquad (5.3.57)$$

and

$$F_n'(b, a) = Y_n'(k_0 a) J_n'(k_0 b) - J_n'(k_0 a) Y_n'(k_0 b). \qquad (5.3.58)$$

• *Annular sectors*

The Green's functions for annular sectors (see Figure 5.3.4), where $a < r, \rho < b$, and $0 < \theta, \theta' < \alpha$, are found following the same technique as we used for annular rings. Here, the mutually orthogonal eigenfunctions that satisfy Dirichlet conditions along its edges are

$$u_{m\nu}(r, \theta) = [Y_\nu(k_{m\nu} a) J_\nu(k_{m\nu} r) - J_\nu(k_{m\nu} a) Y_\nu(k_{m\nu} r)] \sin(\nu\theta), \quad (5.3.59)$$

where $\nu = n\pi/\alpha$, $m = 1, 2, 3, \ldots$, and $n = 1, 2, 3, \ldots$ The eigenvalues $k_{m\nu}$ are computed from

$$\frac{J_\nu(k_{m\nu} a)}{Y_\nu(k_{m\nu} a)} = \frac{J_\nu(k_{m\nu} b)}{Y_\nu(k_{m\nu} b)}. \qquad (5.3.60)$$

Proceeding as we did for the annular rings, we obtain the Green's function

$$g(r,\theta|\rho,\theta') = \frac{2}{\pi} \sum_{n=1}^{\infty} \sum_{m=1}^{\infty} \frac{F_{m\nu}(\rho)F_{m\nu}(r)\sin(\nu\theta)\sin(\nu\theta')}{[b^2 F_{m\nu}'^2(b) - a^2 F_{m\nu}'^2(a)](k_{m\nu}^2 - k_0^2)}, \quad (5.3.61)$$

where

$$F_{m\nu}(r) = Y_\nu(k_{m\nu}a)J_\nu(k_{m\nu}r) - J_\nu(k_{m\nu}a)Y_\nu(k_{m\nu}r). \quad (5.3.62)$$

In the case of Neumann boundary conditions, the eigenfunctions are

$$u_{m\nu}(r,\theta) = [Y_\nu'(k_{m\nu}a)J_\nu(k_{m\nu}r) - J_\nu'(k_{m\nu}a)Y_\nu(k_{m\nu}r)]\cos(\nu\theta). \quad (5.3.63)$$

The eigenvalues $k_{m\nu}$ are found by solving

$$\frac{J_\nu'(k_{m\nu}a)}{Y_\nu'(k_{m\nu}a)} = \frac{J_\nu'(k_{m\nu}b)}{Y_\nu'(k_{m\nu}b)}. \quad (5.3.64)$$

We have the additional eigenvalue $k_{00} = 0$ when $n = 0$.

Proceeding as before, we obtain the Green's function[24]

$$\begin{aligned}
g(r,\theta|\rho,\theta') = &-\frac{2\ell}{\pi(b^2 - a^2)k_0^2} \\
&+ \frac{4\ell}{\pi} \sum_{n=0}^{\infty} \sum_{m=1}^{\infty} \frac{F_{m\nu}(\rho)F_{m\nu}(r)}{\epsilon_\nu \left[(b^2 - \frac{n^2\ell^2}{k_{m\nu}^2})F_{m\nu}^2(b) - (a^2 - \frac{n^2\ell^2}{k_{m\nu}^2})F_{m\nu}^2(a) \right]} \\
&\qquad\qquad \times \frac{\cos(\nu\theta)\cos(\nu\theta')}{k_{m\nu}^2 - k_0^2},
\end{aligned} \quad (5.3.65)$$

where

$$F_{m\nu}(r) = Y_\nu'(k_{m\nu}a)J_\nu(k_{m\nu}r) - J_\nu'(k_{m\nu}a)Y_\nu(k_{m\nu}r). \quad (5.3.66)$$

In the derivation of (5.3.61) and (5.3.65), Chadha and Gupta required that π/α must be a positive integer. However, recent work by Alhargan and Judah[25] showed that this condition is unnecessary.

An alternative expression to (5.3.61) is found by first noting that

$$\delta(\theta - \theta') = \frac{2}{\alpha} \sum_{n=1}^{\infty} \sin\left(\frac{n\pi\theta'}{\alpha}\right)\sin\left(\frac{n\pi\theta}{\alpha}\right), \quad (5.3.67)$$

---

[25] Alhargan, F. A., and S. R. Judah, 1994: Circular and annular sector planar components of arbitrary angle for N-way power dividers/combiners. *IEEE Trans. Microwave Theory Tech.*, **MTT-42**, 1617–1623.

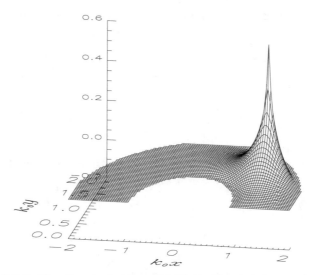

Figure 5.3.5: The Green's function (5.3.74) for Helmholtz's equation within an annular sector $0 < \theta < \pi$ with a Dirichlet boundary conditions on all of the sides when $k_0 a = 1$, $k_0 b = 2$, $\rho = \sqrt{2}$, and $\theta' = \pi/4$.

where $0 < \alpha < 2\pi$. From the form of (5.3.67) we anticipate that

$$g(r,\theta|\rho,\theta') = \frac{2}{\alpha} \sum_{n=1}^{\infty} G_n(r|\rho) \sin\left(\frac{n\pi\theta'}{\alpha}\right) \sin\left(\frac{n\pi\theta}{\alpha}\right), \qquad (5.3.68)$$

where $G_n(r|\rho)$ satisfies the equation

$$\frac{1}{r}\frac{d}{dr}\left(r\frac{dG_n}{dr}\right) - \frac{\nu^2 G_n}{r^2} + k_0^2 G_n = -\frac{\delta(r-\rho)}{r}, \qquad (5.3.69)$$

and $\nu = n\pi/\alpha$. Note that (5.3.68) satisfies the Dirichlet boundary conditions if $G_n(a|\rho) = G_n(b|\rho) = 0$. Solutions to (5.3.69) that satisfy these boundary conditions and the condition that $G_n(\rho^-|\rho) = G_n(\rho^+|\rho)$ are

$$G_n(r|\rho) = A_n F_\nu(k_0 a, k_0 r_<) F_\nu(k_0 b, k_0 r_>), \qquad (5.3.70)$$

where

$$F_\nu(x,y) = J_\nu(x) Y_\nu(y) - Y_\nu(x) J_\nu(y). \qquad (5.3.71)$$

The coefficient A_n is determined by the condition that

$$\left.\frac{dG_n}{dr}\right|_{r=\rho^-}^{r=\rho^+} = -\frac{1}{\rho}. \qquad (5.3.72)$$

This leads to

$$G_n(r|\rho) = \frac{\pi F_\nu(k_0 a, k_0 r_<) F_\nu(k_0 b, k_0 r_>)}{2 F_\nu(k_0 b, k_0 a)}, \tag{5.3.73}$$

and

$$g(r, \theta|\rho, \theta') = \frac{\pi}{\alpha} \sum_{n=1}^{\infty} \frac{F_\nu(k_0 a, k_0 r_<) F_\nu(k_0 b, k_0 r_>)}{F_\nu(k_0 b, k_0 a)} \sin\left(\frac{n\pi\theta'}{\alpha}\right) \sin\left(\frac{n\pi\theta}{\alpha}\right). \tag{5.3.74}$$

We can also find an alternative to (5.3.65), namely

$$g(r, \theta|\rho, \theta') = \frac{\pi}{\alpha} \sum_{n=0}^{\infty} \frac{F_\nu(k_0 a, k_0 r_<) F_\nu(k_0 b, k_0 r_>)}{\epsilon_n F_\nu'(k_0 b, k_0 a)} \cos\left(\frac{n\pi\theta'}{\alpha}\right) \cos\left(\frac{n\pi\theta}{\alpha}\right), \tag{5.3.75}$$

where

$$F_\nu(x, y) = Y_\nu'(x) J_\nu(y) - J_\nu'(x) Y_\nu(y), \tag{5.3.76}$$

$$F_\nu'(x, y) = Y_\nu'(x) J_\nu'(y) - J_\nu'(x) Y_\nu'(y), \tag{5.3.77}$$

$\epsilon_0 = 2$, and $\epsilon_n = 1$ for $n > 0$.

- *Triangular section*

Deriving Green's functions for non-rectangular and non-cylindrical domains is difficult. Chadha and Gupta[26] developed an ingenious technique for triangular shaped regions with Neumann boundary conditions. We illustrate their technique for an isosceles right-angled triangle. The interested reader is referred to their paper for 30°-60° right-angle and equilateral triangles.

The problem hinges on devising an eigenfunction expansion to represent $\delta(x - \xi)\delta(y - \eta)$ for the region shown in Figure 5.3.6. Along each of the three boundaries drawn as solid lines, we want the derivative of this expansion to vanish. Why? The reason lies in the commonly employed technique of using the same eigenfunctions to represent the Green's function and the delta function.

How did they find this expansion? Chadha and Gupta introduced an infinite number of virtual sources whose basic pattern is shown in Figure 5.3.6 and is repeated an infinite number of times in both the x and y directions. Then they used the method of images. Chadha and Gupta found that

$$\delta(x - \xi)\delta(y - \eta) = \frac{1}{2a^2} \sum_{m=-\infty}^{\infty} \sum_{n=-\infty}^{\infty} U(\xi, \eta) U(x, y), \tag{5.3.78}$$

---

[26] Chadha, R., and K. C. Gupta, 1980: Green's functions for triangular segments in planar microwave circuits. *IEEE Trans. Microwave Theory Tech.*, **MTT-28**, 1139–1143. ©1980 IEEE

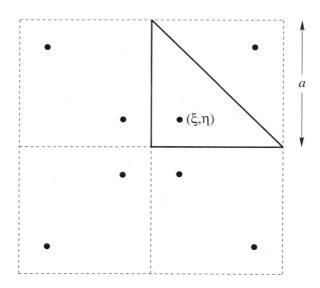

Figure 5.3.6: The location of the various image sources used to find the Green's function within an isosceles right-angled triangular region with Neumann boundary conditions.

where

$$U(x,y) = \cos\left(\frac{m\pi x}{a}\right)\cos\left(\frac{n\pi y}{a}\right) + (-1)^{m+n}\cos\left(\frac{n\pi x}{a}\right)\cos\left(\frac{m\pi y}{a}\right).$$
$$(5.3.79)$$

The first term in (5.3.79) represents the inner four sources, while the second term represents the outer four sources.

Having found the eigenfunction expansion for the delta functions, the corresponding Green's function is

$$g(x,y|\xi,\eta) = \frac{1}{2a^2}\sum_{m=-\infty}^{\infty}\sum_{n=-\infty}^{\infty}A_{mn}U(x,y).\qquad(5.3.80)$$

Recall that these eigenfunctions were chosen to satisfy the Neumann boundary conditions along the boundaries. Therefore, we only have to find A_{mn} so that (5.3.79) satisfies the differential equation. Substituting (5.3.78) and (5.3.80) into (5.3.1), we have that

$$\sum_{m=-\infty}^{\infty}\sum_{n=-\infty}^{\infty}\left\{A_{mn}\left[k_0^2 - \left(\frac{m\pi}{a}\right)^2 - \left(\frac{n\pi}{a}\right)^2\right] + U(\xi,\eta)\right\}U(x,y) = 0.$$
$$(5.3.81)$$

Because the bracketed term must vanish for any x and y, we finally obtain

$$g(x,y|\xi,\eta) = \frac{1}{2}\sum_{m=-\infty}^{\infty}\sum_{n=-\infty}^{\infty}\frac{U(\xi,\eta)U(x,y)}{(m^2+n^2)\pi^2 - k_0^2 a^2}.\qquad(5.3.82)$$

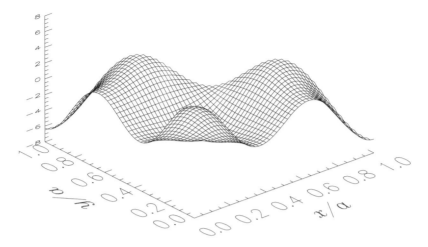

Figure 5.3.7: The Green's function (5.3.82) for Helmholtz's equation over an isosceles, right-angled triangular region with a Neumann boundary condition on each side when $k_0 a = 10$, and $x/a = y/a = 0.25$.

In a similar manner, Terras and Swanson[27] have found all of the domains bounded by linear planes for which the method of images produces Green's functions for Laplace's equation with Dirichlet boundary conditions.

5.4 POISSON'S AND HELMHOLTZ'S EQUATIONS ON A RECTANGULAR STRIP

In this section, we continue our study of the two-dimensional Poisson and Helmholtz equations. Our particular focus will be on domains that are of infinite extent in the y direction but are of finite length in the x direction.

• *Poisson's equation*

We begin by finding the Green's function for Poisson's equation on the semi-infinite strip $0 < x, \xi < L$ and $0 < y, \eta < \infty$ where we solve

$$\frac{\partial^2 g}{\partial x^2} + \frac{\partial^2 g}{\partial y^2} = -\delta(x - \xi)\delta(y - \eta), \qquad (5.4.1)$$

[27] Terras, R., and R. A. Swanson, 1980: Electrostatic image problems with plane boundaries. *Am. J. Phys.*, **48**, 526–531; Terras, R., and R. Swanson, 1980: Image methods for constructing Green's functions and eigenfunctions for domains with plane boundaries. *J. Math. Phys.*, **21**, 2140–2153.

subject to the boundary conditions

$$g(0, y|\xi, \eta) = g(L, y|\xi, \eta) = g(x, 0|\xi, \eta) = 0, \qquad (5.4.2)$$

and

$$\lim_{y \to \infty} |g(x, y|\xi, \eta)| < \infty. \qquad (5.4.3)$$

Cases when $-\infty < y, \eta < \infty$ are left as problems 3, 4, and 5.

Because the Green's function must vanish when $x = 0$ and $x = L$, we express it and $\delta(x - \xi)$ in terms of the Fourier sine expansions

$$\delta(x - \xi) = \frac{2}{L} \sum_{n=1}^{\infty} \sin\left(\frac{n\pi\xi}{L}\right) \sin\left(\frac{n\pi x}{L}\right), \qquad (5.4.4)$$

and

$$g(x, y|\xi, \eta) = \sum_{n=1}^{\infty} G_n(y|\eta) \sin\left(\frac{n\pi\xi}{L}\right) \sin\left(\frac{n\pi x}{L}\right), \qquad (5.4.5)$$

respectively. Substituting (5.4.4) and (5.4.5) into (5.4.1), we find that

$$\frac{d^2 G_n}{dy^2} - \frac{n^2\pi^2}{L^2} G_n = -\delta(y - \eta), \quad 0 < y, \eta < \infty, \qquad (5.4.6)$$

with

$$G_n(0|\eta) = 0, \quad \text{and} \quad \lim_{y \to \infty} |G_n(y|\eta)| < \infty. \qquad (5.4.7)$$

The solution of this boundary-value problem is

$$G_n(y|\eta) = \frac{e^{-n\pi|y-\eta|/L} - e^{-n\pi(y+\eta)/L}}{n\pi}. \qquad (5.4.8)$$

The first term in (5.4.8) is the particular solution to (5.4.6), while the second term is a homogeneous solution, required so that $G_n(0|\eta) = 0$. Therefore, the Green's function governed for (5.4.1)–(5.4.3) is

$$g(x, y|\xi, \eta) = \frac{1}{\pi} \sum_{n=1}^{\infty} \frac{e^{-n\pi|y-\eta|/L} - e^{-n\pi(y+\eta)/L}}{n} \sin\left(\frac{n\pi\xi}{L}\right) \sin\left(\frac{n\pi x}{L}\right). \qquad (5.4.9)$$

Following Melnikov,[28] we can find a closed form for the Green's function. We begin by first rewriting (5.4.9) as

$$g(x, y|\xi, \eta) = \frac{1}{2\pi} \sum_{n=1}^{\infty} \frac{1}{n} \left[e^{-n\pi|y-\eta|/L} - e^{-n\pi(y+\eta)/L} \right]$$

$$\times \left\{ \cos\left[\frac{n\pi(x-\xi)}{L}\right] - \cos\left[\frac{n\pi(x+\xi)}{L}\right] \right\}. \quad (5.4.10)$$

---

[28] Melnikov, Yu. A., 1995: *Green's Functions in Applied Mechanics.* WIT Press/Computational Mechanics Publications, pp. 24–25.

Since

$$\sum_{n=1}^{\infty} \frac{q^n}{n} \cos(n\alpha) = -\ln\left[\sqrt{1 - 2q\cos(\alpha) + q^2}\right], \qquad (5.4.11)$$

provided that $|q| < 1$, and $0 \le \alpha \le 2\pi$, we have that

$$g(x, y|\xi, \eta) = \frac{1}{2\pi} \ln\Bigg\{ \sqrt{\frac{1 - 2e^{-\pi(y+\eta)/L}\cos[\pi(x-\xi)/L] + e^{-2\pi(y+\eta)/L}}{1 - 2e^{-\pi|y-\eta|/L}\cos[\pi(x-\xi)/L] + e^{-2\pi|y-\eta|/L}}}$$

$$\times \sqrt{\frac{1 - 2e^{-\pi|y-\eta|/L}\cos[\pi(x+\xi)/L] + e^{-2\pi|y-\eta|/L}}{1 - 2e^{-\pi(y+\eta)/L}\cos[\pi(x+\xi)/L] + e^{-2\pi(y+\eta)/L}}} \Bigg\}.$$

$$(5.4.12)$$

Multiplying the numerator of the first radial and the denominator of the second radial by $e^{2\pi(y+\eta)/L}$, we arrive at the closed form solution

$$g(x, y|\xi, \eta) = \frac{1}{2\pi} \ln\left[\frac{E(z + \zeta^*)E(z - \zeta^*)}{E(z - \zeta)E(z + \zeta)}\right], \qquad (5.4.13)$$

where $z = y + ix$, $\zeta = \eta + i\xi$, $\zeta^* = \eta - i\xi$, and $E(z) = |e^{\pi z/L} - 1|$.

A similar technique can be used to find the Green's function for (5.4.1) with the boundary conditions

$$g(0, y|\xi, \eta) = g(L, y|\xi, \eta) = g_y(x, 0|\xi, \eta) - \beta g(x, 0|\xi, \eta) = 0, \quad (5.4.14)$$

and (5.4.3), where $\beta \ge 0$. Performing a similar analysis, we find that

$$g(x, y|\xi, \eta) = \frac{1}{\pi} \sum_{n=1}^{\infty} \frac{1}{n} \left[e^{-n\pi|y-\eta|/L} - \frac{\beta L - n\pi}{\beta L + n\pi} e^{-n\pi(y+\eta)/L} \right]$$

$$\times \sin\left(\frac{n\pi\xi}{L}\right) \sin\left(\frac{n\pi x}{L}\right) \qquad (5.4.15)$$

$$= \frac{1}{\pi} \sum_{n=1}^{\infty} \frac{1}{n} \left[e^{-n\pi|y-\eta|/L} + e^{-n\pi(y+\eta)/L} \right]$$

$$\times \sin\left(\frac{n\pi\xi}{L}\right) \sin\left(\frac{n\pi x}{L}\right)$$

$$- \frac{2\beta L}{\pi} \sum_{n=1}^{\infty} \frac{e^{-n\pi(y+\eta)/L}}{n(\beta L + n\pi)} \sin\left(\frac{n\pi\xi}{L}\right) \sin\left(\frac{n\pi x}{L}\right). \quad (5.4.16)$$

$$= \frac{1}{2\pi} \ln\left[\frac{E(z + \zeta)E(z - \zeta^*)}{E(z - \zeta)E(z + \zeta^*)}\right]$$

$$- \frac{2\beta L}{\pi} \sum_{n=1}^{\infty} \frac{e^{-n\pi(y+\eta)/L}}{n(\beta L + n\pi)} \sin\left(\frac{n\pi\xi}{L}\right) \sin\left(\frac{n\pi x}{L}\right). \quad (5.4.17)$$

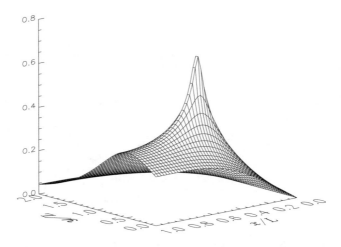

Figure 5.4.1: The Green's function (5.4.20) for the planar Poisson equation over the semi-infinite strip $0 < x < L$ and $0 < y$ subject to the boundary conditions that $g(0, y|\xi, \eta) = g_x(0, y|\xi, \eta) = 0$, and $g_y(x, 0|\xi, \eta) - \beta g(x, 0|\xi, \eta) = 0$ when $\xi/L = \eta/L = 0.5$, and $\beta L = 1$.

The advantage of (5.4.17) over (5.4.15) is its uniform convergence. We obtain the logarithmic term in the same manner as we derived it in (5.4.13).

Finally we solve (5.4.1) subject to the boundary conditions

$$g(0, y|\xi, \eta) = g_x(L, y|\xi, \eta) = 0, \qquad (5.4.18)$$

and

$$g_y(x, 0|\xi, \eta) - \beta g(x, 0|\xi, \eta) = 0, \qquad (5.4.19)$$

with (5.4.3). The Green's function[29] is

$$g(x, y|\xi, \eta) = \frac{1}{2\pi} \ln\left[\frac{E_1(z + \zeta^*)E_1(z - \zeta)E_2(z + \zeta)E_2(z - \zeta^*)}{E_1(z + \zeta)E_1(z - \zeta^*)E_2(z - \zeta)E_2(z + \zeta^*)}\right]$$
$$- \frac{2\beta}{L} \sum_{n=1}^{\infty} \frac{e^{-\nu(y+\eta)}}{\nu(\beta + \nu)} \sin(\nu\xi)\sin(\nu x), \qquad (5.4.20)$$

where $z = y + ix$, $\zeta = \eta + i\xi$, $\zeta^* = \eta - i\xi$, $\nu = (2n - 1)\pi/(2L)$, $E_1(z) = \left|e^{\pi z/(2L)} + 1\right|$, and $E_2(z) = \left|e^{\pi z/(2L)} - 1\right|$.

• *Helmholtz's equation*

Consider the Helmholtz equation

$$\frac{\partial^2 g}{\partial x^2} + \frac{\partial^2 g}{\partial y^2} + k_0^2 g = -\delta(x - \xi)\delta(y - \eta), \qquad (5.4.21)$$

---

[29] *Ibid.*, p. 28.

where $0 < x, \xi < L$, and $-\infty < y, \eta < \infty$, subject to the boundary conditions that

$$g(0, y|\xi, \eta) = g(L, y|\xi, \eta) = 0, \tag{5.4.22}$$

and

$$\lim_{|y| \to \infty} |g(x, y|\xi, \eta)| < \infty. \tag{5.4.23}$$

We begin by introducing the Fourier transform

$$G(x, \ell|\xi, \eta) = \int_{-\infty}^{\infty} g(x, y|\xi, \eta) e^{-i\ell y} \, dy, \tag{5.4.24}$$

and

$$g(x, y|\xi, \eta) = \frac{1}{2\pi} \int_{-\infty}^{\infty} G(x, \ell|\xi, \eta) e^{i\ell y} \, d\ell, \tag{5.4.25}$$

where $\text{Re}(\ell) > 0$. We chose this form of the Fourier transform so that we have outwardly propagating waves if the time dependence is $e^{-i\omega t}$. Equation (5.4.21) then becomes the ordinary differential equation

$$\frac{d^2 G}{dx^2} + \left(k_0^2 - \ell^2\right) G = -\delta(x - \xi) e^{-i\ell \eta}, \quad 0 < x, \xi < L, \tag{5.4.26}$$

with

$$G(0, \ell|\xi, \eta) = G(L, \ell|\xi, \eta) = 0. \tag{5.4.27}$$

To solve (5.4.26), we assume that

$$G(x, \ell|\xi, \eta) = \sum_{n=1}^{\infty} G_n \sin\left(\frac{n\pi x}{L}\right), \tag{5.4.28}$$

with

$$\delta(x - \xi) = \frac{2}{L} \sum_{n=1}^{\infty} \sin\left(\frac{n\pi \xi}{L}\right) \sin\left(\frac{n\pi x}{L}\right). \tag{5.4.29}$$

Substituting (5.4.28) and (5.4.29) into (5.4.26), and matching similar harmonics,

$$\left(k_0^2 - \ell^2 - \frac{n^2\pi^2}{L^2}\right) G_n = -\frac{2}{L} \sin\left(\frac{n\pi \xi}{L}\right) e^{-i\ell \eta}, \tag{5.4.30}$$

and

$$g(x, y|\xi, \eta) = \frac{1}{\pi L} \sum_{n=1}^{\infty} \sin\left(\frac{n\pi \xi}{L}\right) \sin\left(\frac{n\pi x}{L}\right) \int_{-\infty}^{\infty} \frac{e^{i\ell(y-\eta)}}{\ell^2 + \kappa^2} \, d\ell, \tag{5.4.31}$$

where $\kappa = \sqrt{n^2\pi^2/L^2 - k_0^2}$.

We now integrate (5.4.31) via contour integration. By Jordan's lemma, we must close the line integration from $(-\infty, 0)$ to $(\infty, 0)$ with a semicircle of infinite radius in the *upper* half of the ℓ-plane if $y > \eta$; if $y < \eta$, then the semicircle will be in the lower half-plane. Performing the integration via the residue theorem, we obtain

$$g(x, y | \xi, \eta) = \frac{1}{L} \sum_{n=1}^{\infty} \sin\left(\frac{n\pi\xi}{L}\right) \sin\left(\frac{n\pi x}{L}\right) \frac{e^{-\sqrt{n^2\pi^2/L^2 - k_0^2}\, |y-\eta|}}{\sqrt{n^2\pi^2/L^2 - k_0^2}}.$$

(5.4.32)

What happens if $n\pi/L < k_0$ for some n? Then $\sqrt{n^2\pi^2/L^2 - k_0^2} = i\sqrt{k_0^2 - n^2\pi^2/L^2}$. We must have this particular root so that the waves decay *away* from the source at $y = \eta$.

In summary, the Green's function for the two-dimensional Helmholtz equation where the domain is the infinite strip $0 < x < L$ consists of a superposition of two types of waves. For $n\pi/L < k_0$, we have propagating waves radiating away from the source. For sufficiently large n, then the waves become *evanescent* modes that are trapped near the source.

Let us retrace our steps and find another form of the Green's function by the method of images. We begin at the point where we have taken the Fourier transform (5.4.26). In §5.1 we showed that the free-space Green's function that satisfies (5.4.26) is

$$G(x, \ell | \xi, \eta) = \frac{i}{2k} e^{ik|x-\xi| - i\ell y},$$ (5.4.33)

where $k^2 = k_0^2 - \ell^2$. Although (5.4.33) satisfies (5.4.26), it does not satisfy any of the boundary conditions. On the other hand, if we modify (5.4.33) to read

$$G(x, \ell | \xi, \eta) = \frac{i}{2k} e^{ik|x-\xi| - i\ell y} - \frac{i}{2k} e^{ik|x+\xi| - i\ell y},$$ (5.4.34)

a quick check shows that (5.4.34) does satisfy the boundary condition at $x = 0$. Although the second term appears to be a wave that radiates from the source point $(-\xi, \eta)$, it is really the reflection of a wave emanating from (ξ, η) off the surface at $x = 0$.

The reason that (5.4.34) is not the Green's function that we are seeking is its failure to satisfy the boundary condition at $x = L$. Let us modify it to read

$$G(x, \ell | \xi, \eta) = \frac{i}{2k} e^{ik|x-\xi| - i\ell y} - \frac{i}{2k} e^{ik|x+\xi| - i\ell y}$$
$$+ \frac{i}{2k} e^{ik|x-\xi-2L| - i\ell y} - \frac{i}{2k} e^{ik|x+\xi-2L| - i\ell y}.$$ (5.4.35)

Although (5.4.35) now satisfies the boundary conditions at $x = L$, it also violates the one at $x = 0$. Consequently, we must add in more free-space Green's functions so that we again satisfy the boundary condition at $x = 0$. Clearly, this repeated introduction of free-space Green's functions to satisfy both boundary conditions leads to the series:

$$G(x, \ell | \xi, \eta) = \frac{i}{2k} \left[e^{ik|x-\xi|-i\ell y} + \sum_{\substack{n=-\infty \\ n \neq 0}}^{\infty} e^{ik|x-\xi-2nL|-i\ell y} \right.$$

$$\left. - \sum_{n=-\infty}^{\infty} e^{ik|x+\xi+2nL|-i\ell y} \right]. \qquad (5.4.36)$$

Upon using (5.1.22) on each term in (5.4.36), we find that

$$g(x, y | \xi, \eta) = \frac{i}{4} \sum_{n=-\infty}^{\infty} \left\{ H_0^{(1)} \left[k_0 \sqrt{(x - \xi - 2nL)^2 + (y - \eta)^2} \right] \right.$$

$$\left. - H_0^{(1)} \left[k_0 \sqrt{(x + \xi + 2nL)^2 + (y - \eta)^2} \right] \right\}.$$

$$(5.4.37)$$

We can convert (5.4.37) into (5.4.32) through the use of Poisson's summation formula.

What is the physical interpretation of (5.4.37)? Recall that the Helmholtz equation describes the steady-state radiation of waves from the source point (ξ, η). Consequently, (5.4.37) appears to be describing the constructive and destructive interference of waves emanating from an infinite number of sources. Actually, these additional terms are due to multiple reflections of the original wave emanating from (ξ, η) off the boundaries. For this reason, (5.4.37) is often referred to as a *multiple-reflective wave series*.

Our third and final method for solving (5.4.21) involves separation of variables. We begin by considering the homogeneous version of (5.4.21), namely

$$\frac{\partial^2 g}{\partial x^2} + \frac{\partial^2 g}{\partial y^2} + k_0^2 g = 0. \qquad (5.4.38)$$

Assuming that $g(x, y) = X(x)Y(y)$, the homogeneous Helmholtz equation becomes

$$\frac{1}{X} \frac{d^2 X}{dx^2} + \frac{1}{Y} \frac{d^2 Y}{dy^2} + k_0^2 = 0, \qquad (5.4.39)$$

or

$$\frac{d^2 X}{dx^2} + \lambda_x X = 0, \qquad (5.4.40)$$

and

$$\frac{d^2Y}{dy^2} + \lambda_y Y = 0, \tag{5.4.41}$$

with $\lambda_x + \lambda_y = k_0^2$.

Let us associate with (5.4.40) and (5.4.41), a corresponding one-dimensional Green's function problem:

$$\frac{d^2X}{dx^2} + \lambda_x X = -\delta(x - \xi), \qquad 0 < x, \xi < L, \tag{5.4.42}$$

with $X(0|\xi) = X(L|\xi) = 0$, and

$$\frac{d^2Y}{dy^2} + \lambda_y Y = -\delta(y - \eta), \qquad -\infty < y, \eta < \infty, \tag{5.4.43}$$

where $Y(y|\eta)$ must satisfy the radiation condition. The solution to (5.4.42) follows directly from the eigenfunction method presented in §2.3 and equals

$$X(x|\xi) = -\frac{2}{L} \sum_{n=1}^{\infty} \frac{\sin(n\pi\xi/L)\sin(n\pi x/L)}{\lambda_x - n^2\pi^2/L^2}. \tag{5.4.44}$$

Because the y-domain is infinite, we employ Fourier transforms in that direction and find that

$$Y(y|\eta) = \frac{i}{2\sqrt{\lambda_y}} e^{i\sqrt{\lambda_y}\,|y-\eta|} \tag{5.4.45}$$

with $\mathrm{Im}(\sqrt{\lambda_y}) > 0$.

Consider now the following Green's function:

$$g(x, y|\xi, \eta) = -\frac{1}{2\pi i} \oint_{C_x} X(x|\xi) Y(y|\eta)\, d\lambda_x \tag{5.4.46}$$

$$= -\frac{1}{2\pi i} \oint_{C_y} X(x|\xi) Y(y|\eta)\, d\lambda_y, \tag{5.4.47}$$

where C_x encloses the singularities of $X(x|\xi)$ and excludes those of $Y(y|\eta)$; C_y, on the other hand, encloses only the singularities of $Y(y|\eta)$. Here, the relationship $\lambda_x + \lambda_y = k_0^2$ was used to eliminate λ_y in (5.4.46) and λ_x in (5.4.47). Furthermore,

$$\frac{\partial^2 g}{\partial x^2} + \frac{\partial^2 g}{\partial y^2} + k_0^2 g = \frac{\partial^2 g}{\partial x^2} + \lambda_x g + \frac{\partial^2 g}{\partial y^2} + \lambda_y g \tag{5.4.48}$$

$$= -\frac{1}{2\pi i} \oint_{C_x} \left(\frac{\partial^2}{\partial x^2} + \lambda_x + \frac{\partial^2}{\partial y^2} + \lambda_y \right) X(x|\xi) Y(y|\eta)\, d\lambda_x$$

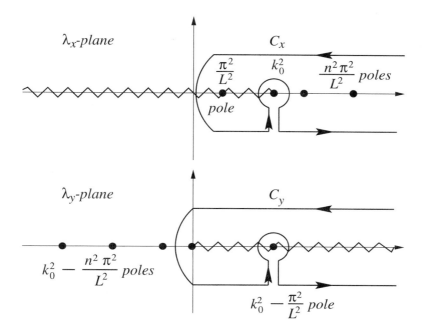

Figure 5.4.2: The contours C_x and C_y used in finding the solutions (5.4.52) and (5.4.54) of the Helmholtz equation.

$$(5.4.49)$$

$$= -\frac{1}{2\pi i} \oint_{C_x} [-\delta(x-\xi)Y(y|\eta) - X(x|\xi)\delta(y-\eta)] \, d\lambda_x$$

$$(5.4.50)$$

$$= \frac{\delta(y-\eta)}{2\pi i} \oint_{C_x} X(x|\xi) \, d\lambda_x = -\delta(x-\xi)\delta(y-\eta)$$

$$(5.4.51)$$

by (2.4.34). The integral (5.4.51) follows from (5.4.50) because C_x excludes the singularities of $Y(y|\eta)$; consequently, $\oint_{C_x} Y(y|\eta) \, d\lambda_x = 0$ by Cauchy's theorem. A similar demonstration proves (5.4.47).

If we use (5.4.46), then

$$g(x,y|\xi,\eta) = \frac{1}{2\pi i} \oint_{C_x} \frac{ie^{i\sqrt{k_0^2-\lambda_x}\,|y-\eta|}}{L\sqrt{k_0^2-\lambda_x}} \sum_{n=1}^{\infty} \frac{\sin(n\pi\xi/L)\sin(n\pi x/L)}{\lambda_x - (n\pi/L)^2} \, d\lambda_x$$

$$(5.4.52)$$

$$= \frac{1}{L} \sum_{n=1}^{\infty} \frac{\sin(n\pi\xi/L)\sin(n\pi x/L)}{\sqrt{(n\pi/L)^2-k_0^2}} e^{-\sqrt{(n\pi/L)^2-k_0^2}\,|y-\eta|},$$

$$(5.4.53)$$

where $\text{Im}[\sqrt{(n\pi/L)^2 - k_0^2}] > 0$. In Figure 5.4.2 we illustrate C_x. The poles are located at $\lambda_x = (n\pi/L)^2$, while the branch cut associated with $\sqrt{k_0^2 - \lambda_x}$ is taken along the real axis from the branch point k_0^2 to $-\infty$. As the figure shows, C_x encloses the poles but not the branch point k_0^2.

If we had used (5.4.47), then the solution would be

$$g(x, y|\xi, \eta) = \frac{1}{\pi L} \int_{-\infty}^{\infty} \sum_{n=1}^{\infty} \sin\left(\frac{n\pi\xi}{L}\right) \sin\left(\frac{n\pi x}{L}\right) \frac{e^{i\omega|y-\eta|}}{\omega^2 + n^2\pi^2/L^2 - k_0^2} \, d\omega,$$

$$(5.4.54)$$

which is a branch cut integral and equals an integral over the continuous spectrum of $Y(y|\eta)$ rather than a sum over the discrete spectrum of $X(x|\xi)$. The contour C_y is illustrated in Figure 5.4.2, where it was assumed that $\pi/L < k_0 < 2\pi/L$. This contour encloses the branch cut for $Y(y|\eta)$ but not the poles of $X(x|\xi)$ which occur now when $\lambda_y = k_0^2 - (n\pi/L)^2$.

In this section, we have shown that (5.4.32), (5.4.37) and (5.4.54) give the Green's function for the two-dimensional Helmholtz equation over the infinite strip $0 < x < L$. Which one should we use? Because the rate of convergence for each series depends upon the particular (x, y), so does the choice. For example, near the source point, (5.4.37) is best because the lead term explicitly resolves the singularity at (ξ, η) and the remaining terms in the series tend to zero rapidly.

5.5 THREE-DIMENSIONAL PROBLEMS IN A HALF-SPACE

In this section, we find the Green's functions for Poisson's and Helmholtz's equations within the half-space $z > 0$.

• *Poisson's equation*

In this case, the Green's function $g(x, y, z|\xi, \eta, \zeta)$ satisfies the partial differential equation

$$\frac{\partial^2 g}{\partial x^2} + \frac{\partial^2 g}{\partial y^2} + \frac{\partial^2 g}{\partial z^2} = -\delta(x - \xi)\delta(y - \eta)\delta(z - \zeta) \qquad (5.5.1)$$

over the half-space $z > 0$, with the boundary condition

$$\alpha g(x, y, 0|\xi, \eta, \zeta) - \beta \frac{\partial g(x, y, 0|\xi, \eta, \zeta)}{\partial z} = 0, \qquad (5.5.2)$$

where α and β are constants. There are three possible cases: (1) the Dirichlet problem with $\alpha = 1$ and $\beta = 0$, (2) the Neumann problem with $\alpha = 0$ and $\beta = -1$, and (3) the Robin problem with $\alpha = h$ and $\beta = 1$ with $h > 0$. Finally, we require that the Green's function vanishes at infinity.

We begin by noting that the most general solution consists of a sum of the free-space Green's function

$$g_1(x, y, z | \xi, \eta, \zeta) = \frac{1}{4\pi} \left[(x - \xi)^2 + (y - \eta)^2 + (z - \zeta)^2 \right]^{-1/2} \quad (5.5.3)$$

found in §5.1 and a homogeneous solution to (5.5.1), g_2. Consequently, from (5.5.2) the boundary condition on g_2 is

$$\alpha g_2(x, y, 0 | \xi, \eta, \zeta) - \beta \frac{\partial g_2(x, y, 0 | \xi, \eta, \zeta)}{\partial z}$$
$$= -\alpha g_1(x, y, 0 | \xi, \eta, \zeta) + \beta \frac{\partial g_1(x, y, 0 | \xi, \eta, \zeta)}{\partial z}. \quad (5.5.4)$$

Let us now introduce an image point at $(\xi, \eta, -\zeta)$, which is the reflection of (ξ, η, ζ) in the plane $z = 0$. Then the free-space Green's function corresponding to this image point is

$$\widehat{g}_2(x, y, z | \xi, \eta, \zeta) = \frac{1}{4\pi} \left[(x - \xi)^2 + (y - \eta)^2 + (z + \zeta)^2 \right]^{-1/2}. \quad (5.5.5)$$

Note, however, that $\widehat{g}_2(x, y, z | \xi, \eta, \zeta)$ is harmonic within the domain $z > 0$. Can we use (5.5.5) to find $g_2(x, y, z | \xi, \eta, \zeta)$? Yes! Because $\widehat{g}_2(x, y, 0 | \xi, \eta, \zeta) = g_1(x, y, 0 | \xi, \eta, \zeta)$, $g_2(x, y, z | \xi, \eta, \zeta) = -\widehat{g}_2(x, y, z | \xi, \eta, \zeta)$ in the case of a Dirichlet condition. Similarly, since

$$\frac{\partial \widehat{g}_2(x, y, 0 | \xi, \eta, \zeta)}{\partial z} = -\frac{(z + \zeta)}{4\pi} \left[(x - \xi)^2 + (y - \eta)^2 + (z + \zeta)^2 \right]^{-3/2} \Big|_{z=0}$$
$$\quad (5.5.6)$$
$$= -\frac{\partial g_1(x, y, 0 | \xi, \eta, \zeta)}{\partial z}, \quad (5.5.7)$$

$g_2(x, y, z | \xi, \eta, \zeta) = \widehat{g}_2(x, y, z | \xi, \eta, \zeta)$ in the case of the Neumann problem. Thus, from the method of images, the Green's function for Poisson's equation within the half-space $z > 0$ is

$$g(x, y, z | \xi, \eta, \zeta) = \frac{1}{4\pi} \left[(x - \xi)^2 + (y - \eta)^2 + (z - \zeta)^2 \right]^{-1/2}$$
$$- \frac{1}{4\pi} \left[(x - \xi)^2 + (y - \eta)^2 + (z + \zeta)^2 \right]^{-1/2}, \quad (5.5.8)$$

when we have a Dirichlet boundary condition and

$$g(x, y, z | \xi, \eta, \zeta) = \frac{1}{4\pi} \left[(x - \xi)^2 + (y - \eta)^2 + (z - \zeta)^2 \right]^{-1/2}$$
$$+ \frac{1}{4\pi} \left[(x - \xi)^2 + (y - \eta)^2 + (z + \zeta)^2 \right]^{-1/2}, \quad (5.5.9)$$

when we have a Neumann boundary condition.

If we try to apply this technique for a Robin boundary condition, we fail. We cannot obtain $g_2(x, y, z | \xi, \eta, \zeta)$ solely in terms of an image source point at $(\xi, \eta, -\zeta)$. Motivated by Example 4.1.2 (recall that Poisson's equation is the steady-state version of the three-dimensional heat equation), we introduce an entire line of image sources on the line $x = \xi$ and $y = \eta$ with z extending from $z = -\zeta$ to $z = -\infty$. Let us try

$$
g_2(x, y, z | \xi, \eta, \zeta) = \frac{1}{4\pi} \left[(x - \xi)^2 + (y - \eta)^2 + (z + \zeta)^2 \right]^{-1/2}
$$
$$
+ \frac{1}{4\pi} \int_{-\infty}^{-\zeta} \frac{\rho(s)}{\sqrt{(x - \xi)^2 + (y - \eta)^2 + (z - s)^2}} \, ds,
$$
$$
\text{(5.5.10)}
$$

where $\rho(s)$ is a presently unknown source density. When $h = 0$ in the Robin boundary condition, (5.5.10) must reduce to the solution for the Neumann problem. Therefore, $\rho(s) = 0$ when $h = 0$. This is why we introduced the free-space Green's function (5.5.5) in (5.5.10).

If $\rho(s)$ decays sufficiently rapidly at infinity so that the integral in (5.5.10) converges and that differentiation under the integral sign is permitted, then g_2 is harmonic for $z > 0$ because all of the singular points in (5.5.10) occur in the lower half-space $z < 0$. Substituting (5.5.10) into (5.5.4),

$$
\frac{\partial g_2(x, y, 0 | \xi, \eta, \zeta)}{\partial z} - h g_2(x, y, 0 | \xi, \eta, \zeta)
$$
$$
= -\frac{\zeta}{4\pi} \left[(x - \xi)^2 + (y - \eta)^2 + \zeta^2 \right]^{-3/2}
$$
$$
- \frac{1}{4\pi} \int_{-\infty}^{-\zeta} \rho(s) \frac{\partial}{\partial s} \left[(x - \xi)^2 + (y - \eta)^2 + s^2 \right]^{-1/2} ds
$$
$$
- \frac{h}{4\pi} \left[(x - \xi)^2 + (y - \eta)^2 + \zeta^2 \right]^{-1/2}
$$
$$
- \frac{h}{4\pi} \int_{-\infty}^{-\zeta} \frac{\rho(s)}{\sqrt{(x - \xi)^2 + (y - \eta)^2 + s^2}} \, ds \quad \text{(5.5.11)}
$$
$$
= -\frac{\partial g_1(x, y, 0 | \xi, \eta, \zeta)}{\partial z} + h g_1(x, y, 0 | \xi, \eta, \zeta) \quad \text{(5.5.12)}
$$
$$
= -\frac{\zeta}{4\pi} \left[(x - \xi)^2 + (y - \eta)^2 + \zeta^2 \right]^{-3/2}
$$
$$
+ \frac{h}{4\pi} \left[(x - \xi)^2 + (y - \eta)^2 + \zeta^2 \right]^{-1/2}. \quad \text{(5.5.13)}
$$

Because the operator $\partial/\partial z$ has the same effect as $-\partial/\partial s$ at $z = 0$, and the use of $\partial/\partial s$ in the integral term allows us to integrate by parts, we obtain

$$\int_{-\infty}^{-\zeta} \rho(s) \frac{\partial}{\partial s} \left[(x-\xi)^2 + (y-\eta)^2 + s^2\right]^{-1/2} ds$$

$$= \rho(-\zeta) \left[(x-\xi)^2 + (y-\eta)^2 + \zeta^2\right]^{-1/2} ds$$

$$- \int_{-\infty}^{-\zeta} \rho'(s) \left[(x-\xi)^2 + (y-\eta)^2 + s^2\right]^{-1/2} ds. \quad (5.5.14)$$

Substituting (5.5.14) into (5.5.11), we find that

$$\left[-\frac{h}{2\pi} - \frac{\rho(-\zeta)}{4\pi}\right] \left[(x-\xi)^2 + (y-\eta)^2 + \zeta^2\right]^{-1/2}$$

$$+ \frac{1}{4\pi} \int_{-\infty}^{-\zeta} \frac{\rho'(s) - h\,\rho(s)}{\sqrt{(x-\xi)^2 + (y-\eta)^2 + s^2}} ds = 0. \quad (5.5.15)$$

Equation (5.5.15) is satisfied if we set

$$\rho'(s) - h\,\rho(s) = 0, \qquad s < -\zeta, \qquad (5.5.16)$$

with

$$\rho(-\zeta) = -2h. \qquad (5.5.17)$$

The solution of the ordinary differential equation (5.5.16)–(5.5.17) is

$$\rho(s) = -2h\, e^{h(s+\zeta)}. \qquad (5.5.18)$$

Note that $\rho(s)$ vanishes for $h = 0$ and that it decays exponentially as $s \to -\infty$.

Thus, the Green's function for Poisson's equation within the half-space $z > 0$ with Robin boundary condition is

$$g(x, y, z|\xi, \eta, \zeta) = \frac{1}{4\pi} \left[(x-\xi)^2 + (y-\eta)^2 + (z-\zeta)^2\right]^{-1/2}$$

$$+ \frac{1}{4\pi} \left[(x-\xi)^2 + (y-\eta)^2 + (z+\zeta)^2\right]^{-1/2}$$

$$- \frac{h}{2\pi} \int_{-\infty}^{-\zeta} \frac{e^{h(s+\zeta)}}{\sqrt{(x-\xi)^2 + (y-\eta)^2 + (z-s)^2}} ds. \quad (5.5.19)$$

• *Helmholtz's equation*

Let us now find the Green's function for the three-dimensional Helmholtz equation (5.1.1) in the space $z > 0$ with the boundary condition that $g(x, y, 0|\xi, \eta, \zeta) = 0$. Using the method of images, the desired Green's function is

$$g(x, y, z|\xi, \eta, \zeta) = \frac{e^{ik_0\sqrt{(z-\zeta)^2+\rho^2}}}{4\pi\sqrt{(z-\zeta)^2+\rho^2}} - \frac{e^{ik_0\sqrt{(z+\zeta)^2+\rho^2}}}{4\pi\sqrt{(z+\zeta)^2+\rho^2}}, \quad (5.5.20)$$

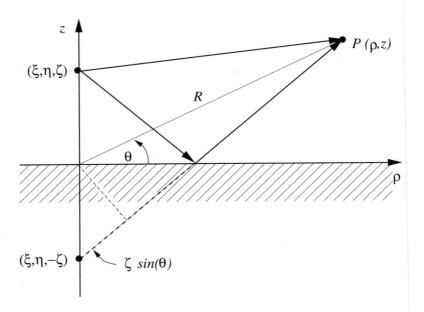

Figure 5.5.1: The solution of the Helmholtz equation for the half-space $z > 0$ using the method of images.

where $\rho^2 = (x - \xi)^2 + (y - \eta)^2$. The first term in (5.1.20) corresponds to the free-space Green's function resulting from the point source (ξ, η, ζ) in the region $z > 0$. Consequently, it satisfies the nonhomogeneous Helmholtz equation (5.1.1). On the other hand, the second term represents the free-space Green's function for a (negative) mirror image about $z = 0$. Because its source point lies at $(\xi, \eta, -\zeta)$, it satisfies the *homogeneous* Helmholtz equation. Its purpose is to ensure that (5.5.20) satisfies the boundary condition at $z = 0$.

Equation (5.5.20) arises in acoustics and gives the complicated interference pattern due to the presence of a wall. We can see this by examining (5.5.20) when the distance from the origin to the point (x, y, z) is much greater than the source depth ζ. Denoting the declination angle by θ, our far-field assumption leads to the following approximations

$$\sqrt{(z - \zeta)^2 + \rho^2} \approx R - \zeta \sin(\theta), \qquad (5.5.21)$$

and

$$\sqrt{(z + \zeta)^2 + \rho^2} \approx R + \zeta \sin(\theta), \qquad (5.5.22)$$

where $R = \sqrt{\rho^2 + z^2}$. See Figure 5.5.1. Assuming further that the ranges appearing in the denominator of both terms of (5.5.20) can be replaced by the slant range R because the amplitude decays slowly with

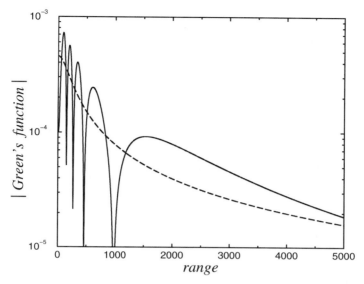

Figure 5.5.2: Surface-interference (Lloyd mirror) solution for a point source in a homogeneous halfspace $z > 0$. The solid line gives $|g(x, y, 200\text{ m}|\xi, \eta, 25\text{ m})|$ from (5.5.20) with $k_0 = 2\pi/(10\text{ m})$ as a function of range ρ in meters. The dashed line is $1/[4\pi\sqrt{\rho^2 + (200\text{ m} - 25\text{ m})^2}]$.

range, we obtain

$$g(x, y, z|\xi, \eta, \zeta) = \frac{1}{4\pi R} \left\{ e^{ik_0[R - \zeta\sin(\theta)]} - e^{ik_0[R + \zeta\sin(\theta)]} \right\} \quad (5.5.23)$$

$$= \frac{e^{ik_0 R}}{4\pi R} \left[e^{-ik_0\zeta\sin(\theta)} - e^{ik_0\zeta\sin(\theta)} \right]. \quad (5.5.24)$$

The two exponentials can be combined to yield sine and (5.5.24) becomes

$$g(x, y, z|\xi, \eta, \zeta) = -\frac{i}{2\pi R} \sin\left[k_0\zeta\sin(\theta)\right] e^{ik_0 R}, \quad (5.5.25)$$

which means that the amplitude of the Green's function varies as

$$|g(x, y, z|\xi, \eta, \zeta)| = \frac{1}{2\pi R} \left| \sin\left[k_0\zeta\sin(\theta)\right] \right|. \quad (5.5.26)$$

For a point source in free space we would have a spherically expanding wave with $|g(x, y, z|\xi, \eta, \zeta)| = 1/(4\pi R)$. Therefore, the presence of the reflecting surface generates a directional pattern with maxima and minima given by

$$|g(x, y, z|\xi, \eta, \zeta)|_{\max} = \frac{1}{2\pi R} \quad \text{for} \quad \sin(\theta) = (2m-1)\frac{\pi}{2k_0\zeta}, \quad m = 1, 2, \dots$$
$$(5.5.27)$$

and

$$|g(x,y,z|\xi,\eta,\zeta)|_{\min} = 0 \quad \text{for} \quad \sin(\theta) = (m-1)\frac{\pi}{k_0\zeta}, \quad m = 1, 2, \ldots$$
$$(5.5.28)$$

This is the classic surface-image or *Lloyd-mirror* interference pattern.[30] Note that the maximum is twice the value that a single source would have (constructive interference), while the minimum is zero (destructive interference). Figure 5.5.2 shows the variation of the $|g(x,y,z|\xi,\eta,\zeta)|$ as a function of range ρ.

5.6 THREE-DIMENSIONAL POISSON'S EQUATION IN A CYLINDRICAL DOMAIN

In this section, we present Green's functions for various limited cylindrical domains. We assume Dirichlet boundary conditions along any surface. In all cases, the governing equation is

$$\frac{1}{r}\frac{\partial}{\partial r}\left(r\frac{\partial g}{\partial r}\right) + \frac{1}{r^2}\frac{\partial^2 g}{\partial \theta^2} + \frac{\partial^2 g}{\partial z^2} = -\frac{\delta(r-\rho)\delta(\theta-\theta')\delta(z-\zeta)}{r}. \quad (5.6.1)$$

We illustrate several simple geometries; Gray and Mathews[31] and Dougall[32] present solutions for more complicated configurations.

• *Space bounded by two parallel plates*

In Example 5.1.2, the three-dimensional free-space Green's function in cylindrical coordinates was found. Although this solution satisfies the differential equation, it does not satisfy most boundary conditions. For example, in the case of Dirichlet boundary conditions along the planes $z = 0$ and $z = c$, we must find a homogeneous solution to (5.6.1) such that it plus the free-space Green's function satisfy the boundary conditions.[33] A quick check shows that

$$V(x,y,z) = -\int_0^\infty \left\{ \frac{\sinh(kz)}{\sinh(kc)}e^{-k(c-\rho)} + \frac{\sinh[k(c-z)]}{\sinh(kc)}e^{-k\rho} \right\} J_0(kR)\, dk,$$
$$(5.6.2)$$

---

[30] In 1833 Humphrey Lloyd (1800–1881) published the results of an experiment that confirmed conical refraction as predicted by William Rowan Hamilton (1805–1865) from Fresnel's wave theory of light.

[31] Gray, A., and G. B. Mathews, 1966: *Treatise on Bessel Functions and Their Applications to Physics.* Dover Publications, Chapter IX, §3.

[32] Dougall, J., 1900: The determination of Green's functions by means of cylindrical and spherical harmonics. *Proc. Edinburgh Math. Soc.*, **18**, 33–83.

[33] Bates [Bates, J. W., 1997: On toroidal Green's functions. *J. Math. Phys.*, **38**, 3679–3691.] used a similar technique to find the Green's functions for Laplace's and the biharmonic equations with Dirichlet boundary conditions along the entire boundary.

where $R = \sqrt{r^2 + \rho^2 - 2r\rho\cos(\theta - \theta')}$, is this homogeneous solution, since its combination with free-space Green's function (5.1.54) yields the Green's function

$$g(r, \theta, z | \rho, \theta', \zeta) = \frac{1}{2\pi} \int_0^\infty \frac{\sinh[k(c - z_>)]\sinh(kz_<)}{\sinh(kc)} J_0(kR)\, dk,$$

(5.6.3)

which satisfies Dirichlet boundary conditions along $z = 0$ and $z = c$.

• *Infinitely long tube*

In the case of an infinitely long tube, we must solve (5.6.1) with the boundary conditions

$$\lim_{r \to 0} |g(r, \theta, z | \rho, \theta', \zeta)| < \infty, \qquad g(a, \theta, z | \rho, \theta', \zeta)| = 0, \qquad (5.6.4)$$

for $0 \le \theta, \theta' \le 2\pi$. The Green's function must be periodic in θ.

We begin by noting that

$$\frac{\delta(r - \rho)\delta(\theta - \theta')}{r} = \frac{1}{\pi a^2} \sum_{n=-\infty}^{\infty} \sum_{m=1}^{\infty} \frac{J_n(k_{nm}\rho) J_n(k_{nm}r)}{J'^2_n(k_{nm}a)} \cos[n(\theta - \theta')],$$

(5.6.5)

where k_{nm} is the mth root of $J_n(ka) = 0$. The form of the Green's function is then

$$g(r, \theta, z | \rho, \theta', \zeta) = \sum_{n=-\infty}^{\infty} \sum_{m=1}^{\infty} G_{nm}(z | \zeta) J_n(k_{nm}r) \cos[n(\theta - \theta')]. \quad (5.6.6)$$

Substitution of (5.6.5) and (5.6.6) into (5.6.1) yields the differential equation

$$\frac{d^2 G_{nm}}{dz^2} - k_{nm}^2 G_{nm} = -\frac{J_n(k_{nm}\rho)}{\pi a^2 J'^2_n(k_{nm}a)} \delta(z - \zeta). \qquad (5.6.7)$$

The solution to (5.6.7) that remains bounded as $|z| \to \infty$ is

$$G_{nm}(z | \zeta) = \frac{J_n(k_{nm}\rho)}{2\pi a^2 k_{nm} J'^2_n(k_{nm}a)} e^{-k_{nm}|z - \zeta|}. \qquad (5.6.8)$$

Consequently, the Green's function for an infinitely long tube where the solution vanishes along the cylindrical side at $r = a$ is

$$g(r, \theta, z | \rho, \theta', \zeta) = \frac{1}{2\pi a^2} \sum_{n=-\infty}^{\infty} \sum_{m=1}^{\infty} \frac{J_n(k_{nm}\rho) J_n(k_{nm}r)}{k_{nm} J'^2_n(k_{nm}a)} \cos[n(\theta - \theta')]$$

$$\times e^{-k_{nm}|z - \zeta|}. \qquad (5.6.9)$$

- *Cylindrical pill box*

When the tube is of finite length and has the additional boundary conditions that

$$g(r, \theta, 0 | \rho, \theta', \zeta) = g(r, \theta, L | \rho, \theta', \zeta) = 0, \qquad (5.6.10)$$

we must solve (5.6.7) subject to the boundary conditions $G_{nm}(0|\zeta) = G_{nm}(L|\zeta) = 0$. The solution is

$$G_{nm}(z|\zeta) = \frac{J_n(k_{nm}\rho) \sinh[k_{nm}(L - z_>)] \sinh(k_{nm}z_<)}{\pi a^2 k_{nm} \sinh(k_{nm}L) J'^2_n(k_{nm}a)}. \qquad (5.6.11)$$

Substituting (5.6.11) into (5.6.6), we obtain

$$g(r, \theta, z | \rho, \theta', \zeta) = \frac{1}{\pi a^2} \sum_{n=-\infty}^{\infty} \sum_{m=1}^{\infty} \frac{J_n(k_{nm}\rho) J_n(k_{nm}r)}{J'^2_n(k_{nm}a)} \cos[n(\theta - \theta')]$$

$$\times \frac{\sinh[k_{nm}(L - z_<)] \sinh(k_{nm}z_>)}{k_{nm} \sinh(k_{nm}L)}. \qquad (5.6.12)$$

In §2.4 we showed that the Green's function for ordinary differential equations such as (5.6.7) can be written as an eigenfunction expansion. In the present case,

$$G_{nm}(z|\zeta) = \frac{2L}{\pi a^2} \frac{J_n(k_{nm}\rho)}{J'^2_n(k_{nm}a)} \sum_{\ell=1}^{\infty} \frac{\sin(\ell\pi\zeta/L) \sin(\ell\pi z/L)}{k^2_{nm}L^2 + \ell^2\pi^2}. \qquad (5.6.13)$$

Thus, an alternative to (5.6.12) is

$$g(r, \theta, z | \rho, \theta', \zeta) = \frac{2L}{\pi a^2} \sum_{n=-\infty}^{\infty} \sum_{m=1}^{\infty} \sum_{\ell=1}^{\infty} \frac{J_n(k_{nm}\rho) J_n(k_{nm}r)}{J'^2_n(k_{nm}a)} \cos[n(\theta - \theta')]$$

$$\times \frac{\sin(\ell\pi\zeta/L) \sin(\ell\pi z/L)}{k^2_{nm}L^2 + \ell^2\pi^2}. \qquad (5.6.14)$$

5.7 POISSON'S EQUATION FOR A SPHERICAL DOMAIN

Finally, we turn to the closed domain of a sphere of radius a with its center at the origin and find the Green's function for Poisson's equation within its interior. Our method of solution works for the Dirichlet problem[34] but not for the Neumann and Robin problems.

---

[34] Additional Green's functions for more complicated geometries are given by Dougall, J., 1900: The determination of Green's functions by means of cylindrical and spherical harmonics. *Proc. Edinburgh Math. Soc.*, **18**, 33–83.

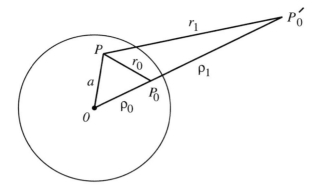

Figure 5.7.1: Diagram used in finding the Green's function for Poisson's equation in a spherical domain.

We begin by introducing a point P located *on* the surface of the sphere. The source point P_0 is located within the sphere at the point (ξ, η, ζ). As shown in Figure 5.7.1, we also introduce an image source point P_0' at (ξ', η', ζ'). The point P_0' lies on the radial line extending from the origin O through the source point P_0. Its distance from the origin equals ρ_1, and the distance of P_0 from the origin is ρ_0. These distances are related by $\rho_0 \rho_1 = a^2$. This process for obtaining P_0' from P_0 is called *inversion with respect to the sphere*. Note that if the radius $a = 1$, we have that $\rho_1 = 1/\rho_0$ and the distances ρ_0 and ρ_1 are inverse to one other.

The triangles $\triangle OPP_0$ and $\triangle OPP_0'$ in Figure 5.7.1 are similar because they share the common angle $\angle POP_0$ and proportional sides $\overline{OP_0}/\overline{OP} = \overline{OP}/\overline{OP_0'}$. This follows because $\overline{OP} = a$, $\overline{OP_0} = \rho_0$, $\overline{OP_0'} = \rho_1$, and $(\overline{OP_0})(\overline{OP_0'}) = \rho_0 \rho_1 = (\overline{OP})^2 = a^2$. The similarity of the triangles implies that all three sides are proportional and we have $\rho_0/a = a/\rho_1 = r_0/r_1$.

Let us now break g into two parts: g_1 and g_2, where g_1 is the free-space Green's function (5.1.53) with $k_0 = 0$ and g_2 is a constant multiple c of the free-space Green's function with the source point at P_0'. Therefore,

$$g(x, y, z | \xi, \eta, \zeta) = \frac{1}{4\pi} \left[(x - \xi)^2 + (y - \eta)^2 + (z - \zeta)^2 \right]^{-1/2}$$
$$+ \frac{c}{4\pi} \left[(x - \xi')^2 + (y - \eta')^2 + (z - \zeta')^2 \right]^{-1/2}. \quad \textbf{(5.7.1)}$$

Because P_0' lies outside the sphere, the second term in (5.7.1) is a harmonic function (a solution to Laplace's equation) within the sphere. Thus, (5.7.1) satisfies the differential equation governing the Green's function. On the surface of the sphere, we therefore have that

$$g(x, y, z|\xi, \eta, \zeta) = \frac{1}{4\pi}\left(\frac{1}{r_0} + \frac{c}{r_1}\right) = \frac{1}{4\pi r_0}\left(1 + \frac{c\rho_0}{a}\right). \qquad (5.7.2)$$

Thus, if $c = -a/\rho_0$, the Green's function g vanishes on the sphere. Therefore, we rewrite (5.7.2) as

$$g(x, y, z|\xi, \eta, \zeta) = \frac{1}{4\pi}\left(\frac{1}{r_0} - \frac{a}{\rho_0 r_1}\right). \qquad (5.7.3)$$

Let us now use this solution to solve Poisson's equation within a sphere of volume V and radius a:

$$\frac{\partial^2 u}{\partial x^2} + \frac{\partial^2 u}{\partial y^2} + \frac{\partial^2 u}{\partial z^2} = -f(x, y, z), \qquad (5.7.4),$$

with the boundary condition that

$$u = B(x, y, z) \qquad (5.7.5)$$

on the surface S of the sphere. We now prove that the solution at any point (ξ, η, ζ) *inside* the sphere is

$$u(\xi, \eta, \zeta) = \iiint_V f(x, y, z)g(x, y, z|\xi, \eta, \zeta)\,dx\,dy\,dz - \oiint_S \frac{\partial g}{\partial n}B(x, y, z)\,dS,$$
$$(5.7.6)$$

where

$$\frac{\partial g}{\partial n} = \frac{1}{4\pi}\left[\frac{\partial}{\partial n}\left(\frac{1}{r_0}\right) - \frac{a}{\rho_0}\frac{\partial g}{\partial n}\left(\frac{1}{r_1}\right)\right] \qquad (5.7.7)$$

$$= \frac{1}{4\pi}\left[\frac{\partial}{\partial r_0}\left(\frac{1}{r_0}\right)\frac{\partial r_0}{\partial n} - \frac{a}{\rho_0}\frac{\partial}{\partial r_1}\left(\frac{1}{r_1}\right)\frac{\partial r_1}{\partial n}\right] \qquad (5.7.8)$$

$$= \frac{1}{4\pi}\left[-\frac{1}{r_0^2}\frac{\partial r_0}{\partial n} + \frac{a}{\rho_0}\frac{1}{r_1^2}\frac{\partial r_1}{\partial n}\right]. \qquad (5.7.9)$$

Recalling that $r_0^2 = (x-\xi)^2 + (y-\eta)^2 + (z-\zeta)^2$, and $r_1^2 = (x-\xi')^2 + (y-\eta')^2 + (z-\zeta')^2$, where (x, y, z) is a point on the sphere, and introducing coordinate systems with their origins at (ξ, η, ζ) and (ξ', η', ζ'), we find that

$$\frac{\partial r_0}{\partial n} = \cos(\theta_0), \qquad \text{and} \qquad \frac{\partial r_1}{\partial n} = \cos(\theta_1). \qquad (5.7.10)$$

The angle θ_0 is the angle between the exterior unit normal vector to the sphere at the point P and vector extending from the source point P_0 to

P. Similarly, θ_1 is the angle between the normal vector at P and the vector from P_0' to P. From Figure 5.7.1 and using the law of cosines,

$$\cos(\theta_0) = \frac{a^2 + r_0^2 - \rho_0^2}{2ar_0}, \tag{5.7.11}$$

and

$$\cos(\theta_1) = \frac{a^2 + r_1^2 - \rho_1^2}{2ar_1}. \tag{5.7.12}$$

Noting that $\rho_1 = a^2/\rho_0$, and $r_1 = ar_0/\rho_0$ in (5.7.12), we have that

$$\cos(\theta_1) = \frac{a^2 + (a^2/\rho_0^2)r_0^2 - (a^4/\rho_0^2)}{2a(a/\rho_0)r_0} = \frac{\rho_0^2 + r_0^2 - a^2}{2\rho_0 r_0}. \tag{5.7.13}$$

Substituting these results into (5.7.9), we finally obtain

$$\frac{\partial g}{\partial n} = \frac{1}{4\pi}\left[-\frac{1}{r_0^2}\left(\frac{a^2 + r_0^2 - \rho_0^2}{2ar_0}\right) + \frac{\rho_0^2}{a^2 r_0^2}\left(\frac{a}{\rho_0}\right)\left(\frac{\rho_0^2 + r_0^2 - a^2}{2\rho_0 r_0}\right)\right]$$
$$\tag{5.7.14}$$

$$= -\frac{1}{4\pi a}\frac{a^2 - \rho_0^2}{r_0^3}. \tag{5.7.15}$$

Thus, (5.7.6) becomes

$$u(\xi, \eta, \zeta) = \frac{1}{4\pi}\iiint_V f(x, y, z)\left(\frac{1}{r_0} - \frac{a}{\rho_0 r_1}\right)dx\,dy\,dz$$
$$+ \frac{1}{4\pi a}\oiint_S\left(\frac{a^2 - \rho_0^2}{r_0^3}\right)B(x, y, z)\,dS. \tag{5.7.16}$$

By transforming to spherical coordinates with their center at the origin, the second integral in (5.7.16) can be written

$$\frac{1}{4\pi a}\oiint_S\left(\frac{a^2 - \rho_0^2}{r_0^3}\right)B(x, y, z)\,dS$$
$$= \frac{a}{4\pi}\int_0^{2\pi}\int_0^{\pi}\frac{a^2 - \rho_0^2}{[a^2 - 2a\rho_0\cos(\gamma) + \rho_0^2]^{3/2}}B(a, \varphi, \theta)\sin(\theta)\,d\theta\,d\varphi,$$
$$\tag{5.7.17}$$

where $\gamma = \angle POP_0$. Equation (5.7.17) is known as *Poisson's integral* for the sphere. We also could have obtained this expression by using separation of variables with Laplace's equation in a spherical domain. It gives the solution to the Dirichlet problem for Laplace equation because in that case $f(x, y, z) = 0$.

• **Example 5.7.1: Green's function for Laplace's equation when the source is located on surface of the sphere**

Let us find the Green's function[35] for Laplace's equation in the space exterior to a sphere of radius a when the source is located on the surface at (a, θ', φ'). We solve the case for a Neumann boundary condition, while we leave the Dirichlet case as problem 35.

This problem is equivalent to solving

$$\frac{1}{r^2} \frac{\partial}{\partial r} \left(r^2 \frac{\partial g}{\partial r} \right) + \frac{1}{r^2 \sin(\theta)} \frac{\partial}{\partial \theta} \left(\sin(\theta) \frac{\partial g}{\partial \theta} \right) + \frac{1}{r^2 \sin^2(\theta)} \frac{\partial^2 g}{\partial \varphi^2} = 0 \quad (\mathbf{5.7.18})$$

subject to the boundary condition

$$\left. \frac{\partial g}{\partial r} \right|_{r=a} = -\frac{\delta(\theta - \theta')\delta(\varphi - \varphi')}{a^2 \sin(\theta')}, \quad (\mathbf{5.7.19})$$

where $a < r < \infty$, $0 \leq \theta, \theta' \leq \pi$, and $0 \leq \varphi, \varphi' \leq 2\pi$. Since

$$\frac{\delta(\theta - \theta')\delta(\varphi - \varphi')}{\sin(\theta')} = \sum_{\ell=0}^{\infty} \sum_{m=0}^{\ell} \frac{[M_{\ell m} \cos(m\varphi) + N_{\ell m} \sin(m\varphi)]}{2\pi} P_\ell^m[\cos(\theta)],$$

$$(\mathbf{5.7.20})$$

where

$$\begin{cases} M_{\ell m} \\ N_{\ell m} \end{cases} = \frac{2\ell + 1}{1 + \delta_{m0}} \frac{(\ell - m)!}{(\ell + m)!} P_\ell^m[\cos(\theta')] \begin{cases} \cos(m\varphi') \\ \sin(m\varphi') \end{cases}, \quad (\mathbf{5.7.21})$$

and δ_{ij} is the Kronecker delta function, separation of variables leads to

$$g(r, \theta, \varphi | a, \theta', \varphi') = \frac{1}{2\pi a} \sum_{\ell=0}^{\infty} \sum_{m=0}^{\ell} \frac{[M_{\ell m} \cos(m\varphi) + N_{\ell m} \sin(m\varphi)]}{\ell + 1}$$

$$\times \left(\frac{a}{r} \right)^{\ell+1} P_\ell^m[\cos(\theta)], \quad (\mathbf{5.7.22})$$

or

$$g(r, \theta, \varphi | a, \theta', \varphi') = \frac{1}{4\pi a} \sum_{\ell=0}^{\infty} \frac{2\ell + 1}{\ell + 1} \left(\frac{a}{r} \right)^{\ell+1} \left\{ P_\ell[\cos(\theta)]P_\ell[\cos(\theta')] \right.$$

$$\left. + 2 \sum_{m=1}^{\ell} \frac{(\ell - m)!}{(\ell + m)!} P_\ell^m[\cos(\theta)]P_\ell^m[\cos(\theta')] \cos[m(\varphi - \varphi')] \right\}.$$

$$(\mathbf{5.7.23})$$

---

[35] Reprinted with permission from Nemenman, I. M., and A. S. Silbergleit, 1999: Explicit Green's function of a boundary value problem for a sphere and trapped flux analysis in Gravity Probe B experiment. *J. Appl. Phys.*, **86**, 614–624. ©American Institute of Physics, 1999.

Upon applying the addition theorem for Legendre functions, we rewrite (5.7.23) as

$$g(r, \theta, \varphi | a, \theta', \varphi') = \frac{1}{4\pi a} \sum_{\ell=0}^{\infty} \frac{2\ell + 1}{\ell + 1} \left(\frac{a}{r}\right)^{\ell+1} P_\ell[\cos(\gamma)] \qquad (5.7.24)$$

$$= \frac{1}{4\pi a} \left\{ 2 \sum_{\ell=0}^{\infty} \left(\frac{a}{r}\right)^{\ell+1} P_\ell[\cos(\gamma)] \right.$$

$$\left. - \sum_{\ell=0}^{\infty} \frac{1}{\ell + 1} \left(\frac{a}{r}\right)^{\ell+1} P_\ell[\cos(\gamma)] \right\}, (5.7.25)$$

where

$$\cos(\gamma) = \cos(\theta)\cos(\theta') + \sin(\theta)\sin(\theta')\cos(\varphi - \varphi'). \qquad (5.7.26)$$

The first series in (5.7.25) is the generating function for Legendre polynomials; the second series is an integral of it, namely

$$\sum_{\ell=0}^{\infty} \frac{\eta^{\ell+1}}{\ell + 1} P_\ell(\zeta) = \int_0^\eta \sum_{\ell=0}^{\infty} \tau^\ell P_\ell(\zeta)\, d\tau = \int_0^\eta \frac{d\tau}{\sqrt{1 - 2\zeta\tau + \tau^2}} \qquad (5.7.27)$$

$$= \ln\left(\frac{\eta - \zeta + \sqrt{1 - 2\zeta\eta + \eta^2}}{1 - \zeta}\right). \qquad (5.7.28)$$

Combining these results, we obtain the Green's function

$$g(r, \theta, \varphi | a, \theta', \varphi') = \frac{1}{2\pi}\left[\frac{1}{|\mathbf{r} - \mathbf{r}_0|} \right.$$

$$\left. - \frac{1}{2a} \ln\left(\frac{a^2 - \mathbf{r} \cdot \mathbf{r}_0 + a|\mathbf{r} - \mathbf{r}_0|}{ar - \mathbf{r} \cdot \mathbf{r}_0}\right)\right], \qquad (5.7.29)$$

where $\mathbf{r}_0 = (a, \theta', \varphi')$ is the position vector of the source. The first term is one half of the potential of a point charge, while the second term is the contribution from the curved boundary.

If we were interested in the Green's function for the *interior* of a sphere of radius a when the charge lies on the surface, we would find that

$$g(r, \theta, \varphi | a, \theta', \varphi') = \frac{1}{2\pi}\left[\frac{1}{|\mathbf{r} - \mathbf{r}_0|} \right.$$

$$\left. - \frac{1}{2r} \ln\left(\frac{r^2 - \mathbf{r} \cdot \mathbf{r}_0 + r|\mathbf{r} - \mathbf{r}_0|}{ar - \mathbf{r} \cdot \mathbf{r}_0}\right)\right], \qquad (5.7.30)$$

Equation (5.7.30) follows from our inversion technique,[36] where we replace r in (5.7.29) with a^2/r and multiplied (5.7.29) by a/r.

5.8 IMPROVING THE CONVERGENCE RATE OF GREEN'S FUNCTIONS

So far in this chapter, we have been content with finding any analytic expression that gives us the Green's function. However, when a Green's function is applied to an actual physical problem, numerous evaluations of the Green's function may be required. Unfortunately, because most Green's functions are series, these expressions may converge so slowly that the Green's function approach must be abandoned. The purpose of this section[37] is to suggest methods to dramatically decrease the number of computations necessary to accurately evaluate a given Green's function.

To illustrate the possible techniques, consider a Green's function that we found in §5.4. There we showed that the Green's function for Helmholtz's equation with Dirichlet boundary conditions on the rectangular strip $0 < x < a$ and $-\infty < y < \infty$ is

$$
g(x, y|\xi, \eta) = \frac{i}{4} \sum_{n=-\infty}^{\infty} \left\{ H_0^{(1)} \left[k_0 \sqrt{(x - \xi - 2na)^2 + (y - \eta)^2} \right] \right.
$$
$$
\left. - H_0^{(1)} \left[k_0 \sqrt{(x + \xi + 2na)^2 + (y - \eta)^2} \right] \right\},
$$

$$
\text{(5.8.1)}
$$

or

$$
g(x, y|\xi, \eta) = \frac{1}{a} \sum_{n=1}^{\infty} \sin\left(\frac{n\pi\xi}{a}\right) \sin\left(\frac{n\pi x}{a}\right) \frac{e^{-\sqrt{n^2\pi^2/a^2 - k_0^2}\,|y-\eta|}}{\sqrt{n^2\pi^2/a^2 - k_0^2}}.
$$

$$
\text{(5.8.2)}
$$

Under certain conditions (5.8.1) or (5.8.2) may converge slowly and we must seek an alternative expression if we wish to evaluate it efficiently.

---

[36] See, for example, Wallace, P. R., 1984: *Mathematical Analysis of Physical Problems*. Dover Publications, Inc., pp. 132–134.

[37] An excellent review of these methods as they apply to the two-dimensional Helmholtz equation in periodic domains is given by Linton, C. M., 1998: The Green's function for the two-dimensional Helmholtz equation in periodic domains. *J. Engng. Math.*, **33**, 377–402. See also Dienstfrey, A., F. Hang, and J. Huang, 2001: Lattice sums and the two-dimensional, periodic Green's function for the Helmholtz equation. *Proc. Royal Soc. London*, **A457**, 67–86.

• *Poisson's summation formula*

Poisson's summation formula can sometimes be used to convert a slowly converging series into a rapidly converging one. The idea here is the reciprocal spreading property of the Fourier transform: If a function $f(x)$ tends slowly to zero as $|x| \to \infty$, then its Fourier transform will tend rapidly to zero in the frequency domain. Poisson's summation formula can be written

$$\sum_{n=-\infty}^{\infty} f(t + nT) = \frac{1}{T} \sum_{n=-\infty}^{\infty} F(n\omega)e^{in\omega t}, \qquad (5.8.3)$$

where $\omega = 2\pi/T$, and $F(\ )$ is the Fourier transform of $f(\ )$.

Let us apply Poisson's summation formula to (5.8.1). A direct application of (5.8.3) yields

$$g(x, y|\xi, \eta) = \frac{i}{4} \sum_{n=-\infty}^{\infty} \left\{ H_0^{(1)} \left[k_0 \sqrt{(x - \xi - 2na)^2 + (y - \eta)^2} \right] \right.$$
$$\left. - H_0^{(1)} \left[k_0 \sqrt{(x + \xi + 2na)^2 + (y - \eta)^2} \right] \right\}$$

$$(5.8.4)$$

$$= \frac{1}{4a} \sum_{n=-\infty}^{\infty} \frac{e^{-\sqrt{n^2\pi^2/a^2 - k_0^2}\ |y-\eta|}}{\sqrt{n^2\pi^2/a^2 - k_0^2}} e^{-n\pi(x-\xi)i/a}$$
$$- \frac{e^{-\sqrt{n^2\pi^2/a^2 - k_0^2}\ |y-\eta|}}{\sqrt{n^2\pi^2/a^2 - k_0^2}} e^{-n\pi(x+\xi)i/a} \qquad (5.8.5)$$

$$= \frac{1}{a} \sum_{n=1}^{\infty} \sin\left(\frac{n\pi\xi}{a}\right) \sin\left(\frac{n\pi x}{a}\right) \frac{e^{-\sqrt{n^2\pi^2/a^2 - k_0^2}\ |y-\eta|}}{\sqrt{n^2\pi^2/a^2 - k_0^2}}, \qquad (5.8.6)$$

and we obtain (5.8.2).

• *Kummer's transformation*

Kummer's transformation applies the principle that the rate of convergence of a series is governed by the asymptotic form of that series. Therefore, this method divides the slowly convergent series into two parts. One series has the same asymptotic behavior as the original series. The other series contains the original series minus the asymptotic series; this new series should converge very rapidly. Sometimes, the asymptotic series can be summed to give a known closed-form expression. If not, the convergence of the asymptotic series can be accelerated by applying Poisson's summation formula.[38] Mathematically, we express Kummer's transformation as

$$\sum_{n=-\infty}^{\infty} f(n) = \sum_{n=-\infty}^{\infty} [f(n) - f_a(n)] + \sum_{n=-\infty}^{\infty} f_a(n), \qquad (5.8.7)$$

where $f_a(n)$ is asymptotic to the function $f(n)$.

Applying the technique to our pet problem, we have

$$
\begin{aligned}
g(x, y|\xi, \eta) = \frac{1}{a} \sum_{n=1}^{\infty} &\left[\frac{e^{-\sqrt{n^2\pi^2/a^2 - k_0^2}\,|y-\eta|}}{\sqrt{n^2\pi^2/a^2 - k_0^2}} \right. \\
&\left. - \frac{ae^{-n\pi|y-\eta|/a}}{n\pi}\left(1 + \frac{k_0^2|y-\eta|a}{2n\pi}\right) \right] \\
&\times \sin\left(\frac{n\pi\xi}{a}\right)\sin\left(\frac{n\pi x}{a}\right) \\
+ \sum_{n=1}^{\infty} &\frac{e^{-n\pi|y-\eta|/a}}{n\pi}\left(1 + \frac{k_0^2|y-\eta|a}{2n\pi}\right)\sin\left(\frac{n\pi\xi}{a}\right)\sin\left(\frac{n\pi x}{a}\right).
\end{aligned}
$$
$$(5.8.8)$$

Our choice of adding and subtracting the series

$$\sum_{n=1}^{\infty} \frac{e^{-n\pi|y-\eta|/a}}{n\pi}\left(1 + \frac{k_0^2|y-\eta|a}{2n\pi}\right)\sin\left(\frac{n\pi\xi}{a}\right)\sin\left(\frac{n\pi x}{a}\right) \qquad (5.8.9)$$

is motivated by the fact that

$$\left(\frac{n^2\pi^2}{a^2} - k_0^2\right)^{1/2} = \frac{n\pi}{a}\left(1 - \frac{k_0^2 a^2}{2n^2\pi^2}\right) + O\left(n^{-3}\right), \qquad (5.8.10)$$

$$\left(\frac{n^2\pi^2}{a^2} - k_0^2\right)^{-1/2} = \frac{a}{n\pi}\left(1 + \frac{k_0^2 a^2}{2n^2\pi^2}\right) + O\left(n^{-5}\right), \qquad (5.8.11)$$

and

$$e^{-\sqrt{n^2\pi^2/a^2 - k_0^2}\,|y-\eta|} = e^{-n\pi|y-\eta|/a}\left(1 + \frac{k_0^2|y-\eta|a}{2n\pi}\right) + O\left(n^{-2}\right), \qquad (5.8.12)$$

as $n \to \infty$.

---

[38] See, for example, Lampe, R., P. Klock, and P. Mayes, 1985: Integral transforms useful for the accelerated summation of periodic, free-space Green's functions. *IEEE Trans. Microwave Theory Tech.*, **MTT-33**, 734–736.

Next, we introduce the polylogarithm function Li_s, which is defined as follows:

$$\mathrm{Li}_s(z) = \sum_{n=1}^{\infty} \frac{z^n}{n^s}, \qquad \mathrm{Li}_1(z) = -\log(1-z), \tag{5.8.13}$$

which converges for $|z| < 1$. This function is useful because it can be evaluated quickly and with high accuracy.[39] Applying this function to our analysis, we have that

$$\sum_{n=1}^{\infty} \frac{e^{-n\pi|y-\eta|/a}}{n\pi} = -\frac{1}{4\pi}\left[\mathrm{Li}_1\left(e^{-\pi z_+/a}\right) + \mathrm{Li}_1\left(e^{-\pi z_+^*/a}\right) \right.$$
$$\left. - \mathrm{Li}_1\left(e^{-\pi z_-/a}\right) - \mathrm{Li}_1\left(e^{-\pi z_-^*/a}\right) \right], \tag{5.8.14}$$

and

$$\sum_{n=1}^{\infty} \frac{e^{-n\pi|y-\eta|/a}}{n^2\pi^2} = -\frac{1}{4\pi^2}\left[\mathrm{Li}_2\left(e^{-\pi z_+/a}\right) + \mathrm{Li}_2\left(e^{-\pi z_+^*/a}\right) \right.$$
$$\left. - \mathrm{Li}_2\left(e^{-\pi z_-/a}\right) - \mathrm{Li}_2\left(e^{-\pi z_-^*/a}\right) \right], \tag{5.8.15}$$

where $z_{\pm} = |y-\eta| + i(x \pm \xi)$. Equations (5.8.14)–(5.8.15) are then used to evaluate the second series in (5.8.8), while the first summation converges very rapidly.

• Example 5.8.1: Melnikov's method

Melnikov[40] developed a variation on Kummer's transformation to accelerate the convergence of Green's function representations for Laplace's equation. To illustrate this technique, consider the Green's function that we found for Laplace's equation in a rectangular domain $(0 < x < a, 0 < y < b)$ when Dirichlet boundary conditions are imposed along the entire boundary. We showed in §5.2 that the Green's function is

$$g(x,y|\xi,\eta) = \frac{2}{\pi} \sum_{n=1}^{\infty} \frac{\sinh[n\pi(a-x_>)/b]\sinh(n\pi x_</b)}{n\;\sinh(n\pi a/b)}$$
$$\times \sin\left(\frac{n\pi\eta}{b}\right)\sin\left(\frac{n\pi y}{b}\right). \tag{5.8.16}$$

---

[39] See the appendix in Nicorovici, N. A., R. C. McPhedran, and R. Petit, 1994: Efficient calculation of the Green's function for electromagnetic scattering by gratings. *Phys. Review E*, **49**, 4563–4577.

[40] Melnikov, Yu. A., 1995: *Green's Functions in Applied Mechanics*. WIT Press/Computational Mechanics Publications, §1.2.

We begin by first rewriting (5.8.16) as

$$
g(x, y|\xi, \eta) = \frac{1}{\pi} \sum_{n=1}^{\infty} \frac{1}{n} \left[e^{-n\pi|x-\xi|/b} - e^{-n\pi(x+\xi)/b} \right] \sin\left(\frac{n\pi\eta}{b}\right) \sin\left(\frac{n\pi y}{b}\right)
$$

$$
- \frac{2}{\pi} \sum_{n=1}^{\infty} \frac{\sinh(n\pi\xi/b) \sinh(n\pi x/b)}{n\, e^{n\pi a/b} \sinh(n\pi a/b)} \sin\left(\frac{n\pi\eta}{b}\right) \sin\left(\frac{n\pi y}{b}\right).
$$

(5.8.17)

The first summation can be re-expressed in terms of logarithms using a technique given in §5.4. Once this done, we can rewrite (5.8.17) as

$$
g(x, y|\xi, \eta) = \frac{1}{2\pi} \ln\left[\frac{E(z - \zeta^*)E(z + \zeta^*)}{E(z - \zeta)E(z + \zeta)} \right]
$$

$$
- \frac{2}{\pi} \sum_{n=1}^{\infty} \frac{\sinh(n\pi\xi/b) \sinh(n\pi x/b)}{n\, e^{n\pi a/b} \sinh(n\pi a/b)} \sin\left(\frac{n\pi\eta}{b}\right) \sin\left(\frac{n\pi y}{b}\right),
$$

(5.8.18)

where $z = x + iy$, $\zeta = \xi + i\eta$, $\zeta^* = \xi - i\eta$ and $E(z) = |e^{\pi z/b} - 1|$.

Although (5.8.18) clearly shows the logarithmic singularity of the Green's function, it converges slowly when both x and ξ lie near the edge $x = a$. To understand why, we rewrite the Fourier coefficient in (5.8.18) as follows:

$$
\frac{\sinh(n\pi\xi/b) \sinh(n\pi x/b)}{n\, e^{n\pi a/b} \sinh(n\pi a/b)} = \frac{\cosh[n\pi(x + \xi)/b] - \cosh[n\pi(x - \xi)/b]}{2n\, e^{n\pi a/b} \sinh(n\pi a/b)}
$$

(5.8.19)

$$
= \frac{\cosh[n\pi(x + \xi)/b] - \cosh[n\pi(x - \xi)/b]}{n\, (e^{2n\pi a/b} - 1)}
$$

(5.8.20)

$$
= \frac{\cosh[n\pi(x + \xi)/b]}{n\, (e^{2n\pi a/b} - 1)} - \frac{\cosh[n\pi(x - \xi)/b]}{n\, (e^{2n\pi a/b} - 1)}.
$$

(5.8.21)

Let us look more closely at the first term in (5.8.21). If $x + \xi \approx 2a$, then this term behaves like n^{-1} and the Fourier series in (5.8.18) converges slowly. To eliminate this problem, we break the Fourier coefficient into two parts; the first part contains the troublesome factor, while the second part contains everything else. Therefore, we write the Fourier coefficient as

$$
\frac{\sinh(n\pi\xi/b) \sinh(n\pi x/b)}{n\, e^{n\pi a/b} \sinh(n\pi a/b)} = \frac{\cosh[n\pi(x + \xi)/b]}{n\, (e^{2n\pi a/b} - 1)} - \frac{\cosh[n\pi(x + \xi)/b]}{n\, e^{2n\pi a/b}}
$$

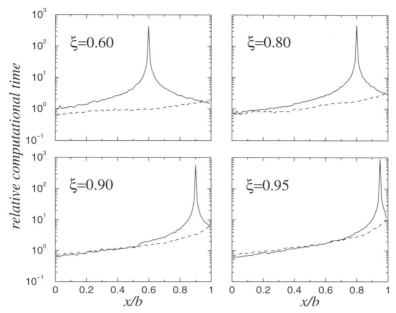

Figure 5.8.1: The relative computational time to compute the Green's functions (5.8.16), (5.8.18) and (5.8.30) with a relative error less than 10^{-6} as a function of x for various ξ's. The solid line gives the relative timing for (5.8.16) against (5.8.30), while the dashed line gives (5.8.18). The timings are based on 99 points in the y direction with $a = b$ and $\xi = \eta$. The point $x = \xi$ and $y = \eta$ was excluded.

$$+ \frac{\cosh[n\pi(x + \xi)/b]}{n\, e^{2n\pi a/b}} - \frac{\cosh[n\pi(x - \xi)/b]}{n\left(e^{2n\pi a/b} - 1\right)}$$

$$\text{(5.8.22)}$$

$$= \frac{\cosh[n\pi(x + \xi)/b]}{n\, e^{2n\pi a/b}} + S_n(x, \xi), \quad \text{(5.8.23)}$$

where

$$S_n(x, \xi) = \frac{e^{n\pi x/b}\sinh[n\pi(\xi - a)/b] - e^{-n\pi x/b}\sinh[n\pi(\xi + a)/b]}{2n\, e^{2n\pi a/b}\sinh(n\pi a/b)}.$$

$$\text{(5.8.24)}$$

Therefore,

$$\frac{2}{\pi}\sum_{n=1}^{\infty}\frac{\sinh(n\pi\xi/b)\sinh(n\pi x/b)}{n\, e^{n\pi a/b}\sinh(n\pi a/b)}\sin\left(\frac{n\pi\eta}{b}\right)\sin\left(\frac{n\pi y}{b}\right)$$

$$= \frac{2}{\pi}\sum_{n=1}^{\infty}\frac{\cosh[n\pi(x + \xi)/b]}{n\, e^{2n\pi a/b}}\sin\left(\frac{n\pi\eta}{b}\right)\sin\left(\frac{n\pi y}{b}\right)$$

$$+ \frac{2}{\pi}\sum_{n=1}^{\infty} S_n(x, \xi)\sin\left(\frac{n\pi\eta}{b}\right)\sin\left(\frac{n\pi y}{b}\right). \quad \text{(5.8.25)}$$

The second term in (5.8.25) converges uniformly and we need not consider it further. On the other hand, the first term in (5.8.25) can be summed by first noting that

$$\frac{2}{\pi} \sum_{n=1}^{\infty} \frac{e^{n\pi(x+\xi)/b} + e^{-n\pi(x+\xi)/b}}{2n\, e^{2n\pi a/b}} \sin\left(\frac{n\pi\eta}{b}\right) \sin\left(\frac{n\pi y}{b}\right)$$

$$= \frac{1}{2\pi} \sum_{n=1}^{\infty} \frac{1}{n} \left[e^{n\pi(x+\xi-2a)/b} + e^{-n\pi(x+\xi+2a)/b} \right]$$

$$\times \left\{ \cos\left[\frac{n\pi(y-\eta)}{b}\right] - \cos\left[\frac{n\pi(y+\eta)}{b}\right] \right\}. \quad (\mathbf{5.8.26})$$

Upon applying

$$\sum_{n=1}^{\infty} \frac{q^n}{n} \cos(n\alpha) = -\ln\left[\sqrt{1 - 2q\cos(\alpha) + q^2}\right], \quad (\mathbf{5.8.27})$$

the right side of (5.8.26) becomes

$$\frac{1}{2\pi} \ln\left\{ \sqrt{\frac{1 - 2e^{\pi(x+\xi-2a)/b}\cos[\pi(y+\eta)/b] + e^{2\pi(x+\xi-2a)/b}}{1 - 2e^{\pi(x+\xi-2a)/b}\cos[\pi(y-\eta)/b] + e^{2\pi(x+\xi-2a)/b}}} \right.$$

$$\left. \times \sqrt{\frac{1 - 2e^{-\pi(x+\xi+2a)/b}\cos[\pi(y+\eta)/b] + e^{-2\pi(x+\xi+2a)/b}}{1 - 2e^{-\pi(x+\xi+2a)/b}\cos[\pi(y-\eta)/b] + e^{-2\pi(x+\xi+2a)/b}}} \right\}.$$

$$(\mathbf{5.8.28})$$

If we now multiply the numerator and denominator of the second radical by $e^{2\pi(x+\xi+2a)/b}$, (5.8.26) becomes

$$\frac{2}{\pi} \sum_{n=1}^{\infty} \frac{\cosh[n\pi(x+\xi)/b]}{n\, e^{2n\pi a/b}} \sin\left(\frac{n\pi\eta}{b}\right) \sin\left(\frac{n\pi y}{b}\right)$$

$$= \frac{1}{2\pi} \ln\left[\frac{E(z_1 + \zeta_1)E(z_2 + \zeta_2)}{E(z_1 + \zeta_1^*)E(z_2 + \zeta_2^*)} \right], \quad (\mathbf{5.8.29})$$

where $z_1 = (x+a)+iy$, $z_2 = (x-a)+iy$, $\zeta_1 = (\xi+a)+i\eta$, $\zeta_2 = (\xi-a)+i\eta$, $\zeta_1^* = (\xi+a)-i\eta$, $\zeta_2^* = (\xi-a)-i\eta$. Finally, if we substitute (5.8.25) and (5.8.29) into (5.8.18), we obtain the final result

$$g(x,y|\xi,\eta) = \frac{1}{2\pi} \ln\left[\frac{E(z-\zeta^*)E(z+\zeta^*)E(z_1+\zeta_1^*)E(z_2+\zeta_2^*)}{E(z-\zeta)E(z+\zeta)E(z_1+\zeta_1)E(z_2+\zeta_2)} \right]$$

$$- \frac{2}{\pi} \sum_{n=1}^{\infty} S_n(x,\xi) \sin\left(\frac{n\pi\eta}{b}\right) \sin\left(\frac{n\pi y}{b}\right). \quad (\mathbf{5.8.30})$$

To illustrate our analysis, the relative computational times to calculate the Green's functions given by (5.8.16), (5.8.18) and (5.8.30) have been computed as a function of x for various ξ's. As Figure 5.8.1 shows, the traditional Green's function (5.8.16) is often at a competitive disadvantage compared with (5.8.18) or (5.8.30), which explicitly resolve the logarithmic singularity. Finally, (5.8.30) is superior when both x and ξ are near the boundary $x = a$, as expected.

• *Integral representation*

As we just showed, we can transform a series representation of a Green's function by Kummer's transformation into one where the sum converges more rapidly. Our third technique uses the integral representation of special functions to convert the Green's function into an improper integral whose integrand exhibits exponential decay.

Turning again to our example of the Green's function for a two-dimensional Helmholtz equation over the rectangular strip $0 < x < a$, we begin by employing (5.4.37) or

$$g(x, y|\xi, \eta) = g_0(x, y|\xi, \eta) - g_0(x, y|-\xi, \eta), \tag{5.8.31}$$

where

$$g_0(x, y|\xi, \eta) = \frac{i}{4} \sum_{n=-\infty}^{\infty} H_0^{(1)}\left[k_0\sqrt{(x - \xi - 2na)^2 + (y - \eta)^2}\right]. \tag{5.8.32}$$

Using the result that

$$e^{-ib} H_0^{(1)}\left(\sqrt{a^2 + b^2}\right) = -\frac{2i}{\pi} \int_0^{\infty} \frac{e^{-bu} \cos\left(a\sqrt{u^2 - 2iu}\right)}{\sqrt{u^2 - 2iu}} \, du, \quad b \geq 0, \tag{5.8.33}$$

with $a = k_0|y - \eta|$, and $b = k_0[2na - (x - \xi)]$, then

$$\sum_{n=1}^{\infty} H_0^{(1)}\left[k_0\sqrt{(x - \xi - 2na)^2 + (y - \eta)^2}\right]$$

$$= -\frac{2i}{\pi} \sum_{n=1}^{\infty} e^{ik_0[2na-(x-\xi)]}$$

$$\times \int_0^{\infty} \frac{e^{-k_0[2na-(x-\xi)]u} \cos\left(k_0|y - \eta|\sqrt{u^2 - 2iu}\right)}{\sqrt{u^2 - 2iu}} \, du \tag{5.8.34}$$

$$= -\frac{2i}{\pi} e^{-ik_0(x-\xi)} \int_0^{\infty} \frac{e^{k_0(x-\xi-2a)u} \cos\left(k_0|y - \eta|\sqrt{u^2 - 2iu}\right)}{\left(e^{-2ik_0a} - e^{-2k_0au}\right)\sqrt{u^2 - 2iu}} \, du, \tag{5.8.35}$$

if $2a \geq x - \xi$. Similarly,

$$\sum_{n=-\infty}^{-1} H_0^{(1)}\left[k_0\sqrt{(x-\xi-2na)^2+(y-\eta)^2}\right]$$

$$= \sum_{m=1}^{\infty} H_0^{(1)}\left[k_0\sqrt{(x-\xi+2ma)^2+(y-\eta)^2}\right] \qquad (5.8.36)$$

$$= -\frac{2i}{\pi}e^{ik_0(x-\xi)}\int_0^\infty \frac{e^{-k_0(x-\xi+2a)u}\cos\left(k_0|y-\eta|\sqrt{u^2-2iu}\right)}{\left(e^{-2ik_0a}-e^{-2k_0au}\right)\sqrt{u^2-2iu}}\,du, \qquad (5.8.37)$$

if $x - \xi \geq -2a$. Therefore, for $-2a < x - \xi < 2a$,

$$g_0(x,y|\xi,\eta) = \frac{i}{4}H_0^{(1)}\left[k_0\sqrt{(x-\xi)^2+(y-\eta)^2}\right]$$

$$+ \frac{1}{2\pi}\int_0^\infty \frac{e^{k_0(x-\xi)(u-i)}\cos\left(k_0|y-\eta|\sqrt{u^2-2iu}\right)}{\left[e^{2k_0a(u-i)}-1\right]\sqrt{u^2-2iu}}\,du,$$

$$+ \frac{1}{2\pi}\int_0^\infty \frac{e^{-k_0(x-\xi)(u-i)}\cos\left(k_0|y-\eta|\sqrt{u^2-2iu}\right)}{\left[e^{2k_0a(u-i)}-1\right]\sqrt{u^2-2iu}}\,du. \qquad (5.8.38)$$

Substituting (5.8.38) into (5.8.31), we obtain

$$g(x,y|\xi,\eta) = \frac{i}{4}H_0^{(1)}\left[k_0\sqrt{(x-\xi)^2+(y-\eta)^2}\right]$$

$$- \frac{i}{4}H_0^{(1)}\left[k_0\sqrt{(x+\xi)^2+(y-\eta)^2}\right]$$

$$- \frac{2}{\pi}\int_0^\infty \frac{\sinh[k_0x(u-i)]\sinh[k_0\xi(u-i)]}{e^{2k_0a(u-i)}-1}$$

$$\times \frac{\cos\left(k_0|y-\eta|\sqrt{u^2-2iu}\right)}{\sqrt{u^2-2iu}}\,du. \qquad (5.8.39)$$

The integral in (5.8.39) has an integrable singularity at $u = 0$ and the integrand decays exponentially as $u \to \infty$ provided $|x - \xi| < 2a$.

An alternative to (5.8.39) can be found by setting $t = \sqrt{u^2-2iu}$ or $u = i + \gamma$, where $\gamma = (t^2-1)^{1/2} = -i(1-t^2)^{1/2}$. The limits of integration change from 0 to ∞ into 0 to $\infty - i$. Substitution yields

$$g(x,y|\xi,\eta) = \frac{i}{4}H_0^{(1)}\left[k_0\sqrt{(x-\xi)^2+(y-\eta)^2}\right]$$

$$- \frac{i}{4}H_0^{(1)}\left[k_0\sqrt{(x+\xi)^2+(y-\eta)^2}\right]$$

$$-\frac{2}{\pi}\oint_0^\infty \frac{\sinh(k_0\gamma x)\sinh(k_0\gamma\xi)}{\gamma\left(e^{2k_0\gamma a}-1\right)}\cos(k_0|y-\eta|t)\,dt \quad \textbf{(5.8.40)}$$

$$=\frac{1}{\pi}\oint_0^\infty \left[\cosh(k_0\gamma x_>) - \coth(k_0\gamma a)\sinh(k_0\gamma x_>)\right]$$

$$\times \sinh(k_0\gamma x_<)\frac{\cos(k_0|y-\eta|t)}{\gamma}\,dt. \quad \textbf{(5.8.41)}$$

The special integral sign denotes an integration along the real t-axis except at the poles where it passes beneath them. This special contour arises because we deformed the original contour from a line between 0 and $\infty - i$ back to the real axis. Finally, (5.8.41) was derived using the relationship that

$$H_0^{(1)}\left(k_0\sqrt{a^2+b^2}\right) = -\frac{2i}{\pi}\int_0^\infty \frac{e^{-k_0\gamma|b|}}{\gamma}\cos(k_0at)\,dt. \quad \textbf{(5.8.42)}$$

• *Ewald's method*

In 1986, Jordan *et al.*[41] introduced a new method, originally used by P. P. Ewald[42] to improve the convergence of power series arising in the analysis of crystal lattices, to accelerate the rate of convergence of three-dimensional, periodic Green's functions. This technique has been subsequently applied by several investigators on problems involving three-dimensional waveguides. One crucial step in these analyses is the replacement of e^{ik_0R}/R with the integral representation

$$\frac{2}{\sqrt{\pi}}\int_0^\infty \exp\left(-R^2s^2 + \frac{k_0^2}{4s^2}\right)ds$$

with a suitably chosen path of integration. This would appear to limit Ewald's technique to three-dimensional problems where the eigenfunction expansion can be rewritten to take advantage of this integral representation.

Recently, Linton[43] showed how we can obtain Ewald sums by recasting the problem as a heat conduction problem. This method originated with Strain,[44] who applied it to Laplace's equation within a closed

[41] Jordan, K. E., G. R. Richter, and P. Sheng, 1986: An efficient numerical evaluation of the Green's function for the Helmholtz operator on periodic structures. *J. Comput. Phys.*, **63**, 222–235.

[42] Ewald, P. P., 1921: Die Berechnung optischer und elektrischer Gitterpotentiale. *Ann. Phys., 4te Folge*, **64**, 253–287.

[43] Linton, C. M., 1999: Rapidly convergent representations for Green's functions for Laplace's equation. *Proc. Royal Soc. London*, **A455**, 1767–1797.

[44] Strain, J., 1992: Fast potential theory. II. Layer potentials and discrete sums. *J. Comput. Phys.*, **99**, 251–270.

domain that has a Robin boundary condition. In his paper, Linton re-
stricted himself to Laplace's equation, but it is clear that his analysis
can be applied just as well to Helmholtz's equation. It is this approach
that we will use to illustrate Ewald's method.

Consider the initial-value problem

$$u_t = \nabla^2 u + k_0^2 u, \qquad 0 < t, \qquad (5.8.43)$$

where ∇ is the three-dimensional gradient operator. We take as our
initial condition

$$u(\mathbf{x}, 0) = \delta(\mathbf{x} - \mathbf{x}_0), \qquad (5.8.44)$$

where $\mathbf{x}$ is the position vector and $\mathbf{x}_0$ is the point of excitation.

Following Linton, we presently assume that the domain is unbounded in at least one direction; the technique would work just as
well on a closed domain. If the domain is of limited extent in other directions, the corresponding boundary conditions are time independent.

Linton's first insight was to observe that the Green's function governed by

$$\nabla^2 g + k_0^2 g = -\delta(\mathbf{x} - \mathbf{x}_0) \qquad (5.8.45)$$

for the same domain and boundary conditions that govern (5.8.43)
equals

$$g(\mathbf{x}|\mathbf{x}_0) = \int_0^\infty u(\mathbf{x}, t)\, dt, \qquad (5.8.46)$$

provided that the integral exists. Why have we introduced (5.8.43)–
(5.8.44)? This new initial-boundary-value problem is generally more
difficult to solve than (5.8.45). This leads us to Linton's second insight.
Let us split (5.8.46) into two parts:

$$g(\mathbf{x}|\mathbf{x}_0) = \int_0^{b^2} u(\mathbf{x}, t)\, dt + \int_{b^2}^\infty u(\mathbf{x}, t)\, dt, \qquad (5.8.47)$$

where b^2 is a free parameter. If we could find two representations of
u such that the first one, u_1, converges rapidly for small values of t
and the second one, u_2, converges rapidly for large t, then the resulting
expansion would converge more rapidly than using a single u for all t.
It is this splitting of (5.8.47) that will lead to Ewald's sums. In general,
we will find that u_1 equals a sum of free-space Green's functions (image
series) while u_2 equals the eigenfunction expansion. As we will see,
there is an optimal value for b^2 that minimizes the total number of
terms needed for a desired accuracy.

Let us illustrate this method by finding Ewald summations for the
Green's function for Helmholtz's equation over the rectangular strip
$0 < x < a$ and $-\infty < y < \infty$; this is the problem that we posed at the
beginning of §5.4.

We begin by first considering the initial-value problem

$$\frac{\partial u}{\partial t} = \frac{\partial^2 u}{\partial x^2} + \frac{\partial^2 u}{\partial y^2} + k_0^2 u, \quad 0 < x < a, \quad -\infty < y < \infty, \quad \textbf{(5.8.48)}$$

with the boundary conditions that

$$u(0, y, t) = u(a, y, t) = 0, \quad -\infty < y < \infty, \quad 0 < t, \quad \textbf{(5.8.49)}$$

and

$$\lim_{|y| \to \infty} |u(x, y, t)| < \infty, \quad 0 < x < a, \quad 0 < t, \quad \textbf{(5.8.50)}$$

and the initial condition

$$u(x, y, 0) = \delta(x - \xi)\delta(y - \eta), \quad 0 < \xi < a, \quad -\infty < \eta < \infty. \quad \textbf{(5.8.51)}$$

This system (5.8.48)–(5.8.51) can be solved by using Laplace transforms in time, Fourier series in the x direction and Fourier transforms in the y direction. The solution is

$$u(x, y, t) = \frac{1}{a} \sum_{n=1}^{\infty} \sin\left(\frac{n\pi\xi}{a}\right) \sin\left(\frac{n\pi x}{a}\right)$$
$$\times \frac{\exp[-(n^2\pi^2/a^2 - k_0^2)t - (y - \eta)^2/(4t)]}{\sqrt{\pi t}}. \quad \textbf{(5.8.52)}$$

Here, we assume that $\pi/a > k_0$. For those cases when $k_0 > \pi/a$, a certain number of the terms in (5.8.52) must be modified to take into account the negative argument of the radical. In any case, we find $g(x, y|\xi, \eta)$ by integrating in time,

$$g(x, y|\xi, \eta) = \frac{1}{a} \sum_{n=1}^{\infty} \sin\left(\frac{n\pi\xi}{a}\right) \sin\left(\frac{n\pi x}{a}\right)$$
$$\times \int_0^{\infty} \frac{\exp[-(n^2\pi^2/a^2 - k_0^2)t - (y - \eta)^2/(4t)]}{\sqrt{\pi t}} \, dt \quad \textbf{(5.8.53)}$$
$$= \frac{2}{a\sqrt{\pi}} \sum_{n=1}^{\infty} \sin\left(\frac{n\pi\xi}{a}\right) \sin\left(\frac{n\pi x}{a}\right)$$
$$\times \int_0^{\infty} \exp\left[-(n^2\pi^2/a^2 - k_0^2)\tau^2 - \frac{(y - \eta)^2}{4\tau^2}\right] d\tau, \quad \textbf{(5.8.54)}$$

where $t = \tau^2$. Upon carrying out the integration,[45] we obtain the eigenfunction expansion (5.4.32).

[45] Gradshteyn, I. S., and I. M. Ryzhik, 1965: *Tables of Integrals, Series, and Products*. Academic Press, formula 3.325.

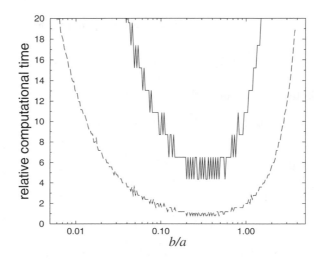

Figure 5.8.2: The relative computational time (in percentage) between using the Ewald sums as opposed to using the eigenfunction expansion as a function of b. The solid and dashed lines correspond to $|y - \eta|/a = 0.001$ and 0.0001, respectively. Here $k_0 a = 1$, and $\xi/a = 0.5$.

Let us examine (5.8.52) more closely. As t increases, the number of terms needed to accurately compute $u(x, y, t)$ decreases dramatically because each term behaves like $e^{-n^2 t}$ as $n \to \infty$. However, as $t \to 0$, we must include many terms when $|y - \eta| \ll 1$. Therefore, although (5.8.52) satisfies our needs for u_2, we must find an alternative expression that converges rapidly for small t for use as u_1.

We begin by noting that (5.8.52) can be rewritten

$$
u(x,y,t) = \sum_{n=-\infty}^{\infty} \frac{\exp[n\pi(x-\xi)i/a - (n^2\pi^2/a^2 - k_0^2)t - (y-\eta)^2/(4t)]}{4a\sqrt{\pi t}}
$$
$$
- \sum_{n=-\infty}^{\infty} \frac{\exp[n\pi(x+\xi)i/a - (n^2\pi^2/a^2 - k_0^2)t - (y-\eta)^2/(4t)]}{4a\sqrt{\pi t}}.
$$

$$(5.8.55)$$

The summations

$$
\sum_{n=-\infty}^{\infty} e^{n\pi(x\pm\xi)i/a - n^2\pi^2 t/a^2}
$$

in (5.8.55) can be replaced by

$$
\frac{a}{\sqrt{\pi t}} \sum_{n=-\infty}^{\infty} \exp\left[-\frac{(x \pm \xi - 2na)^2}{4t} \right]
$$

by Poisson's summation formula. Thus, we rewrite (5.8.53) as

$$g(x,y|\xi,\eta) = \frac{1}{4\pi} \sum_{n=-\infty}^{\infty} \int_{1/b^2}^{\infty} e^{k_0^2/\tau - [(x-\xi-2na)^2 + (y-\eta)^2]\tau} \frac{d\tau}{\tau}$$

$$- \frac{1}{4\pi} \sum_{n=-\infty}^{\infty} \int_{1/b^2}^{\infty} e^{k_0^2/\tau - [(x+\xi-2na)^2 + (y-\eta)^2]\tau} \frac{d\tau}{\tau}$$

$$+ \frac{2}{a\sqrt{\pi}} \sum_{n=1}^{\infty} \sin\left(\frac{n\pi\xi}{a}\right) \sin\left(\frac{n\pi x}{a}\right)$$

$$\times \int_b^{\infty} \exp\left[-\left(\frac{n^2\pi^2}{a^2} - k_0^2\right)\tau^2 - \frac{(y-\eta)^2}{4\tau^2}\right] d\tau,$$

$$(5.8.56)$$

where we have used $\tau = 1/t$ in the first two sums. Carrying out the integrations, we obtain

$$g(x,y|\xi,\eta) = \frac{1}{4\pi} \sum_{n=-\infty}^{\infty} \sum_{m=0}^{\infty} \frac{(k_0 b)^{2m}}{m!} E_{m+1}\left[\frac{(x-\xi-2na)^2 + (y-\eta)^2}{4b^2}\right]$$

$$- \frac{1}{4\pi} \sum_{n=-\infty}^{\infty} \sum_{m=0}^{\infty} \frac{(k_0 b)^{2m}}{m!} E_{m+1}\left[\frac{(x+\xi-2na)^2 + (y-\eta)^2}{4b^2}\right]$$

$$+ \frac{1}{2a} \sum_{n=1}^{\infty} \frac{\sin(n\pi\xi/a)\sin(n\pi x/a)}{\sqrt{n^2\pi^2/a^2 - k_0^2}}$$

$$\times \left[e^{|y-\eta|\sqrt{n^2\pi^2/a^2 - k_0^2}} \operatorname{erfc}\left(b\sqrt{\frac{n^2\pi^2}{a^2} - k_0^2} + \frac{|y-\eta|}{2b}\right)\right.$$

$$\left. + e^{-|y-\eta|\sqrt{n^2\pi^2/a^2 - k_0^2}} \operatorname{erfc}\left(b\sqrt{\frac{n^2\pi^2}{a^2} - k_0^2} - \frac{|y-\eta|}{2b}\right)\right],$$

$$(5.8.57)$$

where erfc( ) denotes the complementary error function and $E_m(\ )$ is the exponential integral.

We show in Figure 5.8.2 the relative computational time between using the Ewald sums (5.8.57) as opposed to using the eigenfunction expansion (5.4.52) as a function of b. To obtain the ratio, we averaged the values given at 100 sample points between $0 < x < a$.

• *Mittag-Leffler's expansion theorem*

Consider the meromorphic function $F(z)$. Mittag-Leffler's expansion theorem[46] states that

$$F(z) = F(0) + \sum_{m=1}^{\infty} R_m \left(\frac{1}{z - z_m} - \frac{1}{z_m} \right), \qquad (5.8.58)$$

where z_m and R_m denote the location and residue of the mth pole of $F(z)$, respectively. Alhargan and Judah[19] used this theorem to reduce the double Fourier series that often arise in modal expansions for Green's function for Helmholtz's equation down to a single series. An additional benefit is the elimination of the eigenvalue problem. This technique is a generalization of a method developed by Carslaw[47] of using Fourier-Bessel eigenfunction expansions of simple analytical functions to simplify various three-dimensional Green's functions.

To illustrate this procedure, consider the complex function

$$F_n(z) = \frac{\pi a z J_n(r_< z) f(r_> z, az)}{4 J_n'(az)} + \frac{\delta_n}{az}, \qquad (5.8.59)$$

where $\delta_0 = 1$, $\delta_n = 0$ if $n > 0$, and

$$f(rz, az) = Y_n'(az) J_n(rz) - J_n'(az) Y_n(rz). \qquad (5.8.60)$$

From now on, let $\rho \neq 0$. Therefore, from Mittag-Leffler's expansion theorem,

$$F_n(z) = F_n(0) + \sum_{m=1}^{\infty} R_m \left(\frac{2z}{z^2 - k_{nm}^2} \right), \qquad (5.8.61)$$

because $F(z)$ has simple poles at $z = \pm k_{nm}$, where k_{nm} is the mth zero of $J_n'(k_{nm} a) = 0$.

Computing $F_n(0)$ and R_m, we find that $F_n(0) = 0$, and

$$R_m = \lim_{z \to k_{nm}} \frac{\pi a z J_n(r_< z) f(r_> z, az)(z - k_{nm})}{4 J_n'(az)} + \frac{(z - k_{nm}) \delta_n}{az}, \quad (5.8.62)$$

$$= \frac{\pi k_{nm} J_n(k_{nm} r_<) f(k_{nm} r_>, a k_{nm})}{4 J_n''(k_{nm} a)}. \qquad (5.8.63)$$

However, because

$$J_n''(k_{nm} a) = \frac{J_n(k_{nm} a)}{a^2} \left(\frac{n^2}{k_{nm}^2} - a^2 \right), \qquad (5.8.64)$$

[46] Morse, P. M., and H. Feshbach, 1953: *Methods of Theoretical Physics.* McGraw-Hill Book Co., 1978 pp. See p. 383.

[47] Carslaw, H. S., 1916: The Green's function for the equation $\nabla^2 u + \kappa^2 u = 0$. *Proc. London Math. Soc., Ser. 2*, **16**, 84–93.

and

$$f(k_{nm}r_>, k_{nm}a) = \frac{2J_n(k_{nm}r_>)}{\pi a k_{nm} J_n(k_{nm}a)},\qquad (5.8.65)$$

we have

$$R_m = -\frac{aJ_n(k_{nm}r)J_n(k_{nm}\rho)}{2(a^2 - n^2/k_{nm}^2)J_n^2(k_{nm}a)}.\qquad (5.8.66)$$

Therefore,

$$\frac{J_n(k_0 r_<)f(k_0 r_>, k_0 a)}{4J_n'(k_0 a)} = -\frac{\delta_n}{\pi k_0^2 a^2}$$

$$+ \frac{1}{\pi}\sum_{m=1}^{\infty}\frac{J_n(k_{nm}r)J_n(k_{nm}\rho)}{(a^2 - n^2/k_{nm}^2)(k_{nm}^2 - k_0^2)J_n^2(k_{nm}a)}.$$

$$(5.8.67)$$

Substituting (5.8.67) into (5.3.34) yields (5.3.35).

Problems

1. Using Fourier sine transforms, find the Green's function governed by

$$\frac{\partial^2 g}{\partial x^2} + \frac{\partial^2 g}{\partial y^2} = -\delta(x - \xi)\delta(y - \eta),$$

for the quarter space $0 < x, y$, with the boundary conditions

$$g(0, y|\xi, \eta) = g(x, 0|\xi, \eta) = 0,$$

and

$$\lim_{x,y\to\infty}|g(x, y|\xi, \eta)| < \infty.$$

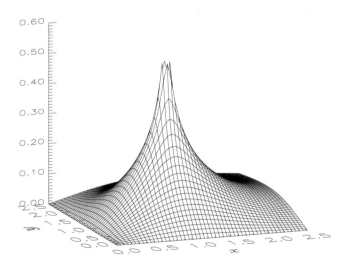

Problem 1

Step 1: Taking the Fourier sine transform in the x direction, show that the partial differential equation reduces to the ordinary differential equation

$$\frac{d^2 G}{dy^2} - k^2 G = -\sin(k\xi)\delta(y - \eta),$$

with the boundary conditions

$$G(k, 0|\xi, \eta) = 0, \quad \text{and} \quad \lim_{y \to \infty} |G(k, y|\xi, \eta)| < \infty.$$

Step 2: Show that

$$G(k, y|\xi, \eta) = \frac{\sin(k\xi)}{2k} \left[e^{-k|y-\eta|} - e^{-k(y+\eta)} \right].$$

Step 3: Taking the inverse, show that

$$g(x, y|\xi, \eta) = \frac{1}{\pi} \int_0^\infty \left[e^{-k|y-\eta|} - e^{-k(y+\eta)} \right] \sin(k\xi) \, \sin(kx) \, \frac{dk}{k}.$$

Step 4: Performing the integration, show that

$$g(x, y|\xi, \eta) = -\frac{1}{4\pi} \ln \left\{ \frac{[(x - \xi)^2 + (y - \eta)^2][(x + \xi)^2 + (y + \eta)^2]}{[(x - \xi)^2 + (y + \eta)^2][(x + \xi)^2 + (y - \eta)^2]} \right\}.$$

The Green's function $g(x, y|\xi, \eta)$ is graphed in the figure labeled Problem 1 when $\xi = \eta = 1$.

Step 5: Use the method of images to check your answer.

2. Use the method of images to show that the Green's function governed by the planar Poisson equation

$$\frac{\partial^2 g}{\partial x^2} + \frac{\partial^2 g}{\partial y^2} = -\delta(x - \xi)\delta(y - \eta),$$

for the quarter space $0 < x, y$, with the boundary conditions

$$g_x(0, y|\xi, \eta) = g_y(x, 0|\xi, \eta) = 0,$$

is

$$g(x, y|\xi, \eta) = -\frac{1}{4\pi} \ln \left\{ [(x - \eta)^2 + (y - \eta)^2][(x - \eta)^2 + (y + \eta)^2] \right.$$

$$\left. \times [(x + \eta)^2 + (y - \eta)^2][(x + \eta)^2 + (y + \eta)^2] \right\}.$$

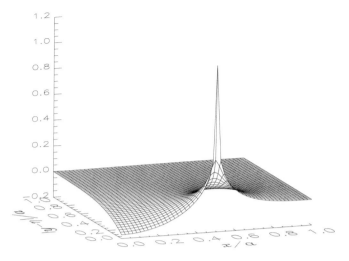

Problem 4

What is the behavior to the Green's function as x and y tend to infinity? Could you have found this Green's function using transform methods?

3. Construct the Green's function governed by the planar Poisson equation

$$\frac{\partial^2 g}{\partial x^2} + \frac{\partial^2 g}{\partial y^2} = -\delta(x - \xi)\delta(y - \eta), \quad 0 < x, \xi < a, \quad -\infty < y, \eta < \infty,$$

with the Dirichlet boundary conditions

$$g(0, y|\xi, \eta) = g(a, y|\xi, \eta) = 0, \quad -\infty < y < \infty,$$

and the conditions at infinity that

$$\lim_{|y| \to \infty} |g(x, y|\xi, \eta)| < \infty, \quad 0 < x < a.$$

4. Construct the Green's function governed by the planar Poisson equation

$$\frac{\partial^2 g}{\partial x^2} + \frac{\partial^2 g}{\partial y^2} = -\delta(x - \xi)\delta(y - \eta), \quad 0 < x, \xi < a, \quad -\infty < y, \eta < \infty,$$

with the Neumann and Dirichlet boundary conditions

$$g_x(0, y|\xi, \eta) = g(a, y|\xi, \eta) = 0, \quad -\infty < y < \infty,$$

and the conditions at infinity that

$$\lim_{|y| \to \infty} |g(x, y|\xi, \eta)| < \infty, \quad 0 < x < a.$$

The Green's function $g(x, y|\xi, \eta)$ is illustrated in the figure captioned Problem 4 when $\xi = 0.5$.

5. Construct the Green's function governed by the planar Poisson equation

$$\frac{\partial^2 g}{\partial x^2} + \frac{\partial^2 g}{\partial y^2} = -\delta(x - \xi)\delta(y - \eta), \quad 0 < x, \xi < a, \quad -\infty < y, \eta < \infty,$$

with the Neumann boundary conditions

$$g_x(0, y|\xi, \eta) = g_x(a, y|\xi, \eta) = 0, \quad -\infty < y < \infty,$$

and the condition at infinity that

$$\lim_{|y| \to \infty} |g(x, y|\xi, \eta)| = O(|y - \eta|^n), \quad 0 < x < a.$$

6. Find the Green's function[48] governed by Poisson's equation

$$\frac{\partial^2 g}{\partial x^2} + \frac{\partial^2 g}{\partial y^2} = -\delta(x - \xi)\delta(y - \eta), \quad -\infty < x, \xi < \infty, \quad 0 < y, \eta < \infty,$$

subject to the boundary conditions

$$g_y(x, 0|\xi, \eta) + \alpha g(x, 0|\xi, \eta) = 0, \quad \lim_{y \to \infty} g_y(x, y|\xi, \eta) \to 0,$$

and

$$\lim_{|x| \to \infty} |g(x, y|\xi, \eta)| < \infty,$$

where α is real.

Step 1: Taking the Fourier transform of the governing partial differential equation, show that it becomes the ordinary differential equation

$$\frac{d^2 G}{dy^2} - k^2 G = -e^{-ik\xi}\delta(y - \eta),$$

with the boundary conditions

$$G'(k, 0|\xi, \eta) + \alpha G(k, 0|\xi, \eta) = 0, \quad \text{and} \quad \lim_{y \to \infty} G'(k, y|\xi, \eta) \to 0.$$

Step 2: Show that

$$G(k, y|\xi, \eta) = -\frac{e^{-ik\xi}}{2|k|}e^{-|k||y-\eta|} + \frac{e^{-ik\xi}}{2|k|}e^{-|k|(y+\eta)} - \frac{e^{-ik\xi}}{|k| - \alpha}e^{-|k|(y+\eta)}.$$

---

[48] Taken from Meylan, M., and V. A. Squire, 1994: The response of ice floes to ocean waves. *J. Geophys. Res.*, **99**, 891–900. ©1994 by the American Geophysical Union.

Step 3: Complete the problem by showing that

$$
\begin{aligned}
g(x, y | \xi, \eta) =\ & -\frac{1}{4\pi} \int_{-\infty}^{\infty} e^{ik(x-\xi)} e^{-|k||y-\eta|} \frac{dk}{|k|} \\
& + \frac{1}{4\pi} \int_{-\infty}^{\infty} e^{ik(x-\xi)} e^{-|k|(y+\eta)} \frac{dk}{|k|} \\
& - \frac{1}{2\pi} \int_{-\infty}^{\infty} e^{ik(x-\xi)} e^{-|k|(y+\eta)} \frac{dk}{|k| - \alpha} \\
=\ & \frac{1}{4\pi} \ln \left[\frac{(x-\xi)^2 + (y-\eta)^2}{(x-\xi)^2 + (y+\eta)^2} \right] \\
& - \frac{1}{2\pi} \int_{-\infty}^{\infty} e^{ik(x-\xi)} e^{-|k|(y+\eta)} \frac{dk}{|k| - \alpha}.
\end{aligned}
$$

7. Find the free-space Green's function[49] governed by

$$
\frac{\partial^2 g}{\partial x^2} + \frac{\partial^2 g}{\partial y^2} - g = -\delta(x - \xi)\delta(y - \eta), \qquad -\infty < x, y, \xi, \eta < \infty.
$$

Step 1: Show that the Fourier transform of the governing partial differential equation is

$$
\frac{d^2 G}{dy^2} - \left(k^2 + 1\right) G = -e^{-ik\xi} \delta(y - \eta).
$$

Step 2: Show that

$$
G(k, y | \xi, \eta) = \frac{e^{-ik\xi}}{2\pi} \int_{-\infty}^{\infty} \frac{e^{i\ell(y-\eta)}}{k^2 + \ell^2 + 1} \, d\ell.
$$

Step 3: Complete the problem by showing that

$$
\begin{aligned}
g(x, y | \xi, \eta) &= \frac{1}{4\pi^2} \int_{-\infty}^{\infty} \int_{-\infty}^{\infty} \frac{e^{ik(x-\xi)} e^{i\ell(y-\eta)}}{k^2 + \ell^2 + 1} \, d\ell \, dk \\
&= \frac{1}{2\pi} \int_{0}^{\infty} \frac{J_0(r\kappa)}{\kappa^2 + 1} \kappa \, d\kappa = \frac{K_0(r)}{2\pi},
\end{aligned}
$$

where $k = \kappa \cos(\theta)$, $\ell = \kappa \sin(\theta)$, $x - \xi = r \cos(\varphi)$, $y - \eta = r \sin(\varphi)$, and $r = \sqrt{(x-\xi)^2 + (y-\eta)^2}$.

---

[49] For its use, see Geisler, J. E., 1970: Linear theory of the response of a two layer ocean to a moving hurricane. *Geophys. Fluid Dyn.*, **1**, 249–272.

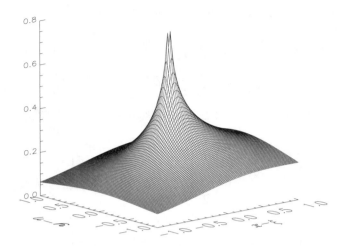

Problem 8

8. Find the free-space Green's function governed by

$$\frac{\partial^2 g}{\partial x^2} + \frac{\partial^2 g}{\partial y^2} - \frac{\partial g}{\partial x} = -\delta(x - \xi)\delta(y - \eta), \qquad -\infty < x, y, \xi, \eta < \infty.$$

Step 1: By introducing $\varphi(x, y|\xi, \eta)$ such that

$$g(x, y|\xi, \eta) = e^{x/2}\varphi(x, y|\xi, \eta),$$

show that the partial differential equation for $g(x, y|\xi, \eta)$ becomes

$$\frac{\partial^2 \varphi}{\partial x^2} + \frac{\partial^2 \varphi}{\partial y^2} - \frac{\varphi}{4} = -e^{-\xi/2}\delta(x - \xi)\delta(y - \eta).$$

Step 2: After taking the Fourier transform with respect to x of the partial differential equation in Step 1, show that it becomes the ordinary differential equation

$$\frac{d^2 \Phi}{dy^2} - \left(k^2 + \tfrac{1}{4}\right)\Phi = -e^{-\xi/2 - ik\xi}\delta(y - \eta).$$

Step 3: Using the methods presented in the previous problem, show that

$$\Phi(k, y|\xi, \eta) = \frac{e^{-\xi/2 - ik\xi}}{2\pi}\int_{-\infty}^{\infty}\frac{e^{i\ell(y - \eta)}}{k^2 + \ell^2 + \tfrac{1}{4}}\,d\ell,$$

and

$$\varphi(x,y|\xi,\eta) = \frac{e^{-\xi/2}}{2\pi} K_0(\tfrac{1}{2}r),$$

where $r = \sqrt{(x-\xi)^2 + (y-\eta)^2}$.

Step 4: Complete the analysis and show that

$$g(x,y|\xi,\eta) = \frac{e^{(x-\xi)/2}}{2\pi} K_0(\tfrac{1}{2}r).$$

This Green's function $g(x,y|\xi,\eta)$ is plotted in the figure captioned Problem 8.

9. Show that the Green's function for Poisson's equation over the strip $0 < \text{Im}(z) < 1$ is

$$g(x,y|\xi,\eta) = \ln\left|\frac{e^{\pi z} - e^{\pi z_0^*}}{e^{\pi z} - e^{\pi z_0}}\right|,$$

where $z = x + iy$ and $z_0 = \xi + i\eta$. What boundary condition does it satisfy?

10. Find the Green's function governed by the planar Poisson equation

$$\frac{\partial^2 g}{\partial x^2} + \frac{\partial^2 g}{\partial y^2} = -\delta(x-\xi)\delta(y-\eta) + \frac{1}{ab}, \quad 0 < x,\xi < a, \quad 0 < y,\eta < b,$$

with Neumann boundary conditions

$$g_x(0,y|\xi,\eta) = g_x(a,y|\xi,\eta) = 0, \quad 0 < y < b,$$

and

$$g_y(x,0|\xi,\eta) = g_y(x,b|\xi,\eta) = 0, \quad 0 < x < a.$$

Marshall[50] has found computationally useful expressions for calculating this Green's function.

11. Show that the Green's function governed by the planar Poisson equation

$$\frac{\partial^2 g}{\partial x^2} + \frac{\partial^2 g}{\partial y^2} = -\delta(x-\xi)\delta(y-\eta), \quad 0 < x,\xi < a, \quad 0 < y,\eta < b,$$

---

[50] Marshall, S. L., 1999: A rapidly convergent modified Green's function for Laplace's equation in a rectangular region. *Proc. R. Soc. London*, **A455**, 1739–1766.

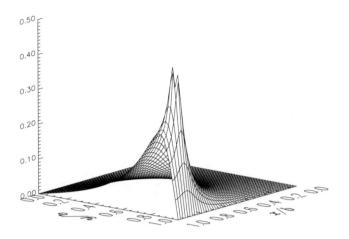

Problem 11

with the boundary conditions

$$g(x,0|\xi,\eta) = g(x,b|\xi,\eta) = 0,$$

and

$$g(0,y|\xi,\eta) = g_x(a,y|\xi,\eta) + \beta g(a,y|\xi,\eta) = 0, \qquad \beta \geq 0,$$

is

$$g(x,y|\xi,\eta) = \sum_{n=1}^{\infty} \frac{\sinh(\nu x_<)\,\{\nu \cosh\left[\nu(a-x_>)\right] + \beta \sinh\left[\nu(a-x_>)\right]\}}{\nu^2 \cosh(\nu a) + \beta \nu \sinh(\nu a)}$$
$$\times \sin\left(\frac{n\pi\eta}{b}\right) \sin\left(\frac{n\pi y}{b}\right),$$

where $\nu = n\pi/b$.

Following Melnikov,[51] as illustrated in §5.4, show that an equivalent expression is

$$g(x,y|\xi,\eta) = \frac{1}{2\pi} \ln\left[\frac{E(z-\zeta^*)E(z+\zeta^*)}{E(z-\zeta)E(z+\zeta)}\right]$$
$$- \frac{2}{\pi} \sum_{n=1}^{\infty} \frac{(\beta b - n\pi)e^{-n\pi a/b}\sinh(n\pi\xi/b)\sinh(n\pi x/b)}{n[\beta b \sinh(n\pi a/b) + n\pi \cosh(n\pi a/b)]}$$
$$\times \sin\left(\frac{n\pi\eta}{b}\right) \sin\left(\frac{n\pi y}{b}\right),$$

[51] Melnikov, Yu. A., 1995: *Green's Functions in Applied Mechanics*. WIT Press/Computational Mechanics Publications, pp. 33.

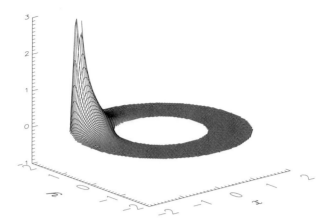

Problem 12

where $z = x + iy$, $\zeta = \xi + i\eta$, $\zeta^* = \xi - i\eta$, and $E(z) = \left| e^{\pi z/b} - 1 \right|$.

Melnikov[52] further showed that $g(x, y | \xi, \eta)$ can be transformed into

$$g(x, y | \xi, \eta) = \frac{1}{2\pi} \ln \left[\frac{E(z - \zeta^*)E(z + \zeta^*)E(z_1 + \zeta_1)E(z_2 + \zeta_2)}{E(z - \zeta)E(z + \zeta)E(z_1 - \zeta_1^*)E(z_2 + \zeta_2^*)} \right]$$
$$- \frac{2}{\pi} \sum_{n=1}^{\infty} T_n(x, \xi) \sin\left(\frac{n\pi\eta}{b}\right) \sin\left(\frac{n\pi y}{b}\right),$$

where

$$T_n(x, \xi) = \frac{2b\beta \sinh(n\pi x/b) \sinh(n\pi\xi/b)}{\pi n^2 \, e^{2n\pi a/b}} + \frac{\cosh[n\pi(x - \xi)/b]}{n \, e^{2n\pi a/b}}$$
$$- \frac{(\beta b - n\pi) \sinh(n\pi x/b) \sinh(n\pi\xi/b)}{\pi n^2 \, e^{n\pi a/b}}$$
$$\times \left\{ \frac{2}{e^{n\pi a/b} \left(e^{2n\pi a/b} + 1 \right)} \right.$$
$$\left. + \frac{b\beta \sinh(n\pi a/b)}{\cosh(n\pi a/b) \left[n\pi \cosh(n\pi a/b) + \beta b \sinh(n\pi a/b) \right]} \right\},$$

$z_1 = (x + a) + iy$, $z_2 = (x - a) + iy$, $\zeta_1 = (\xi + a) + i\eta$, $\zeta_2 = (\xi - a) + i\eta$, $\zeta_1^* = (\xi + a) - i\eta$, and $\zeta_2^* = (\xi - a) - i\eta$. This particular form of the Green's function has the advantage that it converges uniformly. The figure captioned Problem 11 illustrates $g(x, y | \xi, \eta)$ when $\xi/b = \eta/b = 0.9$, $a/b = 1$, and $\beta b = 1$.

12. Construct the Green's function governed by the planar Poisson equation

$$\frac{1}{r} \frac{\partial}{\partial r} \left(r \frac{\partial g}{\partial r} \right) + \frac{1}{r^2} \frac{\partial^2 g}{\partial \theta^2} = -\frac{\delta(r - \rho)\delta(\theta - \theta')}{r},$$

---

[52] *Ibid.*, pp. 33–35.

where $a < r, \rho < b$, and $0 \le \theta, \theta' \le 2\pi$, subject to the boundary conditions $g(a, \theta|\rho, \theta') = g(b, \theta|\rho, \theta') = 0$ and periodicity in θ. The Green's function $g(r, \theta|\rho, \theta')$ is illustrated in the figure captioned Problem 12 for $\rho = 1.3\sqrt{2}$, $\theta' = 3\pi/4$, $a = 1$, and $b = 2$.

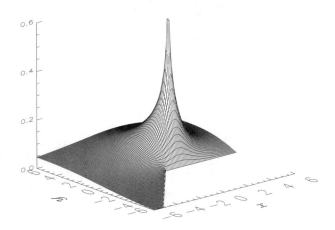

Problem 13

13. Construct the Green's function governed by the planar Poisson equation

$$\frac{1}{r}\frac{\partial}{\partial r}\left(r\frac{\partial g}{\partial r}\right) + \frac{1}{r^2}\frac{\partial^2 g}{\partial \theta^2} = -\frac{\delta(r - \rho)\delta(\theta - \theta')}{r},$$

where $0 < r, \rho < \infty$, and $0 < \theta, \theta' < \beta$, subject to the boundary conditions that $g(r, 0|\rho, \theta') = g(r, \beta|\rho, \theta') = 0$ for all r. This Green's function $g(r, \theta|\rho, \theta')$ is plotted in the figure captioned Problem 13 when $\rho = 1.35\sqrt{2}$, $\theta' = \pi/4$, and $\beta = 4$. Hint:

$$\delta(\theta - \theta') = \frac{2}{\beta}\sum_{n=1}^{\infty}\sin\left(\frac{n\pi\theta'}{\beta}\right)\sin\left(\frac{n\pi\theta}{\beta}\right).$$

14. Construct the Green's function governed by the planar Poisson equation

$$\frac{1}{r}\frac{\partial}{\partial r}\left(r\frac{\partial g}{\partial r}\right) + \frac{1}{r^2}\frac{\partial^2 g}{\partial \theta^2} = -\frac{\delta(r - \rho)\delta(\theta - \theta')}{r},$$

where $0 < r, \rho < a$, and $0 < \theta, \theta' < \beta$, subject to the boundary conditions $g(r, 0|\rho, \theta') = g(r, \beta|\rho, \theta') = g(a, \theta|\rho, \theta') = 0$. This Green's function

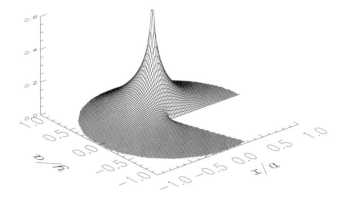

Problem 14

$g(r, \theta | \rho, \theta')$ is graphed in the figure captioned Problem 14 when $\rho = 0.35$, $\theta' = 1.43$, and $\beta = 5$. Hint:

$$\delta(\theta - \theta') = \frac{2}{\beta} \sum_{n=1}^{\infty} \sin\left(\frac{n\pi\theta'}{\beta}\right) \sin\left(\frac{n\pi\theta}{\beta}\right).$$

15. Construct the free-space Green's function[53] for the planar Poisson equation

$$\frac{1}{r}\frac{\partial}{\partial r}\left(r\kappa\frac{\partial g}{\partial r}\right) + \frac{\kappa}{r^2}\frac{\partial^2 g}{\partial \theta^2} = -\frac{\delta(r - \rho)\delta(\theta - \theta')}{r},$$

where $0 < r, \rho < \infty$, and $\kappa = h^2/(1 + h^2 r^2)$, subject to the boundary conditions that

$$\lim_{r \to 0}\left| r\kappa\frac{\partial g}{\partial r} \right| < \infty, \qquad \lim_{r \to \infty}\left| r\kappa\frac{\partial g}{\partial r} \right| < \infty,$$

and the Green's function is periodic in θ.

Step 1: Using the fact that

$$\delta(\theta - \theta') = \frac{1}{\pi} + \frac{2}{\pi} \sum_{n=1}^{\infty} \cos[n(\theta - \theta')],$$

---

[53] Taken from Gardner, H. J., R. L. Dewar, and W. N-C. Sy, 1988: The free-boundary equilibrium problem for helically symmetric plasmas. *J. Comput. Phys.*, **74**, 477–487.

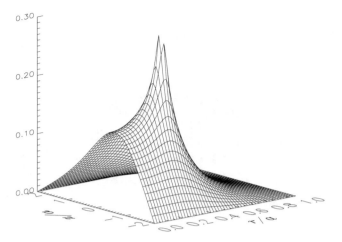

Problem 16

show that the Green's function can be expressed as

$$g(r, \theta | \rho, \theta') = \frac{G_0(r|\rho)}{\pi} + \frac{2}{\pi} \sum_{n=1}^{\infty} G_n(r|\rho) \cos[n(\theta - \theta')],$$

where $G_n(r|\rho)$ is governed by

$$\frac{1}{r} \frac{d}{dr} \left(\frac{rh^2}{1 + h^2 r^2} \frac{dG_n}{dr} \right) - \frac{n^2 h^2}{r^2(1 + h^2 r^2)} G_n = -\frac{\delta(r - \rho)}{r},$$

and $n = 0, 1, 2, \dots..$

Step 2: Show that $G_n(r|\rho)$ is given by

$$G_0(r|\rho) = \left[\ln(r_>) + \tfrac{1}{2} h^2 r_>^2 \right] / (2h^2) - \left[\ln(r_<) + \tfrac{1}{2} h^2 r_<^2 \right] / (2h^2),$$

and

$$G_n(r|\rho) = -\tfrac{1}{2} r\rho I_n'(nhr_<) K_n'(nhr_>).$$

Step 3: Complete the problem by showing that

$$g(r, z|\rho, \zeta) = \left[\ln(r_>) + \tfrac{1}{2} h^2 r_>^2 \right] / (2\pi h^2) - \left[\ln(r_<) + \tfrac{1}{2} h^2 r_<^2 \right] / (2\pi h^2)$$

$$- \frac{r\rho}{\pi} \sum_{n=1}^{\infty} I_n'(nhr_<) K_n'(nhr_>) \cos[n(\theta - \theta')].$$

16. Find the Green's function governed by the axisymmetric Poisson equation

$$\frac{\partial^2 g}{\partial r^2} + \frac{1}{r}\frac{\partial g}{\partial r} + \frac{\partial^2 g}{\partial z^2} = -\frac{\delta(r-\rho)\delta(z-\zeta)}{2\pi r},$$

where $0 < r, \rho < a$, and $0 < z, \zeta < L$, subject to the boundary conditions

$$g(r, 0|\rho, \zeta) = g(r, L|\rho, \zeta) = 0, \quad 0 < r < a,$$

and

$$\lim_{r \to 0} |g(r, z|\rho, \zeta)| < \infty, \quad g(a, z|\rho, \zeta) = 0, \quad 0 < z < L.$$

The Green's function $ag(r, z|\rho, \zeta)$ is illustrated in the figure captioned Problem 16 when $\rho = 0.3$, $\zeta = 0.3$, and $L = a$.

17. Find the Green's function[54] governed by the axisymmetric Poisson equation

$$\frac{\partial^2 g}{\partial r^2} + \frac{1}{r}\frac{\partial g}{\partial r} - \frac{g}{r^2} + \frac{\partial^2 g}{\partial z^2} = -\frac{\delta(r-\rho)\delta(z-\zeta)}{2\pi r},$$

where $0 < r, \rho < \infty$, and $0 < z, \zeta < \infty$, subject to the boundary conditions

$$g_z(r, 0|\rho, \zeta) = 0, \quad \lim_{z \to \infty} |g(r, z|\rho, \zeta)| < \infty, \quad 0 < r < \infty,$$

and

$$\lim_{r \to 0} |g(r, z|\rho, \zeta)| < \infty, \quad \lim_{r \to \infty} |g(r, z|\rho, \zeta)| < \infty, \quad 0 < z < \infty.$$

Step 1: Take the Hankel transform of the partial differential equation and show that it becomes the ordinary differential equation

$$\frac{d^2 G}{dz^2} - k^2 G = -\frac{\delta(z-\zeta)}{2\pi} J_1(k\rho), \quad 0 < z < \infty,$$

with

$$G'(k, 0|\rho, \zeta) = 0, \quad \text{and} \quad \lim_{z \to \infty} |G(k, z|\rho, \zeta)| < \infty.$$

Step 2: Solve the ordinary differential equation in Step 1 and show that

$$G(k, z|\rho, \zeta) = \frac{J_1(k\rho)}{2\pi k} e^{-kz_>} \cosh(kz_<) = \frac{J_1(k\rho)}{4\pi k}\left[e^{-k|z-\zeta|} + e^{-k(z+\zeta)}\right].$$

---

[54] Reprinted from *Int. J. Solids Struct.*, **37**, M. Rahman, The Reissner-Sagoci problem for a half-space under buried torsional forces, 1119–1132, ©2000, with permission from Elsevier Science.

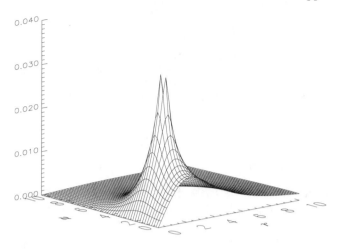

Problem 17

Step 3: Show that

$$g(r, z | \rho, \zeta) = \frac{1}{4\pi} \int_0^\infty \left[e^{-k|z-\zeta|} + e^{-k(z+\zeta)} \right] J_1(k\rho) J_1(kr) \, dk.$$

Step 4: Using Neumann's addition theorem

$$J_1(k\rho) J_1(kr) = \frac{1}{\pi} \int_0^\pi J_0(Rk) \cos(\varphi) \, d\varphi, \quad R = \sqrt{r^2 + \rho^2 - 2r\rho \cos(\varphi)},$$

show that

$$g(r, z | \rho, \zeta) = \frac{1}{4\pi^2} \int_0^\pi \left[S(R, |z - \zeta|) + S(R, z + \zeta) \right] \cos(\varphi) \, d\varphi$$

$$= \frac{1}{4\pi^2} \int_0^\pi \left[\frac{1}{R_1} + \frac{1}{R_2} \right] \cos(\varphi) \, d\varphi,$$

where

$$S(r, z) = \int_0^\infty e^{-sz} J_0(rs) \, ds = \left(r^2 + z^2 \right)^{-1/2},$$

$$R_1 = \sqrt{r^2 + \rho^2 + (z - \zeta)^2 - 2r\rho \cos(\varphi)},$$

and

$$R_2 = \sqrt{r^2 + \rho^2 + (z + \zeta)^2 - 2r\rho \cos(\varphi)}.$$

This Green's function is illustrated in the figure captioned Problem 17 when $\rho = \zeta = 2$. Simpson's rule was used to evaluate the integral.

18. Construct the Green's function governed by the axisymmetric Poisson equation

$$\frac{\partial^2 g}{\partial r^2} + \frac{1}{r}\frac{\partial g}{\partial r} - \frac{g}{r^2} + \frac{\partial^2 g}{\partial z^2} = -\frac{\delta(r-\rho)\delta(z-\zeta)}{2\pi r},$$

where $0 < r, \rho < \infty$, and $0 < z, \zeta < L$, subject to the boundary conditions

$$g_z(r,0|\rho,\zeta) = g_z(r,L|\rho,\zeta) = 0, \quad 0 < r < \infty,$$

and

$$\lim_{r\to 0} |g(r,z|\rho,\zeta)| < \infty, \quad \lim_{r\to\infty} |g(r,z|\rho,\zeta)| < \infty, \quad 0 < z < L.$$

Step 1: Show that

$$g(r,z|\rho,\zeta) = -\frac{G_0(r|\rho)}{L} - \frac{2}{L}\sum_{n=1}^{\infty} G_n(r|\rho)\cos\left(\frac{n\pi\zeta}{L}\right)\cos\left(\frac{n\pi z}{L}\right),$$

where

$$\frac{d^2 G_n}{dr^2} + \frac{1}{r}\frac{dG_n}{dr} - \frac{G_n}{r^2} - n^2 G_n = \frac{\delta(r-\rho)}{2\pi r}, \quad 0 < r < \infty.$$

Step 2: Show that

$$G_0(r|\rho) = \frac{r_<}{4\pi r_>},$$

and

$$G_n(r|\rho) = \frac{I_1(n\pi r_</L)K_1(n\pi r_>/L)}{2\pi}.$$

Step 3: Show that

$$g(r,z|\rho,\zeta) = -\frac{r_<}{4\pi r_> L}$$
$$-\frac{1}{\pi L}\sum_{n=1}^{\infty} I_1\left(\frac{n\pi r_<}{L}\right)K_1\left(\frac{n\pi r_>}{L}\right)\cos\left(\frac{n\pi\zeta}{L}\right)\cos\left(\frac{n\pi z}{L}\right).$$

19. Construct the Green's function[55] governed by the axisymmetric Poisson equation

$$\frac{\partial^2 g}{\partial r^2} + \frac{1}{r}\frac{\partial g}{\partial r} - \frac{g}{r^2} + a^2\frac{\partial^2 g}{\partial z^2} + \lambda a^2\frac{\partial g}{\partial z} = -\frac{\delta(r-\rho)\delta(z-\zeta)}{2\pi r},$$

---

[55] Adapted from material in Tarn, J.-Q., and Y.-M. Wang, 1986: Fundamental solutions for torsional problems of nonhomogeneous transversely isotropic media. *J. Chinese Inst. Engrs.*, **9**, 187–194.

where $0 < r, \rho < \infty$, and $-\infty < z, \zeta < \infty$, subject to the boundary conditions

$$\lim_{r \to 0} |g(r, z|\rho, \zeta)| < \infty, \quad \lim_{r \to \infty} |g(r, z|\rho, \zeta)| < \infty, \quad -\infty < z < \infty,$$

and

$$\lim_{|z| \to \infty} |g(r, z|\rho, \zeta)| < \infty, \quad 0 < r < \infty.$$

Step 1: Show that the Hankel transform of the governing equation is

$$\frac{d^2 G}{dz^2} + \lambda \frac{dG}{dz} - \frac{k^2}{a^2} G = -\frac{J_1(k\rho)\delta(z - \zeta)}{2\pi},$$

where $G(k, z|\rho, \zeta)$ is the Hankel transform of $g(r, z|\rho, \zeta)$:

$$G(k, z|\rho, \zeta) = \int_0^\infty g(r, z) J_1(kr) \, r \, dr,$$

with the boundary conditions that

$$\lim_{|z| \to \infty} |G(k, z|\rho, \zeta)| < \infty, \quad 0 \le k < \infty.$$

Step 2: Find $G(k, z|\rho, \zeta)$ by solving the ordinary differential equation in Step 1 and show that

$$G(k, z|\rho, \zeta) = \frac{e^{-\lambda(z-\zeta)/2 - \beta|z-\zeta|}}{4\pi\beta} J_1(k\rho),$$

where $\beta = \sqrt{k^2/a^2 + \lambda^2/4}$.

Step 3: By inverting the Hankel transform, show that

$$g(r, z|\rho, \zeta) = \frac{e^{-\lambda(z-\zeta)/2}}{4\pi} \int_0^\infty \frac{J_1(k\rho) J_1(kr)}{\sqrt{k^2/a^2 + \lambda^2/4}} e^{-|z-\zeta|\sqrt{k^2/a^2 + \lambda^2/4}} \, k \, dk.$$

20. Find the Green's function[56] for Poisson's equation in parabolic coordinates

$$\frac{1}{\sigma} \frac{\partial}{\partial \sigma} \left(\sigma \frac{\partial g}{\partial \sigma} \right) + \frac{1}{\tau} \frac{\partial}{\partial \tau} \left(\tau \frac{\partial g}{\partial \tau} \right) = -\frac{\delta(\sigma - \xi)\delta(\tau - \eta)}{2\pi\sigma\tau},$$

---

[56] Reprinted with permission from Dorofeev, I. A., 1997: Energy balance in the tip-sample system of a tunneling microscope in the surface modification regime. *Tech. Phys.*, **42**, 1305–1311. ©American Institute of Physics, 1997.

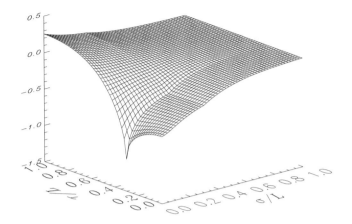

Problem 20

where $0 < \sigma, \xi < a$, and $0 < \tau, \eta < L$, subject to the boundary conditions

$$\lim_{\sigma \to 0} |g(\sigma, \tau|\xi, \eta)| < \infty, \qquad g(a, \tau|\xi, \eta) = 0, \qquad 0 < \tau < L,$$

and

$$\lim_{\tau \to 0} |g(\sigma, \tau|\xi, \eta)| < \infty, \qquad g_\tau(\sigma, L|\xi, \eta) = 0, \qquad 0 < \sigma < a.$$

Step 1: Assuming that the Green's function can be written as

$$g(\sigma, \tau|\xi, \eta) = \sum_{n=1}^{\infty} \Sigma_n(\sigma|\xi) J_0(k_n \tau/L),$$

show that the partial differential equation reduces to the ordinary differential equation

$$\frac{1}{\sigma} \frac{d}{d\sigma}\left(\sigma \frac{d\Sigma_n}{d\sigma}\right) - \frac{k_n^2}{L^2}\Sigma_n = -\frac{J_0(k_n \eta/L)}{\pi L^2 J_0^2(k_n)}\frac{\delta(\sigma - \xi)}{\sigma}, \qquad 0 < \sigma < a,$$

with

$$\lim_{\sigma \to 0} |\Sigma_n(\sigma|\xi)| < \infty, \qquad \Sigma_n(a|\xi) = 0,$$

where k_n is the nth root of $J_1(k) = 0$. Why did we assume this particular eigenfunction expansion?

Step 2: Show that

$$\Sigma_n(\sigma|\xi) = \frac{J_0(k_n\eta/L)}{\pi L^2 J_0^2(k_n)} I_0\left(\frac{k_n\sigma_<}{L}\right)$$
$$\times \left[K_0\left(\frac{k_n\sigma_>}{L}\right) - \frac{K_0(k_na/L)}{I_0(k_na/L)} I_0\left(\frac{k_n\sigma_>}{L}\right)\right],$$

where $\sigma_< = \min(\sigma, \xi)$, and $\sigma_> = \max(\sigma, \xi)$.

Step 3: Show that

$$g(\sigma, \tau|\xi, \eta) = \frac{1}{\pi L^2} \sum_{n=1}^{\infty} \frac{J_0(k_n\eta/L)}{J_0^2(k_n)} J_0\left(\frac{k_n\tau}{L}\right) I_0\left(\frac{k_n\sigma_<}{L}\right)$$
$$\times \left[K_0\left(\frac{k_n\sigma_>}{L}\right) - \frac{K_0(k_na/L)}{I_0(k_na/L)} I_0\left(\frac{k_n\sigma_>}{L}\right)\right].$$

Why do we not have a term corresponding to the $k = 0$ root? This Green's function (multiplied by πL^2) is illustrated in the figure captioned Problem 20 when $a = L$, and $\xi/L = \eta/L = 0.3$.

21. Construct the Green's function[57] governed by the planar Helmholtz equation

$$\frac{\partial^2 g}{\partial x^2} + \frac{\partial^2 g}{\partial y^2} + k_0^2 g = -\delta(x - \xi)\delta(y - \eta), \quad 0 < x, \xi < a, \quad 0 < y, \eta < b,$$

subject to the Neumann boundary conditions

$$g_x(0, y|\xi, \eta) = g_x(a, y|\xi, \eta) = 0, \quad 0 < y < b,$$

and

$$g_y(x, 0|\xi, \eta) = g_y(x, b|\xi, \eta) = 0, \quad 0 < x < a.$$

22. Find the *surface* Green's function[58] governed by the planar Helmholtz equation

$$\frac{\partial^2 g}{\partial x^2} + \frac{\partial^2 g}{\partial y^2} + k_0^2 g = 0, \quad -\infty < x, \xi < \infty, \quad 0 \le y \le 1,$$

[57] Kulkarni *et al.* [Kulkarni, S., F. G. Leppington, and E. G. Broadbent, 2001: Vibrations in several interconnected regions: A comparison of SEA, ray theory and numerical results. *Wave Motion*, **33**, 79–96.] solved this problem when the domain has two different, constant k_0^2's.

[58] Taken from Pompei, A., M. A. Rigano, and M. A. Sumbatyan, 1999: On efficient treatment of high-frequency SH-wave fields in the elastic layer. *Zeit. Angew. Math. Mech.*, **79**, 781–785.

subject to the boundary conditions

$$g(x, 0|\xi) = 0, \quad g_y(x, 1|\xi) = \delta(x - \xi), \quad -\infty < x < \infty,$$

and

$$\lim_{|x| \to \infty} |g(x, y|\xi)| < \infty, \quad 0 \le y \le 1.$$

The temporal behavior of the wave is $e^{-i\omega t}$.

Step 1: Taking the complex Fourier transform in the x direction, show that the partial differential equation reduces to the ordinary differential equation

$$\frac{d^2 G}{dy^2} - \gamma^2 G = 0, \quad 0 \le y \le 1,$$

where $\gamma^2 = k^2 - k_0^2$, and k is the transform variable, with $G(k, 0|\xi) = 0$, and $G'(k, 1|\xi) = e^{-ik\xi}$.

Step 2: Show that

$$g(x, y|\xi) = \frac{1}{2\pi} \int_{-\infty}^{\infty} \frac{\sinh(\gamma y)}{\gamma \cosh(\gamma)} e^{ik(x-\xi)} \, dk.$$

Step 3: Show that

$$g(x, y|\xi) = \sum_{n=1}^{\infty} (-1)^{n-1} \frac{\sin\left[\left(n - \frac{1}{2}\right)\pi y\right]}{\sqrt{\left(n - \frac{1}{2}\right)^2 \pi^2 - k_0^2}}$$

$$\times \exp\left[-|x - \xi|\sqrt{\left(n - \frac{1}{2}\right)^2 \pi^2 - k_0^2}\right]$$

$$= \frac{1}{2\pi} \int_{-\infty}^{\infty} \frac{e^{-\gamma(1-y)} - e^{-\gamma(1+y)}}{1 + e^{-2\gamma}} e^{ik(x-\xi)} \frac{dk}{\gamma}$$

$$= G_0(x, 1 - y) - G_0(x, 1 + y),$$

where

$$G_0(x, y) = \frac{1}{2\pi} \sum_{n=0}^{\infty} (-1)^n \int_{-\infty}^{\infty} e^{-\gamma(2n+y)} e^{ik(x-\xi)} \frac{dk}{\gamma}$$

$$= \frac{i}{2} \sum_{n=0}^{\infty} (-1)^n H_0^{(1)}\left[k_0 \sqrt{(2n + y)^2 + (x - \xi)^2}\right].$$

Give a physical interpretation of $G_0(x, y)$.

23. Construct the free-space Green's function[59] governed by the two-dimensional Helmholtz equation

$$\frac{\partial^2 g}{\partial x^2} + \frac{\partial^2 g}{\partial y^2} + \frac{k_0^2}{(1+ay)^2} g = -\delta(x-\xi)\delta(y-\eta).$$

Step 1: Introducing the Fourier cosine transform

$$g(x,y|\xi,\eta) = \frac{1}{\pi} \int_0^\infty G(k,y|\xi,\eta) \cos[k(x-\xi)]\, dk,$$

show that the problem reduces to solving the ordinary differential equation

$$\frac{d^2 G}{dy^2} + \left[\frac{k_0^2}{(1+ay)^2} - k^2 \right] G = -\delta(y-\eta).$$

Step 2: Introducing $z = (1+ay)/a$, and $k_1 = k_0/a$, show that the differential equation in Step 1 can be transformed into

$$\frac{d^2 G}{dz^2} + \left(\frac{k_1^2}{z^2} - k^2 \right) G = -\delta(z-\zeta),$$

where $\zeta = (1+a\eta)/a$.

Step 3: Show that the solution to the ordinary differential equation in Step 2 is

$$G(k,z|\xi,\zeta) = z_>^{1/2} z_<^{1/2} I_{i\nu}(kz_<) K_{i\nu}(kz_>),$$

where $\nu = -(k_1^2 - \frac{1}{4})^{1/2}$.

Step 4: Complete the analysis by showing that

$$g(x,y|\xi,\eta) = \frac{1}{\pi} \int_0^\infty z_>^{1/2} z_<^{1/2} I_{i\nu}(kz_<) K_{i\nu}(kz_>) \cos[k(x-\xi)]\, dk$$

$$= \frac{1}{2\pi} Q_{i\nu-\frac{1}{2}} \left[\frac{z_<^2 + z_>^2 + (x-\xi)^2}{2 z_< z_>} \right]$$

$$= \frac{1}{2\pi} Q_{i\nu-\frac{1}{2}} \left[1 + \frac{a^2[(x-\xi)^2 + (y-\eta)^2]}{2(1+ay)(1+a\eta)} \right],$$

[59] Reprinted with permission from Li, Y. L., S. J. Franke, and C. H. Liu, 1993: Wave scattering from a ground with a Gaussian bump or trough in an inhomogeneous medium. *J. Acoust. Soc. Am.*, **94**, 1067–1075. ©Acoustical Society of America, 1993.

where $Q_p(\ )$ is the Legendre function of the second kind of order p. Hint: There is a mathematical identity involving modified Bessel functions.

24. Construct the free-space Green's function for the axisymmetric Helmholtz equation

$$\frac{\partial^2 g}{\partial r^2} + \frac{1}{r}\frac{\partial g}{\partial r} - \frac{g}{r^2} + \frac{\partial^2 g}{\partial z^2} + k_0^2 g = -\frac{\delta(r-\rho)\delta(z-\zeta)}{r},$$

where $0 < r, \rho < \infty$, and $-\infty < z, \zeta < \infty$.

Step 1: Taking the Fourier transform of the partial differential equation, show that it reduces to the ordinary differential equation

$$\frac{d^2 G}{dr^2} + \frac{1}{r}\frac{dG}{dr} - \frac{G}{r^2} + \kappa^2 G = -\frac{\delta(r-\rho)}{r}e^{-ik\zeta},$$

where $\kappa = \sqrt{k_0^2 - k^2}$ with $\text{Im}(\kappa) \geq 0$.

Step 2: Show that

$$G(r,k|\rho,\zeta) = \frac{\pi i}{2}J_1(\kappa r_<)H_1^{(1)}(\kappa r_>),$$

where $H_1^{(1)}(\ )$ is a Hankel function of order 1 and the first kind.

25. Let us find the Green's function[60] governed by the Helmholtz-like axisymmetric equation

$$\frac{\partial^2 g}{\partial r^2} + \frac{1}{r}\frac{\partial g}{\partial r} + \frac{\partial^2 g}{\partial z^2} - k_0^2 g = -\frac{\delta(r-\rho)\delta(z-\zeta)}{2\pi r},$$

where $0 < r, \rho < \infty$, and $0 < z, \zeta < L$, subject to the boundary conditions

$$\lim_{r\to 0}|g(r,z|\rho,\zeta)| < \infty, \quad \lim_{r\to\infty}|g(r,z|\rho,\zeta)| < \infty, \quad 0 < z < L,$$

$$g(r,0|\rho,\zeta) - ag_z(r,0|\rho,\zeta) = 0, \quad 0 < r < \infty,$$

and

$$g(r,L|\rho,\zeta) + ag_z(r,L|\rho,\zeta) = 0, \quad 0 < r < \infty,$$

where $a > 0$.

---

[60] Taken from Reynolds, L., C. Johnson, and A. Ishimaru, 1976: Diffuse reflectance from a finite blood medium: Applications to the modeling of fiber optic catheters. *Appl. Optics*, **15**, 2059–2067.

Step 1: Consider the regular Sturm-Liouville problem

$$\varphi'' + k^2\varphi = 0, \qquad 0 < z < L,$$

with

$$\varphi(0) - a\varphi'(0) = 0, \qquad \text{and} \qquad \varphi(L) + a\varphi'(L) = 0.$$

Show that the eigenfunctions are

$$\varphi_n(z) = \sin(k_n z + \gamma_n), \qquad n = 1, 2, 3, \ldots,$$

provided

$$\tan(k_n L) = \frac{2ak_n}{a^2 k_n^2 - 1}, \qquad \text{and} \qquad \gamma_n = \tan^{-1}(ak_n).$$

Step 2: Using the results from Step 1, show that

$$\delta(z - \zeta) = 4 \sum_{n=1}^{\infty} \frac{k_n \, \varphi_n(\zeta) \, \varphi_n(z)}{2k_n L + \sin(2\gamma_n) - \sin\left[2\left(k_n L + \gamma_n\right)\right]}.$$

Step 3: Assuming that

$$g(r, z | \rho, \zeta) = \frac{2}{\pi} \sum_{n=1}^{\infty} \frac{k_n \, \varphi_n(\zeta) \, \varphi_n(z)}{2k_n L + \sin(2\gamma_n) - \sin\left[2\left(k_n L + \gamma_n\right)\right]} R_n(r|\rho),$$

show that $R_n(r|\rho)$ satisfies the ordinary differential equation

$$\frac{1}{r}\frac{d}{dr}\left(r\frac{dR_n}{dr}\right) - \lambda_n^2 R_n = -\frac{\delta(r - \rho)}{r}, \qquad 0 < r, \rho < \infty,$$

with the boundary conditions

$$\lim_{r \to 0} |R_n(r|\rho)| < \infty, \qquad \text{and} \qquad \lim_{r \to \infty} |R_n(r|\rho)| < \infty,$$

where $\lambda_n^2 = k_0^2 + k_n^2$.

Step 4: Conclude the problem by showing that

$$g(r, z | \rho, \zeta) = \frac{2}{\pi} \sum_{n=1}^{\infty} \frac{k_n \, \varphi_n(\zeta) \, \varphi_n(z)}{2k_n L + \sin(2\gamma_n) - \sin\left[2\left(k_n L + \gamma_n\right)\right]}$$
$$\times I_0(\lambda_n r_<) K_0(\lambda_n r_>).$$

26. Use the method of images to show that the Green's function to the three-dimensional Poisson equation

$$\frac{\partial^2 g}{\partial x^2} + \frac{\partial^2 g}{\partial y^2} + \frac{\partial^2 g}{\partial z^2} = -\delta(x - \xi)\delta(y - \eta)\delta(z - \zeta),$$

for the domain $0 < x, y, \xi, \eta < \infty$, and $0 < z, \zeta < L$, is

$$g(x, y, z|\xi, \eta, \zeta) = \sum_{n=-\infty}^{\infty} g_0(r, z - \zeta + 2nL) + g_0(r, z + \zeta + 2nL),$$

where

$$g_0(r, z) = \frac{1}{4\pi R}, \quad R^2 = r^2 + (z - \zeta)^2, \quad r^2 = (x - \xi)^2 + (y - \eta)^2,$$

if the boundary conditions are

$$\lim_{|x|,|y| \to \infty} |g(x, y, z|\xi, \eta, \zeta)| < \infty,$$

and

$$g_z(x, y, 0|\xi, \eta, \zeta) = g_z(x, y, L|\xi, \eta, \zeta) = 0.$$

Now show that the corresponding eigenfunction expansion is

$$g(x, y, z|\xi, \eta, \zeta) = -\frac{\ln(r/L_0)}{2\pi L} + \frac{1}{\pi L} \sum_{n=1}^{\infty} K_0\left(\frac{n\pi r}{L}\right) \cos\left(\frac{n\pi\zeta}{L}\right) \cos\left(\frac{n\pi z}{L}\right),$$

where L_0 is an arbitrary length constant.

27. Construct the Green's function governed by the three-dimensional Poisson equation

$$\frac{\partial^2 g}{\partial x^2} + \frac{\partial^2 g}{\partial y^2} + \frac{\partial^2 g}{\partial z^2} = -\delta(x - \xi)\delta(y - \eta)\delta(z - \zeta),$$

for $0 < x, \xi < a$, $0 < y, \eta < b$, and $0 < z, \zeta < c$, subject to the Dirichlet boundary conditions

$$g(0, y, z|\xi, \eta, \zeta) = g(a, y, z|\xi, \eta, \zeta) = 0, \quad 0 < y < b, \quad 0 < z < c,$$

$$g(x, 0, z|\xi, \eta, \zeta) = g(x, b, z|\xi, \eta, \zeta) = 0, \quad 0 < x < a, \quad 0 < z < c,$$

and

$$g(x, y, 0|\xi, \eta, \zeta) = g(x, y, c|\xi, \eta, \zeta) = 0, \quad 0 < x < a, \quad 0 < y < b.$$

28. Find the free-space Green's function[61] governed by

$$\frac{\partial^2 g}{\partial x^2} + \frac{\partial^2 g}{\partial y^2} + \frac{\partial^2 g}{\partial z^2} + \gamma \frac{\partial g}{\partial x} = \delta(x - \xi)\delta(y - \eta)\delta(z - \zeta),$$

where $-\infty < x, y, z, \xi, \eta, \zeta < \infty$.

Step 1: Setting $g(x, y, z | \xi, \eta, \zeta) = e^{-\gamma x/2}\varphi(x, y, z | \xi, \eta, \zeta)$, show that the original partial differential equation becomes

$$\frac{\partial^2 \varphi}{\partial x^2} + \frac{\partial^2 \varphi}{\partial y^2} + \frac{\partial^2 \varphi}{\partial z^2} - \frac{\gamma^2}{4}\varphi = e^{\gamma\xi/2}\delta(x - \xi)\delta(y - \eta)\delta(z - \zeta).$$

Step 2: After taking the Fourier transform with respect to x of the partial differential equation in Step 1, show that

$$\frac{\partial^2 \Phi}{\partial y^2} + \frac{\partial^2 \Phi}{\partial z^2} - \left(k^2 + \frac{\gamma^2}{4}\right)\Phi = e^{\gamma\xi/2 - ik\xi}\delta(y - \eta)\delta(z - \zeta).$$

Step 3: Following problem 8, show that

$$\Phi(k, y, z | \xi, \eta, \zeta) = \frac{e^{\gamma\xi/2 - ik\xi}}{2\pi} K_0\left[\sqrt{k^2 + \gamma^2/4}\sqrt{(y - \eta)^2 + (z - \zeta)^2}\right].$$

Step 4: Complete the analysis and show that

$$g(x, y, z | \xi, \eta, \zeta)$$
$$= \frac{e^{-\gamma(x-\xi)/2}}{4\pi^2} \int_{-\infty}^{\infty} K_0\left[\sqrt{k^2 + \gamma^2/4}\sqrt{(y - \eta)^2 + (z - \zeta)^2}\right]$$
$$\times e^{ik(x-\xi)}\, dk$$
$$= \frac{e^{-\gamma(x-\xi)/2}}{2\pi^2} \int_0^{\infty} K_0\left[\sqrt{k^2 + \gamma^2/4}\sqrt{(y - \eta)^2 + (z - \zeta)^2}\right]$$
$$\times \cos[k(x - \xi)]\, dk$$
$$= \frac{\exp\left[-\gamma(x - \xi)/2 + \sqrt{(x - \xi)^2 + (y - \eta)^2 + (z - \zeta)^2}\right]}{2\pi^2\sqrt{(x - \xi)^2 + (y - \eta)^2 + (z - \zeta)^2}}.$$

---

[61] Reprinted from *Deep-Sea Res.*, **39**, N. R. McDonald, Flows caused by mass forcing in a stratified ocean, 1767–1790, ©1992, with permission from Elsevier Science.

29. Construct the Green's function governed by the three-dimensional Poisson equation

$$\frac{1}{r}\frac{\partial}{\partial r}\left(r\frac{\partial g}{\partial r}\right) + \frac{1}{r^2}\frac{\partial^2 g}{\partial \theta^2} + \frac{\partial^2 g}{\partial z^2} = -\frac{\delta(r-\rho)\delta(\theta-\theta')\delta(z-\zeta)}{r},$$

where $0 < r, \rho < \infty$, $0 \le \theta, \theta' \le 2\pi$, and $0 < z, \zeta < L$, subject to the boundary conditions

$$\lim_{r \to 0} |g(r,\theta,z|\rho,\theta',\zeta)| < \infty, \quad \lim_{r \to \infty} |g(r,\theta,z|\rho,\theta',\zeta)| < \infty,$$

$$g(r,\theta,0|\rho,\theta',\zeta) = g(r,\theta,L|\rho,\theta',\zeta) = 0,$$

and periodicity in θ.

Step 1: Because

$$\delta(\theta - \theta')\delta(z - \zeta) = \frac{1}{\pi L} \sum_{n=-\infty}^{\infty} \sum_{m=1}^{\infty} \sin\left(\frac{m\pi\zeta}{L}\right) \sin\left(\frac{m\pi z}{L}\right) e^{in(\theta-\theta')},$$

show that

$$g(r,\theta,z|\rho,\theta',\zeta) = \frac{1}{\pi L} \sum_{n=-\infty}^{\infty} \sum_{m=1}^{\infty} G_{nm}(r|\rho) \sin\left(\frac{m\pi\zeta}{L}\right) \sin\left(\frac{m\pi z}{L}\right)$$
$$\times\, e^{in(\theta-\theta')},$$

where

$$\frac{1}{r}\frac{d}{dr}\left(r\frac{dG_{nm}}{dr}\right) - \left(\frac{m^2\pi^2}{L^2} + \frac{n^2}{r^2}\right)G_{nm} = -\frac{\delta(r-\rho)}{r}.$$

Step 2: Show that

$$G_{nm}(r|\rho) = I_n(m\pi r_</L)K_n(m\pi r_>/L).$$

Step 3: Conclude the problem by showing that

$$g(r,\theta,z|\rho,\theta',\zeta) = \frac{1}{\pi L} \sum_{n=-\infty}^{\infty} \sum_{m=1}^{\infty} \sin\left(\frac{m\pi\zeta}{L}\right) \sin\left(\frac{m\pi z}{L}\right) e^{in(\theta-\theta')}$$
$$\times\, I_n(m\pi r_</L)K_n(m\pi r_>/L).$$

Step 4: Show that an alternative form of the Green's function is

$$g(r, \theta, z | \rho, \theta', \zeta) = 2 \sum_{n=-\infty}^{\infty} \int_0^{\infty} e^{in(\theta - \theta')} J_n(kr) J_n(k\rho)\, dk$$

$$\times \begin{cases} \sinh(kz) \sinh[k(L - \zeta)]/\sinh(kL), & 0 \le z \le \zeta, \\ \sinh(k\zeta) \sinh[k(L - z)]/\sinh(kL), & \zeta \le z \le L. \end{cases}$$

30. Construct the Green's function[62] governed by the three-dimensional Poisson equation

$$\frac{1}{r} \frac{\partial}{\partial r} \left(r \frac{\partial g}{\partial r} \right) + \frac{1}{r^2} \frac{\partial^2 g}{\partial \theta^2} + \frac{\partial^2 g}{\partial z^2} = -\frac{\delta(r - \rho)\delta(\theta - \theta')\delta(z - \zeta)}{r},$$

where $a < r, \rho < \infty$, $0 \le \theta, \theta' \le 2\pi$, and $0 < z, \zeta < L$, subject to the boundary conditions

$$\frac{\partial g(a, \theta, z | \rho, \theta', \zeta)}{\partial r} = 0,$$

$$\lim_{r \to \infty} \sqrt{r} \left[\frac{\partial g(r, \theta, z | \rho, \theta', \zeta)}{\partial r} - i\kappa g(r, \theta, z | \rho, \theta', \zeta) \right] = 0,$$

$$\frac{\partial g(r, \theta, 0 | \rho, \theta', \zeta)}{\partial z} + A g(r, \theta, 0 | \rho, \theta', \zeta) = 0, \qquad \frac{\partial g(r, \theta, L | \rho, \theta', \zeta)}{\partial z} = 0,$$

and periodicity in θ.

Step 1: Show that

$$\delta(\theta - \theta') = \frac{1}{2\pi} \sum_{n=-\infty}^{\infty} \cos[n(\theta - \theta')],$$

and

$$\delta(z - \zeta) = \sum_{m=0}^{\infty} Z_m(k_m \zeta) Z_m(k_m z),$$

where

$$Z_0(k_m z) = \cosh[k_0(z - L)]/\sqrt{N_0}, \quad Z_m(k_m z) = \cos[k_m(z - L)]/\sqrt{N_m},$$

$$N_0 = \frac{L(k_0^2 - A^2) + A}{2(k_0^2 - A^2)}, \qquad N_m = \frac{L(k_m^2 + A^2) - A}{2(k_m^2 + A^2)},$$

---

[62] Taken from Chau, F. P., and R. E. Taylor, 1992: Second-order wave diffraction by a vertical cylinder. *J. Fluid Mech.*, **240**, 571–599. Reprinted with the permission of Cambridge University Press.

$$k_0 \tanh(k_0 L) = A, \qquad \text{and} \qquad k_m \tan(k_m L) = -A.$$

Hint: Consider the eigenfunction expansion given by the eigenfunctions $\varphi_n(z)$ from the regular Sturm-Liouville problem

$$\varphi'' + k^2 \varphi = 0, \qquad \varphi'(0) + A\varphi(0) = 0, \qquad \varphi'(L) = 0.$$

Step 2: If we write the Green's function as

$$g(r, \theta, z | \rho, \theta', \zeta) = \frac{1}{2\pi} \sum_{n=-\infty}^{\infty} \sum_{m=0}^{\infty} G_{nm}(r|\rho) Z_m(k_m\zeta) Z_m(k_m z)$$

$$\times \cos[n(\theta - \theta')],$$

show that G_{nm} is governed by

$$\frac{1}{r} \frac{d}{dr}\left(r \frac{dG_{nm}}{dr}\right) - \left(k_m^2 + \frac{n^2}{r^2}\right) G_{nm} = -\frac{\delta(r - \rho)}{r},$$

with

$$G'_{nm}(a|\rho) = 0,$$

and

$$\lim_{r \to \infty} \sqrt{r} \left[G'_{nm}(r|\zeta) - i\kappa G_{nm}(r|\rho)\right] = 0.$$

Step 3: Show that

$$G_{nm}(r|\rho) = \frac{K_n(k_m r_>)}{K'_n(k_m a)}[I_n(k_m r_<)K'_n(k_m a) - I'_n(k_m a)K_n(k_m r_<)].$$

Step 4: Conclude the problem by showing that

$$g(r, \theta, z | \rho, \theta', \zeta) = \frac{1}{2\pi} \sum_{n=-\infty}^{\infty} \sum_{m=0}^{\infty} Z_m(k_m\zeta) Z_m(k_m z) \cos[n(\theta - \theta')]$$

$$\times \frac{K_n(k_m r_>)}{K'_n(k_m a)}[I_n(k_m r_<)K'_n(k_m a)$$

$$- I'_n(k_m a)K_n(k_m r_<)].$$

31. Construct the Green's function governed by the three-dimensional Helmholtz equation

$$\frac{\partial^2 g}{\partial x^2} + \frac{\partial^2 g}{\partial y^2} + \frac{\partial^2 g}{\partial z^2} + k_0^2 g = -\delta(x - \xi)\delta(y - \eta)\delta(z - \zeta),$$

for $0 < x, \xi < a$, $0 < y, \eta < b$, and $0 < z, \zeta < c$, subject to the Dirichlet boundary conditions

$$g(0, y, z|\xi, \eta, \zeta) = g(a, y, z|\xi, \eta, \zeta) = 0, \quad 0 < y < b, \quad 0 < z < c,$$

$$g(x, 0, z|\xi, \eta, \zeta) = g(x, b, z|\xi, \eta, \zeta) = 0, \quad 0 < x < a, \quad 0 < z < c,$$

and

$$g(x, y, 0|\xi, \eta, \zeta) = g(x, y, c|\xi, \eta, \zeta) = 0, \quad 0 < x < a, \quad 0 < y < b.$$

32. Construct the Green's function governed by the three-dimensional Helmholtz equation

$$\frac{\partial^2 g}{\partial x^2} + \frac{\partial^2 g}{\partial y^2} + \frac{\partial^2 g}{\partial z^2} + k_0^2 g = -\delta(x - \xi)\delta(y - \eta)\delta(z - \zeta),$$

for $0 < x, \xi < a$, $0 < y, \eta < b$, and $0 < z, \zeta < c$, subject to the Neumann boundary conditions

$$g_x(0, y, z|\xi, \eta, \zeta) = g_x(a, y, z|\xi, \eta, \zeta) = 0, \quad 0 < y < b, \quad 0 < z < c,$$

$$g_y(x, 0, z|\xi, \eta, \zeta) = g_y(x, b, z|\xi, \eta, \zeta) = 0, \quad 0 < x < a, \quad 0 < z < c,$$

and

$$g_z(x, y, 0|\xi, \eta, \zeta) = g_z(x, y, c|\xi, \eta, \zeta) = 0, \quad 0 < x < a, \quad 0 < y < b.$$

33. Show[63] that the Green's function governed by biharmonic equation

$$\frac{\partial^4 g}{\partial x^4} + 2\frac{\partial^4 g}{\partial x^2 \partial y^2} + \frac{\partial^4 g}{\partial y^4} - k_0^4 g = \delta(x - \xi)\delta(y - \eta),$$

with the boundary conditions

$$g(0, y|\xi, \eta) = g(1, y|\xi, \eta) = g_{xx}(0, y|\xi, \eta) = g_{xx}(1, y|\xi, \eta) = 0,$$

and

$$g(x, 0|\xi, \eta) = g(x, 1|\xi, \eta) = g_{yy}(x, 0|\xi, \eta) = g_{yy}(x, 1|\xi, \eta) = 0,$$

---

[63] Taken from Nicholson, J. W., and L. A. Bergman, 1985: On the efficacy of the modal series representation for the Green functions of vibrating continuous structures. *J. Sound Vibr.*, **98**, 299–304; see §3 of Bergman, L. A., J. K. Hall, G. G. G. Lueschen, and D. M. McFarland, 1993: Dynamic Green's functions for Levy plates. *J. Sound Vibr.*, **162**, 281–310. Published by Academic Press Ltd., London, U.K.

where $0 < x, y, \xi, \eta < 1$, is

$$g(x, y|\xi, \eta) = \frac{1}{k_0^2} \sum_{n=1}^{\infty} \Phi_n(y|\eta) \sin(n\pi\xi) \sin(n\pi x),$$

where

$$\Phi_n(y|\eta) = \frac{\Psi_2(1 - \eta) \Psi_2(y)}{\Psi_2(1)} - \frac{\Psi_1(1 - \eta) \Psi_1(y)}{\Psi_1(1)},$$

$$\Psi_1(y) = \frac{\sinh\left(y\sqrt{n^2\pi^2 + k_0^2}\right)}{\sqrt{n^2\pi^2 + k_0^2}},$$

and

$$\Psi_2(y) = \begin{cases} \sinh\left(y\sqrt{n^2\pi^2 - k_0^2}\right) \Big/ \sqrt{n^2\pi^2 - k_0^2}, & n\pi > k_0, \\[2mm] y, & k = n\pi, \\[2mm] \sin\left(y\sqrt{k_0^2 - n^2\pi^2}\right) \Big/ \sqrt{k_0^2 - n^2\pi^2}, & k_0 > n\pi, \end{cases}$$

or

$$g(x, y|\xi, \eta) = 4 \sum_{n=1}^{\infty} \sum_{m=1}^{\infty} \frac{\sin(n\pi\xi) \sin(n\pi x) \sin(m\pi\eta) \sin(m\pi y)}{\pi^4(n^2 + m^2)^2 - k_0^4}.$$

34. Find the Green's function governed by

$$\frac{\partial^4 g}{\partial x^2 \partial z^2} + a^2\left(\frac{\partial^2 g}{\partial x^2} + \frac{\partial^2 g}{\partial y^2}\right) + b^2\frac{\partial^2 g}{\partial z^2} = -\delta(x - \xi)\delta(y - \eta)\delta(z - \zeta),$$

where $-\infty < x, y, \xi, \eta < \infty$, and $0 < z, \zeta < 1$. The boundary conditions are

$$g(x, y, 0|\xi, \eta, \zeta) = g(x, y, 1|\xi, \eta, \zeta) = 0,$$

$$\lim_{|x|\to\infty} g(x, y, z|\xi, \eta, \zeta) \to 0,$$

and

$$\lim_{|y|\to\infty} g(x, y, z|\xi, \eta, \zeta) \to 0.$$

Step 1: Show that if we expand $g(x, y, z|\xi, \eta, \zeta)$ as the half-range Fourier expansion

$$g(x, y, z|\xi, \eta, \zeta) = \sum_{n=1}^{\infty} g_n(x, y|\xi, \eta) \sin(n\pi z),$$

then the partial differential equation governing $g_n(x, y|\xi, \eta)$ is

$$(a^2 - n^2\pi^2)\frac{\partial^2 g_n}{\partial x^2} + a^2\frac{\partial^2 g_n}{\partial y^2} - n^2\pi^2 b^2 g_n = -2\sin(n\pi\zeta)\delta(x - \xi)\delta(y - \eta).$$

Step 2: Defining the Fourier transform of $g_n(x, y|\xi, \eta)$ by

$$G_n(x, \ell|\xi, \eta) = \int_{-\infty}^{\infty} g_n(x, y|\xi, \eta)e^{-i\ell y}\, dy,$$

show that the Fourier transform of the partial differential equation in Step 1 is

$$(a^2 - n^2\pi^2)\frac{d^2 G_n}{dx^2} - (a^2\ell^2 + n^2\pi^2 b^2)G_n = -2e^{-i\ell\eta}\sin(n\pi\zeta)\delta(x - \xi).$$

Step 3: Let us denote the first n for which $n^2\pi^2 > a^2$ by n_0. Now show that

$$G_n(x, \ell|\xi, \eta) = e^{-i\ell\eta}\begin{cases} \dfrac{\sin(n\pi\zeta)\exp\left(-\sqrt{\frac{a^2\ell^2 + n^2\pi^2 b^2}{a^2 - n^2\pi^2}}\,|x - \xi|\right)}{\sqrt{a^2 - n^2\pi^2}\sqrt{a^2\ell^2 + n^2\pi^2 b^2}}, n < n_0, \\[4ex] \dfrac{2\sin(n\pi\zeta)\sin\left[\sqrt{\frac{a^2\ell^2 + n^2\pi^2 b^2}{n^2\pi^2 - a^2}}\,(x - \xi)\right]}{\sqrt{n^2\pi^2 - a^2}\sqrt{a^2\ell^2 + n^2\pi^2 b^2}}H(\xi - x), \\ \hspace{6cm} n \geq n_0. \end{cases}$$

Step 4: Invert $G_n(x, \ell|\xi, \eta)$ and show that

$$g(x, y, z|\xi, \eta, \zeta) = \frac{1}{\pi a}\sum_{n=1}^{n_0 - 1}\frac{\sin(n\pi\zeta)\sin(n\pi z)}{\sqrt{a^2 - n^2\pi^2}}$$

$$\times K_0\left[\frac{n\pi b}{a}\sqrt{\frac{a^2(x - \xi)^2}{a^2 - n^2\pi^2} + (y - \eta)^2}\right]$$

$$+ \frac{1}{\pi a}\sum_{n=n_0}^{\infty}\frac{\sin(n\pi\zeta)\sin(n\pi z)}{\sqrt{n^2\pi^2 - a^2}}$$

$$\times J_0\left[\frac{n\pi b}{a}\sqrt{\frac{a^2(x - \xi)^2}{n^2\pi^2 - a^2} - (y - \eta)^2}\right]$$

$$\times H(\xi - x)\,H\left[\frac{a^2(x - \xi)^2}{n^2\pi^2 - a^2} - (y - \eta)^2\right].$$

35.   Following Example 5.7.1, show that the Green's function[64] for Laplace's equation is

$$g(r, \theta, \varphi | a, \theta', \varphi') = \frac{r^2 - a^2}{4\pi |\mathbf{r} - \mathbf{r_0}|^3}$$

for the region exterior to a sphere of radius a when the source is located on the surface at $(a, \theta' \varphi')$ and given by

$$g(a, \theta, \varphi | a, \theta', \varphi') = \frac{\delta(\theta - \theta') \delta(\varphi - \varphi')}{a \sin(\theta')}.$$

Here $\mathbf{r}$ and $\mathbf{r_0}$ denote the position vectors of a point (r, θ, φ) inside the sphere and the source, respectively. What should be the Green's function for the spherical domain $0 < r < a$ if the same surface Dirichlet boundary condition occurs? Hint:

$$\frac{1 - t^2}{(1 - 2tx + t^2)^{3/2}} = \sum_{n=0}^{\infty} (2n + 1) P_n(x) t^n.$$

36. Use (D.5) to establish the identity that

$$e^{ikr \cos(\theta)} = \sum_{n=0}^{\infty} i^n (2n + 1) j_n(kr) P_n[\cos(\theta)].$$

---

[64] Reprinted with permission from Nemenman, I. M., and A. S. Silbergleit, 1999: Explicit Green's function of a boundary value problem for a sphere and trapped flux analysis in Gravity Probe B experiment. *J. Appl. Phys.*, **86**, 614–624. ©American Institute of Physics, 1999.

Appendix A:
The Fourier Transform

We review here some of the basic properties of the Fourier transform and how they are used in solving differential equations. We employ these techniques throughout the text.

A.1 DEFINITION AND PROPERTIES OF FOURIER TRANSFORMS

The Fourier transform is the natural extension of Fourier series to a function $f(t)$ of infinite period. The Fourier transform is defined in terms of a pair of integrals:

$$f(t) = \frac{1}{2\pi} \int_{-\infty}^{\infty} F(\omega) e^{i\omega t} d\omega, \qquad (\mathbf{A.1.1})$$

and

$$F(\omega) = \int_{-\infty}^{\infty} f(t) e^{-i\omega t} dt. \qquad (\mathbf{A.1.2})$$

Equation (A.1.2) is the *Fourier transform* of $f(t)$ while (A.1.1) is the *inverse Fourier transform*, which converts a Fourier transform back to $f(t)$. Hamming[1] has suggested the following analog in understanding

---

[1] Hamming, R. W., 1977: *Digital Filters*. Prentice-Hall, p. 136.

Table A.1.1: Some General Properties of Fourier Transforms

	function, f(t)	Fourier transform, F(ω)
1. Linearity	$c_1 f(t) + c_2 g(t)$	$c_1 F(\omega) + c_2 G(\omega)$
2. Complex conjugate	$f^*(t)$	$F^*(-\omega)$
3. Scaling	$f(\alpha t)$	$F(\omega/\alpha)/|\alpha|$
4. Delay	$f(t - \tau)$	$e^{-i\omega\tau} F(\omega)$
5. Frequency translation	$e^{i\omega_0 t} f(t)$	$F(\omega - \omega_0)$
6. Duality in time and frequency	$F(t)$	$2\pi f(-\omega)$
7. Time differentiation	$f'(t)$	$i\omega F(\omega)$

the Fourier transform. Let us imagine that $f(t)$ is a light beam. Then the Fourier transform, like a glass prism, breaks up the function into its component frequencies ω, each of intensity $F(\omega)$. In optics, the various frequencies are called colors; by analogy the Fourier transform gives us the color spectrum of a function. On the other hand, the inverse Fourier transform blends a function's spectrum to give back the original function.

• **Example A.1.1**

Repeatedly in this book, we must find the Fourier transform of the derivative of a function $f(t)$ that is differentiable for all t and vanishes as $t \to \pm\infty$. From the definition of the Fourier transform,

$$\mathcal{F}[f'(t)] = \int_{-\infty}^{\infty} f'(t) e^{-i\omega t}\, dt \qquad (A.1.3)$$

$$= f(t) e^{-i\omega t}\Big|_{-\infty}^{\infty} + i\omega \int_{-\infty}^{\infty} f(t) e^{-i\omega t}\, dt \qquad (A.1.4)$$

$$= i\omega F(\omega), \qquad (A.1.5)$$

where $F(\omega)$ is the Fourier transform of $f(t)$. Similarly,

$$\mathcal{F}[f''(t)] = -\omega^2 F(\omega). \qquad (A.1.6)$$

In principle, we can compute any Fourier transform from the definition. However, it is far more efficient to derive some simple relationships that relate transforms to each other. Some of the most useful properties are given in Table A.1.1.

A.2 INVERSION OF FOURIER TRANSFORMS

Although we can find the inverse by direct integration or partial fractions, in many instances, the Fourier transform does not lend itself to these techniques. On the other hand, if we view the inverse Fourier transform as a line integral along the real axis in the complex ω-plane, then we can use complex variable theory to evaluate the integral. To this end, we rewrite the inversion integral (A.1.1) as

$$f(t) = \frac{1}{2\pi} \int_{-\infty}^{\infty} F(\omega)e^{it\omega}\, d\omega = \frac{1}{2\pi} \oint_C F(z)e^{itz}\, dz - \frac{1}{2\pi} \int_{C_R} F(z)e^{itz}\, dz,$$

$$\textbf{(A.2.1)}$$

where C denotes a closed contour consisting of the entire real axis plus a new contour C_R that joins the point $(\infty, 0)$ to $(-\infty, 0)$. There are countless possibilities for C_R. For example, it could be the loop $(\infty, 0)$ to (∞, R) to $(-\infty, R)$ to $(-\infty, 0)$ with $R > 0$. However, any choice of C_R must be such that we can compute $\int_{C_R} F(z)e^{itz}\, dz$. When we take that constraint into account, the number of acceptable contours decreases to just a few. The best is given by *Jordan's lemma*:

Jordan's lemma: *Suppose that, on a circular arc C_R with radius R and center at the origin, $f(z) \to 0$ uniformly as $R \to \infty$. Then*

$$(1) \qquad \lim_{R\to\infty} \int_{C_R} f(z)e^{imz}dz = 0, \qquad (m > 0) \qquad \textbf{(A.2.2)}$$

if C_R lies in the first and/or second quadrant;

$$(2) \qquad \lim_{R\to\infty} \int_{C_R} f(z)e^{-imz}dz = 0, \qquad (m > 0) \qquad \textbf{(A.2.3)}$$

if C_R lies in the third and/or fourth quadrant;

$$(3) \qquad \lim_{R\to\infty} \int_{C_R} f(z)e^{mz}dz = 0, \qquad (m > 0) \qquad \textbf{(A.2.4)}$$

if C_R lies in the second and/or third quadrant; and

$$(4) \qquad \lim_{R\to\infty} \int_{C_R} f(z)e^{-mz}dz = 0, \qquad (m > 0) \qquad \textbf{(A.2.5)}$$

if C_R lies in the first and/or fourth quadrant.

The proof is given elsewhere. Consider now the following inversions of Fourier transforms:

• **Example A.2.1**

Let us find the inverse for

$$F(\omega) = \frac{1}{\omega^2 - 2ib\omega - a^2 - b^2}, \qquad a, b > 0. \tag{A.2.6}$$

From the inversion integral,

$$f(t) = \frac{1}{2\pi} \int_{-\infty}^{\infty} \frac{e^{it\omega}}{\omega^2 - 2ib\omega - a^2 - b^2} \, d\omega, \tag{A.2.7}$$

or

$$f(t) = \frac{1}{2\pi} \oint_C \frac{e^{itz}}{z^2 - 2ibz - a^2 - b^2} \, dz - \frac{1}{2\pi} \int_{C_R} \frac{e^{itz}}{z^2 - 2ibz - a^2 - b^2} \, dz, \tag{A.2.8}$$

where C denotes a closed contour consisting of the entire real axis plus C_R. Because $f(z) = 1/(z^2 - 2ibz - a^2 - b^2)$ tends to zero uniformly as $|z| \to \infty$ and $m = t$, the second integral in (A.2.8) vanishes by Jordan's lemma if C_R is a semicircle of infinite radius in the upper half of the z-plane when $t > 0$ and a semicircle in the lower half of the z-plane when $t < 0$.

Next, we must find the location and nature of the singularities. They are located at

$$z^2 - 2ibz - a^2 - b^2 = 0, \tag{A.2.9}$$

or

$$z = \pm a + bi. \tag{A.2.10}$$

Therefore we can rewrite (A.2.8) as

$$f(t) = \frac{1}{2\pi} \oint_C \frac{e^{itz}}{(z - a - bi)(z + a - bi)} \, dz. \tag{A.2.11}$$

Thus, all of the singularities are simple poles.

Consider now $t > 0$. As stated earlier, we close the line integral with an infinite semicircle in the upper half-plane. See Figure A.2.1. Inside this closed contour there are two singularities: $z = \pm a + bi$. For these poles,

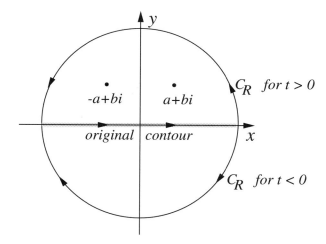

Figure A.2.1: Contour used to find the inverse of the Fourier transform (A.2.6). The contour C consists of the line integral along the real axis plus C_R.

$$\text{Res}\left(\frac{e^{itz}}{z^2 - 2ibz - a^2 - b^2}; a + bi\right)$$

$$= \lim_{z \to a+bi} (z - a - bi)\frac{e^{itz}}{(z - a - bi)(z + a - bi)} \qquad \text{(A.2.12)}$$

$$= \frac{e^{iat}e^{-bt}}{2a} = \frac{e^{-bt}}{2a}[\cos(at) + i\sin(at)], \qquad \text{(A.2.13)}$$

where we used Euler's formula to eliminate e^{iat}. Similarly,

$$\text{Res}\left(\frac{e^{itz}}{z^2 - 2ibz - a^2 - b^2}; -a + bi\right) = -\frac{e^{-bt}}{2a}[\cos(at) - i\sin(at)].$$
$$\text{(A.2.14)}$$

Consequently, the inverse Fourier transform follows from (A.2.11) after applying the residue theorem and equals

$$f(t) = -\frac{e^{-bt}}{2a}\sin(at) \qquad \text{(A.2.15)}$$

for $t > 0$.

For $t < 0$ the semicircle is in the lower half-plane because the contribution from the semicircle vanishes as $R \to \infty$. Because there are no singularities within the closed contour, $f(t) = 0$. Therefore, we can write in general that

$$f(t) = -\frac{e^{-bt}}{2a}\sin(at)H(t). \qquad \text{(A.2.16)}$$

• **Example A.2.2**

Let us find the inverse of the Fourier transform

$$F(\omega) = \frac{e^{-\omega i}}{\omega^2 + a^2}, \tag{A.2.17}$$

where a is real and positive.

From the inversion integral,

$$f(t) = \frac{1}{2\pi} \int_{-\infty}^{\infty} \frac{e^{i(t-1)\omega}}{\omega^2 + a^2} \, d\omega \tag{A.2.18}$$

$$= \frac{1}{2\pi} \oint_C \frac{e^{i(t-1)z}}{z^2 + a^2} \, dz - \frac{1}{2\pi} \int_{C_R} \frac{e^{i(t-1)z}}{z^2 + a^2} \, dz, \tag{A.2.19}$$

where C denotes a closed contour consisting of the entire real axis plus C_R. The contour C_R is determined by Jordan's lemma because $1/(z^2 + a^2) \to 0$ uniformly as $|z| \to \infty$. Since $m = t - 1$, the semicircle C_R of infinite radius lies in the upper half-plane if $t > 1$ and in the lower half-plane if $t < 1$. Thus, if $t > 1$,

$$f(t) = \frac{1}{2\pi}(2\pi i)\text{Res}\left[\frac{e^{i(t-1)z}}{z^2 + a^2}; ai\right] = \frac{e^{-a(t-1)}}{2a}, \tag{A.2.20}$$

whereas for $t < 1$,

$$f(t) = \frac{1}{2\pi}(-2\pi i)\text{Res}\left[\frac{e^{i(t-1)z}}{z^2 + a^2}; -ai\right] = \frac{e^{a(t-1)}}{2a}. \tag{A.2.21}$$

The minus sign in front of the $2\pi i$ arises from the clockwise direction or negative sense of the contour. We can write the inverse as the single expression

$$f(t) = \frac{e^{-a|t-1|}}{2a}. \tag{A.2.22}$$

A.3 SOLUTION OF ORDINARY DIFFERENTIAL EQUATIONS

As with Laplace transforms, we can use Fourier transforms to solve differential equations. However, this method gives only the particular solution and we must find the complementary solution separately.

Consider the differential equation

$$y' + y = \tfrac{1}{2}e^{-|t|}, \qquad -\infty < t < \infty. \tag{A.3.1}$$

Taking the Fourier transform of both sides of (A.3.1),

$$i\omega Y(\omega) + Y(\omega) = \tfrac{1}{2} \int_{-\infty}^{0} e^{t-i\omega t} \, dt + \tfrac{1}{2} \int_{0}^{\infty} e^{-t-i\omega t} \, dt = \frac{1}{\omega^2 + 1}, \quad (\mathbf{A.3.2})$$

where we used the time differentiation rule from Table A.1.1 to obtain the transform of y' and $Y(\omega) = \mathcal{F}[y(t)]$. Therefore,

$$Y(\omega) = \frac{1}{(\omega^2 + 1)(1 + \omega i)}. \quad (\mathbf{A.3.3})$$

Applying the inversion integral to (A.3.3),

$$y(t) = \frac{1}{2\pi} \int_{-\infty}^{\infty} \frac{e^{it\omega}}{(\omega^2 + 1)(1 + \omega i)} \, d\omega. \quad (\mathbf{A.3.4})$$

We evaluate (A.3.4) by contour integration. For $t > 0$ we close the line integral with an infinite semicircle in the upper half of the ω-plane. The integration along this arc equals zero by Jordan's lemma. Within this closed contour we have a second-order pole at $z = i$. Therefore,

$$\mathrm{Res}\left[\frac{e^{itz}}{(z^2+1)(1+zi)}; i\right] = \lim_{z\to i} \frac{d}{dz}\left[(z-i)^2 \frac{e^{itz}}{i(z-i)^2(z+i)}\right] (\mathbf{A.3.5})$$

$$= \frac{te^{-t}}{2i} + \frac{e^{-t}}{4i}, \quad (\mathbf{A.3.6})$$

and

$$y(t) = \frac{1}{2\pi}(2\pi i)\left[\frac{te^{-t}}{2i} + \frac{e^{-t}}{4i}\right] = \frac{e^{-t}}{4}(2t+1). \quad (\mathbf{A.3.7})$$

For $t < 0$, we again close the line integral with an infinite semicircle, but this time it is in the lower half of the ω-plane. The contribution from the line integral along the arc vanishes by Jordan's lemma. Within the contour, we have a simple pole at $z = -i$. Therefore,

$$\mathrm{Res}\left[\frac{e^{itz}}{(z^2+1)(1+zi)}; -i\right] = \lim_{z\to -i}(z+i)\frac{e^{itz}}{i(z+i)(z-i)^2} = -\frac{e^t}{4i}, \quad (\mathbf{A.3.8})$$

and

$$y(t) = \frac{1}{2\pi}(-2\pi i)\left(-\frac{e^t}{4i}\right) = \frac{e^t}{4}. \quad (\mathbf{A.3.9})$$

The minus sign in front of the $2\pi i$ results from the clockwise direction or negative sense in which the closed contour is taken. Using the Heaviside step function, we can combine (A.3.7) and (A.3.9) into the single expression

$$y(t) = \tfrac{1}{4}e^{-|t|} + \tfrac{1}{2}te^{-t}H(t). \quad (\mathbf{A.3.10})$$

Note that we found only the particular or forced solution to (A.3.1). The most general solution therefore requires that we add the complementary solution Ae^{-t}, yielding

$$y(t) = Ae^{-t} + \tfrac{1}{4}e^{-|t|} + \tfrac{1}{2}te^{-t}H(t). \qquad \textbf{(A.3.11)}$$

The arbitrary constant A would be determined by the initial condition, which we have not specified.

A.4 SOLUTION OF PARTIAL DIFFERENTIAL EQUATIONS

Just as we can solve ordinary differential equations by Fourier transforms, similar considerations hold for partial differential equations. To illustrate this technique, let us calculate the sound waves[2] radiated by a sphere of radius a whose surface expands radially with an impulsive acceleration $v_0\delta(t)$.

If we assume radial symmetry, the wave equation is

$$\frac{1}{r^2}\frac{\partial}{\partial r}\left(r^2\frac{\partial u}{\partial r}\right) = \frac{1}{c^2}\frac{\partial^2 u}{\partial t^2}, \qquad \textbf{(A.4.1)}$$

subject to the boundary condition

$$\frac{\partial u}{\partial r} = -\rho v_0\delta(t) \qquad \textbf{(A.4.2)}$$

at $r = a$, where $u(r,t)$ is the pressure field, c is the speed of sound, and ρ is the average density of the fluid. Assuming that the pressure field possesses a Fourier transform, we may re-express it by the Fourier integral

$$u(r,t) = \frac{1}{2\pi}\int_{-\infty}^{\infty} U(r,\omega)e^{i\omega t}d\omega. \qquad \textbf{(A.4.3)}$$

Substituting into (A.4.1)–(A.4.2), they become

$$\frac{1}{2\pi}\int_{-\infty}^{\infty}\left[\frac{1}{r^2}\frac{d}{dr}\left(r^2\frac{dU}{dr}\right) + k_0^2 U\right]e^{i\omega t}d\omega = 0, \qquad \textbf{(A.4.4)}$$

and

$$\frac{1}{2\pi}\int_{-\infty}^{\infty}\left[\frac{dU(a,\omega)}{dr} + \rho v_0\right]e^{i\omega t}d\omega = 0, \qquad \textbf{(A.4.5)}$$

---

[2] Taken from Hodgson, D. C., and J. E. Bowcock, 1975: Billet expansion as a mechanism for noise production in impact forming machines. *J. Sound Vib.*, **42**, 325–335. Published by Academic Press Ltd., London, U.K.

with $k_0 = \omega/c$. Because (A.4.4) and (A.4.5) must be true for any time t, the bracketed quantities must vanish and we have

$$\frac{1}{r^2}\frac{d}{dr}\left(r^2\frac{dU}{dr}\right) + k_0^2 U = 0, \qquad (\textbf{A.4.6})$$

and

$$\frac{dU(a,\omega)}{dr} = -\rho v_0. \qquad (\textbf{A.4.7})$$

The most general solution of (A.4.6) is

$$U(r,\omega) = A(\omega)\frac{e^{ik_0 r}}{r} + B(\omega)\frac{e^{-ik_0 r}}{r}, \qquad (\textbf{A.4.8})$$

and

$$u(r,t) = \frac{1}{2\pi}\int_{-\infty}^{\infty}\left[A(\omega)\frac{e^{i\omega t + ik_0 r}}{r} + B(\omega)\frac{e^{i\omega t - ik_0 r}}{r}\right]d\omega. \qquad (\textbf{A.4.9})$$

At this point, we note that the first and second terms on the right side of (A.4.9) represent inwardly and outwardly propagating waves, respectively. Because there is no source of energy at infinity, the inwardly propagating wave is nonphysical and we discard it.

Upon substituting (A.4.8) into (A.4.7) with $A(\omega) = 0$,

$$U(r,\omega) = \frac{\rho a^2 v_0}{2\pi r}\frac{e^{-i\omega(r-a)/c}}{1 + i\omega a/c}, \qquad (\textbf{A.4.10})$$

or

$$u(r,t) = \frac{1}{2\pi}\int_{-\infty}^{\infty} U(r,\omega)e^{i\omega t}\,d\omega = \frac{\rho a^2 v_0}{2\pi r}\int_{-\infty}^{\infty}\frac{e^{i\omega[t-(r-a)/c]}}{1 + i\omega a/c}\,d\omega.$$

$$(\textbf{A.4.11})$$

To evaluate (A.4.11), we employ the residue theorem. For $t < (r-a)/c$ Jordan's lemma dictates that we close the line integral by an infinite semicircle in the lower half-plane. For $t > (r-a)/c$, we close the contour with an semicircle in the upper half-plane. The final result is

$$u(r,t) = \frac{\rho a c v_0}{r}\exp\left[-\frac{c}{a}\left(t - \frac{r-a}{c}\right)\right]H\left(t - \frac{r-a}{c}\right). \qquad (\textbf{A.4.12})$$

In this example, we eliminated the temporal dependence by using Fourier transforms. The ordinary differential equation was then solved using homogeneous solutions. In certain cases, an alternative would be to solve the ordinary differential equation by Fourier or Hankel transforms. This is particularly true in the case of Green's functions because the forcing function is a delta function. In the two-dimensional case, we obtain an algebraic equation that we solve to find the *joint transform*. We then compute the inverses successively. The order in which the transforms are inverted depends upon the problem. This repeated application of transforms or Fourier series to a linear partial differential equation to reduce it to an algebraic or ordinary differential equation can be extended to higher spatial dimensions.

Appendix B:
The Laplace Transform

If a function is nonzero only when $t > 0$, we can replace the Fourier transform with the *Laplace transform*. It is particularly useful in solving initial-value problems involving linear, constant coefficient, ordinary and partial differential equations. We summarize the main points here.

B.1 DEFINITION AND ELEMENTARY PROPERTIES

Consider a function $f(t)$ such that $f(t) = 0$ for $t < 0$. Then the *Laplace integral*

$$\mathcal{L}[f(t)] = F(s) = \int_0^\infty f(t)e^{-st}dt \qquad \textbf{(B.1.1)}$$

defines the Laplace transform of $f(t)$, which we write $\mathcal{L}[f(t)]$ or $F(s)$. The Laplace transform of $f(t)$ exists, for sufficiently large s, provided $f(t)$ satisfies the following conditions:

- $f(t) = 0$ for $t < 0$,
- $f(t)$ is continuous or piece-wise continuous in every interval,
- $t^n|f(t)| < \infty$ as $t \to 0$ for some number n, where $n < 1$,
- $e^{-s_0 t}|f(t)| < \infty$ as $t \to \infty$, for some number s_0. The quantity s_0 is called the *abscissa of convergence*.

- **Example B.1.1**

    Let us find the Laplace transform for the *Heaviside step function*:

$$H(t - a) = \begin{cases} 1, & t > a, \\ 0, & t < a, \end{cases} \tag{B.1.2}$$

where $a \geq 0$. The Heaviside step function is essentially a bookkeeping device that gives us the ability to "switch on" and "switch off" a given function. For example, if we want a function $f(t)$ to become nonzero at time $t = a$, we represent this process by the product $f(t)H(t - a)$.

From the definition of the Laplace transform,

$$\mathcal{L}[H(t - a)] = \int_a^\infty e^{-st} dt = \frac{e^{-as}}{s}, \qquad s > 0. \tag{B.1.3}$$

- **Example B.1.2**

    Let us find the Laplace transform of the *Dirac delta function* or *impulse function*. From (1.2.10), the Laplace transform of the delta function is

$$\mathcal{L}[\delta(t - a)] = \int_0^\infty \delta(t - a)e^{-st} dt = \lim_{n \to \infty} \frac{n}{2} \int_{a-1/n}^{a+1/n} e^{-st} dt \tag{B.1.4}$$

$$= \lim_{n \to \infty} \frac{n}{2s} \left(e^{-as+s/n} - e^{-as-s/n} \right) \tag{B.1.5}$$

$$= \lim_{n \to \infty} \frac{n e^{-as}}{2s} \left(1 + \frac{s}{n} + \frac{s^2}{2n^2} + \cdots - 1 + \frac{s}{n} - \frac{s^2}{2n^2} + \cdots \right) \tag{B.1.6}$$

$$= e^{-as}. \tag{B.1.7}$$

- **Example B.1.3**

    Although we could compute (B.1.1) for every function that has a Laplace transform, these results have already been tabulated and are given in many excellent tables.[1] However, there are four basic transforms that the reader should memorize. They are

$$\mathcal{L}(e^{at}) = \int_0^\infty e^{at} e^{-st} dt = \int_0^\infty e^{-(s-a)t} dt \tag{B.1.8}$$

$$= -\left. \frac{e^{-(s-a)t}}{s - a} \right|_0^\infty = \frac{1}{s - a}, \qquad s > a, \tag{B.1.9}$$

---

[1] The most complete set is given by Erdélyi, A., W. Magnus, F. Oberhettinger, and F. G. Tricomi, 1954: *Tables of Integral Transforms, Vol I*. McGraw-Hill Co., 391 pp.

Table B.1.1: Some General Properties of Laplace Transforms with $a > 0$

	function, f(t)	**Laplace transform, F(s)**
1. Linearity	$c_1 f(t) + c_2 g(t)$	$c_1 F(s) + c_2 G(s)$
2. Scaling	$f(t/a)/a$	$F(as)$
3. Multiplication by e^{bt}	$e^{bt} f(t)$	$F(s - b)$
4. Translation	$f(t - a)H(t - a)$	$e^{-as} F(s)$
5. Differentiation	$f^{(n)}(t)$	$s^n F(s) - s^{n-1} f(0)$ $-s^{n-2} f'(0) - \cdots$ $-f^{(n-1)}(0)$
6. Integration	$\int_0^t f(\tau)\, d\tau$	$F(s)/s$
7. Convolution	$\int_0^t f(t - \tau)g(\tau)\, d\tau$	$F(s)G(s)$

$$
\mathcal{L}[\sin(at)] = \int_0^\infty \sin(at)e^{-st}dt = - \left. \frac{e^{-st}}{s^2 + a^2}[s\sin(at) + a\cos(at)] \right|_0^\infty
$$
(**B.1.10**)

$$
= \frac{a}{s^2 + a^2}, \qquad s > 0,
$$
(**B.1.11**)

$$
\mathcal{L}[\cos(at)] = \int_0^\infty \cos(at)e^{-st}dt = \left. \frac{e^{-st}}{s^2 + a^2}[-s\cos(at) + a\sin(at)] \right|_0^\infty
$$
(**B.1.12**)

$$
= \frac{s}{s^2 + a^2}, \qquad s > 0
$$
(**B.1.13**)

and

$$
\mathcal{L}(t^n) = \int_0^\infty t^n e^{-st}dt = n!e^{-st} \left. \sum_{m=0}^n \frac{t^{n-m}}{(n-m)!s^{m+1}} \right|_0^\infty = \frac{n!}{s^{n+1}}, \qquad s > 0,
$$
(**B.1.14**)

where n is a positive integer.

The Laplace transform inherits two important properties from its integral definition. First, the transform of a sum equals the sum of the transforms:

$$\mathcal{L}[c_1 f(t) + c_2 g(t)] = c_1 \mathcal{L}[f(t)] + c_2 \mathcal{L}[g(t)]. \qquad \textbf{(B.1.15)}$$

This linearity property holds with complex numbers and functions as well.

The second important property deals with derivatives. Suppose $f(t)$ is continuous and has a piece-wise continuous derivative $f'(t)$. Then

$$\mathcal{L}[f'(t)] = \int_0^\infty f'(t) e^{-st} dt = e^{-st} f(t)\big|_0^\infty + s \int_0^\infty f(t) e^{-st} dt \quad \textbf{(B.1.16)}$$

by integration by parts. If $f(t)$ is of exponential order,[2] $e^{-st} f(t)$ tends to zero as $t \to \infty$, for large enough s, so that

$$\mathcal{L}[f'(t)] = sF(s) - f(0). \qquad \textbf{(B.1.17)}$$

Similarly, if $f(t)$ and $f'(t)$ are continuous, $f''(t)$ is piece-wise continuous, and all three functions are of exponential order, then

$$\mathcal{L}[f''(t)] = s\mathcal{L}[f'(t)] - f'(0) = s^2 F(s) - sf(0) - f'(0). \qquad \textbf{(B.1.18)}$$

In general,

$$\mathcal{L}[f^{(n)}(t)] = s^n F(s) - s^{n-1} f(0) - \cdots - sf^{(n-2)}(0) - f^{(n-1)}(0) \quad \textbf{(B.1.19)}$$

on the assumption that $f(t)$ and its first $n-1$ derivatives are continuous, $f^{(n)}(t)$ is piece-wise continuous, and all are of exponential order so that the Laplace transform exists.

B.2 THE SHIFTING THEOREMS

Consider the transform of the function $e^{-at} f(t)$, where a is any real number. Then, by definition,

$$\mathcal{L}\left[e^{-at} f(t)\right] = \int_0^\infty e^{-st} e^{-at} f(t)\, dt = \int_0^\infty e^{-(s+a)t} f(t)\, dt, \quad \textbf{(B.2.1)}$$

or

$$\mathcal{L}\left[e^{-at} f(t)\right] = F(s + a). \qquad \textbf{(B.2.2)}$$

---

[2] By exponential order we mean that there exist some constants, M and k, for which $|f(t)| \leq M e^{kt}$ for all $t > 0$.

Equation (B.2.2) is known as the *first shifting theorem* and states that if $F(s)$ is the transform of $f(t)$ and a is a constant, then $F(s+a)$ is the transform of $e^{-at}f(t)$.

- **Example B.2.1**

Let us find the Laplace transform of $f(t) = e^{-at}\sin(bt)$. Because the Laplace transform of $\sin(bt)$ is $b/(s^2 + b^2)$,

$$\mathcal{L}\left[e^{-at}\sin(bt)\right] = \frac{b}{(s+a)^2 + b^2}, \tag{B.2.3}$$

where we have simply replaced s by $s + a$ in the transform for $\sin(bt)$.

- **Example B.2.2**

Let us find the inverse of the Laplace transform

$$F(s) = \frac{s+2}{s^2 + 6s + 1}. \tag{B.2.4}$$

Rearranging terms,

$$F(s) = \frac{s+2}{s^2 + 6s + 1} = \frac{s+2}{(s+3)^2 - 8} \tag{B.2.5}$$

$$= \frac{s+3}{(s+3)^2 - 8} - \frac{1}{2\sqrt{2}}\frac{2\sqrt{2}}{(s+3)^2 - 8}. \tag{B.2.6}$$

Immediately, from the first shifting theorem,

$$f(t) = e^{-3t}\cosh(2\sqrt{2}t) - \frac{e^{-3t}}{2\sqrt{2}}\sinh(2\sqrt{2}t). \tag{B.2.7}$$

The *second shifting theorem* states that if $F(s)$ is the transform of $f(t)$, then $e^{-bs}F(s)$ is the transform of $f(t-b)H(t-b)$, where b is real and positive. To show this, consider the Laplace transform of $f(t-b)H(t-b)$. Then, from the definition,

$$\mathcal{L}[f(t-b)H(t-b)] = \int_0^\infty f(t-b)H(t-b)e^{-st}\,dt \tag{B.2.8}$$

$$= \int_b^\infty f(t-b)e^{-st}\,dt = \int_0^\infty e^{-bs}e^{-sx}f(x)\,dx \tag{B.2.9}$$

$$= e^{-bs}\int_0^\infty e^{-sx}f(x)\,dx, \tag{B.2.10}$$

or

$$\mathcal{L}[f(t-b)H(t-b)] = e^{-bs}F(s), \qquad \textbf{(B.2.11)}$$

where we have set $x = t - b$. This theorem is of fundamental importance because it allows us to write down the transforms for "delayed" time functions. These functions "turn on" b units after the initial time.

● **Example B.2.3**

Let us find the inverse of the transform $e^{-\pi s}/[s^2(s^2+1)]$. Because

$$\frac{e^{-\pi s}}{s^2(s^2+1)} = \frac{e^{-\pi s}}{s^2} - \frac{e^{-\pi s}}{s^2+1}, \qquad \textbf{(B.2.12)}$$

$$
\begin{aligned}
\mathcal{L}^{-1}\left[\frac{e^{-\pi s}}{s^2(s^2+1)}\right] &= \mathcal{L}^{-1}\left(\frac{e^{-\pi s}}{s^2}\right) - \mathcal{L}^{-1}\left(\frac{e^{-\pi s}}{s^2+1}\right) && \textbf{(B.2.13)} \\
&= (t-\pi)H(t-\pi) - \sin(t-\pi)H(t-\pi) && \textbf{(B.2.14)} \\
&= (t-\pi)H(t-\pi) + \sin(t)H(t-\pi), && \textbf{(B.2.15)}
\end{aligned}
$$

since $\mathcal{L}^{-1}(1/s^2) = t$, and $\mathcal{L}^{-1}[1/(s^2+1)] = \sin(t)$.

B.3 CONVOLUTION

In this section, we turn to a fundamental concept in Laplace transforms: convolution. We begin by formally introducing the mathematical operation of the *convolution product*:

$$f(t) * g(t) = \int_0^t f(t-x)g(x)\,dx = \int_0^t f(x)g(t-x)\,dx. \qquad \textbf{(B.3.1)}$$

● **Example B.3.1**

Let us find the convolution between $\cos(t)$ and $\sin(t)$.

$$
\begin{aligned}
\cos(t) * \sin(t) &= \int_0^t \sin(t-x)\cos(x)\,dx && \textbf{(B.3.2)} \\
&= \tfrac{1}{2}\int_0^t [\sin(t) + \sin(t-2x)]\,dx && \textbf{(B.3.3)} \\
&= \tfrac{1}{2}\int_0^t \sin(t)\,dx + \tfrac{1}{2}\int_0^t \sin(t-2x)\,dx && \textbf{(B.3.4)} \\
&= \tfrac{1}{2}\sin(t)\,x\big|_0^t + \tfrac{1}{4}\cos(t-2x)\big|_0^t = \tfrac{1}{2}t\sin(t). && \textbf{(B.3.5)}
\end{aligned}
$$

The reason we have introduced convolution derives from the following fundamental theorem (often called *Borel's theorem*[3]). If

$$w(t) = u(t) * v(t) \qquad \text{(B.3.6)}$$

then

$$W(s) = U(s)V(s). \qquad \text{(B.3.7)}$$

In other words, we can invert a complicated transform by convoluting the inverses to two simpler functions.

● **Example B.3.2**

Let us find the inverse of the transform

$$\frac{1}{(s^2 + a^2)^2} = \frac{1}{a^2} \left(\frac{a}{s^2 + a^2} \times \frac{a}{s^2 + a^2} \right) \qquad \text{(B.3.8)}$$

$$= \frac{1}{a^2} \mathcal{L}[\sin(at)]\mathcal{L}[\sin(at)]. \qquad \text{(B.3.9)}$$

Therefore,

$$\mathcal{L}^{-1} \left[\frac{1}{(s^2 + a^2)^2} \right] = \frac{1}{a^2} \int_0^t \sin[a(t - x)] \sin(ax) \, dx \qquad \text{(B.3.10)}$$

$$= \frac{1}{2a^2} \int_0^t \cos[a(t - 2x)] \, dx - \frac{1}{2a^2} \int_0^t \cos(at) \, dx \qquad \text{(B.3.11)}$$

$$= -\frac{1}{4a^3} \sin[a(t - 2x)] \Big|_0^t - \frac{1}{2a^2} \cos(at) \, x \Big|_0^t \qquad \text{(B.3.12)}$$

$$= \frac{1}{2a^3} [\sin(at) - at \cos(at)]. \qquad \text{(B.3.13)}$$

B.4 SOLUTION OF LINEAR ORDINARY DIFFERENTIAL EQUATIONS WITH CONSTANT COEFFICIENTS

The primary use of Laplace transforms is the solution of ordinary, constant coefficient, linear differential equations. For example, let us solve the *initial-value problem*

$$\frac{d^n y}{dt^n} + a_1 \frac{d^{n-1} y}{dt^{n-1}} + \cdots + a_{n-1} \frac{dy}{dt} + a_n y = f(t), \quad t > 0 \qquad \text{(B.4.1)}$$

---

[3] Borel, É., 1901: *Leçons sur les séries divergentes.* Gauthier-Villars, p. 104.

by Laplace transforms, where $a_1, a_2, \ldots$ are constants and we know the value of $y, y', \ldots, y^{(n-1)}$ at $t = 0$. The procedure is as follows. Applying the derivative rule (B.1.19) to (B.4.1), we reduce the *differential* equation to an *algebraic* one involving the constants $a_1, a_2, \ldots, a_n$, the parameter s, the Laplace transform of $f(t)$, and the values of the initial conditions. We then solve for the Laplace transform of $y(t)$, $Y(s)$. Finally, we apply one of the many techniques of inverting a Laplace transform to find $y(t)$.

Similar considerations hold with *systems* of ordinary differential equations. The Laplace transform of the system of ordinary differential equations results in an algebraic set of equations containing $Y_1(s), Y_2(s)$, $\ldots, Y_n(s)$. By some method, we solve this set of equations and invert each transform $Y_1(s), Y_2(s), \ldots, Y_n(s)$ in turn, to give $y_1(t), y_2(t)$, $\ldots, y_n(t)$.

The following example illustrates the details of the process.

● **Example B.4.1**

Let us solve the ordinary differential equation

$$y'' + 2y' = 8t \qquad\qquad (\text{B.4.2})$$

subject to the initial conditions that $y'(0) = y(0) = 0$. Taking the Laplace transform of both sides of (B.4.2),

$$\mathcal{L}(y'') + 2\mathcal{L}(y') = 8\mathcal{L}(t), \qquad\qquad (\text{B.4.3})$$

or

$$s^2 Y(s) - sy(0) - y'(0) + 2sY(s) - 2y(0) = \frac{8}{s^2}, \qquad (\text{B.4.4})$$

where $Y(s) = \mathcal{L}[y(t)]$. Substituting the initial conditions into (B.4.4) and solving for $Y(s)$,

$$Y(s) = \frac{8}{s^3(s+2)} = \frac{A}{s^3} + \frac{B}{s^2} + \frac{C}{s} + \frac{D}{s+2} \qquad (\text{B.4.5})$$

$$= \frac{8}{s^3(s+2)} = \frac{(s+2)A + s(s+2)B + s^2(s+2)C + s^3 D}{s^3(s+2)}. \quad (\text{B.4.6})$$

Matching powers of s in the numerators of (B.4.6), $C + D = 0$, $B + 2C = 0$, $A + 2B = 0$, and $2A = 8$ or $A = 4$, $B = -2$, $C = 1$, and $D = -1$. Therefore,

$$Y(s) = \frac{4}{s^3} - \frac{2}{s^2} + \frac{1}{s} - \frac{1}{s+2}. \qquad\qquad (\text{B.4.7})$$

Finally, performing term-by-term inversion of (B.4.7), the final solution equals

$$y(t) = 2t^2 - 2t + 1 - e^{-2t}. \qquad\qquad (\text{B.4.8})$$

B.5 INVERSION BY CONTOUR INTEGRATION

Usually, we can find the inverse of the Laplace transform $F(s)$ by looking it up in a table. In this section, we show an alternative method that inverts Laplace transforms through the powerful method of contour integration.

Consider the piece-wise differentiable function $f(x)$ that vanishes for $x < 0$. We can express the function $e^{-cx}f(x)$ by the complex Fourier representation of

$$f(x)e^{-cx} = \frac{1}{2\pi} \int_{-\infty}^{\infty} e^{i\omega x} \left[\int_0^{\infty} e^{-ct}f(t)e^{-i\omega t}dt \right] d\omega, \qquad \text{(B.5.1)}$$

for any value of the real constant c, where the integral

$$I = \int_0^{\infty} e^{-ct}|f(t)|\,dt \qquad \text{(B.5.2)}$$

exists. By multiplying both sides of (B.5.1) by e^{cx} and bringing it inside the first integral,

$$f(x) = \frac{1}{2\pi} \int_{-\infty}^{\infty} e^{(c+\omega i)x} \left[\int_0^{\infty} f(t)e^{-(c+\omega i)t}dt \right] d\omega. \qquad \text{(B.5.3)}$$

With the substitution $z = c + \omega i$, where z is a new, complex variable of integration,

$$f(x) = \frac{1}{2\pi i} \int_{c-\infty i}^{c+\infty i} e^{zx} \left[\int_0^{\infty} f(t)e^{-zt}dt \right] dz. \qquad \text{(B.5.4)}$$

The quantity inside the square brackets is the Laplace transform $F(z)$. Therefore, we can express $f(t)$ in terms of its transform by the complex contour integral:

$$f(t) = \frac{1}{2\pi i} \int_{c-\infty i}^{c+\infty i} F(z)e^{tz}dz. \qquad \text{(B.5.5)}$$

This line integral, *Bromwich's integral*,[4] runs along the line $x = c$ parallel to the imaginary axis and c units to the right of it, the so-called

---

[4] Bromwich, T. J. I'A., 1916: Normal coordinates in dynamical systems. *Proc. London Math. Soc.*, Ser. 2, **15**, 401–448.

Bromwich contour. We select the value of c sufficiently large so that the integral (B.5.2) exists; subsequent analysis shows that this occurs when c is larger than the real part of any of the singularities of $F(z)$.

We must now evaluate the contour integral. Because of the power of the *residue* theorem in complex variables, the contour integral is usually transformed into a closed contour through the use of *Jordan's lemma*, given in Appendix A. The following examples illustrate the proper use of (B.5.5).

• Example B.5.1

Let us invert

$$F(s) = \frac{e^{-3s}}{s^2(s-1)}. \tag{B.5.6}$$

From Bromwich's integral,

$$f(t) = \frac{1}{2\pi i} \int_{c-\infty i}^{c+\infty i} \frac{e^{(t-3)z}}{z^2(z-1)} \, dz \tag{B.5.7}$$

$$= \frac{1}{2\pi i} \oint_C \frac{e^{(t-3)z}}{z^2(z-1)} \, dz - \frac{1}{2\pi i} \int_{C_R} \frac{e^{(t-3)z}}{z^2(z-1)} \, dz, \tag{B.5.8}$$

where C_R is a semicircle of infinite radius in either the right or left half of the z-plane and C is the closed contour that includes C_R and Bromwich's contour. See Figure B.5.1.

Our first task is to choose an appropriate contour so that the integral along C_R vanishes. By Jordan's lemma, this requires a semicircle in the right half-plane if $t - 3 < 0$ and a semicircle in the left half-plane if $t - 3 > 0$. Consequently, by considering these two separate cases, we force the second integral in (B.5.8) to zero and the inversion simply equals the closed contour.

Consider the case $t < 3$ first. Because Bromwich's contour lies to the right of any singularities, there are no singularities within the closed contour and $f(t) = 0$.

Consider now the case $t > 3$. Within the closed contour in the left half-plane, there is a second-order pole at $z = 0$ and a simple pole at $z = 1$. Therefore,

$$f(t) = \text{Res}\left[\frac{e^{(t-3)z}}{z^2(z-1)}; 0\right] + \text{Res}\left[\frac{e^{(t-3)z}}{z^2(z-1)}; 1\right], \tag{B.5.9}$$

where

$$\text{Res}\left[\frac{e^{(t-3)z}}{z^2(z-1)}; 0\right] = \lim_{z \to 0} \frac{d}{dz}\left[z^2 \frac{e^{(t-3)z}}{z^2(z-1)}\right] \tag{B.5.10}$$

$$= \lim_{z \to 0} \left[\frac{(t-3)e^{(t-3)z}}{z-1} - \frac{e^{(t-3)z}}{(z-1)^2}\right] \tag{B.5.11}$$

$$= 2 - t, \tag{B.5.12}$$

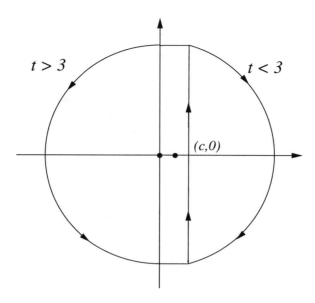

Figure B.5.1: Contours used in the inversion of (B.5.6).

and

$$\text{Res}\left[\frac{e^{(t-3)z}}{z^2(z-1)}; 1\right] = \lim_{z \to 1} (z-1)\frac{e^{(t-3)z}}{z^2(z-1)} = e^{t-3}. \qquad \textbf{(B.5.13)}$$

Taking our earlier results into account, the inverse equals

$$f(t) = \left[e^{t-3} - (t-3) - 1\right] H(t-3), \qquad \textbf{(B.5.14)}$$

which we would have obtained from the second shifting theorem and tables.

● **Example B.5.2**

For our second example of the inversion of Laplace transforms by complex integration, let us find the inverse of

$$F(s) = \frac{1}{s \sinh(as)}, \qquad \textbf{(B.5.15)}$$

where a is real. From Bromwich's integral,

$$f(t) = \frac{1}{2\pi i} \int_{c-\infty i}^{c+\infty i} \frac{e^{tz}}{z \sinh(az)} \, dz. \qquad \textbf{(B.5.16)}$$

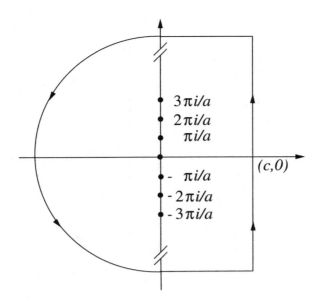

Figure B.5.2: Contours used in the inversion of (B.5.15).

Here c is greater than the real part of any of the singularities in (B.5.15). Using the infinite product for the hyperbolic sine,[5]

$$\frac{e^{tz}}{z\sinh(az)} = \frac{e^{tz}}{az^2[1 + a^2z^2/\pi^2][1 + a^2z^2/(4\pi^2)][1 + a^2z^2/(9\pi^2)]\cdots}.$$
(B.5.17)

Thus, we have a second-order pole at $z = 0$ and simple poles at $z_n = \pm n\pi i/a$, where $n = 1, 2, 3, \ldots$

We convert the line integral (B.5.16), with the Bromwich contour lying parallel and slightly to the right of the imaginary axis, into a closed contour using Jordan's lemma through the addition of an infinite semicircle joining $i\infty$ to $-i\infty$, as shown in Figure B.5.2. We now apply the residue theorem. For the second-order pole at $z = 0$,

$$\text{Res}\left[\frac{e^{tz}}{z\sinh(az)}; 0\right] = \frac{1}{1!}\lim_{z\to 0}\frac{d}{dz}\left[\frac{(z-0)^2 e^{tz}}{z\sinh(az)}\right]$$
(B.5.18)

$$= \lim_{z\to 0}\frac{d}{dz}\left[\frac{ze^{tz}}{\sinh(az)}\right]$$
(B.5.19)

$$= \lim_{z\to 0}\left[\frac{e^{tz}}{\sinh(az)} + \frac{zte^{tz}}{\sinh(az)} - \frac{az\cosh(az)e^{tz}}{\sinh^2(az)}\right]$$

---

[5]   Gradshteyn, I. S. and Ryzhik, I. M., 1965: *Table of Integrals, Series and Products.* Academic Press, §1.431, formula 2.

$$(\textbf{B.5.20})$$

$$= \frac{t}{a} \qquad (\textbf{B.5.21})$$

after using $\sinh(az) = az + O(z^3)$. For the simple poles $z_n = \pm n\pi i/a$,

$$\text{Res}\left[\frac{e^{tz}}{z\sinh(az)}; z_n\right] = \lim_{z\to z_n} \frac{(z-z_n)e^{tz}}{z\sinh(az)} \qquad (\textbf{B.5.22})$$

$$= \lim_{z\to z_n} \frac{e^{tz}}{\sinh(az) + az\cosh(az)} \qquad (\textbf{0.10.23})$$

$$= \frac{\exp(\pm n\pi it/a)}{(-1)^n(\pm n\pi i)}, \qquad (\textbf{B.5.24})$$

because $\cosh(\pm n\pi i) = \cos(n\pi) = (-1)^n$. Thus, summing up all of the residues gives

$$f(t) = \frac{t}{a} + \sum_{n=1}^{\infty} \frac{(-1)^n \exp(n\pi it/a)}{n\pi i} - \sum_{n=1}^{\infty} \frac{(-1)^n \exp(-n\pi it/a)}{n\pi i}$$

$$(\textbf{B.5.25})$$

$$= \frac{t}{a} + \frac{2}{\pi}\sum_{n=1}^{\infty} \frac{(-1)^n}{n} \sin(n\pi t/a). \qquad (\textbf{B.5.26})$$

B.6 SOLUTION OF PARTIAL DIFFERENTIAL EQUATIONS

The solution of linear partial differential equations by transform methods is the most commonly employed analytic technique after the method of separation of variables.

Consider the case where we wish to solve a partial differential equation that depends on x and t. If the solution is denoted by $u(x, t)$, then its Laplace transform is

$$U(x, s) = \int_0^{\infty} u(x, t)e^{-st}\, dt, \qquad (\textbf{B.6.1})$$

because the transform consists solely of an integration with respect to time. Partial derivatives involving time have transforms similar to those that we encounter in the case of functions of a single variable. They include

$$\mathcal{L}[u_t(x, t)] = sU(x, s) - u(x, 0), \qquad (\textbf{B.6.2})$$

and

$$\mathcal{L}[u_{tt}(x, t)] = s^2 U(x, s) - su(x, 0) - u_t(x, 0). \qquad (\textbf{B.6.3})$$

These transforms introduce the initial conditions such as $u(x,0)$ and $u_t(x,0)$. On the other hand, derivatives involving x become

$$\mathcal{L}[u_x(x,t)] = \frac{d}{dx}\{\mathcal{L}[u(x,t)]\} = \frac{dU(x,s)}{dx}, \qquad \text{(B.6.4)}$$

and

$$\mathcal{L}[u_{xx}(x,t)] = \frac{d^2}{dx^2}\{\mathcal{L}[u(x,t)]\} = \frac{d^2U(x,s)}{dx^2}. \qquad \text{(B.6.5)}$$

Because the transformation eliminates the time variable t, only $U(x,s)$ and its derivatives remain in the equation. Consequently, we have transformed the partial differential equation into a boundary-value problem for an ordinary differential equation. Because this equation is often easier to solve than a partial differential equation, the use of Laplace transforms has considerably simplified the original problem. Of course, the Laplace transforms must exist for this technique to work.

To summarize this method, we have constructed the following schematic:

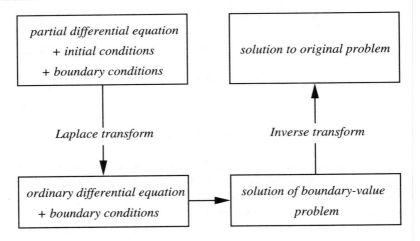

We illustrate these concepts by finding the sound waves that arise when a sphere of radius a begins to pulsate at time $t = 0$. The symmetric wave equation in spherical coordinates is

$$\frac{1}{r}\frac{\partial^2(ru)}{\partial r^2} = \frac{1}{c^2}\frac{\partial^2 u}{\partial t^2}, \qquad \text{(B.6.6)}$$

where c is the speed of sound, $u(r,t)$ is the velocity potential and $-\partial u/\partial r$ gives the velocity of the parcel of air. At the surface of the sphere $r = a$, the radial velocity must equal the velocity of the pulsating sphere

$$-\frac{\partial u}{\partial r} = \frac{d\xi}{dt}, \qquad \text{(B.6.7)}$$

where $\xi(t)$, the displacement of the surface of the pulsating sphere, equals $B \sin(\omega t) H(t)$. The air is initially at rest.

From (B.6.2)–(B.6.5) the Laplace transform of (B.6.6) is

$$\frac{d^2}{dr^2}\left[r\,U(r,s)\right] - \frac{s^2}{c^2}r\,U(r,s) = 0. \tag{B.6.8}$$

The solution of (B.6.8) is

$$r\,U(r,s) = A\exp(-rs/c). \tag{B.6.9}$$

We have discarded the $\exp(rs/c)$ solution because it becomes infinite in the limit of $r \to \infty$. After substituting (B.6.9) into the Laplace transformed (B.6.7),

$$-\frac{d}{dr}\left(A\frac{e^{-sr/c}}{r}\right)\Bigg|_{r=a} = \frac{\omega B s}{s^2 + \omega^2} = Ae^{-as/c}\left(\frac{1}{a^2} + \frac{s}{ac}\right). \tag{B.6.10}$$

Therefore,

$$r\,U(r,s) = \frac{\omega B a^2 cs}{(s^2 + \omega^2)(as + c)}e^{-s(r-a)/c} \tag{B.6.11}$$

$$= \frac{\omega B a^2 c}{a^2\omega^2 + c^2}e^{-s(r-a)/c}\left(\frac{cs + \omega^2 a}{s^2 + \omega^2} - \frac{c}{s + c/a}\right). \tag{B.6.12}$$

Applying the second shifting theorem and tables, the inversion of (B.6.12) follows directly

$$ru(r,t) = \frac{\omega B a^2 c^2}{a^2\omega^2 + c^2}\left\{\cos\left[\omega\left(t - \frac{r-a}{c}\right)\right] + \frac{\omega a}{c}\sin\left[\omega\left(t - \frac{r-a}{c}\right)\right]\right.$$
$$\left. - \exp\left[-\frac{c}{a}\left(t - \frac{r-a}{c}\right)\right]\right\}H\left(t - \frac{r-a}{c}\right). \tag{B.6.13}$$

In this example, we eliminated the temporal dependence by using Laplace transforms. The ordinary differential equation was then solved using homogeneous solutions. In certain cases, an alternative would be to solve the ordinary differential equation by Fourier or Hankel transforms. This is particularly true in the case of Green's functions because the forcing function is a delta function. For a Green's function problem that depends upon a spatial dimension and time, we obtain an algebraic equation that we solve to find the *joint transform*. The inverses would then be found successively. Whether we invert the Laplace or the spatial transform first depends upon the problem. This repeated application of transforms or Fourier series to a linear partial differential equation to reduce it to an algebraic or ordinary differential equation can be extended to higher spatial dimensions.

Appendix C:
Bessel Functions

In the solution of problems involving cylindrical geometries, we will repeatedly encounter *Bessel functions*. In this appendix, we highlight their most useful properties.

C.1 BESSEL FUNCTIONS AND THEIR PROPERTIES

Bessel's equation[1] is

$$x^2 y'' + x y' + (\mu^2 x^2 - n^2) y = 0, \qquad \text{(C.1.1)}$$

or

$$\frac{d}{dx}\left(x \frac{dy}{dx} \right) + \left(\mu^2 x - \frac{n^2}{x} \right) y = 0. \qquad \text{(C.1.2)}$$

Because we cannot write down the solution to (C.1.1) or (C.1.2) in a simple closed form and since Bessel's equation is singular at $x = 0$, the solution is sought as a power series given by the method of Frobenius.

---

[1] The classic reference on Bessel functions is Watson, G. N., 1966: *A Treatise on the Theory of Bessel Functions.* Cambridge University Press, 804 pp.

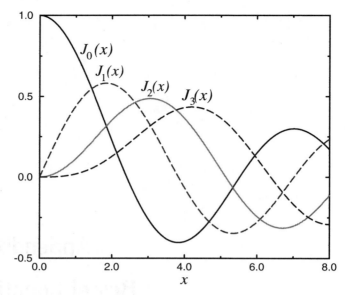

Figure C.1.1: The first four Bessel functions of the first kind over $0 \leq x \leq 8$.

Since n is usually an integer in this book, we present the results for this case.

Frobenius' method yields two linearly independent series that are written down elsewhere but can be treated as tabulated functions.[2] The first solution to (C.1.1)–(C.1.2) is called a Bessel function of the first kind of order n, is denoted by $J_n(x)$, and behaves as x^n, as $x \to 0$. The second general solution is Neumann's Bessel function of the second kind of order n, $Y_n(x)$. The Neumann functions $Y_0(x)$ and $Y_n(x)$ behave as $\ln(x)$ and $-x^{-n}$, respectively, as $x \to 0$ and $n > 0$. Consequently, the general solution to (C.1.1) is

$$y(x) = AJ_n(\mu x) + BY_n(\mu x). \tag{C.1.3}$$

Figure C.1.1 illustrates the functions $J_0(x)$, $J_1(x)$, $J_2(x)$, and $J_3(x)$ while Figure C.1.2 gives $Y_0(x)$, $Y_1(x)$, $Y_2(x)$, and $Y_3(x)$.

The Bessel functions J_n and Y_n enjoy certain *recurrence relations* where functions of order n are related to functions of order $n + 1$ and $n - 1$. If we denote J_n or Y_n by $\mathcal{C}_n$, the most useful relationships are

$$\mathcal{C}_{n-1}(z) + \mathcal{C}_{n+1}(z) = 2n\mathcal{C}_n(z)/z, \qquad n = 1, 2, 3, \ldots, \tag{C.1.4}$$

$$\mathcal{C}_{n-1}(z) - \mathcal{C}_{n+1}(z) = 2\mathcal{C}'_n(z), \qquad n = 1, 2, 3, \ldots, \tag{C.1.5}$$

---

[2] See Press, W. H., B. P. Flannery, S. A. Teukolsky, and W. T. Vetterling, 1986: *Numerical Recipes: The Art of Scientific Computing*. Cambridge University Press, §6.4.

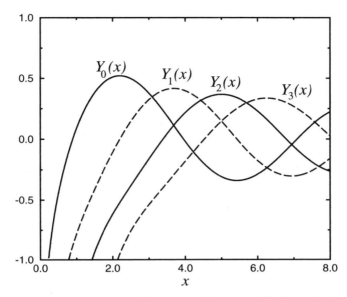

Figure C.1.2: The first four Bessel functions of the second kind over $0 \leq x \leq 8$.

$$\frac{d}{dz}\left[z^n C_n(z)\right] = z^n C_{n-1}(z), \qquad n = 1, 2, 3, \ldots, \qquad \text{(C.1.6)}$$

and

$$\frac{d}{dz}\left[z^{-n} C_n(z)\right] = -z^{-n} C_{n+1}(z), \qquad n = 1, 2, 3, \ldots, \qquad \text{(C.1.7)}$$

with the special case of $J_0'(z) = -J_1(z)$, and $Y_0'(z) = -Y_1(z)$.

An equation that is very similar to (C.1.1) is

$$x^2 \frac{d^2 y}{dx^2} + x \frac{dy}{dx} - (n^2 + x^2)y = 0. \qquad \text{(C.1.8)}$$

If we substitute $ix = t$ (where $i = \sqrt{-1}$) into (C.1.8), it becomes Bessel's equation:

$$t^2 \frac{d^2 y}{dt^2} + t \frac{dy}{dt} + (t^2 - n^2)y = 0. \qquad \text{(C.1.9)}$$

Consequently, we can immediately write the solution to (C.1.8) as

$$y(x) = c_1 J_n(ix) + c_2 Y_n(ix), \qquad \text{(C.1.10)}$$

if n is an integer. Traditionally the solution to (C.1.10) has been written

$$y(x) = c_1 I_n(x) + c_2 K_n(x) \qquad \text{(C.1.11)}$$

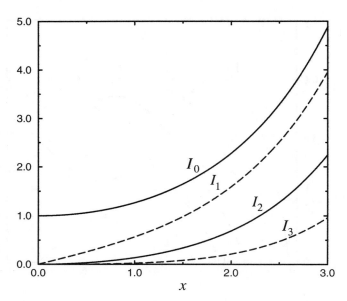

Figure C.1.3: The first four modified Bessel functions of the first kind over $0 \leq x \leq 3$.

rather than in terms of $J_n(ix)$ and $Y_n(ix)$ The function $I_n(x)$ is the modified Bessel function of the first kind, of order n, while $K_n(x)$ is the modified Bessel function of the second kind, of order n. Figure C.1.3 illustrates $I_0(x)$, $I_1(x)$, $I_2(x)$, and $I_3(x)$, while in Figure C.1.4 $K_0(x)$, $K_1(x)$, $K_2(x)$, and $K_3(x)$ have been graphed. Again, both I_n and K_n can be treated as tabulated functions.[3] Note that $K_n(x)$ has no real zeros while $I_n(x)$ equals zero only at $x = 0$ for $n > 0$.

As our derivation suggests, modified Bessel functions are related to ordinary Bessel functions via complex variables. In particular, $J_n(iz) = i^n I_n(z)$ and $I_n(iz) = i^n J_n(z)$ for z complex.

As in the case of J_n and Y_n, there are *recurrence relations* where functions of order n are related to functions of order $n + 1$ and $n - 1$. If we denote by C_n either I_n or $e^{n\pi i} K_n$, we have

$$C_{n-1}(z) - C_{n+1}(z) = 2nC_n(z)/z, \qquad n = 1, 2, 3, \ldots, \qquad \textbf{(C.1.12)}$$

$$C_{n-1}(z) + C_{n+1}(z) = 2C'_n(z), \qquad n = 1, 2, 3, \ldots, \qquad \textbf{(C.1.13)}$$

$$\frac{d}{dz}\left[z^n C_n(z)\right] = z^n C_{n-1}(z), \qquad n = 1, 2, 3, \ldots, \qquad \textbf{(C.1.14)}$$

and

$$\frac{d}{dz}\left[z^{-n} C_n(z)\right] = z^{-n} C_{n+1}(z), \qquad n = 1, 2, 3, \ldots, \qquad \textbf{(C.1.15)}$$

---

[3] *Ibid.*, §6.5.

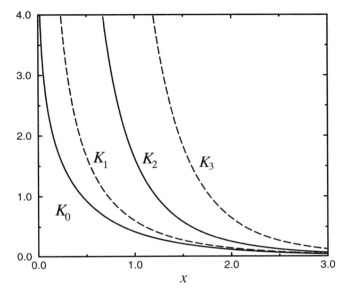

Figure C.1.4: The first four modified Bessel functions of the second kind over $0 \le x \le 3$.

with the special case of $I_0'(z) = I_1(z)$, and $K_0(z) = -K_1(z)$.

C.2 HANKEL FUNCTIONS

Hankel functions are an alternative to J_n and Y_n. Here we define them and show why they are useful.

One of the fundamental equations studied in any differential equations course is

$$y'' + k^2 y = 0. \tag{C.2.1}$$

Its solution is generally written

$$y(x) = A \cos(kx) + B \sin(kx), \tag{C.2.2}$$

where A and B are arbitrary constants. Another, less often quoted, solution is

$$y(x) = C e^{ikx} + D e^{-ikx}, \tag{C.2.3}$$

where C and D are generally complex constants. The advantage of (C.2.3) over (C.2.2) is the convenience of dealing with exponentials rather than sines and cosines.

Let us return to (C.1.1):

$$x^2 y'' + x y' + \left(x^2 - n^2\right) y = 0, \tag{C.2.4}$$

Table C.2.1: Some Useful Recurrence Relations for Hankel Functions.

$$\frac{d}{dx}\left[x^n H_n^{(p)}(x)\right] = x^n H_{n-1}^{(p)}(x), \ n = 1, 2, \ldots; \frac{d}{dx}\left[H_0^{(p)}(x)\right] = -H_1^{(p)}(x)$$

$$\frac{d}{dx}\left[x^{-n} H_n^{(p)}(x)\right] = -x^{-n} H_{n+1}^{(p)}(x), \qquad n = 0, 1, 2, 3, \ldots$$

$$H_{n-1}^{(p)}(x) + H_{n+1}^{(p)}(x) = \frac{2n}{x} H_n^{(p)}(x), \qquad n = 1, 2, 3, \ldots$$

$$H_{n-1}^{(p)}(x) - H_{n+1}^{(p)}(x) = 2\frac{dH_n^{(p)}(x)}{dx}, \qquad n = 1, 2, 3, \ldots$$

where n is a positive (including zero) integer. Its general solution is most often written

$$y(x) = A\, J_n(x) + B\, Y_n(x). \tag{C.2.5}$$

Again A and B are arbitrary constants. However, another, equivalent solution is

$$y(x) = CH_n^{(1)}(x) + DH_n^{(2)}(x), \tag{C.2.6}$$

where

$$H_n^{(1)}(x) = J_n(x) + iY_n(x), \tag{C.2.7}$$

and

$$H_n^{(2)}(x) = J_n(x) - iY_n(x). \tag{C.2.8}$$

These functions $H_n^{(1)}(x), H_n^{(2)}(x)$ are referred to as Bessel functions of the third kind or *Hankel functions*, after the German mathematician Hermann Hankel (1839–1873). Some of their properties are listed in Table C.2.1.

The similarity between the solution (C.2.3) and (C.2.6) is clarified by the asymptotic expansions for the Hankel functions:

$$H_n^{(1)}(z) \sim \sqrt{\frac{2}{\pi z}}\, e^{i(z - n\pi/2 - \pi/4)}, \tag{C.2.9}$$

and

$$H_n^{(2)}(z) \sim \sqrt{\frac{2}{\pi z}}\, e^{-i(z - n\pi/2 - \pi/4)}. \tag{C.2.10}$$

C.3 FOURIER-BESSEL SERIES

A Fourier-Bessel series is a Fourier series where $J_n(\ )$ is used in place of the sine and cosine. The exact form of the expansion depends upon the boundary condition at $x = L$. There are three possible cases. One of them is that $y(L) = 0$ and results in the condition that $J_n(\mu_k L) = 0$. Another condition is $y'(L) = 0$ and gives $J'_n(\mu_k L) = 0$. Finally, if $hy(L) + y'(L) = 0$, then $hJ_n(\mu_k L) + \mu_k J'_n(\mu_k L) = 0$. In all of these cases, the eigenfunction expansion is the same, namely

$$f(x) = \sum_{k=1}^{\infty} A_k J_n(\mu_k x), \qquad \text{(C.3.1)}$$

where μ_k is the kth positive solution of either $J_n(\mu_k L) = 0$, $J'_n(\mu_k L) = 0$, or $hJ_n(\mu_k L) + \mu_k J'_n(\mu_k L) = 0$.

We now need a mechanism for computing A_k. We begin by multiplying (C.3.1) by $xJ_n(\mu_m x)\, dx$ and integrate from 0 to L. This yields

$$\sum_{k=1}^{\infty} A_k \int_0^L xJ_n(\mu_k x)J(\mu_m x)\, dx = \int_0^L xf(x)J_n(\mu_m x)\, dx. \qquad \text{(C.3.2)}$$

From the general orthogonality condition

$$\int_0^L xJ_n(\mu_k x)J_n(\mu_m x)\, dx = 0 \qquad \text{(C.3.3)}$$

if $k \neq m$, (C.3.2) simplifies to

$$A_m \int_0^L xJ_n^2(\mu_m x)\, dx = \int_0^L xf(x)J_n(\mu_m x)\, dx, \qquad \text{(C.3.4)}$$

or

$$A_k = \frac{1}{C_k} \int_0^L xf(x)J_n(\mu_k x)\, dx, \qquad \text{(C.3.5)}$$

where

$$C_k = \int_0^L xJ_n^2(\mu_k x)\, dx \qquad \text{(C.3.6)}$$

and k has replaced m in (C.3.4).

The factor C_k depends upon the boundary conditions at $x = L$. It can be shown that

$$C_k = \tfrac{1}{2}L^2 J_{n+1}^2(\mu_k L), \tag{C.3.7}$$

if $J_n(\mu_k L) = 0$. If $J_n'(\mu_k L) = 0$, then

$$C_k = \frac{\mu_k^2 L^2 - n^2}{2\mu_k^2} J_n^2(\mu_k L). \tag{C.3.8}$$

Finally,

$$C_k = \frac{\mu_k^2 L^2 - n^2 + h^2 L^2}{2\mu_k^2} J_n^2(\mu_k L), \tag{C.3.9}$$

if $\mu_k J_n'(\mu_k L) = -h J_n(\mu_k L)$.

All of the preceding results must be slightly modified when $n = 0$ and the boundary condition is $J_0'(\mu_k L) = 0$ or $\mu_k J_1(\mu_k L) = 0$. This modification arises from the additional eigenvalue $\mu_0 = 0$ being present and we must add the extra term A_0 to the expansion. For this case, (C.3.1) becomes

$$f(x) = A_0 + \sum_{k=1}^{\infty} A_k J_0(\mu_k x), \tag{C.3.10}$$

where the equation for finding A_0 is

$$A_0 = \frac{2}{L^2} \int_0^L f(x)\, x\, dx. \tag{C.3.11}$$

Equations (C.3.5) and (C.3.8) with $n = 0$ give the remaining coefficients.

• **Example C.3.1**

Let us expand $f(x) = x$, $0 < x < 1$, in the series

$$f(x) = \sum_{k=1}^{\infty} A_k J_1(\mu_k x), \tag{C.3.12}$$

where μ_k denotes the kth zero of $J_1(\mu)$. From (C.3.5) and (C.3.7),

$$A_k = \frac{2}{J_2^2(\mu_k)} \int_0^1 x^2 J_1(\mu_k x)\, dx. \tag{C.3.13}$$

However, from (C.1.6),

$$\frac{d}{dx}\left[x^2 J_2(x)\right] = x^2 J_1(x), \tag{C.3.14}$$

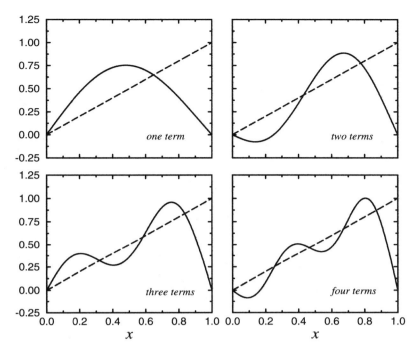

Figure C.3.1: The Fourier-Bessel series representation (C.3.16) for $f(x) = x$, $0 \leq x < 1$, when we truncate the series so that it includes only the first, first two, first three, and first four terms.

if $n = 2$. Therefore, (C.3.13) becomes

$$A_k = \frac{2x^2 J_2(x)}{\mu_k^3 J_2^2(\mu_k)} \bigg|_0^{\mu_k} = \frac{2}{\mu_k J_2(\mu_k)}, \qquad \text{(C.3.15)}$$

and the resulting expansion is

$$x = 2 \sum_{k=1}^{\infty} \frac{J_1(\mu_k x)}{\mu_k J_2(\mu_k)}, \qquad 0 \leq x < 1. \qquad \text{(C.3.16)}$$

Figure C.3.1 shows the Fourier-Bessel expansion of $f(x) = x$ in truncated form when we only include one, two, three, and four terms.

● **Example C.3.2**

Let us expand the function $f(x) = x^2$, $0 < x < 1$, in the series

$$f(x) = \sum_{k=1}^{\infty} A_k J_0(\mu_k x), \qquad \text{(C.3.17)}$$

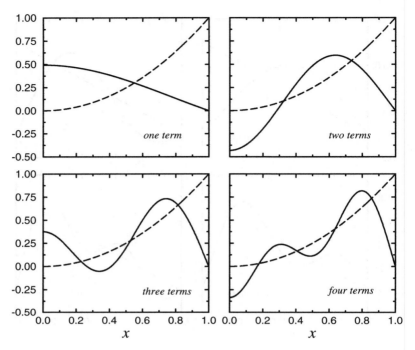

Figure C.3.2: The Fourier-Bessel series representation (C.3.27) for $f(x) = x^2$, $0 < x < 1$, when we truncate the series so that it includes only the first, first two, first three, and first four terms.

where μ_k denotes the kth positive zero of $J_0(\mu)$. From (C.3.5) and (C.3.7),

$$A_k = \frac{2}{J_1^2(\mu_k)} \int_0^1 x^3 J_0(\mu_k x) \, dx. \qquad \textbf{(C.3.18)}$$

If we let $t = \mu_k x$, the integration (C.3.18) becomes

$$A_k = \frac{2}{\mu_k^4 J_1^2(\mu_k)} \int_0^{\mu_k} t^3 J_0(t) \, dt. \qquad \textbf{(C.3.19)}$$

We now let $u = t^2$ and $dv = tJ_0(t) \, dt$ so that integration by parts gives

$$A_k = \frac{2}{\mu_k^4 J_1^2(\mu_k)} \left[t^3 J_1(t) \big|_0^{\mu_k} - 2 \int_0^{\mu_k} t^2 J_1(t) \, dt \right] \qquad \textbf{(C.3.20)}$$

$$= \frac{2}{\mu_k^4 J_1^2(\mu_k)} \left[\mu_k^3 J_1(\mu_k) - 2 \int_0^{\mu_k} t^2 J_1(t) \, dt \right], \qquad \textbf{(C.3.21)}$$

because $v = tJ_1(t)$ from (C.1.6). If we integrate by parts once more, we find that

$$A_k = \frac{2}{\mu_k^4 J_1^2(\mu_k)} \left[\mu_k^3 J_1(\mu_k) - 2\mu_k^2 J_2(\mu_k) \right] \qquad \textbf{(C.3.22)}$$

$$= \frac{2}{J_1^2(\mu_k)} \left[\frac{J_1(\mu_k)}{\mu_k} - \frac{2J_2(\mu_k)}{\mu_k^2} \right]. \tag{C.3.23}$$

However, from (C.1.4) with $n = 1$,

$$J_1(\mu_k) = \tfrac{1}{2}\mu_k \left[J_2(\mu_k) + J_0(\mu_k) \right], \tag{C.3.24}$$

or

$$J_2(\mu_k) = \frac{2J_1(\mu_k)}{\mu_k}, \tag{C.3.25}$$

because $J_0(\mu_k) = 0$. Therefore,

$$A_k = \frac{2(\mu_k^2 - 4)J_1(\mu_k)}{\mu_k^3 J_1^2(\mu_k)}, \tag{C.3.26}$$

and

$$x^2 = 2\sum_{k=1}^{\infty} \frac{(\mu_k^2 - 4)J_0(\mu_k x)}{\mu_k^3 J_1(\mu_k)}, \qquad 0 < x < 1. \tag{C.3.27}$$

Figure C.3.2 shows the representation of x^2 by the Fourier-Bessel series (C.3.27) when we truncate it so that it includes only one, two, three, or four terms. As we add each additional term in the orthogonal expansion, the expansion fits $f(x)$ better in the "least squares" sense.

Appendix D:
Relationship between Solutions of Helmholtz's and Laplace's Equations in Cylindrical and Spherical Coordinates

In §5.1 we showed how solutions to the Helmholtz or scalar wave equation in one coordinate system can be re-expressed as a superposition (integral) of solutions in another coordinate system. We extend these thoughts as they apply to product solutions involving Laplace's and Helmholtz's equations.

Consider the three-dimensional Helmholtz equation in cylindrical and spherical coordinates

$$\frac{\partial^2 u}{\partial \rho^2} + \frac{1}{\rho}\frac{\partial u}{\partial \rho} + \frac{1}{\rho^2}\frac{\partial^2 u}{\partial \varphi^2} + \frac{\partial^2 u}{\partial z^2} + k_0^2 u = 0, \qquad (\mathbf{D.1})$$

and

$$\frac{\partial^2 u}{\partial r^2} + \frac{2}{r}\frac{\partial u}{\partial r} + \frac{1}{r^2 \sin(\theta)}\frac{\partial}{\partial \theta}\left[\sin(\theta)\frac{\partial u}{\partial \theta}\right] + \frac{1}{r^2 \sin^2(\theta)}\frac{\partial^2 u}{\partial \varphi^2} + k_0^2 u = 0. \quad (\mathbf{D.2})$$

Separation of variables[1] leads to the product solutions

$$Z_{m,\theta_0}(\rho, \varphi, z) = e^{ik_0 z \cos(\theta_0)} J_m[k_0 \rho \sin(\theta_0)] e^{im\varphi}, \qquad (\mathbf{D.3})$$

---

[1] See Stratton, J. A., 1941: *Electromagnetic Theory*. McGraw-Hill, p. 395, Eq. (21) with $h = k_0 \cos(\theta_0)$.

and

$$\Psi_n^m(r, \varphi, \theta) = j_n(k_0 r) P_n^m[\cos(\theta)]e^{im\varphi} \qquad \textbf{(D.4)}$$

for (D.1) and (D.2), respectively, where $j_n(\ )$ is a spherical Bessel function of order n, and $P_n^m(\ )$ is an associated Legendre polynomial of the first kind, order n and type m.

The first relationship is given in Stranton's book,[2] namely that

$$Z_{m,\theta_0}(\rho, \varphi, z) = \sum_{n=m}^{\infty} \frac{(2n+1)i^{n-m}(n-m)!}{(n+m)!} P_n^m[\cos(\theta_0)]\Psi_n^m(r, \varphi, \theta),$$

$$\textbf{(D.5)}$$

where n and m are *nonnegative* integers and $0 \leq \theta_0 \leq \pi$. To obtain Ψ_n^m in terms of Z_{m,θ_0}, we multiply (D.5) by $P_n^m[\cos(\theta_0)] \sin(\theta_0) d\theta_0$, integrate from 0 to π and obtain that

$$\Psi_n^m(r, \varphi, \theta) = \frac{i^{m-n}}{2} \int_0^{\pi} P_n^m[\cos(\theta_0)]Z_{m,\theta_0}(\rho, \varphi, z) \sin(\theta_0) \, d\theta_0.$$

$$\textbf{(D.6)}$$

Let us apply[3] (D.5) to establish the relationship between solutions of Laplace's equation in cylindrical and spherical coordinates. Setting $\theta_0 = it$ and $k_0 = 2\lambda e^{-t}$, then

$$e^{i\lambda(1+e^{-2t})z} J_m \left[\rho\lambda i \left(1 - e^{-2t}\right)\right] e^{im\varphi}$$

$$= \sum_{n=m}^{\infty} \frac{(2n+1)i^{n-m}(n-m)!}{(n+m)!} P_n^m \left(\frac{e^t + e^{-t}}{2}\right)$$

$$\times j_n \left(\frac{2\lambda r}{e^t}\right) P_n^m[\cos(\theta)]e^{im\varphi}. \qquad \textbf{(D.7)}$$

---

[2] *Ibid.*, p. 413, Eq. (82).

[3] Erofeenko, V. T., 1972: Relation between the basic solutions of the Helmholtz and Laplace equations in cylindrical and spherical coordinates (in Russian). *Izv. Akad. Nauk BSSR, Ser. Fiz.-Mat. Nauk*, **No. 4**, 42–46.

In the limit of $t \to \infty$,

$$P_n^m \left(\frac{e^t + e^{-t}}{2} \right) \approx \frac{(-1)^m (2n)! e^{(n-m)t}}{2^n n! (n-m)! 2^{n-m}}, \tag{D.8}$$

and

$$j_n \left(\frac{2\lambda r}{e^t} \right) \approx \frac{2^n n! (2r\lambda)^n}{(2n+1)! e^{nt}}, \tag{D.9}$$

so that

$$e^{i\lambda z} J_m(i\rho\lambda) e^{im\varphi} = (-1)^m \sum_{n=m}^{\infty} \frac{(i\lambda)^n}{(n+m)!} \, r^n P_n^m [\cos(\theta)] e^{im\varphi}, \tag{D.10}$$

and

$$J_{m,\lambda}(\rho, \varphi, z) = (-1)^m \sum_{n=m}^{\infty} \frac{(i\lambda)^n}{(n+m)!} \, R_n^m(r, \varphi, \theta), \tag{D.11}$$

where

$$J_{m,\lambda}(\rho, \varphi, z) = e^{i\lambda z} J_m(i\lambda\rho) e^{im\varphi}, \tag{D.12}$$

and

$$R_n^m(r, \varphi, \theta) = r^n P_n^m [\cos(\theta)] e^{im\varphi} \tag{D.13}$$

are product solutions[4] of Laplace's equation in cylindrical and spherical coordinates, respectively.

The final relationship is derived from the expression

$$\frac{1}{(\rho^2 + z^2)^{m+\frac{1}{2}}} = \frac{1}{r^{2m+1}} \tag{D.14}$$

$$= \frac{\sqrt{\pi}}{\pi 2^m \rho^m \Gamma\left(m + \frac{1}{2}\right)} \int_{-\infty}^{\infty} e^{i\lambda z} |\lambda|^m K_m(|\lambda|\rho) \, d\lambda, \tag{D.15}$$

for $m \geq 0$. Therefore,

$$\frac{1}{r^{m+1}} P_m^m [\cos(\theta)] = \frac{1}{r^{m+1}} \frac{(2m)! (-1)^m}{2^m m!} \sin^m(\theta) \tag{D.16}$$

$$= \frac{(2m)! (-1)^m}{2^m m!} \frac{\rho^m}{r^{2m+1}} \tag{D.17}$$

$$= \frac{(-1)^m}{\pi} \int_{-\infty}^{\infty} e^{i\lambda z} |\lambda|^m K_m(|\lambda|\rho) \, d\lambda. \tag{D.18}$$

---

[4] See Stinson, D. C., 1976: *Intermediate Mathematics of Electromagnetics.* Prentice-Hall, pp. 125 and 155.

Because $n > m \geq 0$,

$$
\frac{d^{n-m}}{dz^{n-m}} \left\{ \frac{1}{r^{m+1}} P_m^m[\cos(\theta)] \right\}
$$

$$
= (-1)^{n-m}(n-m)! \frac{1}{r^{n+1}} P_n^m[\cos(\theta)] \tag{D.19}
$$

$$
= \frac{(-1)^m}{\pi} i^{n-m} \int_{-\infty}^{\infty} e^{i\lambda z} \lambda^{n-m} |\lambda|^m K_m(|\lambda|\rho) \, d\lambda. \tag{D.20}
$$

Consequently,

$$
\frac{1}{r^{n+1}} P_n^m[\cos(\theta)] = \frac{(-1)^n i^{n-m}}{\pi(n-m)!} \int_{-\infty}^{\infty} e^{i\lambda z} \lambda^{n-m} |\lambda|^m K_m(|\lambda|\rho) \, d\lambda. \tag{D.21}
$$

The final result is

$$
\boxed{\Phi_n^m(r, \varphi, \theta) = \frac{i^{n-m}(-1)^n}{\pi(n-m)!} \int_{-\infty}^{\infty} \lambda^n K_{m,\lambda}(\rho, \varphi, z) \, d\lambda,} \tag{D.22}
$$

where

$$
K_{m,\lambda}(\rho, \varphi, z) = e^{i\lambda z} K_m(\lambda\rho) e^{im\varphi}, \tag{D.23}
$$

with the modified Bessel function of the second kind

$$
K_m(\lambda) = \begin{cases} K_m(\lambda), & \lambda > 0, \\ (-1)^m K_m(-\lambda), & \lambda < 0, \end{cases} \tag{D.24}
$$

and

$$
\Phi_n^m(r, \varphi, \theta) = r^{-n-1} P_n^m[\cos(\theta)] e^{im\varphi} \tag{D.25}
$$

are product solutions[4] of Laplace's equation in cylindrical and spherical coordinates, respectively.

Answers to
Some of the Problems

Chapter 2

1.
$$g(t|\tau) = e^{-k(t-\tau)} H(t - \tau)$$

2.
$$g(t|\tau) = \tfrac{1}{4} \left[e^{3(t-\tau)} - e^{-(t-\tau)} \right] H(t - \tau)$$

3.
$$g(t|\tau) = \tfrac{1}{2} \left[e^{-(t-\tau)} - e^{-3(t-\tau)} \right] H(t - \tau)$$

4.
$$g(t|\tau) = \tfrac{1}{2} \sin[2(t - \tau)]\, e^{t-\tau} H(t - \tau)$$

5.
$$g(t|\tau) = \left[e^{2(t-\tau)} - e^{t-\tau} \right] H(t - \tau)$$

6.
$$g(t|\tau) = (t - \tau) e^{-2(t-\tau)} H(t - \tau)$$

7.

$$g(t|\tau) = \tfrac{1}{6}\left[e^{3(t-\tau)} - e^{-3(t-\tau)}\right]H(t-\tau)$$

8.

$$g(t|\tau) = \sin(t-\tau)H(t-\tau)$$

9.

$$g(t|\tau) = \left[e^{t-\tau} - 1\right]H(t-\tau)$$

10.

$$g(x|\xi) = \frac{(L - x_>)(\alpha + x_<)}{L + \alpha}$$

and

$$g(x|\xi) = 2\sum_{n=1}^{\infty}\frac{(1 + \alpha^2 k_n^2)\sin[k_n(L - \xi)]\sin[k_n(L - x)]}{[\alpha + L(1 + \alpha^2 k_n^2)]k_n^2},$$

where k_n is the nth root of $\tan(kL) = -\alpha k$.

11.

$$g(x|\xi) = (1 + x_<)(L - 1 - x_>)/L$$

and

$$g(x|\xi) = -\frac{2e^{x+\xi}}{e^{2L} - 1} + \frac{2L^3}{\pi^2}\sum_{n=1}^{\infty}\frac{\varphi_n(\xi)\varphi_n(x)}{n^2(n^2\pi^2 + 1)},$$

where $\varphi_n(x) = \sin(n\pi x/L) + n\pi\cos(n\pi x/L)/L$.

12.

$$g(x|\xi) = \frac{(1 + x_<)(L + 1 - x_>)}{2 + L}$$

and

$$g(x|\xi) = 2\sum_{n=1}^{\infty}\frac{\varphi_n(\xi)\varphi_n(x)}{(2 + L + k_n^2 L)k_n^2},$$

where $\varphi_n(x) = \sin(k_n x) + k_n\cos(k_n x)$ and k_n is the nth root of $\tan(kL)$ $= 2k/(k^2 - 1)$.

13.

$$g(x|\xi) = \frac{a}{3} - x_> + \frac{x^2 + \xi^2}{2a}$$

14.

$$g(x|\xi) = \frac{\sinh(kx_<)\sinh[k(L - x_>)]}{k\sinh(kL)}$$

and

$$g(x|\xi) = 2L \sum_{n=1}^{\infty} \frac{\sin(n\pi x/L)\sin(n\pi\xi/L)}{k^2 L^2 + n^2\pi^2}$$

15.

$$g(x|\xi) = \frac{\cosh(kx_<)\cosh[k(L-x_>)]}{k\sinh(kL)}$$

and

$$g(x|\xi) = \frac{1}{k^2 L} + 2L \sum_{n=1}^{\infty} \frac{\cos(n\pi x/L)\cos(n\pi\xi/L)}{k^2 L^2 + n^2\pi^2}$$

16.

$$g(x|\xi) = \frac{\sinh(kx_<)\{k\cosh[k(x_> - L)] - \sinh[k(x_> - L)]\}}{k\sinh(kL) + k^2\cosh(kL)}$$

and

$$g(x|\xi) = 2\sum_{n=1}^{\infty} \frac{(1+k_n^2)\sin(k_n\xi)\sin(k_n x)}{(k^2 + k_n^2)[1 + L(1+k_n^2)]},$$

where k_n is the nth root of $\tan(kL) = -k$.

17.

$$g(x|\xi) = \frac{\sinh(kx_<)\{\sinh[k(x_> - L)] + k\cosh[k(x_> - L)]\}}{k^2\cosh(kL) - k\sinh(kL)}$$

and

$$g(x|\xi) = 2\sum_{n=1}^{\infty} \frac{(1+k_n^2)\sin(k_n\xi)\sin(k_n x)}{(k^2 + k_n^2)[L(1+k_n^2) - 1]},$$

where k_n is the nth root of $\tan(kL) = k$.

18.

$$g(x|\xi) = \frac{a\sinh(kx_<) - k\cosh(kx_<)}{k[a\cosh(kL) - k\sinh(kL)]}\cosh[k(L-x_>)]$$

and

$$g(x|\xi) = 2\sum_{n=1}^{\infty} \frac{(a^2 + k_n^2)\cos[k_n(\xi - L)]\cos[k_n(x - L)]}{(k^2 + k_n^2)[L(a^2 + k_n^2) - a]},$$

where k_n is the nth root of $kL\tan(kL) = -aL$.

19.

$$g(x|\xi) = \frac{\{\sinh[k(L-x_>)] - k\cosh[k(L-x_>)]\}}{k[(1+k^2)\sinh(kL) - 2k\cosh(kL)]}$$
$$\times [\sinh(kx_<) - k\cosh(kx_<)]]$$

and

$$g(x|\xi) = 2 \sum_{n=1}^{\infty} \frac{\varphi_n(\xi)\varphi_n(x)}{(k^2 + k_n^2)[(1 + k_n^2)L - 2]},$$

where $\varphi_n(x) = \sin(k_n x) - k_n \cos(k_n x)$ and $\tan(k_n L) = 2k_n/(1 - k_n^2)$.

21.

$$g(x|\xi) = \frac{e^{a(\xi - x)/2} \sinh\left[(\pi - x_>)\sqrt{k^2 + a^2/4}\right] \sinh\left(x_< \sqrt{k^2 + a^2/4}\right)}{\sqrt{k^2 + a^2/4} \, \sinh\left(\pi\sqrt{k^2 + a^2/4}\right)}$$

22.

$$g(x|\xi) = \frac{1}{\sqrt{1 + 4\nu s}} \begin{cases} \exp\left[\frac{1 + \sqrt{1 + 4\nu s}}{4\nu}(x - \xi)\right], & x \le \xi \\[2mm] \exp\left[\frac{1 - \sqrt{1 + 4\nu s}}{4\nu}(x - \xi)\right], & x \ge \xi \end{cases}$$

26.

$$g(x|\xi) = \frac{\pi k^2 L^2}{2} \frac{J'_m(kx_<)[J'_m(kx_>)Y'_m(kL) - J'_m(kL)Y'_m(kx_>)]}{J'_m(kL)}$$

and

$$g(x|\xi) = 2 \sum_{n=1}^{\infty} \frac{k_{nm}^4 J'_m(k_{nm}\xi) J'_m(k_{nm}x)}{(k_{nm}^2 - k^2)(k_{nm}^2 - m^2)J_m^2(k_{nm}L)},$$

where $J'_m(k_{nm}L) = 0$.

27.

$$g(r|\rho) = \frac{a}{2\pi^3 \rho r} \sum_{n=1}^{\infty} \frac{1}{n^2} \sin\left(\frac{n\pi\rho}{a}\right) \sin\left(\frac{n\pi r}{a}\right).$$

28.

$$g(x|\xi) = \frac{kb\cosh(kx_<) - a\sinh(kx_<)}{k(kb - a)} e^{-kx_>}$$

30.

$$g(x|\xi) = \begin{cases} \left[e^{-|x - \xi|\sqrt{s}} - R(s)e^{-(x+\xi)\sqrt{s}}\right]/\sqrt{s}, & x \ge 0, \xi \ge 0, \\[2mm] T(s)e^{x\sqrt{s+\rho} - \xi\sqrt{s}}/\sqrt{s}, & x \le 0, \xi \ge 0, \\[2mm] T(s)e^{\xi\sqrt{s+\rho} - x\sqrt{s}}/\sqrt{s}, & x \ge 0, \xi \le 0, \\[2mm] \left[e^{-|x - \xi|\sqrt{s+\rho}} + R(s)e^{(x+\xi)\sqrt{s+\rho}}\right]/\sqrt{s+\rho}, & x \le 0, \xi \le 0, \end{cases}$$

where

$$R(s) = \frac{\sqrt{s+\rho}-\sqrt{s}}{\sqrt{s+\rho}+\sqrt{s}} \quad \text{and} \quad T(s) = \frac{2\sqrt{s}}{\sqrt{s+\rho}+\sqrt{s}}.$$

31.

$$g(x|\xi) = \frac{(x_> - L)x_<}{6L}\left(x^2 - 2Lx_> + \xi^2\right)$$

$$= 2L^3 \sum_{n=1}^{\infty} \frac{\sin(n\pi\xi/L)\sin(n\pi x/L)}{n^4\pi^4}$$

32.

$$g(x|\xi) = -\frac{1}{4k^3}\left(ie^{ik|x-\xi|} - e^{-k|x-\xi|}\right)$$

33.

$$g(x|\xi) = \frac{\sin[k(L-x_>)]\sin(kx_<)}{2k^3\,\sin(kL)}$$

$$- \frac{\sinh[k(L-x_>)]\sinh(kx_<)}{2k^3\,\sinh(kL)}$$

and

$$g(x|\xi) = 2L^3 \sum_{n=1}^{\infty} \frac{\sin(n\pi\xi/L)\sin(n\pi x/L)}{n^4\pi^4 - k^4L^4}$$

Chapter 3

2.

$$g(x,t|\xi,\tau) = \frac{4}{\pi}H(t-\tau)\sum_{n=1}^{\infty}\frac{1}{2n-1}\sin\left[\frac{(2n-1)\pi\xi}{2L}\right]\sin\left[\frac{(2n-1)\pi x}{2L}\right]$$

$$\times \sin\left[\frac{(2n-1)\pi(t-\tau)}{2L}\right]$$

Chapter 4

20.

$$g(x,t|\xi,\tau) = \frac{2}{L}H(t-\tau)\sum_{n=1}^{\infty}\sin\left(\frac{n\pi\xi}{L}\right)\sin\left(\frac{n\pi x}{L}\right)e^{-n^2\pi^2(t-\tau)/L^2}$$

21.

$$g(x,t|\xi,\tau) = \frac{2}{L}H(t-\tau)\sum_{n=1}^{\infty}\sin\left[\frac{(2n-1)\pi\xi}{2L}\right]\sin\left[\frac{(2n-1)\pi x}{2L}\right]$$

$$\times \exp\left[-\frac{(2n-1)^2\pi^2(t-\tau)}{4L^2}\right]$$

22.

$$g(x,t|\xi,\tau) = \frac{H(t-\tau)}{L}$$

$$\times \left\{1 + 2\sum_{n=1}^{\infty}\cos\left(\frac{n\pi\xi}{L}\right)\cos\left(\frac{n\pi x}{L}\right)e^{-n^2\pi^2(t-\tau)/L^2}\right\}$$

26.

$$g(x,t|\xi,\tau) = \frac{2}{L}H(t-\tau)\sum_{n=1}^{\infty}\sin\left[\frac{(2n-1)\pi\xi}{2L}\right]\sin\left[\frac{(2n-1)\pi x}{2L}\right]$$

$$\times \exp\left[-a^2k^2(t-\tau) - \frac{a^2(2n-1)^2\pi^2(t-\tau)}{4L^2}\right]$$

Chapter 5

2.

$$g(x,y|\xi,\eta) = \sum_{n=1}^{\infty}\frac{\exp\left(-n\pi|y-\eta|/a\right)}{n\pi}\sin\left(\frac{n\pi\xi}{a}\right)\sin\left(\frac{n\pi x}{a}\right)$$

4.

$$g(x,y|\xi,\eta) = \sum_{n=0}^{\infty}\frac{\exp\left[-\left(n+\frac{1}{2}\right)\pi|y-\eta|/a\right]}{\left(n+\frac{1}{2}\right)\pi}$$

$$\times \cos\left[\frac{\left(n+\frac{1}{2}\right)\pi x}{a}\right]\cos\left[\frac{\left(n+\frac{1}{2}\right)\pi\xi}{a}\right]$$

5.

$$g(x,y|\xi,\eta) = \frac{|y-\eta|}{2a} - \sum_{n=1}^{\infty}\frac{\exp\left(-n\pi|y-\eta|/a\right)}{n\pi}\cos\left(\frac{n\pi x}{a}\right)\cos\left(\frac{n\pi\xi}{a}\right)$$

10.

$$g(x,y|\xi,\eta) = \frac{4}{ab} \sum_{n=0}^{\infty} \sum_{m=0}^{\infty} \frac{\epsilon_{nm}}{n^2\pi^2/a^2 + m^2\pi^2/b^2}$$

$$\times \cos\left(\frac{n\pi x}{a}\right) \cos\left(\frac{n\pi\xi}{a}\right) \cos\left(\frac{m\pi y}{b}\right) \cos\left(\frac{m\pi\eta}{b}\right),$$

where $\epsilon_{00} = 0$, $\epsilon_{n0} = \epsilon_{0m} = \frac{1}{2}$, and $\epsilon_{nm} = 1$ if $n > 0$ and $m > 0$.

12.

$$g(r,\theta|\rho,\theta') = 2\frac{\ln(r_</a)\ln(b/r_>)}{\ln(b/a)}$$

$$+ 2\sum_{n=1}^{\infty} \frac{\cos[n(\theta - \theta')]}{n\left[1 - (a/b)^{2n}\right]} \left(r_<^n - a^{2n}/r_<^n\right)\left(1/r_>^n - r_>^n/b^{2n}\right).$$

13.

$$g(r,\theta|\rho,\theta') = \frac{1}{\pi}\sum_{n=1}^{\infty}\frac{1}{n} r_<^{n\pi/\beta} r_>^{-n\pi/\beta} \sin\left(\frac{n\pi\theta'}{\beta}\right)\sin\left(\frac{n\pi\theta}{\beta}\right).$$

14.

$$g(r,\theta|\rho,\theta') = \frac{1}{\pi}\sum_{n=1}^{\infty}\frac{1}{n} r_<^{n\pi/\beta}\left(\frac{1}{r_>^{n\pi/\beta}} - \frac{r_>^{n\pi/\beta}}{a^{2n\pi/\beta}}\right)\sin\left(\frac{n\pi\theta'}{\beta}\right)\sin\left(\frac{n\pi\theta}{\beta}\right).$$

16.

$$g(r,z|\rho,\zeta) = \frac{2}{\pi a^2 L}\sum_{m=1}^{\infty}\sum_{n=1}^{\infty}\frac{J_0(\beta_m r/a)J_0(\beta_m\rho/a)}{\left[(\beta_m/a)^2 + n^2\pi^2/L^2\right]J_1^2(\beta_m)}$$

$$\times \sin\left(\frac{n\pi\zeta}{L}\right)\sin\left(\frac{n\pi z}{L}\right),$$

where β_m is the mth root of $J_0(\beta) = 0$.

21.

$$g(x,y|\xi,\eta) = \frac{2}{b}\sum_{m=0}^{\infty}\alpha_m\frac{\cos(k_m x_<)\cos[k_m(x_> - a)]}{k_m\sin(k_m a)}$$

$$\times \cos\left(\frac{m\pi\eta}{b}\right)\cos\left(\frac{m\pi y}{b}\right)$$

$$= \frac{4}{ab}\sum_{n=0}^{\infty}\sum_{m=0}^{\infty}\alpha_n\alpha_m\frac{\cos(n\pi\xi)\cos(n\pi x)}{n^2\pi^2/a^2 - k_m^2}$$

$$\times \cos\left(\frac{m\pi\eta}{b}\right)\cos\left(\frac{m\pi y}{b}\right),$$

where $k_m^2 = k_0^2 - m^2\pi^2/b^2$, $\alpha_0 = \frac{1}{2}$, and $\alpha_m = 1$ for $m > 0$.

27.

$$g(x,y,z|\xi,\eta,\zeta) = \frac{8}{abc\pi^2} \sum_{m=1}^{\infty} \sum_{n=1}^{\infty} \sum_{p=1}^{\infty} \frac{\sin(p\pi\zeta/c)\sin(p\pi z/c)}{n^2/a^2 + m^2/b^2 + p^2/c^2}$$

$$\times \sin\left(\frac{m\pi\eta}{b}\right)\sin\left(\frac{m\pi y}{b}\right)\sin\left(\frac{n\pi\xi}{a}\right)\sin\left(\frac{n\pi x}{a}\right)$$

31.

$$g(x,y,z|\xi,\eta,\zeta) = \frac{4}{bc} \sum_{m=0}^{\infty} \sum_{n=0}^{\infty} \frac{\sin(k_{mn}x_<)\sin[k_{mn}(a-x_>)]}{k_{mn}\sin(k_{mn}a)}$$

$$\times \sin\left(\frac{m\pi\eta}{b}\right)\sin\left(\frac{m\pi y}{b}\right)\sin\left(\frac{n\pi\zeta}{c}\right)\sin\left(\frac{n\pi z}{c}\right)$$

$$= \frac{8}{abc} \sum_{n=0}^{\infty} \sum_{m=0}^{\infty} \sum_{k=0}^{\infty} \frac{\sin(k\pi\xi)\sin(k\pi x)}{k^2\pi^2/a^2 - k_m^2}$$

$$\times \sin\left(\frac{m\pi\eta}{b}\right)\sin\left(\frac{m\pi y}{b}\right)\sin\left(\frac{n\pi\zeta}{c}\right)\sin\left(\frac{n\pi z}{c}\right),$$

where $k_m^2 = k_0^2 - m^2\pi^2/b^2 - n^2\pi^2/c^2$.

32.

$$g(x,y,z|\xi,\eta,\zeta) = \frac{4}{bc} \sum_{m=0}^{\infty} \sum_{n=0}^{\infty} \alpha_n\alpha_m \frac{\cos(k_{mn}x_<)\cos[k_{mn}(x_> - a)]}{k_{mn}\sin(k_{mn}a)}$$

$$\times \cos\left(\frac{m\pi\eta}{b}\right)\cos\left(\frac{m\pi y}{b}\right)\cos\left(\frac{n\pi\zeta}{c}\right)\cos\left(\frac{n\pi z}{c}\right)$$

$$= \frac{8}{abc} \sum_{n=0}^{\infty} \sum_{m=0}^{\infty} \sum_{k=0}^{\infty} \alpha_n\alpha_m\alpha_k \frac{\cos(k\pi\xi)\cos(k\pi x)}{k^2\pi^2/a^2 - k_m^2}$$

$$\times \cos\left(\frac{m\pi\eta}{b}\right)\cos\left(\frac{m\pi y}{b}\right)\cos\left(\frac{n\pi\zeta}{c}\right)\cos\left(\frac{n\pi z}{c}\right).$$

Here $k_m^2 = k_0^2 - m^2\pi^2/b^2 - n^2\pi^2/c^2$, $\alpha_0 = \frac{1}{2}$, and $\alpha_m = 1$ for $m > 0$.

Index